10th Mathematical Physics Meeting: School and Conference on Modern Mathematical Physics 2019

AA004989

Belgrade, Serbia
9 – 14 September 2019

Editors:

B. Dragovich
I. Salom
M. Vojinovic

ISBN: 978-1-7138-2124-3

Printed from e-media with permission by:

Curran Associates, Inc.
57 Morehouse Lane
Red Hook, NY 12571

Some format issues inherent in the e-media version may also appear in this print version.

Printed with permission by Curran Associates, Inc. (2021)

For permission requests, please contact Belgrade Institute of Physics
at the address below.

Belgrade Institute of Physics
Pregrevica 118
11080 Zemun – Belgrade
Serbia

Phone: +381 (0)11 37 13 000
Fax: +381 (0)11 31 62 190

info@ipb.ac.rs

Additional copies of this publication are available from:

Curran Associates, Inc.
57 Morehouse Lane
Red Hook, NY 12571 USA
Phone: 845-758-0400
Fax: 845-758-2633
Email: curran@proceedings.com
Web: www.proceedings.com

P r o c e e d i n g s

of the

10th MATHEMATICAL PHYSICS MEETING: School and Conference on Modern Mathematical Physics

September 9 – 14, 2019, Belgrade, Serbia

Editors

B. Dragovich, I. Salom and M. Vojinović

Institute of Physics
Belgrade, 2020
SERBIA

10th MATHEMATICAL PHYSICS MEETING:
School and Conference on Modern Mathematical Physics

Belgrade, September 9 - 14, 2019

Organizers

- Institute of Physics, Belgrade, Serbia
- Faculty of Mathematics, Belgrade, Serbia
- Mathematical Institute, Belgrade, Serbia
- Faculty of Science, Kragujevac, Serbia
- Serbian Academy of Sciences and Arts, Serbia

Co-organizers

- Institute of Nuclear Sciences "Vinča", Belgrade, Serbia
- Institute of Physics, Kragujevac, Serbia
- Faculty of Physics, Belgrade, Serbia
- Faculty of Sciences, Novi Sad, Serbia
- School of Electrical Engineering, Belgrade, Serbia
- SEENET-MTP, Niš, Serbia

International Advisory Committee

International Organizing Committee

Vladimir Dobrev (Sofia, Bulgaria)
Radu Constantinescu (Craiova, Romania)
Branko Dragovich (Belgrade, Serbia)
Goran Đorđević (Niš, Serbia)
Zoran Rakić (Belgrade, Serbia)
Đorđe Šijački (Belgrade, Serbia)
Marko Vojinović (Belgrade, Serbia)
Igor Volovich (Moscow, Russia)

Local Organizing Committee

Branko Dragovich (Co-Chairman, Institute of Physics, Belgrade)
Marko Vojinović (Co-Chairman, Institute of Physics, Belgrade)
Sanja Ćirković (Institute of Physics, Belgrade)
Ljubica Davidović (Institute of Physics, Belgrade)
Ivan Dimitrijević (Faculty of Mathematics, Belgrade, Serbia)
Ilija Ivanišević (Institute of Physics, Belgrade)
Danijel Obrić (Institute of Physics, Belgrade)
Tijana Radenković (Institute of Physics, Belgrade)
Igor Salom (Institute of Physics, Belgrade)
Dejan Simić (Institute of Physics, Belgrade)

Sponsors

- Ministry of Education, Science and Technological Development, Republic of Serbia

МИНИСТАРСТВО ПРОСВЕТЕ,
НАУКЕ И ТЕХНОЛОШКОГ РАЗВОЈА

- Telekom Srbija

Telekom Srbija

- Open Access Journal "Symmetry" (MDPI, Switzerland)

symmetry

Media Partner

- Open Access Journal "Entropy" (MDPI, Switzerland)

entropy
an Open Access Journal by MDPI

P R E F A C E

This volume contains some reviews and original research contributions, which are related to the **10th Mathematical Physics Meeting: School and Conference on Modern Mathematical Physics**, organized by the Institute of Physics, Belgrade (Serbia), September 9–14, 2019. The programme of this meeting was mainly oriented towards some recent developments in gravity and cosmology, string and quantum field theory, and some relevant mathematical methods. We hope that articles presented here will be valuable literature not only for the participants of this meeting but also for many other PhD students and researchers in modern mathematical and theoretical physics. We are grateful to all authors for writing their contributions for these proceedings.

The previous nine meetings in this series of schools and conferences on modern mathematical physics were also held in Serbia: Sokobanja 2001, Kopaonik 2002, Zlatibor 2004, Belgrade 2006, 2008, 2010, 2012, 2014, and 2017. The corresponding proceedings of all these meetings were published by the Institute of Physics Belgrade, and are available in the printed form as well as online at the websites. According to an agreement with the journal Symmetry, several papers are published in the special issue "Selected Papers: 10th Mathematical Physics Meeting".

This jubilary tenth meeting took place at two different venues — the opening and the first day of lectures was held in the grand lecture hall of the Serbian Academy of Sciences and Arts, while the lectures for the remaining five days were held at the Mathematical Institute. Both venues are located in Belgrade downtown, across the road of each other. We hope that all attendees of this meeting will recall it as a useful and pleasant event, and will wish to participate again in the future.

We wish to thank all lecturers and other speakers for their interesting and valuable talks. We also thank all participants for their active participation. Financial support of our sponsors, *Ministry of Education, Science and Technological Development of the Republic of Serbia, Belgrade; Telekom Srbija; Open access journal "Symmetry"*, and the support of our media partner, *Open access journal "Entropy"*, were very significant for realization of this activity.

April 2020

E d i t o r s

B. Dragovich
I. Salom
M. Vojinović

CONTENTS

Review and Research Works

Non-Riemannian volume elements dynamically generate inflation

David Benisty*
Physics Department, Ben Gurion University of the Negev
Beer Sheva, Israel

Frankfurt Institute for Advanced Studies (FIAS)
Frankfurt am Main, Germany
Eduardo Guendelman†
Physics Department, Ben Gurion University of the Negev
Beer Sheva, Israel
Frankfurt Institute for Advanced Studies (FIAS)
Frankfurt am Main, Germany
Bahamas Advanced Study Institute and Conferences
Stella Maris, Long Island, The Bahamas

Emil Nissimov and Svetlana Pacheva‡
Institute for Nuclear Research and Nuclear Energy
Bulgarian Academy of Sciences, Sofia, Bulgaria

Abstract

Our primary objective is the formulation of a plausible cosmological inflationary model entirely in terms of a pure modified gravity without any *a priori* matter couplings within the formalism of non-Riemannian spacetime volume elements. The non-Riemannian volume elements *dynamically* create in the physical Einstein frame a canonical scalar matter field and produce a non-trivial inflationary scalar potential with a large flat region and a low-lying stable minimum corresponding to the late universe stage. This dynamically generated inflationary potential is a substantial generalization of the classic Starobinsky potential. Our model predicts scalar power spectral index and tensor to scalar ratio in accordance with the available observational data.

1. Introduction

The theoretical framework based on the concept of "inflation" in the study of the evolution of the early Universe provides an attractive solution ex-

* e-mail address: benidav@post.bgu.ac.il
† e-mail address: guendel@bgu.ac.il, guendelman@fias.uni-frankfurt.de
‡ e-mail address: nissimov@inrne.bas.bg, svetlana@inrne.bas.bg

plaining the "puzzles" of Big-Bang cosmology (the horizon problem, the flatness problem, the magnetic monople problem, etc. [1]-[5]. Likewise it is an important instrumentarium for treatment of primordial density perturbations [6, 7]. For some recent detailed accounts, see Refs.[8]-[12].

On the other hand, in a parallel development another groundbreaking concept emerged in the last decade or so about the intrinsic necessity to modify (extend) gravity theories beyond the scope of standard Einstein's general relativity with the main motivation to overcome the limitations of the latter coming from: (i) Cosmology – for solving the problems of dark energy and dark matter and explaining the large scale structure of the Universe [13, 14]; (ii) Quantum field theory in curved spacetime – due to renormalization of ultraviolet divergences in higher loops [15]; (iii) Modern string theory – due to the natural appearance of higher-order curvature invariants and scalar-tensor couplings in low-energy effective field theories [16].

Various classes of modified gravity theories have been employed to construct viable inflationary models: $f(R)$-gravity; scalar-tensor gravity; Gauss-Bonnet gravity (see Refs.[17]-[21] for a detailed accounts); recently also based on non-local gravity (Ref.[22] and references therein) or based on brane-world scenarios (Ref.[23] and references therein). The first early successful cosmological model based on the extended $f(R) = R + R^2$-gravity is the classical Starobinsky potential [1].

A further specific broad class of actively developed modified (extended) gravitational theories is based on the formalism of *non-Riemannian spacetime volume elements* (originally proposed in Refs.[24]-[28]; see Refs.[29, 30] for a systematic geometric formulation). This formalism was used as a basis for constructing a series of extended gravity-matter models describing unified dark energy and dark matter scenario [31, 32], quintessential cosmological models with gravity-assisted and inflaton-assisted dynamical suppression (in the "early" universe) or dynamical generation (in the post-inflationary universe) of electroweak spontaneous symmetry breaking and charge confinement [33]-[35], as well as a novel mechanism for dynamical supersymmetric Brout-Englert-Higgs effect in supergravity [29].

In what follows we will describe in some detail the construction of a viable cosmological inflationary model starting from a modified pure gravity involving several independent non-Riemannian volume elements and *without any a priori coupling* to matter fields.

2. Brief Reminder on Non-Riemannian Volume-Forms (Volume Elements)

Let us briefly recall the essence of the non-Riemannian volume-form formalism (cf. Ref.[36]).

In integrals over differentiable manifolds (not necessarily Riemannian one, so *no* metric is needed) volume-forms are given by nonsingular maximal rank differential forms ω:

$$\int_{\mathcal{M}} \omega(\ldots) = \int_{\mathcal{M}} dx^D \, \Omega(\ldots) \;, \quad \omega = \frac{1}{D!}\omega_{\mu_1\ldots\mu_D} dx^{\mu_1} \wedge \ldots \wedge dx^{\mu_D} \;, \tag{1}$$

where $\omega_{\mu_1\ldots\mu_D} = -\varepsilon_{\mu_1\ldots\mu_D}\Omega$ and Ω is the volume element density. Our conventions for the totally anti-symmetric symbols are

$$\varepsilon^{01\ldots D-1} = 1 \;, \; \varepsilon_{01\ldots D-1} = -1.$$

In Riemannian D-dimensional spacetime manifolds a standard generally-covariant volume-form is defined through the "D-bein" (frame-bundle) canonical one-forms $e^A = e^A_\mu dx^\mu$ $(A = 0, \ldots, D-1)$:

$$\omega = e^0 \wedge \ldots \wedge e^{D-1} = \det \|e^A_\mu\| \, dx^{\mu_1} \wedge \ldots \wedge dx^{\mu_D} \;, \tag{2}$$

where the standard Riemannian volume element density reads

$$\Omega = \det \|e^A_\mu\| = \sqrt{-\det \|g_{\mu\nu}\|} \equiv \sqrt{-g}.$$

To construct modified gravitational theories as alternatives to ordinary standard theories in Einstein's general relativity, instead of $\sqrt{-g}$ we can employ one or more alternative *non-Riemannian* volume element(s) as in (1) given by non-singular *exact* D-forms $\omega = dA$, where: $A = \frac{1}{(D-1)!} A_{\mu_1\ldots\mu_{D-1}} dx^{\mu_1} \wedge \ldots \wedge dx^{\mu-1}$ and the corresponding volume element density reads:

$$\Omega \equiv \Phi(A) = \frac{1}{(D-1)!}\varepsilon^{\mu_1\ldots\mu_D} \, \partial_{\mu_1} A_{\mu_2\ldots\mu_D} \,. \tag{3}$$

Thus, non-Riemannian volume element densities $\Phi(A)$ are defined in terms of the (scalar density of the) dual field-strength of auxiliary rank $D-1$ tensor gauge fields $A_{\mu_1\ldots\mu_{D-1}}$.

As an important remark, let us note that in the first-order (Palatini) formalism ($g_{\mu\nu}$ and $\Gamma^\lambda_{\mu\nu}$ *a priori* independent), the auxiliary tensor gauge fields $A_{\mu_1\ldots\mu_{D-1}}$ turn out to be (almost) pure-gauge – *no propagating field degrees of freedom* except for few discrete degrees of freedom with conserved canonical momenta appearing as arbitrary integration constants. See Refs.[30]-[33] (appendices A) for a systematic proof of the latter fact using the standard canonical Hamiltonian treatment of systems with gauge symmetries, *i.e.*, systems with first-class Hamiltonian constraints a'la Dirac [37, 38].

However, in the second-order (metric) formalism (where $\Gamma^\lambda_{\mu\nu}$ is the usual Levi-Civita connection of the metric $g_{\mu\nu}$) the first non-Riemannian volume form $\Phi(A)$, replacing $\sqrt{-g}$ in the modified Einstein-Hilbert part of the action:

$$S = \int d^4x \Phi(A) R + \ldots \,, \tag{4}$$

is *not* any more pure-gauge. The reason is that in the action (4) the scalar curvature R (in the metric formalism) containes *second-order* (time) derivatives (the latter amount to a total derivative in the ordinary case $S = \int d^4x \sqrt{-g} R + \ldots$).

So defining $\chi_1 \equiv \Phi(A)/\sqrt{-g}$, the latter field becomes physical degree of freedom as seen from the equations of motion resulting from varying (4) w.r.t. $g^{\mu\nu}$:

$$R_{\mu\nu} + \frac{1}{\chi_1}\left(g_{\mu\nu}\Box\chi_1 - \nabla_\mu\nabla_\nu\chi_1\right) + \ldots = 0 \tag{5}$$

3. Modified Pure Gravity with Non-Riemannian Volume Elements

Let us now consider the following simple modified gravity model without any couplings to matter fields (we will use "Planck units" $16\pi G_N = 1$):

$$S = \int d^4x \Big\{ \Phi_1(A)\Big[R - 2\Lambda_0 \frac{\Phi_1(A)}{\sqrt{-g}}\Big] + \Phi_2(B)\frac{\Phi_0(C)}{\sqrt{-g}} \Big\} \,. \tag{6}$$

Here R is the scalar curvature in the metric formalism and:

$$\Phi_1(A) \equiv \frac{1}{3!}\varepsilon^{\mu\nu\kappa\lambda}\partial_\mu A_{\nu\kappa\lambda} \,, \; \Phi_2(B) \equiv \frac{1}{3!}\varepsilon^{\mu\nu\kappa\lambda}\partial_\mu B_{\nu\kappa\lambda} \,,$$

$$\Phi_0(C) \equiv \frac{1}{3!}\varepsilon^{\mu\nu\kappa\lambda}\partial_\mu C_{\nu\kappa\lambda} \,, \tag{7}$$

denote three different independent non-Riemannian volume element densities as in (3) for $D = 4$. Λ_0 is dimensionful parameter which will turn out in what follows to play the role of an inflationary scale.

It is important to stress that the form of the action (6) is uniquely specified by the requirement about global Weyl-scale invariance under:

$$g_{\mu\nu} \to \lambda g_{\mu\nu} \,, \; A_{\mu\nu\kappa} \to \lambda A_{\mu\nu\kappa} \,, \; B_{\mu\nu\kappa} \to \lambda^2 B_{\mu\nu\kappa} \,, \; C_{\mu\nu\kappa} \to C_{\mu\nu\kappa} \,. \tag{8}$$

where $\lambda =$ const. Its importance within the context of non-Riemannian volume element formalism has been originally stressed in [26].

The equations of motion from the action (6) w.r.t. the auxiliary gauge fields $A_{\mu\nu\lambda}$, $B_{\mu\nu\lambda}$, $C_{\mu\nu\lambda}$ defining the non-Riemannian volume elements (7) yield, respectively:

$$R - 4\Lambda_0 \frac{\Phi_1(A)}{\sqrt{-g}} = -M_1 \equiv \text{const} \,, \tag{9}$$

$$\frac{\Phi_0(C)}{\sqrt{-g}} = -M_2 \equiv \text{const} \,, \quad \frac{\Phi_2(B)}{\sqrt{-g}} = \chi_2 \equiv \text{const} \,. \tag{10}$$

Here M_1, M_2 and χ_2 are (dimensionful and dimensionless, respectively) free integration constants; M_1, M_2 indicate spontaneous breaking of global Weyl symmetry (8).

Also, it is important to observe that, since the scalar curvature R contains terms with second-order time derivatives on $g_{\mu\nu}$, Eq.(9) is a genuine dynamical equation of motion and *not* a constraint.

The equations of motion w.r.t. $g_{\mu\nu}$ from (6) read:

$$R_{\mu\nu} - \Lambda_0 \chi_1 \, g_{\mu\nu} + \frac{1}{\chi_1}\big(g_{\mu\nu}\Box\chi_1 - \nabla_\mu \nabla_\nu \chi_1\big) - \frac{\chi_2 M_2}{\chi_1} g_{\mu\nu} = 0 \,, \tag{11}$$

with $\chi_1 \equiv \Phi(A)/\sqrt{-g}$. Taking the trace of (11):

$$3\frac{\Box\chi_1}{\chi_1} - \frac{4\chi_2 M_2}{\chi_1} - M_1 = 0 \tag{12}$$

yields a dynamical equation of motion for the composite scalar field χ_1.

4. From Modified Gravity to the Physical Einstein Frame

We now transform Eqs.(11) and (12) to the physical Einstein frame via the conformal transformation $\bar{g}_{\mu\nu} = \chi_1 g_{\mu\nu}$, upon using the well-known (cf. Ref.[39]) conformal transformation formulas (bars indicate magnitudes in the $\bar{g}_{\mu\nu}$-frame):

$$\begin{aligned} R_{\mu\nu}(g) &= R_{\mu\nu}(\bar{g}) - 3\frac{\bar{g}_{\mu\nu}}{\chi_1}\bar{g}^{\kappa\lambda}\partial_\kappa \chi_1^{1/2}\partial_\lambda \chi_1^{1/2} \\ &\quad + \chi_1^{-1/2}\big(\bar{\nabla}_\mu \bar{\nabla}_\nu \chi_1^{1/2} + \bar{g}_{\mu\nu}\bar{\Box}\chi_1^{1/2}\big) \,, \end{aligned} \tag{13}$$

$$\Box\chi_1 = \chi_1\Big(\bar{\Box}\chi_1 - 2\bar{g}^{\mu\nu}\frac{\partial_\mu \chi_1^{1/2}\partial_\nu \chi_1}{\chi_1^{1/2}}\Big) \,. \tag{14}$$

Hereby the transformed equations acquire the standard form of Einstein equations w.r.t. the new "Einstein-frame" metric $\bar{g}_{\mu\nu}$:

$$R_{\mu\nu}(\bar{g}) - \frac{1}{2}\bar{g}_{\mu\nu}R(\bar{g}) = \frac{1}{2}\Big[\partial_\mu u \partial_\nu u - \bar{g}_{\mu\nu}\big(\frac{1}{2}\bar{g}^{\kappa\lambda}\partial_\kappa u \partial_\lambda u + U_{\rm eff}(u)\big)\Big] \,, \tag{15}$$

$$\bar{\Box} u + \frac{\partial U_{\rm eff}}{\partial u} = 0 \,, \tag{16}$$

where we have redefined

$$\Phi_1(A)/\sqrt{-g} \equiv \chi_1 = \exp\,(u/\sqrt{3}) \tag{17}$$

in order to obtain a canonically normalized kinetic term for the scalar field u, and where we have obtained a *dynamically generated effective scalar potential*:

$$U_{\rm eff}(u) = 2\Lambda_0 - M_1 \exp\,(-\frac{u}{\sqrt{3}}) + \chi_2 M_2 \exp\,(-2\frac{u}{\sqrt{3}}) \,. \tag{18}$$

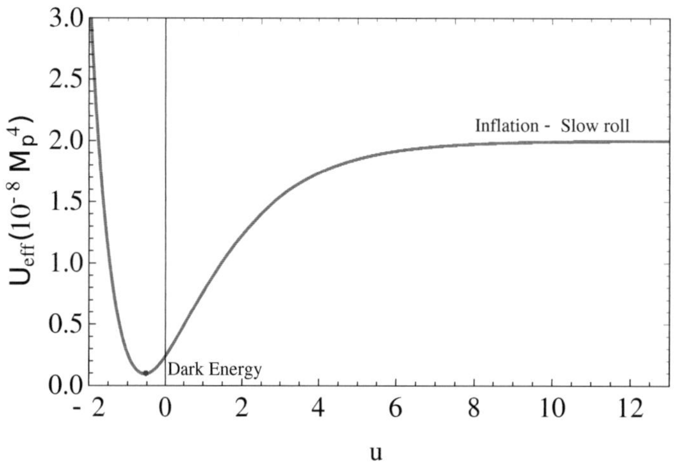

Figure 1: Qualitative shape of the dynamically generated effective scalar potential $U_{\rm eff}$ (18) as function of u. The unit for u is $M_{Planck}/\sqrt{2}$.

$U_{\rm eff}$ (18) is a generalization of the classic *Starobinsky potential* [1]; the latter is a special case of (18) for $\Lambda_0 = \frac{1}{4}M_1 = \frac{1}{2}\chi_2 M_2$.

Accordingly, the corresponding Einstein-frame action reads:

$$S_{\rm EF} = \int d^4x\sqrt{-\bar{g}}\Big[R(\bar{g}) - \frac{1}{2}\bar{g}^{\mu\nu}\partial_\mu u \partial_\nu u - U_{\rm eff}(u)\Big] , \qquad (19)$$

with $U_{\rm eff}$ as in (18).

Let us particularly emphasize that the Einstein-frame action (19) is entirely dynamically generated:

(a) The canonical scalar field u is dynamically created from the ratio of the volume-element densities $\Phi_1(A)/\sqrt{-g}$ (17);

(b) The effective potential $U_{\rm eff}(u)$ (18) is dynamically generated due to the appearance of the free integration constants $M_{1,2}, \chi_2$ in (18) as a result of the specific (constrained) dynamics (9)-(10) of the auxiliary gauge fields $A_{\mu\nu\lambda}, B_{\mu\nu\lambda}, C_{\mu\nu\lambda}$ – constituents of the non-Riemannian volume element densities $\Phi(A), \Phi(B), \Phi(C)$ (7). $U_{\rm eff}(u)$ (18) is graphically depicted on Fig.1.

The dynamically generated potential $U_{\rm eff}(u)$ (18) has two main features relevant for cosmological applications.

First, $U_{\rm eff}(u)$ (18) possesses a long flat region for large positive u and, second, it has a stable minimum for a small finite value $u = u_*$:

- (i) $U_{\rm eff}(u) \simeq 2\Lambda_0$ for large u;

- (ii) $\frac{\partial U_{\rm eff}}{\partial u} = 0$ for $u \equiv u_*$ where:

$$\exp(-\frac{u_*}{\sqrt{3}}) = \frac{M_1}{2\chi_2 M_2} \quad , \quad \frac{\partial^2 U_{\rm eff}}{\partial u^2}\Big|_{u=u_*} = \frac{M_1^2}{6\chi_2 M_2} > 0 \; . \tag{20}$$

The flat region of $U_{\rm eff}(u)$ for large positive u correspond to "early" universe' slow-roll inflationary evolution with energy scale $2\Lambda_0$. On the other hand, the region around the stable minimum at $u = u_*$ (20) correspond to "late" universe' evolution where:

$$U_{\rm eff}(u_*) = 2\Lambda_0 - \frac{M_1^2}{4\chi_2 M_2} \equiv 2\Lambda_{\rm DE} \tag{21}$$

is the dark energy density value dynamically generated through the free integration constants $M_{1,2}$, χ_2.

5. FLRW Reduction and Evolution of the Homogeneous Solution

Let us mow consider the reduction of the Einstein-frame action (19) to the Friedmann-Lemaitre-Robertson-Walker (FLRW) setting with metric $ds^2 = -N^2 dt^2 + a(t)^2 d\vec{x}^2$, and with $u = u(t)$.

The FLRW-reduced action reads:

$$S_{\rm FLRW} = \int d^4x \Big[-6\frac{a\,\dot{a}^2}{N} + N a^3 \Big(\frac{1}{2}\,\dot{u}^2 + M_1 e^{-u/\sqrt{3}} - M_2 \chi_2 e^{-2u/\sqrt{3}} - 2\Lambda_0 \Big) \Big] \; . \tag{22}$$

We will study the evolution of $u = u(t)$ and $a = a(t)$ specified by (22) using the method of autonomous dynamical systems.

The pertinent Friedmann and u-field equations resulting from (22) are given by:

$$H^2 = \frac{1}{6}\rho \; , \quad \rho = \frac{1}{2}\,\dot{u}^2 + U_{\rm eff}(u) \; , \tag{23}$$

$$\dot{H} = -\frac{1}{4}(\rho + p) \; , \quad p = \frac{1}{2}\,\dot{u}^2 - U_{\rm eff}(u) \; , \tag{24}$$

$$\ddot{u} + 3H\,\dot{u} + \frac{\partial U_{\rm eff}}{\partial u} = 0 \; . \tag{25}$$

It is instructive (following Ref.[40]) to rewite the system of Eqs.(23)-(25) in terms of a set of dimensionless variables:

$$x := \frac{\dot{u}}{\sqrt{12}H}, \quad y := \frac{\sqrt{U_{\rm eff}(u) - 2\Lambda_{\rm DE}}}{\sqrt{6}H}, \quad z := \frac{\sqrt{\Lambda_{\rm DE}}}{\sqrt{3}H} \; , \tag{26}$$

with $L_{\rm DE} = \Lambda_0 - \frac{M_1^2}{8\chi_2 M_2}$ as in (21).

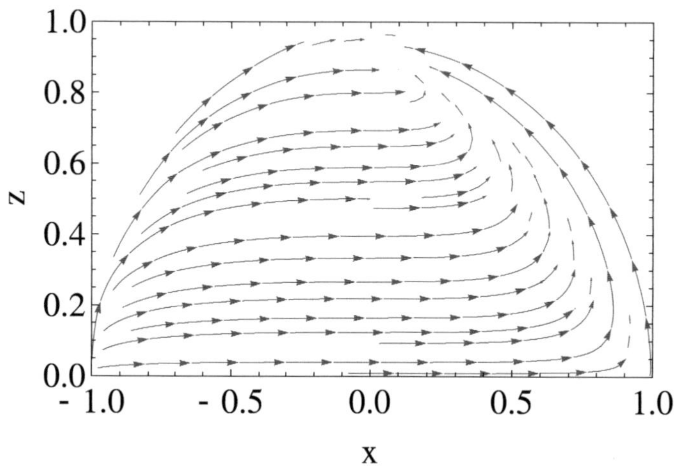

Figure 2: Phase space portrait of the autonomous system (27). The x axis denotes the relative kinetic part of the scalar inflaton, and the z axis denotes the relative part of the dark energy density $\Lambda_{\rm DE}$.

The first Friedman Eq.(23) yields an algebraic constraint $x^2+y^2+z^2=1$, so that the autonomous dynamical system w.r.t. (x,z) reads:

$$x' = \frac{\sqrt{3}}{2\Lambda_{DE}} z^2 \left[-M_1\xi(x,z) + 2M_2\chi_2\xi^2(x,z)\right] - 3x(1-x^2) \,,$$
$$z' = 3zx^2 \quad , \tag{27}$$

where the primes denote derivative w.r.t. the parameter $\mathcal{N} = \log a$ (number of e-foldings), and the function $\xi(x,z)$ is defined as:

$$\xi(x,z) = \frac{M_1}{2\chi_2 M_2}\left[1 - \sqrt{\frac{8\Lambda_0 M_2\chi_2}{M_1^2}\frac{1-x^2-z^2}{z^2}}\right] . \tag{28}$$

Phase space portrait of the autonomous system (27) is depicted numerically on Fig.2.

The autonomous system (27) possesses the following two critical points:

(a) Stable critical point $A\,(x=0, z=1)$ corresponding to the "late" universe de Sitter behavior with a cosmological constant $\Lambda_{\rm DE}$ (21).

(b) Unstable critical point $B\left(x=0, z=\sqrt{\Lambda_{\rm DE}/\Lambda_0}\right)$ corresponding to beginning of evolution in the "early" universe at large u. If the evolution starts at any point close to B, then the dynamics drives the system away from B all the way towards the stable point A at late times.

Numerical solutions of the FLRW system (23)-(25) are graphically presented on Fig.3 for the Hubble parameter $H(t)$, and on Fig.4 for the scalar field $u(t)$.

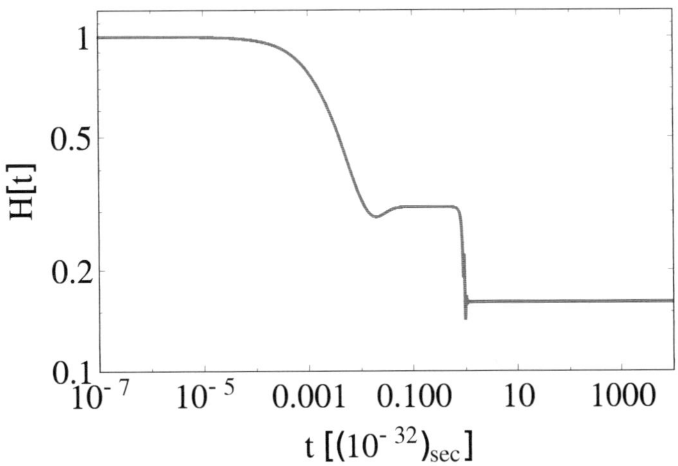

Figure 3: Numerical example of the solution for the Hubble parameter $H(t)$ vs. time. Initially for short times the inflationary Hubble parameter is large and afterwards approaches its cosmological late time value.

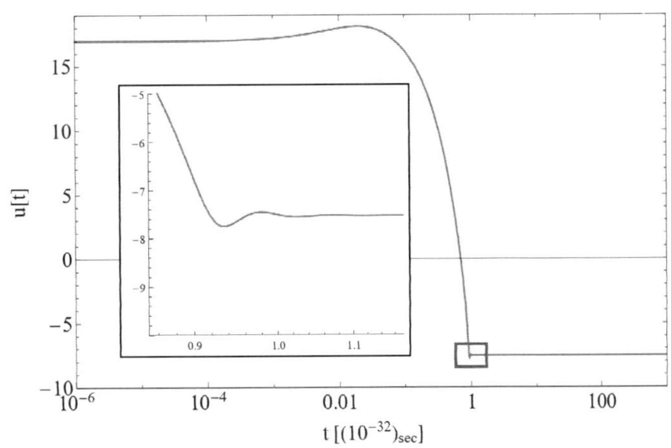

Figure 4: Numerical example of the solution for the scalar field $u(t)$ vs. time. The unit for u is $M_{Planck}/\sqrt{2}$. The blown-up rectangle depicts the oscillations of $u(t)$ around the minimum of $U_{\rm eff}$ (18).

6. Perturbations and Observables

In order to check the viability of our model we will investigate the perturbations of the above FLRW background evolution (23)-(25), in particular focusing on the inflationary observables such as the scalar power spectral index n_s and the tensor-to-scalar ratio r (for definitions, see *e.g.* Ref.[41]). As usual, we introduce the Hubble slow-roll parameters, which in our case using the potential $U_{\rm eff}(u)$ (18) read:

$$\epsilon = \Big(\frac{U'_{\rm eff}(u)}{U_{\rm eff}(u)}\Big)^2 = \frac{4\zeta^2}{3}\frac{(1/2-\zeta)^2}{[(1/2-\zeta)^2+\delta/4]^2} \;, \tag{29}$$

$$|\eta| = 2|\frac{U''_{\rm eff}(u)}{U_{\rm eff}(u)}| = \frac{2\zeta}{3}\frac{(1-4\zeta)}{[(1/2-\zeta)^2+\delta/4]} \;, \tag{30}$$

where:

$$\zeta \equiv \frac{M_2\chi_2}{M_1}e^{-u/\sqrt{3}} \quad , \quad \delta \equiv \frac{8M_2\chi_2}{M_1^2}\Lambda_{\rm DE} \;, \tag{31}$$

with $\Lambda_{\rm DE}$ – the dark energy density (21), and therefore the parameter δ being very small.

Inflation ends when $\epsilon(u_f) = 1$ for some $u = u_f$ whose value (using the short-hand notation $\zeta_f \equiv \frac{M_2\chi_2}{M_1}e^{-u_f/\sqrt{3}}$) is given by:

$$\zeta_f = \frac{1}{2(1+2/\sqrt{3})}\Big[1+\frac{1}{\sqrt{3}} - \sqrt{1/3-(1+2/\sqrt{3})\delta}\Big] \simeq \frac{1}{2(1+2/\sqrt{3})} \;. \tag{32}$$

For the number of e-foldings $\mathcal{N} = \frac{1}{2}\int_{u_i}^{u_f} du\, U_{\rm eff}/U'_{\rm eff}$ we obtain:

$$\mathcal{N} = \frac{3}{8}(1+\delta)\Big(1/\zeta_i - 1/\zeta_f\Big) - \frac{3}{4}(1-\delta)\log\frac{\zeta_f}{\zeta_i} + \frac{3}{4}\delta\,\log\Big(\frac{1-2\zeta_i}{1-2\zeta_f}\Big) \;, \tag{33}$$

where $\zeta_i \equiv \frac{M_2\chi_2}{M_1}e^{-u_i/\sqrt{3}}$ and $u = u_i$ is very large corresponding to the start of the inflation.

Ignoring the very small δ we have for $\mathcal{N}$ approximately:

$$\mathcal{N} \simeq \frac{3M_1}{8M_2\chi_2}e^{u_i/\sqrt{3}} - \frac{\sqrt{3}}{4}u_i - \frac{3}{4}(1+2/\sqrt{3}) + \frac{3}{4}\log\Big(2(1+2/\sqrt{3})\Big) \;. \tag{34}$$

Using the slow-roll parameters (29)-(30), one can calculate the values of the scalar spectral index n_s and the tensor-to-scalar ratio r, respectively, as functions of the e-foldings $\mathcal{N}$:

$$r \approx 16\,\epsilon(u_i(\mathcal{N})) \quad , \quad n_s \approx 1 - 6\,\epsilon(u_i(\mathcal{N})) + 2\eta\,(u_i(\mathcal{N})) \;, \tag{35}$$

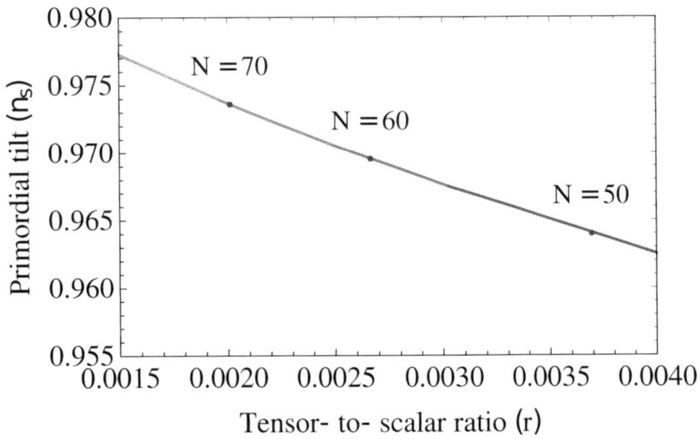

Figure 5: The predicted values of the r and n_s for different number of e-foldings

where $u_i(\mathcal{N})$ is the solution of the transcedental Eq.(34) for u_i as a function of $\mathcal{N}$. From (35), (34), (29), (30) we find:

$$r \simeq \frac{12}{\left[\mathcal{N} + \frac{\sqrt{3}}{4} u_i(\mathcal{N}) + c_0\right]^2} \quad , \quad c_0 \equiv \frac{\sqrt{3}}{2} - \frac{3}{4}\log\Big(2(1 + 2/\sqrt{3})\Big) \; , \tag{36}$$

$$n_s \simeq 1 - \frac{r}{4} - \sqrt{\frac{r}{3}} \; . \tag{37}$$

The numerical results for (36)-(37) are depicted on Fig.5.

The different values of the r and n_s are compatible with the PLANCK observational data ($0.95 < n_s < 0.97$, $r < 0.064$) (cf. Ref.[42]).

Indeed, for the viable example of $\mathcal{N} = 60$ e-foldings until the end of inflation we obtain from (34)-(37):

$$n_s \approx 0.969 \quad , \quad r \approx 0.002 \; . \tag{38}$$

7. Conclusions

- We proposed a very simple modified gravity model without any initial coupling to matter fields in terms of several alternative non-Riemannian spacetime volume elements within the second order (metric) formalism.
- We show how the non-Riemannian volume elements, when passing to the physical Einstein frame, create a canonical scalar field and produce dynamically a non-trivial inflationary-type potential for the

latter possesing a large flat region describing slow-roll inflation and a stable low-lying minimum corresponding to the late universe stage.

- We study the evolution and stability of the cosmological solutions from the point of view of the theory of dynamical systems. Our model predicts scalar spectral index $n_s \approx 0.969$ and tensor-to-scalar ratio $r \approx 0.002$ for 60 e-folds, which is in accordance with the observational data.

Acknowledgments

E.N. and S.P. are sincerely grateful to Prof. Branko Dragovich, Prof. Marko Vojinovich and all the organizers of the *Tenth Meeting in Modern Mathematical Physics* in Belgrade for cordial hospitality. We all gratefully acknowledge support of our collaboration through the academic exchange agreement between the Ben-Gurion University in Beer-Sheva, Israel, and the Bulgarian Academy of Sciences. D.B., E.N. and E.G. have received partial support from European COST actions CA15117, CA16104 and CA18108. E.N. and S.P. are also thankful to Bulgarian National Science Fund for support via research grant DN-18/1.

References

[1] A. Starobinsky, *JETP Lett.*, **30** (1979) 682 [*Pisma Zh. Eksp. Teor. Fiz.*, **30** (1979) 719].

[2] A. Starobinsky, *Phys. Lett.* **91B** (1980) 99.

[3] A. Guth, *Phys. Rev.* **D23** (1981) 347.

[4] A. Linde, *Phys. Lett.* **108B** (1982) 389.

[5] A. Albrecht and P Steinhardt, *Phys. Rev. Lett.* **48** (1982) 1220.

[6] V. Mukhanov and G. Chibisov, *JETP Lett.* **33** (1981) 532 [*Pisma Zh. Eksp. Teor. Fiz.* **33** (1981) 549].

[7] A. Guth and S. Y. Pi, *Phys. Rev. Lett.* **49** (1982) 1110.

[8] S. Weinberg, *Cosmology*, Oxford Univ. Press, 2008.

[9] D. Lyth, *Cosmology for Physicists*, CRC, 2017.

[10] G. Calcagni, *Classical and Quantum Cosmology*, Springer, 2017.

[11] D. Gorbunov and V. Rubakov, *Introduction to the Theory of the Early Universe. Hot Big Bang Theory*, 2nd Ed., World Scientific, 2018.

[12] V. Mukhanov and S. Winitzki, *Introduction to Quantum Effects in Gravity*, Cambridge Univ. Press, 2007.

[13] S. Perlmutter *et al.* [Supernova Cosmology Project Collaboration], Astrophys. J. **517** (1999) 565 (*astro-ph/9812133*).

[14] E. Copeland, M. Sami and S. Tsujikawa, Int. J. Mod. Phys. D **15** (2006) 1753 (*hep-th/0603057*).

[15] S. Weinberg, *Ultraviolet divergences in quantum theories of gravitation*, in *General Relativity. An Einstein Centenary Survey*, pp. 790-831, S. Hawking and W. Israel (eds.), Cambridge Univ. Press, 1979.

[16] M. Green, J. Schwarz and E. Witten, *Superstring Theory*, vol.1, Cambridge Univ. Press, 1988

[17] S. Capozziello and V. Faraoni, *Beyond Einstein Gravity – A Survey of Gravitational Theories for Cosmology and Astrophysics*, Springer, 2011.

[18] S. Capozziello and M. De Laurentis, Phys. Reports **509**, 167 (2011) (*arXiv:1108.6266*).

[19] S. Nojiri and S. Odintsov, Phys. Reports **505**, 59 (2011).

[20] E. Berti *et.al.*, *Class. Quantum Grav.* **32** (2016) 243001 (*arXiv:1501.07274*).

[21] S. Nojiri, S. Odintsov and V. Oikonomou, Phys. Reports **692**, 1 (2017) (*arXiv:1705.11098*).

[22] I.Dimitrijevic, B. Dragovich, Al. Koshelev, Z. Rakic and J. Stankovic, *Phys. Lett.* **797B** (2019) (*arXiv:1906.07560*).

[23] N. Bilic, D. Dimitrijevic, G. Djordjevic, M. Miloevic and M. Stojanovci, *JCAP* **08** (2019) 034 (*arXiv:1809.07216*).

[24] E. I. Guendelman and A. Kaganovich, *Phys. Rev.* **D53** (1996) 7020 (*arXiv:gr-qc/9605026*).

[25] F.Gronwald, U.Muench, A.Macias, F. Hehl, *Phys. Rev.* **D58** (1998) 084021 (*arXiv:gr-qc/9712063*).

[26] E. I. Guendelman, Mod. Phys. Lett. **A14** (1999) 1043-1052 (*arXiv:gr-qc/9901017*).

[27] E. I. Guendelman and A. Kaganovich, Phys. Rev. **D60** (1999) 065004 (*arXiv:gr-qc/9905029*).

[28] E.I. Guendelman and A.B. Kaganovich, *Ann. Phys.* **323** (2008) 866 (*arXiv:0704.1998*).

[29] E. I. Guendelman, E. Nissimov and S. Pacheva, Bulg. J. Phys. **41** (2014) 123 (*arXiv:1404.4733*).

[30] E. I. Guendelman, E. Nissimov and S. Pacheva, *Int. J. Mod. Phys.* **A30** (2015) 1550133 (*arXiv:1504.01031*).

[31] E. I. Guendelman, E. Nissimov and S. Pacheva, *Eur. J. Phys.* **C75**, (2015) 472-479 (*arXiv:1508.02008*).

[32] E. I. Guendelman, E. Nissimov and S. Pacheva, *Eur. J. Phys.* **C76** (2016) 90 (*arXiv:1511.07071*).

[33] E. I. Guendelman, E. Nissimov and S. Pacheva, *Int. J. Mod. Phys.* **D25** (2016) 1644008 (*arXiv:1603.06231*).

[34] E. I. Guendelman, E. Nissimov and S. Pacheva, in *"Quantum Theory and Symmetries with Lie Theory and Its Applications in Physics", vol.2* ed. V. Dobrev, Springer Proceedings in Mathematics and Statistics v.225, Springer, 2018 (*arXiv:1712.09844*).

[35] E. I. Guendelman, E. Nissimov and S. Pacheva, *AIP Conference Proceedings* **2075** (2019) 090030 (*arXiv:1808.03640*).

[36] M. Spivak, *"Calculus On Manifolds – a Modern Approach To Classical Theorems Of Advanced Calculus"*, Ch.5, p.126, CRC Press, 2018.

[37] M. Henneaux and C. Teitelboim, *Quantization of Gauge Systems*, Princeton Univ. Press, 1991.

[38] H. Rothe and K. Rothe, *Classical and Quantum Dynamics of Constrained Hamiltonian Systems*, Ch.3, World Scientific, 2010.

[39] M. Dabrowski, J. Garecki and D. Blaschke, *Ann. Phys.* **18** (2009) 13 (*arXiv:0806.2683*).

[40] S. Bahamonde, C. G. Boehmer, S. Carloni, E. J. Copeland, W. Fang and N. Tamanini, Phys. Rep. **775-777** (2018) 1 (*arXiv:1712.03107*).

[41] S. Nojiri, S. Odintsov and E. Saridakis, *Phys. Lett.* **797B** (2019) 134829 (*arXiv:1904.01345*).

[42] Y. Akrami *et al.* [Planck Collaboration], *arXiv:1807.06211* .

Baryongenesis until dark matter: H-particles proliferation

Francisco Bulnes*
IINAMEI Director, Head of Research Department in Mathematics and Engineering, TESCHA.

Abstract

This paper is a digression on the dark matter and dark energy considering the baryongenesis and their evolution until the actual Universe. The factors of inflation and matter-energy are analyzed, likewise the hadrons as fundamental part in the dark matter production. Some results on torsion in baryongenesis are used to complete the evidences as Baryon acoustic oscillations, the effects due to the cosmic expansion, redshift, space distortions, etcetera.

1. Introduction

The current model of our Universe given by CMB[1], consider the nucleosynthesis as the fundamental process as former of the energy-matter in a long process passing for the leptogenesis and baryongenesis in the Early Universe. In a sometime, the Universe was conformed by the proliferation of leptons, which conformed charges in the Universe. This, in the process of the matter due to the actuation of torsion field [1-3] produce the gravitational field. Then, this conforms an Universe very seemed to the actual Universe with high proliferation of atoms with the extra energy evidenced for the neutrinos and free photons observed for example, in the luminous disk of a galaxy horizon until 10 kpc [2] (see the figure 1). But this indicium is not precise accord with the inhomogeneity detected by the WMAP (redshift-space distortions and microwaves in background) observations in the background of the actual Universe [4]. Precisely, we could consider an extension of the standard model, but what happen with the gravity failure in these cosmic inhomogeneity effects [5]? Well, in it is necessary consider additional studies on torsion evidence extrapolate to the baryongenesis evolution to prove that the cosmic inhomogeneity effects are due the torsion field in a microscopic level of observation [6], for example, in the baryon oscillations inside cosmic expansion (see the figure 2).

* e-mail address: francisco.bulnes@tesch.edu.mx
[1]Cosmic Microwave Background.
[2]kpc = Kilo-Parsec (astronomical distance unit).

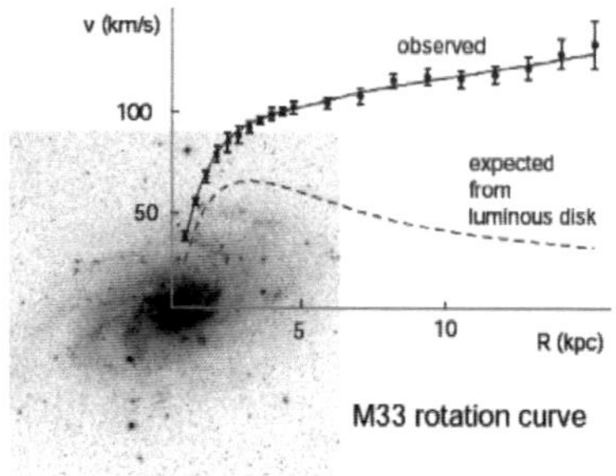

Figure 1: Cosmic expansion in dark matter.

Then possibly will be very useful some aspects of the general relativity and quantum field theory focused to fit with the CMB-model and the super-symmetrical extensions that must be considered in the standard model to action fields [7, 8].

Figure 2: Cosmic expansion in dark matter.

2. Dark matter and torsionful gravity

In the studies of the dark matter have arisen certain questions wanting establish the origin of this. One question is given around gravity, which says: physically the dark matter is a manifestation of gravity or matter-energy?

Remember that we concluded [9] that the torsion field stay univocally determined through **H**$-$ fields considering the established ranges and the Dark matter begins their action on the fermionic dispersion to produce gravity annulling the chirality of the neutrinos/anti-neutrinos interaction and becoming these in oscillations of gravitons or "gravitational waves" registered and measured through spinor frames of the Majorana states [10,11].

Then through this mechanism could be explained the dark matter as vacuum energy with a magnitude. Of fact could use the scalar field theory and the Majorana states registered in fermionic process.

How could be the observation of dark energy affected for inhomogeneity?

The mechanism begins with the right-handed Neutrinos whose inherence meets with the detected in the CMB-model, This could suppose that the gravity involves an inner scattering which can be provoked for the rests of the nucleo-synthesis in the early baryongenesis process with certain anomalies due to the energy density present in that step of gravitational evolution of the space-time.

In other paper [12], the WMAP exploring gives certain data disfavor strongly on the hot dark matter (Neutrinos) having that the energy density of the Universe comply with $\Omega_\nu h^2 < 0.0076$ corresponding to state $m_e < 0.23eV$. In the case of the warm dark matter (gravitino) is disfavored by the evidence to re-ionization at redshift $z \cong 20$.

Then the inhomogeneity can be explained for the torsion effects always and when to the cold dark matter remains axions, super-symmetric dark matter (lightest SUSY particle (LPS)) and permanence super-heavy particles masses of the order $\cong 10^{14\pm5}GeV$.

Figure 3: Dark matter evidence in the Observable Universe. This sidereal object is a galaxy group in the Major Can constellation. The lightening due the high concentration of sterile Neutrinos.

But this can be detected considering from the CDM (Cold Dark Matter), the average energy density $\Omega_{CDM}h^2 < 0.112 \pm 0.006$. Then the dark matter is extended to the standard model with the cold dark matter remains. But this in gravity is the torsion existence until levels where torsionful gravity explains the microscopic inhomogeneity of the space-time (see the figure 4).

The indicium of the inhomogeneity considering the before discussion is measurement of baryon acoustic peak in local Universe (see the figure 5). Here was used the 6-degree Field Galaxy Redshift Survey where relative inflationary value is $H_0 = 67.0 \pm 3,2kms^{-1}Mpc$.

3. Dark energy

The cosmic expansion is equivalent to the baryon oscillations, which produce the dark energy. This fact can be gauged considering an appropriate w- parameter inside the field equations in the framework of the scalar field theory having in count certain values of w, to different physics character-

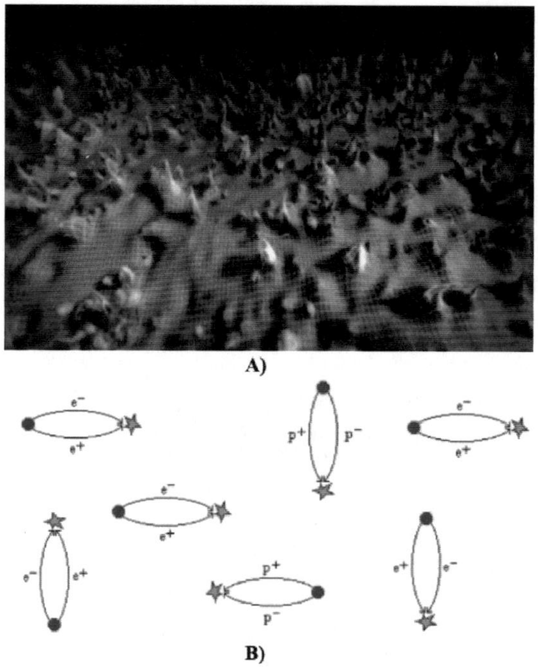

Figure 4: A). Microscopic inhomogeneity of the space-time [13]. B). Torsion mechanism from the baryons evolution [6, 10]. To inhomogeneity from a point of view of gravity, we require the higher dimensional theories, as example superstring theories.

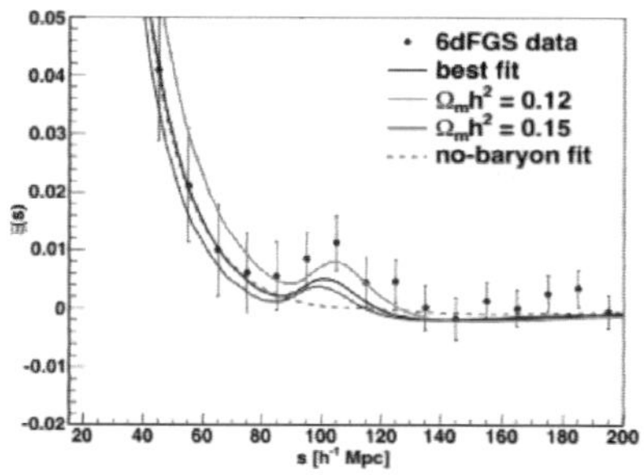

Figure 5: The 6-degree Field Galaxy Redshift Survey. In the graph is showed an inconsistent defined in the range $\Omega_{CDM}h^2 < 0.112 \pm 0.006$.

istics or different key values [3]. For example the values $w = 0$, to matter,

[3]Using baryon acoustic oscillations, it is possible to investigate the effect of dark energy in the history of the Universe, and constrain parameters of the equation of state of dark energy.

$w = 1/3$, to radiation, $w = -1$, as cosmological constant and $w < -1/3$, to accelerating fluids in the Universe.

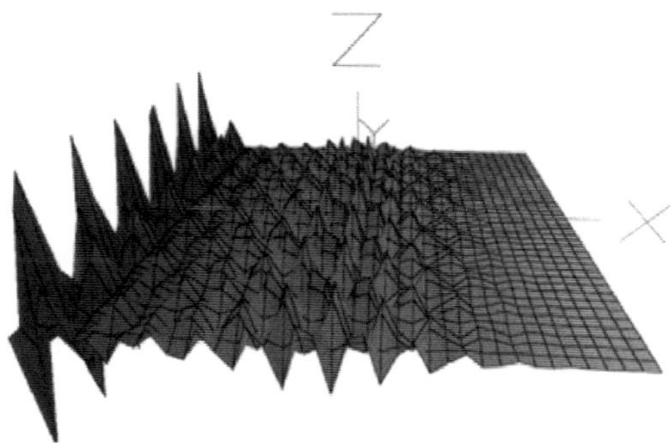

Figure 6: Geometrical model created by [11] to explain the persistency of torsion field in the production of the gravitational waves. The dark energy could be the remains axions in the expansion process. Possibly a key value that can arise in the field equations. Observe the horizon of the 2-dimensional space model and compare with the 6-degree Field Galaxy Redshift Survey given in the figure 5. The inconsistency could arise of the second process or baryongenesis, where dark energy begin in the Universe joined with gravity. This explains the appearing of matter to starting of the nucleon-genesis process [14].

The before values are considered as characters of the corresponding work that is required in each physical entity (matter, radiation,...) considering the latent force of certain "pressure" (tug) of the physical entity contrary to the displacement of certain heavy particle. These values are theoretical and are obtained through dynamical models where is applied the virial energy model of the Universe. Then could be that the dark energy had a negative pressure, contrary to the dark matter, which had a positive pressure. Finally, the variations of the pressure in both physical aspects; dark matter and energy can be detected along the Universe expansion process as baryon oscillations, considering the experiments with heavy particles. Then dark energy is a gravitational + photon energy that full all the space and interacts with the matter producing field effects from these bodies which are observed as maser and laser radiation. Of fact in this step of the Universe is given a proliferation of

H-particles[4], which explains the major quantity of hydrogen in the Universe, as fundamental chemical element to the composing of the interstellar and sidereal bodies (stars, star globular, nebulas, galaxies...).

[4]Hydrogen atoms.

TABLE I
PHYSICS OF DARK ENERGY WITH W-PARAMETER AND SCALAR FACTOR a.

#	Physics	Equation
1	State (Radiation, temperature, scattering)	$P = w\rho$
2	Conservation of Energy	$dE = d(\rho a^3) = -pd(a^3)$
3	Re-arranging	$\rho \propto a^{-3(1+w)}$
4	Friedmann equation (thermal inflation, expansion, matter variation)	$da / dt \propto a^{-(1+3w)/2}$

4. Baryon oscillations and proliferation of H-particle sources in dark energy

The dark energy is in reality the behind of the gravitational and electromagnetic energy in the Universe in the microscopic fundamental levels and the application of the Einstein field equations can be useful until certain level, on all in the explanations of the state and thermal phenomena of the Universe. But no explains the gravity energy and the inconsistencies in the inflation behavior of the Universe comparing the poor existence matter in the space-time and their expansion of himself. Also the forming of the H-particles in the Universe, their proliferation necessity and interacting with the photon sources extended these in the widest vacuum regions of the Universe. To this is necessary first recognize the step or moment in the Universe evolution where is given the proliferation of the H-particles after of the Baryongenesis (in a short time after this Nuclogenesis) [7].

Then arises the matter, but in the initial state with the interaction of wide photon sources extended in the Universe beyond of the de Sitter Universe.

Theorem 4. 1. (F. Bulnes). The dark energy is given beyond of the fundamental interacting of the Baryongenesis and arises with the proliferation of the H-particles in extended photon sources.

Proof. If the dark energy happens beyond of the Baryongenesis then in a photon horizon, with the interaction with the first hydrogenic atoms are produced the instabilities which can be indirectly detected or measured for the variation produced to $\Omega_{CDM}h^2$ (see the figure 5).

This can be proved since CDM are axions, that play an important role to the inhomogeneity [7, 8]. This inhomogeneity, which can be, explained for the torsion effects and the super-symmetric dark matter (lightest SUSY particle (LPS)) permanencies as super-heavy particles masses of the order $\cong 10^{14\pm5}GeV$, which produce a hydrogenic particle scale, whose field has effective sources for emission and absorption of H-particles and photons. However, not explains the high radiation that can be produced of the inhomogeneity.

However, the initial description of H-particles when is assumed their stability only is true when is considered a very short scale time interval. The two particles, H-particle and photon are recombining to shape a single H-particle. Thus, a description of the coupling between weak, causally arranged H-particle sources that does not refer to the real existence of two or more particles propagating between them is physically incomplete.

Then this argumentation does to suppose that this "incompleteness" produce an instability, which provokes high radiation, that comes from the H-fields, and which are hydrogenic particles in a next short time interval, which come from the Baryongenesis. Indeed, of Baryon acoustic oscillations or maser are the consequences of consider the interaction between H-particles and photons along of a large time interval in the process. But, this carries a proliferation of H-particles in extended photon sources, where these last sources are themself resulted of the set of heavy charged particles of photons where their residuals are hadrons (strong interactions of hadrons emerging as residuals) whose forces are stronger magnetic forces. This last to explain in the macroscopic level the existence of magnetic intersidereal fields, as is the case, for example in the galaxies that are extended until the decaying of the dark energy in the limits of the proper sidereal object (see the figure 7). These sidereal matter bodies and all the other bodies are enveloped by the energy of this proliferation of H-particles in the extended photon sources. $\diamond$.

The before arguments used in the demonstration of the theorem 4. 1., can be enforced considering the radiation gauge that substitute the metric tensor $g^{\mu\nu}$, given for the sum $\sum\limits_{\lambda} e^{\mu}_{k\lambda} e^{\nu}_{k\lambda}{}^{*}$, which could have a consistency of dark energy that comes from the strong interacting of hadrons, where each residual given along a λ, can be expressed as transverse parts of multiplying currents

$$J = i \int (d\xi)(d\xi') J_1^{\mu}(\xi) D_+(\xi - \xi') J_{2\mu}(\xi'), \tag{1}$$

then

$$\sum_{\lambda} e^{\mu}_{k\lambda} e^{\nu}_{k\lambda}{}^{*} = 1 - \frac{\mathbf{kk}}{(k^0)^2}, \tag{2}$$

when these multiplying currents are the unit then begins the matter energy and thus the gravitational fields in their early state as hydrogenic particles.

If we consider the unit 1, as the Universe extended to infinitum then all amplitude of the extended photon sources immerse of the Universe must satisfy to their fields that:

$$\langle 0_+ \mid 0_- \rangle^{JK} = 1 + i \int (d\xi)(d\xi') J_1^{\mu}(\xi) D_+(\xi - \xi') J_{2\mu}(\xi') \times$$

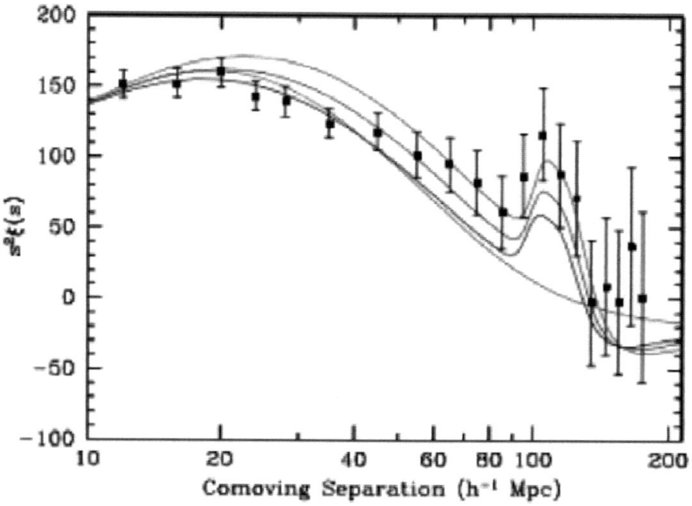

Figure 7: Magnetic intersidereal field decaying in the limits of the proper galactic horizon. Observe how the perturbation is evident when are considered proliferation of the H-particles in the extended photon sources.

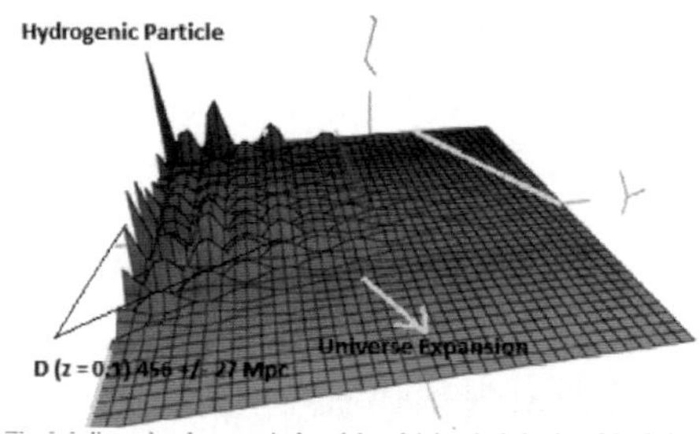

Figure 8: 2-dimensional geometrical model explaining the behavior of the dark matter and dark energy in the Universe expansion. The gravitational field indicium are the early gravitational waves, which until our days are detected [8].

$$i\int(d\chi)(d\chi')K_1(\chi)\Delta_+(\chi-\chi')K_2(\chi')+\ldots \tag{3}$$

where has been used the mathematical property of the mutual actions of two particles [15], where one is an H-particle and the other a photon. These processes multiplies at infinitum.

5. Experiments

Now in our days the experiments reveal strong interactions between hadrons and mesons, which reveals presence of dark matter particles, but not necessarily physics as is suggested in the theorem 4. 1. Then arise indirect detections, excess of gamma ray annihilation products seen in the center of our galaxy, although this is pure speculation and could have other function that not is the dark matter production or some physics relative of the interactions in this site, for example gravitational energy as waves.

In the Large Hadron Collider, should be produced an abundance of dark matter due to the principal idea of collider the Hydrogenic particles with UV light (see the figure 9). These products can be, for the theorem 4.1, the supermassive particles or WIMPS, considering the corresponding step analyzed in the theorem 4. 1, and that come from of the dark energy, such as is mentioned.

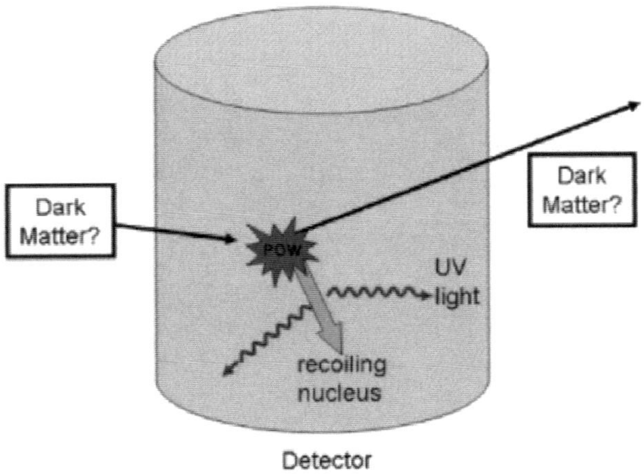

Figure 9: Fundamental scheme of the Hadrons collider (see the figure 11).

Then WIMPs can be detected directly, indirectly or at colliders, within the grasp of current experiments, that is to say, residuals as the transverse currents components given in (1).

The limitations of experiments are remitted for the high interaction of energy and abundance of sterile Neutrinos. However, a good study on super-symmetric extensions to the standard model of particle physics could provide a good theoretical framework on the base of the excess gamma rays. Although in particle physics, this could have a connotation on the lightest super-symmetric particle (LSP) which are additional hypothetical particles found in super-symmetric models.

In models with R-parity conservation, the LSP is stable; in other words, the LSP cannot decay into any Standard Model particle, since all SM particles have the opposite R-parity [19]. There is extensive observational

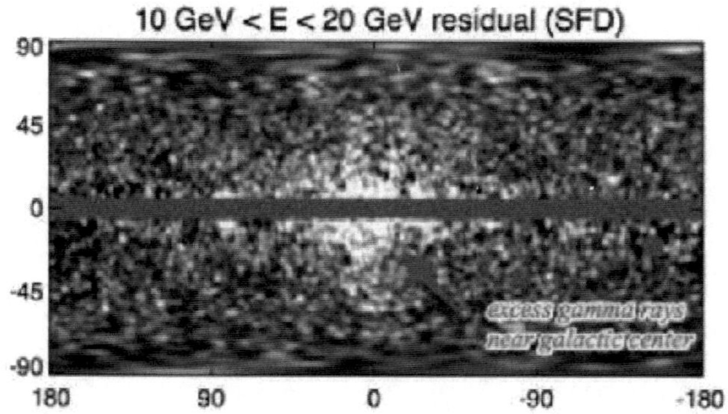

Figure 10: Physics evidence of the excess gamma rays near galactic center, which suggest the dark matter formation [18].

evidence for an additional component of the matter density in the Universe that can be the dark matter.

Finally, the perspective in this study that comes from an extension of the standard model, is projected on the necessity to reproduce observed oscillations from the right-handed Majorana neutrinos [16]. Then the oscillations can be represented by Dirac equations, whose solutions are the planted by the Bianchi cosmologies, that is to say gravitation on supersymmetries.

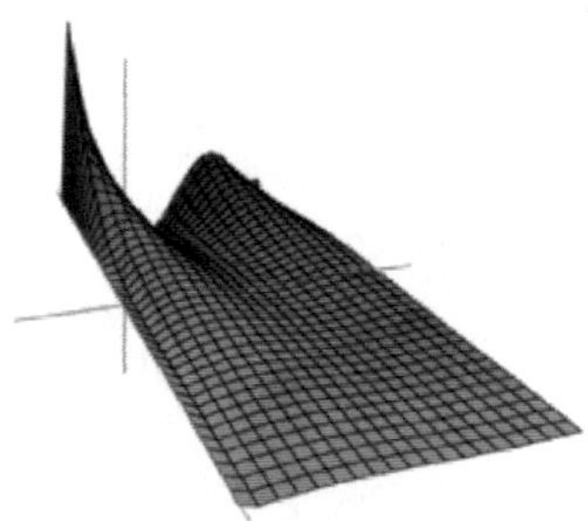

Figure 11: 2-Dimensional model of inconsistency given in the red region for the proliferation of H-particles along of the large interval (recoiling nucleous) (see the figure 9) [15, 17].

References

[1] É. Cartan (1926), Espaces à connexion affine, projective et conforme, Acta Math. 48: 1–42, doi:10.1007/BF02629755

[2] É. Cartan, "Sur une généralisation de la notion de courbure de Riemann et les espaces à torsion." C. R. Acad. Sci. (Paris) 174, 593–595 (1922).

[3] N. E. Mavromatos, S. Sarkar, "Neutrinos in the Early Universe, Kalb-Ramond Torsion and Matter-Antimatter Asymmetry," High Energy Physics - Phenomenology (hep-ph); High Energy Physics - Theory (hep-th) KCL-PH-TH/2013-45, LCTS/2013-34 arXiv:1312.5230

[4] WMAP in NASA, http://map.gsfc.nasa.gov/universe/uni_shape.html.

[5] Smith, P.F.; Bennett, J.R.J.; Homer, G.J.; Lewin, J.D.; Walford, H.E.; Smith, W.A. (November 1981). "A search for anomalous hydrogen in enriched D2O, using a time-of-flight spectrometer". Nucl. Phys. B206 (3): 333–348.

[6] M. E. Shaposhnikov, G. R. Farrar (1993). "Baryon Asymmetry of the Universe in the Minimal Standard Model".Physical Review Letters 70 (19): 2833–2836.

[7] Ellis, John R.; Hagelin, J.S.; Nanopoulos, Dimitri V.; Olive, Keith A.; Srednicki, M. (July 1983). ""Supersymmetric Relics from the Big Bang". Nucl. Phys. B238 (2): 453–476.

[8] Jungman, Gerard; Kamionkowski, Marc; Griest, Kim. "Supersymmetric dark matter". Phys. Rep. 267 (5–6): 195–373.

[9] S. Gorbuix, E. Guivenchy, "Torsion Matter and their Corresponding Dirac Equations on Neutrinos and Matter/anti-Matter Asymmetries (Part II: Gravity Insights)," Journal on Photonics and Spintronics, Vol 4, no. 1, 2014, pp15-25.

[10] A-Wollmann Kleinert, F. Bulnes, "Leptons, the subtly Fermions and their Lagrangians for Spinor Fields: Their Integration in the Electromagnetic Strengthening," Journal on Photonics and Spintronics, Vol. 2 no.1 (2013) pp12-21.

[11] Francisco Bulnes (2017). Detection and Measurement of Quantum Gravity by a Curvature Energy Sensor: H-States of Curvature Energy, Recent Studies in Perturbation Theory, Dr. Dimo Uzunov (Ed.), InTech, DOI: 10.5772/68026. Available from: https://www.intechopen.com/books/recent-studies-in-perturbation-theory/detection-and-measurement-of-quantum-gravity-by-a-curvature-energy-sensor-h-states-of-curvature-ener

[12] E. Guivenchy, "Torsion Matter and their Corresponding Dirac Equations on Neutrinos and Matter/anti-Matter Asymmetries (Part I: scattering difference between Neutrinos/anti-Neutrinos)," Journal on Photonics and Spintronics, Vol 3, no. 4, 2014, pp9-18.

[13] Bulnes, F. "Electromagnetic Gauges and Maxwell Lagrangians Applied to the Determination of Curvature in the Space-Time and their Applications," Journal of Electromagnetic Analysis and Applications, Vol. 4 No. 6, 2012, pp. 252-266. doi: 10.4236/jemaa.2012.46035.

[14] Bulnes, F. , Stropovsvky, Y. and Rabinovich, I. (2017) Curvature Energy and Their Spectrum in the Spinor-Twistor Framework: Torsion as Indicium of Gravitational Waves. Journal of Modern Physics, 8, 1723-1736. doi: 10.4236/jmp.2017.810101.

[15] Wilson, F. L. (1968). "Fermi's Theory of Beta Deacay," American Journal of Physics 36 (12): 1150.

[16] Y. Farzan, S. Pascolli, "A minimal model linking two great mysteries: neutrino mass and dark matter," The GGI for Theoretical Physics, PRD77 (2008) 43516.

[17] U. Debnath, B. Mukhopadhyay, N. Dadhich, "Space-time curvature coupling of spinors in early universe: Neutrino asymmetry and a possible source of baryogenesis," Mod.Phys.Lett. A21 (2006) 399-408 DOI: 10.1142/S0217732306019542

[18] M.A. Ramírez, L. Ramírez, O. Ramírez, F. Bulnes, "The Chronotron, the bi-Particle h, *Journal on Photonics and Spintronics*, Vol. 3 (3), pp23-30.

[19] Byrne, Mark; Kolda, Christopher; Regan, Peter (2002). "Bounds on Charged, Stable Superpartners from Cosmic Ray Production". Physical Review D. 66 (7).

Singularity resolution in fuzzy de Sitter cosmology*

Maja Burić†
University of Belgrade, Faculty of Physics, Belgrade, Serbia
Duško Latas‡
University of Belgrade, Faculty of Physics, Belgrade, Serbia

Abstract

We review and discuss properties of the recently introduced fuzzy de Sitter space [1, 2, 3], in particular its predictions in cosmology.

1. Introduction

One of pending problems in theoretical physics is description of the structure of spacetime at small distances, that is (if we believe in Einstein's insight on the relation between geometry and gravity), quantization of gravity. Since straightforward quantization of the gravitational field does not work, other ideas are developed. On the line of physics, it is for example the idea that there is an elementary substructure (like strings) exists, that it can be quantized by the usual methods yielding the 'quanta of spacetime'. On the line of mathematics, the idea that there is yet to be discovered algebraic structure that generalizes the existing notion of geometry, and provides Einstein's gravity in the large scale limit. There are many ideas in between, combining quantum field theory with geometry. Quantum spacetime should, or should have a potential to, solve two main problems of the current description of space, time and matter: singularities in general relativity and divergences in quantum field theory.

We usually belive that spacetime at small scales will have some kind of geometric structure, different from that of a manifold: perhaps discrete. This structure might be effective or emergent, but also fundamental. A discrete structure physicists are very familiar with is that of an algebra, e.g. the algebra of operators in Hilbert space or a Lie algebra. In noncommutative geometry one assumes that spacetime is described by an algebra of operators, more precisely, a C^* algebra. Furthermore, in order to describe

* This work has been supported by ON171031 grant of the MPNTR, Serbia
† e-mail address: majab@ipb.ac.rs
‡ e-mail address: latas@ipb.ac.rs

fields and equations of motion, one introduces derivatives. In principle, the goal is to introduce noncommutative differential geometry (as quantum spacetime), and describe classical and quantized fields on it. There are several different approaches to this: we work in the following within the noncommutative frame formalism of Madore, [4].

A paradigmatic example of a noncommutative space is the fuzzy sphere, [5]. To construct it one uses its symmetry: here we use similar construction, aiming to obtain four-dimensional fuzzy spacetimes with spherical symmetry. This will enable, we believe, to define noncommutative generalizations of important configurations such as black holes or cosmological spacetimes. As the initial step, we define four-dimensional fuzzy de Sitter space, using the algebra of the de Sitter group SO(1,4) and its unitary irreducible representations.

2. Noncommutative frames

A noncommutative space is an algebra $\mathcal{A}$ generated by a set of hermitian coordinates x^μ

$$[x^\mu, x^\nu] = i\hbar J^{\mu\nu}(x). \tag{1}$$

We can either have an abstract position algebra or its concrete representation, ideally we discuss both. The structure of a noncommutative space can for example be described by the spectra of its coordinates x^μ. There are however other ways to characterize a fuzzy space like its symmetries, or a set of coherent states, or very importantly, its commutative/large scale limit. Diffeomorphisms of noncommutative space are functions on the algebra. Obviously, changes of coordinates change their spectrum. Therefore, although it became quite usual to relate the term 'fuzzy' with either discrete spectrum or finite-dimensional representations, we will in the following identify 'fuzzy' with noncommutative, not presuming *a priori* any properties of the spectra.

Differential structure of $\mathcal{A}$ is given by the momentum algebra. Momenta p_α define a set of vector fields e_α – the free falling frame or tetrad – by

$$e_\alpha f = [p_\alpha, f]. \tag{2}$$

Commutator satisfies the Leibniz rule, and therefore e_α are indeed derivations. In many examples of fuzzy spaces one can identify the algebra generated by momenta with the algebra generated by coordinates, and thus assume that $\mathcal{A}$ is given by either of the sets. But as we want to include, as a particular case, the usual commutative geometry, we do not do so. On commutative manifold, moving frame is given by $e_\alpha f = e^\mu_\alpha\,(\partial_\mu f)$, that is

$$p_\alpha = e^\mu_\alpha\,\partial_\mu, \quad e^\mu_\alpha = [p_\alpha, x^\mu]\,. \tag{3}$$

Momenta lie outside the coordinate algebra $\mathcal{A}$, e_α are outer derivatives. The space of vector fields, tangent space, has dimension equal to dimension of spacetime.

In the noncommutative case we can define, in analogy

$$e^{\mu}_{\alpha} = [p_\alpha, x^\mu], \qquad g^{\mu\nu} = e^{\mu}_{\alpha}\, e^{\nu}_{\beta}\, \eta^{\alpha\beta}\,. \tag{4}$$

Geometry is defined by the choice of a set of momenta, which now does not have *a priori* fixed number of elements, as the tangent space of a noncommutative algebra is infinite-dimensional. One can also define the frame 1-forms θ^α dual to e_α, and the differential d

$$df = (e_\alpha f)\,\theta^\alpha, \qquad [\theta^\alpha, \theta^\beta] = 0. \tag{5}$$

There are additional conditions [4], to assure orthonormality of the moving frame and compatibility of the algebraic (1) with the differential structure (2). If we work with an abstract algebra, the Jacobi identities are also imposed as constraints. It is then possible to define differential-geometric quantities like connection, covariant derivative, curvature and torsion, by formulas analogous to those given in the Cartan's description of geometry. Laplacian of a scalar function is defined naturally,

$$\Delta f = \eta^{\alpha\beta}[p_\alpha, [p_\beta, f]]. \tag{6}$$

Thus one can describe curved noncommutative spaces as well as scalar, spinor and gauge fields. One notion that is representation-dependent is that of a trace, and therefore on an abstract noncommutative space the action is defined only formally.

3. Fuzzy de Sitter space

Construction which we describe here shows that it is in fact possible to adapt and generalize the fuzzy sphere, and obtain noncommutative extensions for all homogeneous spaces. We discuss four-dimensional fuzzy de Sitter space. In the commutative case, de Sitter space is defined as the embedding, [6]

$$-v^2 + w^2 + x^2 + y^2 + z^2 = \frac{3}{\Lambda} \tag{7}$$

in the flat five-dimensional space

$$ds^2 = -dv^2 + dw^2 + dx^2 + dy^2 + dz^2. \tag{8}$$

It has maximal symmetry. We (and originally, [7, 8]) define fuzzy de Sitter space using the algebra of its symmetry group SO(1,4). It has ten generators $M_{\alpha\beta}$,

$$[M_{\alpha\beta}, M_{\gamma\delta}] = -i(\eta_{\alpha\gamma}M_{\beta\delta} - \eta_{\alpha\delta}M_{\beta\gamma} - \eta_{\beta\gamma}M_{\alpha\delta} + \eta_{\beta\delta}M_{\alpha\gamma}) \tag{9}$$

$\alpha, \beta, ... = 0,1,2,3,4$; we use signature $\eta_{\alpha\beta} = \mathrm{diag}(+----)$. The SO(1,4) has two Casimir operators, quadratic and quartic

$$\mathcal{Q} = -\frac{1}{2}\, M_{\alpha\beta}M^{\alpha\beta}, \tag{10}$$

$$\mathcal{W} = -W_\alpha W^\alpha, \quad W_\alpha = \frac{1}{8}\, \epsilon_{\alpha\beta\gamma\delta\eta} M^{\beta\gamma} M^{\delta\eta}. \tag{11}$$

Clearly, relation $\mathcal{W}$=const would be analogous to the embedding of (7) of the four-dimensional commutative de Sitter space in five flat dimensions. Therefore, we introduce coordinates as

$$x^\alpha = \ell\, W^\alpha \tag{12}$$

and define fuzzy de Sitter space to be a unitary irreducible representation (UIR) of the de Sitter algebra. The quartic Casimir of SO(1,4) is then related to the cosmological constant, as $\eta_{\alpha\beta}\, x^\alpha x^\beta = \ell^2\, W^\alpha W_\alpha = 3/\Lambda$.

Coordinates x^α are quadratic in the group generators, and in general they do not close into a Lie or quadratic algebra under commutation:

$$[W^\alpha, W^\beta] = -\frac{i}{2}\, \epsilon^{\alpha\beta\gamma\delta\eta}\, W_\gamma M_{\delta\eta}\, . \tag{13}$$

However, one can show [2], that in irreducible representations they generate the whole algebra via

$$i\,\mathcal{W}M^{\rho\sigma} = [W^\rho, W^\sigma] + \frac{1}{2}\, \epsilon^{\alpha\mu\rho\sigma\tau} W_\tau [W_\alpha, W_\mu]. \tag{14}$$

We will see later that this formula is the Fourier transformation.

We have (at least) two choices of momenta that give de Sitter metric in the commutative limit of this fuzzy space, [1]. The simplest is

$$ip_0 = M_{04}\, , \qquad ip_i = M_{i4} + M_{0i}. \tag{15}$$

The frame formalism gives us the line element,

$$ds^2 = d\tau^2 - e^{-2\tau}\, (dx^i)^2 \tag{16}$$

where cosmic time τ is defined by

$$\frac{\tau}{\ell} = \log(W_0 - W_4)\, . \tag{17}$$

In the conformal group notation, $M_{i4} + M_{0i}$ are translations and M_{04} is dilatation. From

$$[\, iM_{04}, W_0 - W_4] = W_0 - W_4 \tag{18}$$

we find that dilatation is canonically conjugate to the cosmic time, i.e, it can be identified with the hamiltonian.

Unitary irreducible representations of de Sitter group are known, found in [9]. They are labelled by two quantum numbers (s, ρ) or (ν, q) and fall into following categories:

- Principal continuous series: $\rho \geq 0$, $s = 0, \frac{1}{2}, 1, \frac{3}{2}, \ldots$
  $\mathcal{Q} = -s(s+1) + \frac{9}{4} + \rho^2, \quad \mathcal{W} = s(s+1)(\frac{1}{4} + \rho^2)$
- Complementary continuous series: $\nu \in R$, $|\nu| < \frac{3}{2}$, $s = 0, 1, 2\ldots$
  $\mathcal{Q} = -s(s+1) + \frac{9}{4} - \nu^2, \quad \mathcal{W} = s(s+1)(\frac{1}{4} - \nu^2)$
- Discrete series: $s = \frac{1}{2}, 1, \frac{3}{2}, 2\ldots$, $q = s, s-1, \ldots 0$ or $\frac{1}{2}$
  $\mathcal{Q} = -s(s+1) - (q+1)(q-2), \quad \mathcal{W} = s(s+1)q(q-1)$.

4. Spatial coordinates

The principal continuous series has Hilbert space representations [10], which are given in the Bargmann-Wigner representation spaces [11] of the unitary irreducible representations of the Poincaré group. The Hilbert space for the $(\rho, s = \frac{1}{2})$ UIR is the space of Dirac bispinors $\psi(\vec{p})$ which satisfy the Dirac equation. In the Dirac representation of γ-matrices,

$$\psi(\vec{p}) = \begin{pmatrix} \varphi(\vec{p}) \\ -\dfrac{\vec{p}\cdot\vec{\sigma}}{p_0 + m}\,\varphi(\vec{p}) \end{pmatrix} . \tag{19}$$

The scalar product is given by

$$(\psi, \psi') = \int \frac{d^3p}{2p_0}\, \psi^\dagger \gamma^0 \psi' = \int \frac{d^3p}{p_0}\, \frac{2m}{p_0 + m}\, \varphi^\dagger \varphi' . \tag{20}$$

Having an explicit representation, we can solve the eigenvalue problems of the coordinates and thus determine properties of the space. Because of the specific scalar product, hermiticity of coordinates is sometimes not completely obvious. Introducing the expressions for the generators, we find

$$W^0 = -\frac{1}{2m} \begin{pmatrix} (\rho - \frac{i}{2}) p_i \sigma^i + i\, p_0^2 \frac{\partial}{\partial p^i} \sigma^i & \epsilon^{ijk} p_0 p_i \frac{\partial}{\partial p^j}\, \sigma_k + \frac{3i}{2} p_0 \\ \epsilon^{ijk} p_0 p_i \frac{\partial}{\partial p^j}\, \sigma_k + \frac{3i}{2} p_0 & (\rho - \frac{i}{2}) p_i \sigma^i + i p_0^2 \frac{\partial}{\partial p^i}\, \sigma^i \end{pmatrix} \tag{21}$$

$$W^4 = -\frac{1}{2} \begin{pmatrix} i p_0 \frac{\partial}{\partial p^i} \sigma^i & \epsilon^{ijk} p_i \frac{\partial}{\partial p^j}\, \sigma_k + \frac{3i}{2} \\ \epsilon^{ijk} p_i \frac{\partial}{\partial p^j}\, \sigma_k + \frac{3i}{2} & i p_0 \frac{\partial}{\partial p^i} \sigma^i \end{pmatrix} . \tag{22}$$

To find the spectrum of the embedding time W^0 we do not have to solve a differential equation: from the matrix elements of $M_{\alpha\beta}$ one can easily find that the spectrum of W^0 is discrete in all UIR's, with eigenvalues $k(k+1) - k'(k'+1)$.

Because of symmetry, the spectra of spatial coordinates W^i and W^4 are the same: as simpler, we solve the differential equation for W^4. It commutes with spatial rotations so we can take the Ansatz of the form

$$\varphi(\vec{p}) = \frac{f(p)}{p}\,\varphi_{jm} + \frac{h(p)}{p}\,\chi_{jm}\,, \tag{23}$$

where φ_{jm} and χ_{jm} are spinor spherical harmonics,

$$\varphi_{jm} = \begin{pmatrix} \sqrt{\frac{j+m}{2j}}\,Y^{m-1/2}_{j-1/2} \\ \sqrt{\frac{j-m}{2j}}\,Y^{m+1/2}_{j-1/2} \end{pmatrix}, \qquad \chi_{jm} = \begin{pmatrix} \sqrt{\frac{j+1-m}{2(j+1)}}\,Y^{m-1/2}_{j+1/2} \\ -\sqrt{\frac{j+1+m}{2(j+1)}}\,Y^{m+1/2}_{j+1/2} \end{pmatrix}, \tag{24}$$

and Y^m_l are spherical harmonics in momentum space, $p = |\vec{p}|$, $j = \frac{1}{2}, \frac{3}{2}, \dots$.

The W_4−eigenvalue equation for bispinor $\tilde{\psi}_{\sigma jm}$, (19), (denoted by tilde to distinguish it from eigenfunctions of τ in the next section)

$$W_4\,\tilde{\psi}_{\sigma jm} = \sigma\tilde{\psi}_{\sigma jm}\,, \tag{25}$$

reduces to two coupled equations for spinors $\tilde{\varphi}_{\sigma jm}$. Introducing

$$\tilde{f} = (x^2-1)^{1/4}\,\tilde{F}\,, \qquad \tilde{h} = (x^2-1)^{1/4}\,\tilde{H}\,, \tag{26}$$

and variable $x = p_0/m \in (1,\infty)$, we obtain a set of Legendre equations

$$(x^2-1)\,\frac{d^2\tilde{F}}{dx^2} + 2x\,\frac{d\tilde{F}}{dx} - \frac{j^2}{x^2-1}\,\tilde{F} = 2i\sigma(2i\sigma-1)\tilde{F}\,, \tag{27}$$

$$(x^2-1)\,\frac{d^2\tilde{H}}{dx^2} + 2x\,\frac{d\tilde{H}}{dx} - \frac{(j+1)^2}{x^2-1}\,\tilde{H} = 2i\sigma(2i\sigma-1)\tilde{H} \tag{28}$$

with a relation between $\tilde{F}$ and $\tilde{H}$. A regular solution to these equations exists for every real σ, and is expressed in terms of the associated Legendre functions:

$$\tilde{f}_{\sigma j} = A\,(x^2-1)^{\frac{1}{4}}\,P^{-j}_{-2i\sigma}(x), \qquad \tilde{h}_{\sigma j} = A\,(2i\sigma-j-1)(x^2-1)^{\frac{1}{4}}\,P^{-j-1}_{-2i\sigma}(x) \tag{29}$$

Eigenfunctions of W^4 are orthogonal and normalized to δ-function,

$$(\tilde{\psi}_{\sigma jm}, \tilde{\psi}_{\sigma' j' m'}) = 2A^*A'\frac{\Gamma(\frac{1}{2}-2i\sigma)\,\Gamma(\frac{1}{2}+2i\sigma')}{\Gamma(j+1-2i\sigma)\,\Gamma(j+1+2i\sigma')}\delta_{mm'}\,\delta_{jj'}\,\delta(\sigma-\sigma')\,. \tag{30}$$

Thus, the spectrum of W^4 (and all W^i) is continuous, the real line.

5. Cosmic time

We can use the same Ansatz to analyze the eigenvalues of the cosmic time τ, $e^{\tau/\ell} = W_0 - W_4$, i.e. to solve

$$(W_0 - W_4)\,\psi_{\lambda jm} = \lambda \psi_{\lambda jm} \,. \tag{31}$$

After some calculation, we find that now the natural variable in which differential equations simplify is $z = \sqrt{\frac{p_0 - m}{p_0 + m}} \in (0,1)$. We introduce

$$f = \left(\frac{2}{1-z^2}\right)^{-i\rho} z^{j+\frac{1}{2}}\, F\,, \quad h = \left(\frac{2}{1-z^2}\right)^{-i\rho} z^{-j-\frac{1}{2}}\, H \tag{32}$$

and obtain the following of Bessel equations

$$z^2 \frac{d^2F}{dz^2} + z\frac{dF}{dz} + \left(4\lambda^2 z^2 - j^2\right) F = 0$$
$$z^2 \frac{d^2H}{dz^2} + z\frac{dH}{dz} + \left(4\lambda^2 z^2 - (j+1)^2\right) H = 0\,.$$

The regular solution to these equations is given by

$$f_{\lambda j} = C\left(\frac{2}{1-z^2}\right)^{-i\rho} \sqrt{z}\, J_j(2\lambda z), \quad h_{\lambda j} = iC\left(\frac{2}{1-z^2}\right)^{-i\rho} \sqrt{z}\, J_{j+1}(2\lambda z). \tag{33}$$

It exists for every real λ, but for λ and $-\lambda$ the functions are proportional. So apparently, $\lambda \in (0,\infty)$, and the spectrum of $W_0 - W_4$ is continuous. However, when we calculate the scalar product

$$(\psi_{\lambda jm}, \psi_{\lambda' j' m'}) \sim \delta_{jj'}\delta_{mm'} \int_0^1 z dz \left(J_j(2\lambda z) J_j(2\lambda' z) + J_{j+1}(2\lambda z) J_{j+1}(2\lambda' z)\right) \tag{34}$$

we find that it is bounded for $\lambda = \lambda'$, i.e. all solutions are normalizable, which is in contradiction with the statement that they belong to a continuous spectrum. The solutions are also not orthogonal for $\lambda \neq \lambda'$.

Thus, not all of formal solutions (33) can be eigenfunctions, that is, $W_0 - W_4$ is not a self-adjoint operator. One can perform a detailed analysis of hermiticity [3], and find that the domain $\mathcal{D}(W_0 - W_4)$ is unequal to the domain of the adjoint operator, $\mathcal{D}(W_0^\dagger - W_4^\dagger)$. The corresponding deficiency indices are equal, $(n_+, n_-) = (1,1)$. That means that it is possible to find self-adjoint extensions by restricting the domain of $W_0^\dagger - W_4^\dagger$, [12].

The appropriate condition which defines the domain is

$$F(0) = H(0) = 0, \qquad H(1) = icF(1)\,, \tag{35}$$

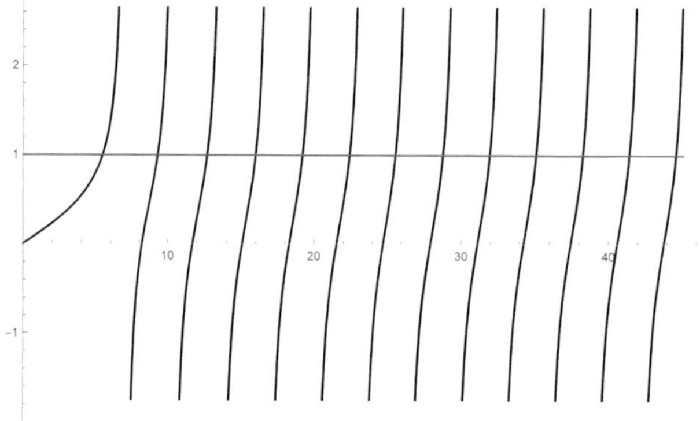

Figure 1: Solutions to Eq. (36) for $j = \frac{7}{2}$, $c = 1$.

One can see that eigenfunctions (33) can satisfy (35). First relation is clearly true, the second gives

$$\frac{J_{j+1}(2\lambda)}{J_j(2\lambda)} = c = \text{const}, \tag{36}$$

that is, an equation for λ. This equation, as seen from Figure 1, has infinitely many solutions for every real c; the set of solutions is discrete. Using the recurrence relations between the Bessel functions, it is possible to show that these solutions orthonormal. We thus conclude thae cosmic time has discrete spectrum.

6. Initial singularity

There are several immediate consequences of the above considerations. We can verify that fuzzy de Sitter space corresponds to an expanding cosmology. The (squared) radius of the universe is given by

$$(x^i)^2 = -\ell^2 W_i W^i \tag{37}$$

and its evolution can be traced by the expectation value $\langle (x^i)^2 \rangle$ in the eigenstates of time. Using the Casimir relation

$$-W_i W^i = \mathcal{W} + W_0^2 - W_4^2 \,, \tag{38}$$

and taking the expectation value in normalized eigenstates $\psi_{\lambda jm}$, we find

$$\langle -W_i W^i \rangle = \mathcal{W} + \langle (W_0 + W_4)(W_0 - W_4) \rangle = \mathcal{W} + \lambda^2 + 2\lambda \, \langle W_4 \rangle \,.$$

The value of $\langle W_4\rangle$ can be estimated: analyzing the explicit expressions we obtain $0\leq\langle W_4\rangle\leq\frac{\lambda}{2}$, and hence

$$\mathcal{W}+\lambda^2\leq\langle -W_iW^i\rangle\leq\mathcal{W}+2\lambda^2\,. \tag{39}$$

The expectation value of the radius of the universe is bounded below by $\ell\sqrt{\mathcal{W}}$: it cannot vanish in physical states. This fact resolves the problem of the initial big bang singularity. The radius grows exponentially with time: for late times we have $\sqrt{\langle -W_iW^i\rangle}\sim\lambda=e^{\langle\tau\rangle/\ell}$.

Discreteness obtained by requiring self-adjointness is known in other cases of quantum spaces too, [13, 14, 15, 16]. It becomes relevant in the 'deep quantum region' $\lambda\to 0$, i.e. $\langle\tau\rangle\to -\infty$, i.e. near the big bang. For values away from the Planck scale time is almost continuous: the difference between its consecutive eigenvalues is macroscopically negligible,

$$\tau_{n+1}-\tau_n\approx\ell\,\log\,(1+\frac{1}{n})\,. \tag{40}$$

Discretization of time means also that the initial classical symmetries of fuzzy de Sitter space near the Planck scale are spontaneously broken, by the choice of the self-adjoint extension. As we can see, it is restored at large scales. It will be important to understand, in the future, the nature of the symmetry breaking, and whether perhaps symmetry remains as deformed or quantum symmetry.

References

[1] M. Buric and J. Madore, Eur. Phys. J. C **75** (2015) no.10, 502 [arXiv:1508.06058 [hep-th]].

[2] M. Buric, D. Latas and L. Nenadovic, Eur. Phys. J. C **78** (2018) no.11, 953 [arXiv:1709.05158 [hep-th]].

[3] M. Buric and D. Latas, Phys. Rev. D **100** (2019) no.2, 024053 doi:10.1103/PhysRevD.100.024053 [arXiv:1903.08378 [hep-th]].

[4] J. Madore, *An Introduction To Noncommutative Differential Geometry And Its Physical Applications*, Lond. Math. Soc. Lect. Note Ser. **257** (2000).

[5] J. Madore, Class. Quant. Grav. **9** (1992) 69, J. Hoppe, "Quantum theory of a massless relativistic surface and a two-dimensional bound state problem", Ph.D. Thesis, MIT, 1982.

[6] S. W. Hawking and G. F. R. Ellis, "Large Scale Structure of Space-Time", Cambridge University Press (1975)

[7] J. P. Gazeau, J. Mourad and J. Queva, [quant-ph/0610222].

[8] D. Jurman and H. Steinacker, JHEP **1401** (2014) 100 [arXiv:1309.1598 [hep-th]].

[9] L. H. Thomas, Ann. Mathematics (1941) 113, T. D. Newton, Ann. Mathematics (1950) 730, J. Dixmier, Bull. Soc. Math. France 89 (1961) 9.

[10] P. Moylan, J. Math. Phys. **24** (1983) 2706, P. Moylan, J. Math. Phys. **26** (1985) 29.

[11] V. Bargmann and E. P. Wigner, Proc. Nat. Acad. Sci. **34** (1948) 211.

[12] N. Dunford and J. T. Schwartz, "Linear Operators", J. Wiley & Sons (1957).

[13] J. Schwenk and J. Wess, Phys. Lett. B **291** (1992) 273.

[14] A. Hebecker, S. Schreckenberg, J. Schwenk, W. Weich and J. Wess, Z. Phys. C **64** (1994) 355.

[15] A. Kempf, G. Mangano and R. B. Mann, Phys. Rev. D **52** (1995) 1108 [hep-th/9412167].

[16] M. Fichtmuller, A. Lorek and J. Wess, Z. Phys. C **71** (1996) 533 [hep-th/9511106].

Dynamical symmetries, coherent states and nonlinear realizations: the SO(2,4) case*

Diego Julio Cirilo-Lombardo †

Consejo Nacional de Investigaciones Cientificas y Tecnicas (CONICET), Universidad de Buenos Aires, National Institute of Plasma Physics (INFIP), Facultad de Ciencias Exactas y Naturales, Ciudad Universitaria, Buenos Aires 1428, Argentina
and
Bogoliubov Laboratory of Theoretical Physics, Joint Institute for Nuclear Research, 141980 Dubna, Russian Federation
and
Andrej B. Arbuzov Bogoliubov Laboratory of Theoretical Physics, Joint Institute for Nuclear Research, Dubna, 141980, Russia and Dubna State University, Dubna, 141980, Russian Federation

Abstract

Nonlinear realizations of the SO(4,2) group are discussed from the point of view of symmetries. Dynamical symmetry breaking is introduced. One linear and one quadratic model in curvature are constructed. Coherent states of the Klauder–Perelomov type are defined for both cases taking into account the coset geometry. A new spontaneous compactification mechanism is defined in the subspace invariant under the stability subgroup. The physical implications of the symmetry breaking in the context of nonlinear realizations and direct gauging are analyzed and briefly discussed

1. Introduction

Studies of higher-dimension theories that involve (spontaneously) broken symmetries and non-commutativity in the quantum case are motivated by searches for a unified theory. Dimensional reduction of such theories is not unique and becomes extremely involved when gravity is included. We believe that the guiding principles for the reduction are provided by the

* DJC-L thanks CONICET-Argentina for financial support, BLTP-JINR for its institutional support where part of this work was developed, and the MPHYS10 conference organization for hospitality and giving us the opportunity to present our research.

† e-mail address: diego777jcl@gmail.com

observed (or desirable) physical field content and by the group theoretical structure itself.

From the technical point of view, we have to extend physical fields into an extra (internal) space with preserving the general noncommutative quantum structure. However the development of a mechanisms that permit us display the set of physical fields in interaction with the corresponding four dimensional world implies that some of the original symmetries of the higher-dimension manifold have been broken. There exist many theoretical attempts to realize the above ideas such as string and brane theories but none of them can be treated as the final answer: formulation of such theories contain serious problems that are still non solved. In spite of the fact that in these theories the solution seems to include a non-commutative structure [1, 2], the concrete implementation of these symmetries in a substructure of any (super) manifold seems to be very complicated from the technical and geometrical viewpoint.

However there exist another way to attack the unification problem that is in the context of gauge theories of gravity [3, 4, 5]. The first model of gauge gravitation theory was suggested by R. Utiyama [6] in 1956 just two years after the birth of gauge theory itself. He was the first who generalized the original gauge model of Yang and Mills for $SU(2)$ to an arbitrary symmetry Lie group and, in particular, to the Lorentz group in order to describe gravity. However, he met the problem of treating general covariant transformations and a pseudo-Riemannian metric which had no partner in the Yang–Mills gauge theory. To eliminate this drawback, representing a tetrad gravitational field as a gauge field of a translation subgroup of the Poincaré group was attempted because, by analogy with gauge potentials in Yang–Mills gauge theory, the indices a of a tetrad field μ were treated as those of a translation group, see [3, 4, 7, 8, 9, 10, 11] and references therein. Since the Poincaré group comes from the Wigner–Inonu contraction of de Sitter groups $SO(2,3)$ and $SO(1,4)$ and it is a subgroup of the conformal group, gauge theories on fibre bundles with these structure groups were also considered [12, 13, 14, 15, 16, 17, 18]. Because these fibre bundles fail to be natural, the lift of the group Diff(X) of diffeomorphisms of the fiber onto the base should be defined [19, 20]. However, these gauging approaches contain the problem with a non-linear (translation) summand of an affine connection being a soldering form, but neither a frame (vierbein) field nor a tetrad field. Thus the latter doesn't have the status of a gauge field [21, 22, 23]. At the same time, gauge theory in the case of spontaneous symmetry breaking also contains classical Higgs fields, besides the gauge and matter ones [24, 25, 26, 27, 28, 29, 30, 31, 32]. Therefore, basing on the mathematical definition of a pseudo-Riemannian metric, some authors formulated gravitation theory as a gauge theory with a reduced Lorentz structure where a metric gravitational field is treated as a Higgs field [33, 34, 35, 36, 37].

The most satisfactory answer to the formulation of gravity as a gauge theory was developed in the pure geometrical context in the works of

D.V. Volkov et al. [38, 39]; in the context of supergravity by Arnowitt and Pran Nath [40]; and finally by Mansouri [41] who was able to solve some of the problems listed before by means of a principal fiber bundle imposing a condition of orthogonality of the generators of the fiber and base manifold. Such conditions that break the symmetry of the original group are implemented by means of a particular choice of the metric tensor. This approach was implemented in a supergroup structure obtaining a gauge theory of supergravity. Notice that the underlying geometry must be reductive (in the Cartan sense) or weakly reductive in the case of supergravity.

As always, even the problem to determine which fields transform as gauge fields and which not, as well as which fields are physical ones and which are redundant, nonetheless remains. Also the relation between the coset factorization (as in the case of the non-linear realization approach [47, 48, 49]) and the specific breaking of the symmetry in the pure topological theories of grand unification (GUT) is still unclear.

2. Coset coherent states

Let us remind the definition of coset coherent states

$$H_0 = \{g \in G \mid \mathcal{U}(g)\, V_0 = V_0\} \subset G. \tag{1}$$

Consequently the orbit is isomorphic to the coset, e.g.

$$\mathcal{O}(V_0) \simeq G/H_0. \tag{2}$$

Analogously, if we remit to the operators, e.g.

$$|V_0\rangle \langle V_0| \equiv \rho_0 \tag{3}$$

then the orbit

$$\mathcal{O}(V_0) \simeq G/H \tag{4}$$

with

$$\begin{aligned} H &= \{g \in G \mid \mathcal{U}(g)\, V_0 = \theta V_0\} \\ &= \left\{g \in G \mid \mathcal{U}(g)\, \rho_0 \mathcal{U}^\dagger(g) = \rho_0\right\} \subset G. \end{aligned} \tag{5}$$

The orbits are identified with cosets spaces of G with respect to the corresponding stability subgroups H_0 and H being the vectors V_0 in the second case defined within a phase. From the quantum viewpoint $|V_0\rangle \in \mathcal{H}$ (the Hilbert space) and $\rho_0 \in \mathcal{F}$ (the Fock space) are V_0 normalized fiducial vectors (embedded unit sphere in $\mathcal{H}$).

3. Symmetry breaking mechanism: the $SO(4,2)$ case

3.1. General considerations

i) Let $a,b,c = 1,2,3,4,5$ and $i,j,k = 1,2,3,4$ (in the six-matrix representation) then the Lie algebra of $SO(2,4)$ is

$$i[J_{ij},J_{kl}] = \eta_{ik}J_{jl} + \eta_{jl}J_{ik} - \eta_{il}J_{jk} - \eta_{jk}J_{il}, \tag{6}$$
$$i[J_{5i},J_{jk}] = \eta_{ik}J_{5j} - \eta_{ij}J_{5k}, \tag{7}$$
$$i[J_{5i},J_{5j}] = -J_{ij}, \tag{8}$$
$$i[J_{6a},J_{bc}] = \eta_{ac}J_{6b} - \eta_{ab}J_{6c}, \tag{9}$$
$$i[J_{6a},J_{6b}] = -J_{ab}. \tag{10}$$

ii) Identifying the first set of commutation relations (6) as the lie algebra of the $SO(1,3)$ with generators $J_{ik} = -J_{ki}$.

iii) The commutation relations (6) plus (7) and (8) are identified as the Lie algebra $SO(2,3)$ with the additional generators J_{5i} and $\eta_{ij} = (1,-1,-1,-1)$.

iv) The commutation relations (6)–(10) is the Lie algebra $SO(2,4)$ written in terms of the Lorentz group $SO(1,3)$ with the additional generators J_{5i}, J_{6b}, and $J_{ab} = -J_{ba}$, where $\eta_{ab} = (1,-1,-1,-1,1)$. It follows that the embedding is given by the chain $SO(1,3) \subset SO(2,3) \subset SO(2,4)$.

From the six dimensional matrix representation we know from that parameterizing the coset $\mathcal{C} = \frac{SO(2,4)}{SO(2,3)}$ and $\mathcal{P} = \frac{SO(2,3)}{SO(1,3)}$, then any element G of $SO(2,4)$ is written as

$$SO(2,4) \approx \frac{SO(2,4)}{SO(2,3)} \times \frac{SO(2,3)}{SO(1,3)} \times SO(1,3), \tag{11}$$

explicitly

$$\begin{aligned} G &= e^{-iz^a(x)J_a} G(H) \\ &= e^{-iz^a(x)J_a} e^{-i\varepsilon^k(x)P_k} H(\Lambda). \end{aligned} \tag{12}$$

Consequently we have $G(H) : H \to G$ is an embedding of an element of $SO(2,3)$ into $SO(2,4)$ where $J_a \equiv \frac{1}{\lambda}J_{6a}$ and $H(\Lambda) : \Lambda \to H$ is an embedding of an element of $SO(1,3)$ into $SO(2,3)$ where $P_k \equiv \frac{1}{m}J_{5k}$ as follows

$$\underbrace{G = e^{-iz^a(x)J_a} e^{-i\varepsilon^k(x)P_k} \underbrace{\begin{pmatrix} \boxed{SO(3,1)} & \mathbf{0} \\ \mathbf{0} & \boxed{I_{2x2}} \end{pmatrix}}_{H(\Lambda)}}_{G(H)} \tag{13}$$

then any element G of $SO(2,4)$ is written as the product of an $SO(2,4)$ boost, an ADS boost, and a Lorentz rotation.

4. Goldstone fields and symmetries

i) Our starting point is to introduce two 6-dimensional vectors V_1 and V_2 being invariant under $SO\,(3,1)$ in a canonical form. Explicitly

$$\left.\underbrace{\begin{pmatrix} 0 \\ 0 \\ 0 \\ 0 \\ A \\ 0 \end{pmatrix}}_{V_1} + \underbrace{\begin{pmatrix} 0 \\ 0 \\ 0 \\ 0 \\ 0 \\ -B \end{pmatrix}}_{V_2} = \underbrace{\begin{pmatrix} 0 \\ 0 \\ 0 \\ 0 \\ A \\ -B \end{pmatrix}}_{V_0}\right\} \textit{invariant under } SO\,(3,1)\,. \tag{14}$$

2) Now we take an element of $Sp\,(2) \subset Mp\,(2)$ embedded in the 6-dimensional matrix representation operating over V as follows

$$\mathcal{M}V \rightarrow \underbrace{\begin{pmatrix} 0 & 0 & 0 & 0 & 0 & 0 \\ 0 & 0 & 0 & 0 & 0 & 0 \\ 0 & 0 & 0 & 0 & 0 & 0 \\ 0 & 0 & 0 & 0 & 0 & 0 \\ 0 & 0 & 0 & 0 & a & b \\ 0 & 0 & 0 & 0 & c & d \end{pmatrix}}_{Sp(2)\subset Mp(2)} \underbrace{\begin{pmatrix} 0 \\ 0 \\ 0 \\ 0 \\ A \\ -B \end{pmatrix}}_{V_0} = \begin{pmatrix} 0 \\ 0 \\ 0 \\ 0 \\ A' \\ -B' \end{pmatrix} = V' \tag{15}$$

where

$$\begin{aligned} A' &= aA - bB, \\ -B' &= cA - dB \end{aligned} \tag{16}$$

consequently we obtain a *Klauder-Perelomov generalized coherent state* with the fiducial vector V_0.

ii) The specific task to be made by the vectors is to perform the breakdown to $SO(3,1)$. Using the transformed vectors above ($Sp(2) \sim Mp(2)$ CS) the symmetry of G can be extended to an internal symmetry as $SU(1,1)$ given by $\widetilde{G}$ below (notice $|\lambda|^2 - |\mu|^2 = 1$):

$$\widetilde{G}V' = \underbrace{e^{-iz^a(x)J_a} e^{-i\varepsilon^k(x)P_k} \underbrace{\begin{pmatrix} \boxed{SO(3,1)} & \mathbf{0} \\ \mathbf{0} & \boxed{\begin{matrix} \lambda & \mu \\ \mu^* & \lambda^* \end{matrix}} \end{pmatrix}}_{\widetilde{H}(\Lambda)}}_{\widetilde{G}(H)} V' = \tag{17}$$

$$= \underbrace{e^{-iz^a(x)J_a} e^{-i\varepsilon^k(x)P_k} \underbrace{\begin{pmatrix} \boxed{SO(3,1)} & \mathbf{0} \\ \mathbf{0} & \boxed{\begin{matrix} \alpha & 0 \\ 0 & \beta \end{matrix}} \end{pmatrix}}_{H(\Lambda)}}_{G(H)} V_0 = GV_0, \tag{18}$$

$$\mathcal{M} = \begin{pmatrix} 0 & 0 & 0 & 0 & 0 & 0 \\ 0 & 0 & 0 & 0 & 0 & 0 \\ 0 & 0 & 0 & 0 & 0 & 0 \\ 0 & 0 & 0 & 0 & 0 & 0 \\ 0 & 0 & 0 & 0 & \lambda^*\alpha & -\mu\beta \\ 0 & 0 & 0 & 0 & -\mu^*\alpha & \lambda\beta \end{pmatrix} \tag{19}$$

and if we also ask for $\mathrm{Det}\mathcal{M} = 1$ then $\alpha\beta = 1$, e.g. the additional phase: it will bring us the 10th Goldstone field. The other nine are given by $z^a(x)$ and $\varepsilon^k(x)$ $(a, b, c = 1, 2, 3, 4, 5$ and $i, j, k = 1, 2, 3, 4)$ coming from the parameterizations of the cosets $\mathcal{C} = \frac{SO(2,4)}{SO(2,3)}$ and $\mathcal{P} = \frac{SO(2,3)}{SO(1,3)}$.

5. Invariant $SO(2,4)$ action and breakdown mechanism

5.1. Linear in R^{AB}

$$S=\int \mu_{AB}\wedge R^{AB} \tag{20}$$

in this case we note at first, that the tensor μ_{AB} $SO(2,4)$-valuated acts as multiplier in S (without any role in dynamics, generally speaking). Having this fact in mind, let us consider the following points:
i) if we have two diffeomorphic (or gauge) nonequivalent $SO(2,4)$-valuated connections, namely Γ^{AB} and $\widetilde{\Gamma}^{AB}$, their difference transforms as a second rank six-tensor under the action of $SO(2,4)$

$$\kappa^{AB}=G^{A}_{C}G^{B}_{D}\kappa^{CD}, \tag{21}$$
$$\kappa^{AB}\equiv\widetilde{\Gamma}^{AB}-\Gamma^{AB}. \tag{22}$$

ii) If we now calculate the curvature from $\widetilde{\Gamma}^{AB}$ we obtain

$$\widetilde{R}^{AB}=R^{AB}+\mathcal{D}\kappa^{AB}, \tag{23}$$

where the $SO(2,4)$ covariant derivative is defined in the usual way

$$\mathcal{D}\kappa^{AB}=d\kappa^{AB}+\Gamma^{A}_{C}\wedge\kappa^{CB}+\Gamma^{B}_{D}\wedge\kappa^{AD}. \tag{24}$$

iii) Redefining the $SO(2,4)$ six vectors as $V_2^A\equiv\psi^A$ and $V_1^B\equiv\varphi^B$ (in order to put all in standard notation), the 2-form κ^{AB} can be constructed as

$$\kappa^{AB}\rightarrow\psi^{[A}\varphi^{B]}dU. \tag{25}$$

Then we introduce all into the $\widetilde{R}^{AB}$ (U scalar function) and get

$$\begin{aligned}\widetilde{R}^{AB}&=R^{AB}+\mathcal{D}\left(\psi^{[A}\varphi^{B]}dU\right)\\&=R^{AB}+\left(\psi^{[A}\mathcal{D}\varphi^{B]}-\varphi^{[A}\mathcal{D}\psi^{B]}\right)\wedge dU.\end{aligned} \tag{26}$$

The next step is to find the specific form of μ_{AB} such that $\widetilde{\mu}_{AB}=\mu_{AB}$ (invariant under tilde transformation) in order to make the splitting of the transformed action $\widetilde{S}$ reductive as follows.

iv) Let us define

$$\widetilde{\theta}^{A}=\widetilde{\mathcal{D}}\varphi^{A} \tag{27}$$

with the connection $\widetilde{\Gamma}^{AB}=\Gamma^{AB}+\kappa^{AB}$, then

$$\widetilde{\theta}^{A}=\underbrace{\mathcal{D}\varphi^{A}}_{\theta^{A}}+\kappa^{A}_{B}\varphi^{B},$$

$$\widetilde{\theta}^{A}=\theta^{A}+\left[\psi^{A}\left(\varphi^{B}\right)^{2}-\varphi^{A}\left(\psi\cdot\varphi\right)\right]\wedge dU, \tag{28}$$

where $(\varphi^B)^2 = (\varphi_B \varphi^B)$ and $(\psi \cdot \varphi) = \psi_B \varphi^B$ etc.

In the same manner we also define

$$\begin{aligned}\widetilde{\eta}^A &= \widetilde{\mathcal{D}}\psi^A,\\ \widetilde{\eta}^A &= \eta^A + \left[\psi_2^A (\psi \cdot \varphi) - \varphi^A (\psi^B)^2\right] \wedge dU.\end{aligned} \tag{29}$$

v) To determine μ_{AB} we propose to cast it in the form

$$\mu_{AB} \propto \rho_s \left[a\psi^F \varphi^E \epsilon_{ABCDEF} \left(\theta^C \wedge \eta^D + \theta^C \wedge \theta^D + \eta^C \wedge \eta^D\right) + b\kappa^{AB}\right] \tag{30}$$

with ρ_s, a, b scalar functions in particular contractions of vectors and bivectors $SO(2,4)$-valuated with ϵ_{ABCDEF}) to be determined. The behaviour under the tilde transformation is

$$\widetilde{\mu}_{AB} \propto \mu_{AB} - \frac{1}{2}\rho_s a \psi^F \varphi^E \epsilon_{ABEF} d\xi \wedge dU, \tag{31}$$

where $\xi = (\psi^A)^2 (\varphi^B)^2 - (\psi \cdot \varphi)^2$.

vi) Finally we must see the behaviour of the transformed action

$$\begin{aligned}\widetilde{S} &= \int \widetilde{\mu}_{AB} \wedge \widetilde{R}^{AB}\\ &= S + \int \frac{1}{2}\rho_s a \kappa_{AB} \wedge R^{AB} \wedge d\xi + \int \mu_{AB} \wedge \mathcal{D}\kappa^{AB}.\end{aligned} \tag{32}$$

We see that till this point, the $SO(2,4)$-valuated six-vectors ψ^F and φ^E are in principle arbitrary. However, under the conditions discussed in the first Section the vectors go to the fiducial ones modulo a phase. Consequently

$$\xi \to A^2 B^2 \tag{33}$$

and the bivector comes to

$$\kappa^{AB} \to \psi^{[A} \varphi^{B]} dU \to \Delta (AB) \epsilon^{\alpha\beta} = \alpha\beta AB \epsilon^{\alpha\beta} = AB\epsilon^{\alpha\beta}, \qquad \alpha, \beta : 5, 6, \tag{34}$$

where we define the 2nd rank antisymmetric tensor $\epsilon^{\alpha\beta}$ and

$$\Delta = Det \begin{pmatrix} \lambda^* \alpha & -\mu\beta \\ -\mu^* \alpha & \lambda\beta \end{pmatrix} = \alpha\beta = 1(\text{unitary transformation}) \tag{35}$$

Below we consider two important cases with respect to the components m and λ.

5.2. $A = m$ and $B = \lambda$

1 If the coefficients $A = m$ and $B = \lambda$ play the role of *constant parameters* we have

$$d\xi \to d\left(\lambda^2 m^2\right) = 0 \tag{36}$$

and

$$\mathcal{D}\kappa^{AB} \to d\left(\lambda m\right)\epsilon^{\alpha\beta} \wedge dU = 0 \tag{37}$$

making the original action S invariant e.g.:

$$\left.\widetilde{S}\right|_{V_0} = \int \widetilde{\mu}_{AB} \wedge \widetilde{R}^{AB} = \int \mu_{AB} \wedge R^{AB} = S \tag{38}$$

being $\left.\widetilde{S}\right|_{V_0}$ the restriction of $\widetilde{S}$ under the subspace generated by V_0 and consequently breaking the symmetry from $SO\left(2,4\right) \to SO\left(1,3\right)$.

2 The connections after the symmetry breaking (when the mentioned conditions with λ and m constants are fulfilled) become

$$\widetilde{\Gamma}^{AB} = \Gamma^{AB} + \kappa^{AB} \Rightarrow b.o.s. \to \widetilde{\Gamma}^{ij} = \Gamma^{ij};\ \widetilde{\Gamma}^{i5} = \Gamma^{i5}, \qquad \widetilde{\Gamma}^{i6} = \Gamma^{i6}, \tag{39}$$

$$\text{but} \qquad \widetilde{\Gamma}^{56} = \Gamma^{56} - \left(\lambda m\right) dU. \tag{40}$$

3 Vectors $\widetilde{\theta}^A$ and $\widetilde{\eta}^A$ after the symmetry breaking and under the same conditions become

$$\begin{aligned}
\widetilde{\theta}^A &= \underbrace{d\varphi^A + \Gamma^A_{\ C} \wedge \varphi^C}_{\theta^A} + \kappa^A_{\ B}\varphi^B \Rightarrow \text{b.o.s.},\\
\widetilde{\theta}^i &= \theta^i = 0 + \Gamma^i_{\ 5} m + 0 \Rightarrow \theta^i = \Gamma^i_{\ 5} m,\\
\widetilde{\theta}^5 &= 0 = 0 + 0 = 0,\\
\widetilde{\eta}^A &= \underbrace{d\psi^A + \Gamma^A_{\ C} \wedge \psi^C}_{\theta^A} + \kappa^A_{\ B}\psi^B \Rightarrow \text{b.o.s.},\\
\widetilde{\eta}^i &= \eta^i = 0 - \Gamma^i_{\ 6}\lambda + 0 \Rightarrow \eta^i = -\Gamma^i_{\ 6}\lambda,\\
\widetilde{\eta}^6 &= \eta^6 = 0
\end{aligned}$$

and evidently $\mu_{i5} = \mu_{i6} = 0$.

4 Consequently from the last points, curvatures becomes to

$$R^{ij} = R^{ij}_{\{\}} + m^{-2}\theta^i \wedge \theta^j + \lambda^{-2}\eta^i \wedge \eta^j, \tag{41}$$

$$R^{i5} = m^{-1}\left[\overbrace{d\theta^i + \omega^i{}_j \wedge \theta^j}^{D\theta^i} + \left(\frac{m}{\lambda}\right)\eta^i \wedge \Gamma^{65}\right]$$

$$= m^{-1}\left[D\theta^i - \frac{m}{\lambda}\eta^i \wedge \Gamma^{65}\right], \tag{42}$$

$$R^{i6} = -\lambda^{-1}\left[D\eta^i - \left(\frac{m}{\lambda}\right)^{-1}\theta^i \wedge \Gamma^{56}\right], \tag{43}$$

$$R^{56} = d\Gamma^{56} + (m\lambda)^{-1}\theta_i \wedge \eta^i, \tag{44}$$

where clearly D is the $SO(1,3)$ covariant derivative.

5 The tensor responsible of the symmetry breaking becomes to

$$\mu_{ij} = -2\rho_s a\lambda m\epsilon_{ijkl}\left(\theta^k \wedge \eta^l + \theta^k \wedge \theta^l + \eta^k \wedge \eta^l\right) \tag{45}$$

$$\mu_{56} = -\rho_s b\epsilon_{56}\lambda m dU. \tag{46}$$

6 Consequently, with all ingredients at hand, the action will be

$$S = \int \mu_{AB} \wedge R^{AB} = \underbrace{\int \mu_{ij} \wedge R^{ij}}_{S_1} + \underbrace{\int \mu_{56} \wedge R^{56}}_{S_2}, \tag{47}$$

where

$$S_1 = -2\int \rho_s a\epsilon_{ijkl}\left(\theta^k \wedge \eta^l + \theta^k \wedge \theta^l + \eta^k \wedge \eta^l\right) \wedge \left(\lambda m R^{ij}_{\{\}} + \frac{\lambda}{m}\theta^i \wedge \theta^j + \frac{m}{\lambda}\eta^i \wedge \eta^j\right)$$

$$= -2\int \rho_s a\epsilon_{ijkl}\left(\theta^k \wedge \eta^l \wedge \lambda m R^{ij}_{\{\}} + \theta^k \wedge \theta^l \wedge \lambda m R^{ij}_{\{\}} + \eta^k \wedge \eta^l \wedge \lambda m R^{ij}_{\{\}}\right)$$

$$-2\int \rho_s a\epsilon_{ijkl}\left(\theta^k \wedge \eta^l \wedge \frac{\lambda}{m}\theta^i \wedge \theta^j + \theta^k \wedge \theta^l \wedge \frac{\lambda}{m}\theta^i \wedge \theta^j + \eta^k \wedge \eta^l \wedge \frac{\lambda}{m}\theta^i \wedge \theta^j\right)$$

$$-2\int \rho_s a\epsilon_{ijkl}\left(\theta^k \wedge \eta^l \wedge \frac{m}{\lambda}\eta^i \wedge \eta^j + \theta^k \wedge \theta^l \wedge \frac{m}{\lambda}\eta^i \wedge \eta^j + \eta^k \wedge \eta^l \wedge \frac{m}{\lambda}\eta^i \wedge \eta^j\right)$$

and

$$S_2 = -\lambda m\int \rho_s b\epsilon_{56} \wedge \left(d\Gamma^{56} + (m\lambda)^{-1}\theta_i \wedge \eta^i\right).$$

7 At this point (the mathematical justification will come later) we can naturally associate the tetrad field with the $\theta - form$

$$\theta^k \sim e^k_a \omega^a \tag{48}$$

consequently a metric can be induced in M_4:

$$\eta_{ab} = g_{jk} e^j_a e^k_b, \qquad g_{jk} = \eta_{ab} e^a_j e^b_k, \qquad e^k_a e^b_k = \delta^a_b, \qquad \text{etc.}, \tag{49}$$

where η_{jk} is the Minkowsky metric. That allows us to up and to down indices, and η^i with the following symmetry typical of a $SU(2,2)$ Clifford structure

$$\eta^k \sim f^k_a \omega^a, \tag{50}$$

$$e^a_j f^k_a g_{lk} = f_{lj} = -f_{jl} \tag{51}$$

that consequently allows us to introduce the electromagnetic field (that will be proportional to f_{lj}) into the model.

8 So we can re-write the action as

$$\begin{aligned} S_1 &= -2\int \rho_s a \epsilon_{ijkl} \left(\theta^k \wedge \eta^l + \theta^k \wedge \theta^l + \eta^k \wedge \eta^l\right) \wedge \left(\lambda m R^{ij}_{\{\}} + \frac{\lambda}{m}\theta^i \wedge \theta^j + \frac{m}{\lambda}\eta^i \wedge \eta^j\right) \\ &= -2\int \rho_s a \left[\lambda m \left(f_{ij} R^{ij}_{\{\}} + \left(g_{ij} + f^k_i f_{kj}\right) R^{ij}_{\{\}}\right) + \left(\frac{\lambda}{m} + \frac{m}{\lambda}\right) f^{kj} f_{kj} \right. \\ &\left. + \left(\frac{\lambda}{m}\sqrt{g} + \frac{m}{\lambda}\sqrt{f}\right)\right] d^4x. \end{aligned} \tag{52}$$

In the above expression we have taken into account the following:
i) terms $\sim \eta \wedge \eta \wedge \eta \wedge \theta$ and $\eta \wedge \theta \wedge \theta \wedge \theta$ vanish;
ii) terms $\sim \eta \wedge \eta \wedge \theta \wedge \theta$ and $\eta \wedge \eta \wedge \theta \wedge \theta$ lead to $\rightarrow f^{kj} f_{kj}$;
iii) term $\sim \epsilon_{ijkl} \theta^k \wedge \eta^l \wedge R^{ij}_{\{\}}$ leads $\rightarrow f_{ij} R^{ij}_{\{\}}$ picking the antisymmetric part of the generalized Ricci tensor (containing torsion);
iv) term $\sim \epsilon_{ijkl}\left(\theta^k \wedge \theta^l + \eta^k \wedge \eta^l\right) R^{ij}_{\{\}}$ leads to $\rightarrow \left(g_{ij} + f^k_i f_{kj}\right) R^{ij}_{\{\}}$ picking the symmetric part of the generalized Ricci tensor (containing Einstein-Hilbert plus quadratic torsion term);
v) terms $\sim \eta \wedge \eta \wedge \eta \wedge \eta$ and $\theta \wedge \theta \wedge \theta \wedge \theta$ lead to the volume elements $\sqrt{f}$ and $\sqrt{g}$, respectively, where we defined as usual $g \equiv Det(g_{lk})$ and $f \equiv Det(f_{lk}) = \left(f^*_{lk} f^{lk}\right)^2$.

5.3. $A = m(x)$ and $B = \lambda(x)$: Spontaneous subspace

If the coefficients $A = m(x)$ and $B = \lambda(x)$ are not *constants* but functions of the coordinates we have

$$d\xi \rightarrow d\left(\lambda^2 m^2\right) = 2d(\lambda m) \tag{53}$$

and

$$\mathcal{D}\kappa^{AB} \rightarrow d(\lambda m)\, \epsilon^{\alpha\beta} \wedge dU. \tag{54}$$

Consequently from the following explicit computations

$$\begin{aligned}\widetilde{S} &= \int \widetilde{\mu}_{AB} \wedge \widetilde{R}^{AB} \\ &= S + \int \frac{1}{2}\rho_s a \kappa_{AB} \wedge R^{AB} \wedge d\xi + \int \mu_{AB} \wedge \mathcal{D}\kappa^{AB} \\ &= S - \int \frac{1}{2}\rho_s a R^{AB} \wedge \kappa_{AB} \wedge d\xi + \int \mu_{AB} \wedge \mathcal{D}\kappa^{AB} \\ &= S - \int \frac{1}{2}\rho_s a R_{\alpha\beta}\epsilon^{\alpha\beta}\lambda m dU \wedge 2d(\lambda m) + \int \mu_{\alpha\beta}\epsilon^{\alpha\beta} d(\lambda m) \wedge dU \\ &= S + \int \frac{1}{2}\rho_s a R_{\alpha\beta}\epsilon^{\alpha\beta}\lambda m 2d(\lambda m) \wedge dU + \int \mu_{\alpha\beta}\epsilon^{\alpha\beta} d(\lambda m) \wedge dU, \\ \widetilde{S} &= S + \int \left[\mu_{\alpha\beta} + \rho_s a R_{\alpha\beta}\lambda m\right]\epsilon^{\alpha\beta} d(\lambda m) \wedge dU.\end{aligned} \tag{55}$$

we obtain the required condition:

$$\begin{aligned}\widetilde{S} &= S \qquad \text{if} \\ \mu_{\alpha\beta} &= -\rho_s a R_{\alpha\beta}\lambda m,\end{aligned} \tag{56}$$

then we see that μ_{AB} takes the place of induced metric and is proportional to the curvature

$$R_{\alpha\beta} = \Lambda \mu_{\alpha\beta} \tag{57}$$

$$\text{with} \qquad \Lambda = -\left(\rho_s a \lambda m\right)^{-1}. \tag{58}$$

Note that we have now a four-dimensional spacetime plus the above "internal" space of a constant curvature. This point is very important as a new compactification-like mechanism

6. Supergravity as a gauge theory and topological QFT

In previous works we have shown, by means of a toy model, that there exists a supersymmetric analog of the above symmetry breaking mechanism coming from the topological QFT. Here we recall some of the above ideas in order to see clearly the analogy between the group structures of the simplest supersymmetric case, $Osp(4)$, and of the classical conformal group $SO(2,4)$.

The starting point is the super $SL(2C)$ superalgebra (strictly speaking $Osp(4)$)

$$\begin{aligned}&[M_{AB}, M_{CD}] = \epsilon_C\left({}_A M_B\right)_D + \epsilon_D\left({}_A M_B\right)_C, \\ &[M_{AB}, Q_C] = \epsilon_C\left({}_A Q_B\right), \qquad \{Q_A, Q_B\} = 2M_{AB}.\end{aligned} \tag{59}$$

Here the indices $A, B, C...$ stay for $\alpha, \beta, \gamma... \left(\dot{\alpha}, \dot{\beta}, \dot{\gamma}...\right)$ spinor indices: $\alpha, \beta \left(\dot{\alpha}, \dot{\beta}\right) = 1, 2 \left(\dot{1}, \dot{2}\right)$ in the Van der Werden spinor notation. We define the superconnection A due the following "gauging"

$$A^p T_p \equiv \omega^{\alpha\dot{\beta}} M_{\alpha\dot{\beta}} + \omega^{\alpha\beta} M_{\alpha\beta} + \omega^{\dot{\alpha}\dot{\beta}} M_{\dot{\alpha}\dot{\beta}} + \omega^{\alpha} Q_{\alpha} - \omega^{\dot{\alpha}} \overline{Q}_{\dot{\alpha}}, \tag{60}$$

where (ωM) define a ten dimensional bosonic manifold and $p \equiv$multi-index, as usual. Analogically the super-curvature is defined by $F \equiv F^p T_p$ with the following detailed structure

$$F(M)^{AB} = d\omega^{AB} + \omega^{A}_{\ C} \wedge \omega^{CB} + \omega^{A} \wedge \omega^{B}, \tag{61}$$

$$F(Q)^{A} = d\omega^{A} + \omega^{A}_{\ C} \wedge \omega^{C}. \tag{62}$$

From (60) is easy to see, that there are a bosonic part and a fermionic one associated with the even and odd generators of the superalgebra. Our proposal for the "toy" action was (as before for $SO(2,4)$) as follows

$$S = \int F^p \wedge \mu_p, \tag{63}$$

where the tensor μ_p (that plays the role of a $Osp(4)$ diagonal metric as in the Mansouri proposal) is *defined* as

$$\mu_{\alpha\dot{\beta}} = \zeta_{\alpha} \wedge \overline{\zeta}_{\dot{\beta}} \quad \mu_{\alpha\beta} = \zeta_{\alpha} \wedge \zeta_{\beta} \quad \mu_{\alpha} = \nu \zeta_{\alpha} \ etc. \tag{64}$$

with $\zeta_{\alpha} \left(\overline{\zeta}_{\dot{\beta}}\right)$ anti-commuting spinors (suitable basis) and ν the parameter of the breaking of super $SL(2C)$ $(Osp(4))$ to $SL(2C)$ symmetry of μ_p. Notice that the introduction of the parameter ν means that we are not take care in the particular dynamics to break the symmetry.

In order to obtain dynamical equations of the theory, we proceed to perform the variation of the proposed action (63)

$$\begin{aligned} \delta S &= \int \delta F^p \wedge \mu_p + F^p \wedge \delta \mu_p \\ &= \int d_A \mu_p \wedge \delta A^p + F^p \wedge \delta \mu_p, \end{aligned} \tag{65}$$

where d_A is the exterior derivative with respect to the super-$SL(2C)$ connection and: $\delta F = d_A \delta A$ have been used. Then, as the result, the dynamics are described by

$$d_A \mu = 0, \qquad F = 0. \tag{66}$$

[3]Corresponding to the number of generators of $SO(4,1)$ or $SO(3,2)$ that define the group manifold

[4]In general this tensor has the same structure that the Cartan-Killing metric of the group under consideration.

The fist equation claims that μ is covariantly constant with respect to the super $SL(2C)$ connection. This fact will be very important when the super $SL(2C)$ symmetry breaks down to $SL(2C)$ because $d_A\mu = d_A\mu_{AB} + d_A\mu_A = 0$, a soldering form will appear. The second equation gives the condition for a super Cartan connection $A = \omega^{AB} + \omega^A$ to be flat, as it is easy to see from the reductive components of above expressions

$$\begin{aligned} F(M)^{AB} &= R^{AB} + \omega^A \wedge \omega^B = 0, \\ F(Q)^A &= d\omega^A + \omega^A_C \wedge \omega^C = d_\omega \omega^A = 0, \end{aligned} \tag{67}$$

where now d_ω is the exterior derivative with respect to the $SL(2C)$ connection and $R^{AB} \equiv d\omega^{AB} + \omega^A_C \wedge \omega^{CB}$ is the $SL(2C)$ curvature. Then

$$F = 0 \Leftrightarrow R^{AB} + \omega^A \wedge \omega^B = 0 \qquad \text{and} \qquad d_\omega \omega^A = 0 \tag{68}$$

the second condition says that the $SL(2C)$ connection is super-torsion free. The first says not that the $SL(2C)$ connection is flat but that it is homogeneous with a cosmological constant related to the explicit structure of the Cartan forms ω^A, as we will see when the super $SL(2C)$ action is reduced to the Volkov-Pashnev model [42].

6.1. The geometrical reduction: extended symplectic super-metrics

6.1.1. Example: Volkov-Pashnev metric

The super-metric under consideration, proposed by Volkov and Pashnev in [42], is the simplest example of symplectic (super) metrics induced by the breaking of symmetry from a pure topological first order action. It can be obtained from the $Osp(4)$ ($superSL(2C)$) action via the following procedure.

i) The Inönu-Wigner contraction [43] in order to pass from $SL(2C)$ to the super-Poincare algebra (corresponding to the original symmetry of the model of refs. [44, 42]) then, the even part of the curvature is split into a $\mathbb{R}^{3,1}$ part $R^{\alpha\dot{\beta}}$ and a $SO(3,1)$ part $R^{\alpha\beta}$ $\left(R^{\dot{\alpha}\dot{\beta}}\right)$ associated with the remaining six generators of the original five dimensional $SL(2C)$ group. This fact is easily realized knowing that the underlying geometry is reductive: $SL(2C) \sim SO(4,1) \to SO(3,1) + \mathbb{R}^{3,1}$. Than we rewrite the superalgebra (59) as

$$\begin{matrix} [M,M] \sim M & [M,\Pi] \sim \Pi & [\Pi,\Pi] \sim M \\ [M,S] \sim S & [\Pi,S] \sim S & \{S,S\} \sim M + \Pi \end{matrix} \tag{69}$$

with $\Pi \sim M_{\alpha\dot{\beta}}$, $M \sim M_{\alpha\beta}\left(M_{\dot{\alpha}\dot{\beta}}\right)$, and re-scale $m^2\Pi = P$ and $mS = Q$. In the limit $m \to 0$, one recovers the super Poincare algebra. Note that one does not re-scale M since one wants to keep $[M,M] \sim M$ Lorentz algebra, that also is a symmetry of (1).

ii) The *spontaneous* breaking of the super $SL(2C)$ down to the $SL(2C)$ symmetry of μ_p (e.g. $\nu \to 0$ in μ_p) of such manner that the even part of the super $SL(2C)$ action $F(M)^{AB}$ remains.

After these evaluations, have been explicitly realized that the even part of the original super $SL(2C)$ action (now a super-Poincare invariant) can be related with the original metric (1) as follows:

$$R(M) + R(P) + \omega^{\alpha}\omega_{\alpha} - \omega^{\dot{\alpha}}\omega_{\dot{\alpha}} \to \omega^{\mu}\omega_{\mu} + \mathbf{a}\omega^{\alpha}\omega_{\alpha} - \mathbf{a}^{*}\omega^{\dot{\alpha}}\omega_{\dot{\alpha}}\,|_{VP}\,. \quad (70)$$

Note that there is mapping $R(M)+R(P) \to \omega^{\mu}\omega_{\mu}\,|_{VP}$ that is well defined and can be realized of different forms, and the map of interest here $\omega^{\alpha}\omega_{\alpha} - \omega^{\dot{\alpha}}\omega_{\dot{\alpha}} \to \mathbf{a}\omega^{\alpha}\omega_{\alpha} - \mathbf{a}^{*}\omega^{\dot{\alpha}}\omega_{\dot{\alpha}}\,|_{VP}$ that associate the Cartan forms of the original super $SL(2C)$ action (63) with the Cartan forms of the Volkov-Pashnev supermodel: $\omega^{\alpha} = (\mathbf{a})^{1/2}\omega^{\alpha}\,|_{VP}$, $\omega^{\dot{\alpha}} = (\mathbf{a}^{*})^{1/2}\omega^{\dot{\alpha}}\,|_{VP}$. Then, the origin of the coefficients $\mathbf{a}$ and $\mathbf{a}^{*}$ becomes clear from the geometrical point of view.

From the first condition in (68) and the association (70) it is not difficult to see that, as in the case of the spacetime cosmological constant $\Lambda : R = \frac{\Lambda}{3} e \wedge e$ $(e \equiv \mathit{space-time\ tetrad})$, there is a cosmological term from the superspace related to the complex parameters $\mathbf{a}$ and $\mathbf{a}^{*}$: $R = -\left(\mathbf{a}\omega^{\alpha}\omega_{\alpha} - \mathbf{a}^{*}\omega^{\dot{\alpha}}\omega_{\dot{\alpha}}\right)$ and is easily to see from the minus sign in above expression, why for supersymmetric (supergravity) models it is more natural to use $SO(3,2)$ instead of $SO(4,1)$.

On the associated spinorial action in the action (63), notice that the role of this part is constrained by the nature of $\nu\zeta_{\alpha}$ in μ_p:

i) If they are of the same nature of the ω^{α} this term is a total derivative, has not influence into the equations of motion, then the action proposed by Volkov and Pashnev in [42] has the correct fermionic form.

ii) If they are not with the same $SL(2C)$ invariance that the ω^{α}, the symmetry of the original model has been modified. In this direction a relativistic supersymmetric model for particles was proposed in ref. [45] considering an N-extended Minkowsky superspace and introducing central charges to the superalgebra. Hence the underlying rigid symmetry gets enlarged to N-extended super-Poincare algebra. Considering for our case similar superextension that in ref. [45] we can introduce the following new action

$$\begin{aligned} S &= -m\int_{\tau 1}^{\tau 2} d\tau\sqrt{\overset{\circ}{\omega}_{\mu}\overset{\circ}{\omega}{}^{\mu} + a\dot{\theta}^{\alpha}\dot{\theta}_{\alpha} - a^{*}\dot{\overline{\theta}}{}^{\dot{\alpha}}\dot{\overline{\theta}}_{\dot{\alpha}} + i(\theta^{\alpha i}A_{ij}\dot{\theta}^{j}_{\alpha} - \overline{\theta}^{\dot{\alpha} i}A_{ij}\dot{\overline{\theta}}{}^{j}_{\dot{\alpha}})} \\ &= \int_{\tau 1}^{\tau 2} d\tau L\left(x, \theta, \overline{\theta}\right) \end{aligned} \quad (71)$$

that is the super-extended version of the superparticle model proposed in [42] with the addition of a first order fermionic part. The matrix tensor

A_{ij} introduce the symplectic structure of such manner that now $\zeta_{\alpha i} \sim A_{ij}\theta^j_\alpha$ is not covariantly constant under d_ω. Note that the "Dirac-like" fermionic part is obviously *inside* the square root because it is a part of the full curvature, fact that was not advertised by the authors in [45] (see also [29]) that not take account on the geometrical origin of the action. An interesting point is to perform the same quantization that in the first part of the research given in [44] in order to obtain and compare the spectrum of physical states with the obtained in ref. [45]. This issue will be presented elsewhere [46].

The spontaneous symmetry breaking happens here because the parameter has no dynamics. But this doesn't happen in the nonlinear realization approach where the parameters have a particular dynamics associated with the spacetime coordinates.

7. Quadratic in R^{AB}

The previous action, linear in the generalized curvature, has some drawbacks that make necessary the introduction of additional "subsidiary conditions" due that the curvatures R^{i5} and R^{i6} play not role into the linear/first order action. Such curvatures have very important information about the dynamics of θ and η fields. In order to simplify the equations of motion we define

$$\Gamma^{56} \equiv A, \tag{72}$$

$$m^{-1}\theta^i \equiv \widetilde{\theta}^i, \tag{73}$$

$$\lambda^{-1}\eta^i \equiv \widetilde{\eta}^i, \tag{74}$$

and as always

$$R^{ij} = R^{ij}_{\{\}} + m^{-2}\theta^i \wedge \theta^j + \lambda^{-2}\eta^i \wedge \eta^j \tag{75}$$

with the $SO(1,3)$ curvature $R^{ij}_{\{\}} = d\omega^{ij} + \omega^i{}_\lambda \wedge \omega^{\lambda j}$. Consequently from the quadratic Lagrangian density

$$S = \int R_{AB} \wedge R^{AB} \tag{76}$$

we obtain the following equations of motion:

$$\frac{\delta\left(R_{AB}\wedge R^{AB}\right)}{\delta\theta^{i}}\rightarrow D\left(D\widetilde{\theta}_{j}\right)+2R_{ij}\wedge\widetilde{\theta}^{i}-\widetilde{\theta}^{i}\wedge\widetilde{\eta}_{i}\wedge\widetilde{\eta}_{j}+\widetilde{\theta}_{j}\wedge A\wedge A=0, \tag{77}$$

$$\frac{\delta\left(R_{AB}\wedge R^{AB}\right)}{\delta\eta^{i}}\rightarrow D\left(D\widetilde{\eta}_{j}\right)+2R_{jk}\wedge\widetilde{\eta}^{k}-\widetilde{\theta}^{i}\wedge\widetilde{\eta}_{i}\wedge\widetilde{\theta}_{j}+\widetilde{\eta}_{j}\wedge A\wedge A=0, \tag{78}$$

$$\frac{\delta\left(R_{AB}\wedge R^{AB}\right)}{\delta\Gamma^{56}}\rightarrow\widetilde{\theta}^{i}\wedge\widetilde{\theta}_{i}=\widetilde{\eta}^{i}\wedge\widetilde{\eta}_{i}, \tag{79}$$

$$\frac{\delta\left(R_{AB}\wedge R^{AB}\right)}{\delta\omega_{j}^{i}}\rightarrow -DR_{kl}+D\widetilde{\theta}_{k}\wedge\widetilde{\theta}_{l}+D\widetilde{\eta}_{k}\wedge\widetilde{\eta}_{l}+\widetilde{\theta}_{k}\wedge\widetilde{\eta}_{l}\wedge A=0. \tag{80}$$

7.1. Maxwell equations and the electromagnetic field

As we claim before we can identify

$$\theta^{i}\equiv e_{\mu}^{i}dx^{\mu}, \tag{81}$$

$$\eta^{i}\equiv f_{\mu}^{i}dx^{\mu} \tag{82}$$

with the symmetries

$$e_{\mu}^{i}e_{i}^{\nu}=\delta_{\mu}^{\nu},e_{\mu}^{i}e_{i\nu}=g_{\mu\nu}=g_{\nu\mu} \tag{83}$$

and

$$f_{\mu}^{i}f_{i}^{\nu}=\delta_{\mu}^{\nu},\qquad e_{i\nu}f_{\mu}^{i}=f_{\mu\nu}=-f_{\nu\mu} \tag{84}$$

such that the geometrical (Bianchi) condition

$$\nabla_{[\rho}f_{\mu\nu]}=\nabla_{\rho}^{*}f^{\rho\nu}=0 \tag{85}$$

or in the language of differential forms

$$D\left(\widetilde{\theta}^{i}\wedge\widetilde{\eta}_{i}\right)=0 \tag{86}$$

holds, enforce to the curvatures R^{i6} and R^{i5} to be null. And conversely if R^{i6} and R^{i5} are zero then $D\left(\widetilde{\theta}^{i}\wedge\widetilde{\eta}_{i}\right)=0$ or equivalently $\nabla_{[\rho}f_{\mu\nu]}=\nabla_{\rho}^{*}f^{\rho\nu}=0$.

Proof. From expressions (42,43), namely: $R^{i5}=\left[D\widetilde{\theta}^{i}-\widetilde{\eta}^{i}\wedge\Gamma^{65}\right]$ and $R^{i6}=\left[-D\widetilde{\eta}^{i}+\widetilde{\theta}^{i}\wedge\Gamma^{56}\right]$ we make

$$R^{i5}\wedge\widetilde{\eta}_{i}+\widetilde{\theta}_{i}\wedge R^{i6}=D\left(\widetilde{\theta}^{i}\wedge\widetilde{\eta}_{i}\right)+\left(\widetilde{\eta}^{i}\wedge\Gamma^{56}\right)\wedge\widetilde{\eta}_{i}+\widetilde{\theta}_{i}\wedge\left(\widetilde{\theta}^{i}\wedge\Gamma^{56}\right), \tag{87}$$

$$R^{i5}\wedge\widetilde{\eta}_{i}+\widetilde{\theta}_{i}\wedge R^{i6}=D\left(\widetilde{\theta}^{i}\wedge\widetilde{\eta}_{i}\right). \tag{88}$$

In the last line we used the constraint given by eq. (79) Consequently if R^{i6} and R^{i5} are zero then $D\left(\widetilde{\theta}^i \wedge \widetilde{\eta}_i\right) = 0$ or equivalently $\nabla_{[\rho} f_{\mu\nu]} = \nabla^*_\rho f^{\rho\nu} = 0$ and vice versa.

Corollary 1 *Note that the vanishing of the R^{56} curvature (that transforms as a Lorentz scalar) does not modify the equation of motion for Γ^{56} and simultaneously defines the electromagnetic field as*

$$R^{56} = d\Gamma^{56} + (m\lambda)^{-1}\,\theta_i \wedge \eta^i = 0, \tag{89}$$
$$\Rightarrow dA - F = 0. \tag{90}$$

■

7.2. Equations of motion in components and symmetries

Let us define

$$R^{ij}_{\{\}\mu\nu} = \partial_\mu \omega^{ij}_\nu - \partial_\nu \omega^{ij}_\mu + \omega^i_{\mu k}\omega^{kj}_\nu - \omega^{kj}_\mu \omega^i_{\nu k}, \tag{91}$$
$$T^i_{\mu\nu} = \partial_\mu e^i_\nu - \partial_\nu e^i_\mu + \omega_\mu{}^i{}_k e^k_\nu - \omega_\nu{}^i{}_k e^k_\mu, \tag{92}$$
$$S^i_{\mu\nu} = \partial_\mu f^i_\nu - \partial_\nu f^i_\mu + \omega_\mu{}^i{}_k f^k_\nu - \omega_\nu{}^i{}_k f^k_\mu. \tag{93}$$

Note that $S^i_{\mu\nu}$ is a totally antisymmetric torsion field due the symmetry of $f^i_\nu dx^\nu \equiv \eta^i$. Consequently the equations of motion in components become

$$\begin{aligned}
&\nabla_\mu\left[\sqrt{|g|}R^{ij\mu\nu}\right] + \sqrt{|g|}\left(-m^{-2}T^{ji\nu} + \lambda^{-2}S^{ji\nu}\right) - \sqrt{|g|}\,(\lambda m)^{-1} f^{[i\nu}A^{i]} = 0,\\
&\nabla_\mu\left[\sqrt{|g|}\left(R^{ij\mu\nu}_{\{\}} - m^{-2}e^{[i\mu}e^{j]\nu} + \lambda^{-2}f^{[i\mu}f^{j]\nu}\right)\right]\\
&\quad + \sqrt{|g|}\left(-m^{-2}T^{ji\nu} + \lambda^{-2}S^{ji\nu}\right) - \sqrt{|g|}\,(\lambda m)^{-1} f^{[i\nu}A^{i]} = 0,\\
&\nabla_\mu\left(\sqrt{|g|}T^{j\mu\nu}\right) + \sqrt{|g|}\left(R^{j\nu}_{\{\}} - m^{-2}e^{j\nu} + A^i A^\nu\right) = 0,\\
&\nabla_\mu\left(\sqrt{|g|}S^{j\mu i}\right) + \sqrt{|g|}\left(R^{ij}_{\{\}} - \lambda^{-2}f^{ij} + A^{[i}A^{j]}\right) = 0,\\
&\nabla_{[\mu}A_{\nu]} = F_{\mu\nu} = (\lambda m)^{-1}F_{\mu\nu},\\
&\nabla_{[\rho}F_{\mu\nu]} = 0.
\end{aligned} \tag{94}$$

8. Nonlinear realizations viewpoint

Notice that in our case eqs. (81,82) identify $\theta^i \sim e^i$ and $\eta^i \sim f^i$ being the table below completely understood. Also the Γ^{65} is identified with the **g** of E. Ivanov and J. Niederle [14, 15].

	this work	[14, 15]
R^{ij}	$R^{ij}_{\{\}} + m^{-2}\theta^i \wedge \theta^j + \lambda^{-2}\eta^i \wedge \eta^j$	$R^{ij}_{\{\}} + 4ge^i \wedge f^j$
R^{i5}	$m^{-1}\left[D\theta^i - \frac{m}{\lambda}\eta^i \wedge \Gamma^{65}\right]$	$De^i + 2ge^i \wedge \mathbf{g}$
R^{i6}	$-\lambda^{-1}\left[D\eta^i - \left(\frac{m}{\lambda}\right)^{-1}\theta^i \wedge \Gamma^{56}\right]$	$Df^i - 2gf^i \wedge \mathbf{g}$
R^{56}	$d\Gamma^{56} + (m\lambda)^{-1}\theta_i \wedge \eta^i$	$d\mathbf{g} + 4ge_i \wedge f^i$
DS/ADS reduction	Yes	No

Algebra and transformations in the case of the work of Ivanov and Niederle are different due different definitions of the generators of the $SO(2,4)$ algebra, however the meaning of $\mathbf{g}$ that it is associated to the connection Γ^{65} remains obscure for us because the second Cartan structure equations R^{i5} and R^{i6}. Notice that, although the group theoretical viewpoint in the case of the simoultaneous nonlinear realization of the affine and conformal group [50] to obtain Einstein gravity are more or less clear, the pure geometrical picture is still hard to recognize due the factorization problem and the orthogonality between coset elements and the corresponding elements of the stability subgroup.

9. Discussion

In this work we introduced two geometrical models: one linear and another one quadratic in curvature. Both models are based on the $SO(2,4)$ group. Dynamical breaking of this symmetry was considered. In both cases we introduced coherent states of the Klauder-Perelomov type, which as defined by the action of a group (generally a Lie group) will be invariant to the stability subgroup of the respective coset being related to the possible extension of the connection which maintains the proposed action invariant.

The linear action, unlike the cases of West or even McDowell and Mansouri[41], uses a symmetry breaking tensor that is dynamic and unrelated to a particular metric. Such a tensor depends on the introduced vectors (i.e. the coherent states) that intervene in the extension of the permissible symmetries of the original connection. Only some components of the curvature defined by the second structure equation of Cartan, are involved in the action, leaving the remaining ones as a system of independent or ignorable equations in the final dynamics. The quadratic action, however, is independent of any additional structure or geometric artifacts and all the curvatures (e.g. all the geometrical equations for the fields) play a role in the final action (Lagrangian of the theory).

With regard to the parameters that come into play $\boldsymbol{\lambda}$ and m (one plays the role of cosmological constant and mass respectively) we saw that in the case of linear action if they are taken dependent on the coordinates and under the conditions of invariance of the action, a new spontaneous compactification mechanism is defined in the subspace invariant under the stability subgroup.

Following this line of research, the steps to follow in terms of possible physical applications would be to point to the scenarios of the traditional GUT theories, the derivation of the symmetries of the standard model together with the gravitational ones and to obtain in a precise way the underlying fundamental theory. This will be important, for example, to solve the problem of hierarchies and fundamental constants, the masses of physical states and their interaction.

Also an important point under research is the noncommutative extensions using the coherent states structure of the connections, due that if there are coherent states a convolutory product (e.g. star product) can be stablished.

References

[1] A. P. Isaev, J. Math. Phys. **35**, 6784 (1994) [hep-th/9402060].

[2] P. Aschieri, L. Castellani and A. P. Isaev, Int. J. Mod. Phys. A **18**, 3555 (2003) [hep-th/0201223].

[3] M. Blagojevic, Gravitation and gauge symmetries, Bristol, UK: IOP (2002) 522 p.

[4] K. Hayashi and T. Shirafuji, Prog. Theor. Phys. **66**, 2258 (1981).

[5] A. B. Borisov, J. Phys. A **11**, 1057 (1978).

[6] R. Utiyama, Phys. Rev. **101**, 1597 (1956).

[7] S. Capozziello and M. De Laurentis, Phys. Rept. **509**, 167 (2011) [arXiv:1108.6266 [gr-qc]].

[8] F. W. Hehl, J. D. McCrea, E. W. Mielke and Y. Ne'eman, Phys. Rept. **258**, 1 (1995) [gr-qc/9402012].

[9] D. Ivanenko and G. Sardanashvily, Phys. Rept. **94**, 1 (1983).

[10] Y. N. Obukhov, Int. J. Geom. Meth. Mod. Phys. **3**, 95 (2006) [gr-qc/0601090].

[11] Y. Ne'eman and T. Regge, Riv. Nuovo Cim. **1N5**, 1 (1978).

[12] S. Gotzes and A. C. Hirshfeld, Annals Phys. **203**, 410 (1990).

[13] T. Shirafuji and M. Suzuki, Prog. Theor. Phys. **80**, 711 (1988).

[14] E. A. Ivanov and J. Niederle, Phys. Rev. D **25**, 976 (1982).

[15] E. A. Ivanov and J. Niederle, Phys. Rev. D **25**, 988 (1982).

[16] M. Leclerc, Annals Phys. **321**, 708 (2006) [gr-qc/0502005].

[17] K. S. Stelle and P. C. West, Phys. Rev. D **21**, 1466 (1980).

[18] A. A. Tseytlin, Phys. Rev. D **26**, 3327 (1982).

[19] E. A. Lord and P. Goswami, J. Math. Phys. **27**, 2415 (1986).

[20] E. A. Lord, J. Math. Phys. **27**, 3051 (1986).

[21] M. Greenberg, Lectures on Algebraic Topology, W.A. Benjamin Inc., Menlo Park, 1971.

[22] S. Kobayashi and K. Nomizu, Foundations of Differential Geometry, John Wiley & Sons, New York - London, 1963.

[23] G. Sardanashvily, Int. J. Geom. Meth. Mod. Phys. **8**, 1869 (2011) [arXiv:1110.1176 [math-ph]].

[24] G.Giachetta, L.Mangiarotti and G.Sardanashvily, Advanced Classical Field Theory, World Scientific, 2009.

[25] I. Kirsch, Phys. Rev. D **72**, 024001 (2005) [hep-th/0503024].

[26] M. Keyl, J. Math. Phys. **32**, 1065 (1991).

[27] L. Nikolova and V. A. Rizov, Rept. Math. Phys. **20**, 287 (1984).

[28] A. Sardanashvily, J. Math. Phys. **33**, 1546 (1992).

[29] G. Sardanashvily, Int. J. Geom. Meth. Mod. Phys. **3**, 139 (2006) [hep-th/0510168].

[30] G. Sardanashvily, Mathematical models of spontaneous symmetry breaking, arXiv:0802.2382 [math-ph].

[31] G. Sardanashvily, Theor. Math. Phys. **181**, 1598 (2014) [arXiv:1602.03818 [math-ph]].

[32] A. Trautman, Differential Geometry For Physicists, Naples, Italy: Bibliopolis (1984) 145 P.

[33] H. B. Lawson and M. L. Michelsohn, Spin geometry, Princeton Univ. Press, Princeton, 1989.

[34] G. Sardanashvily, Phys. Lett. A **75**, 257 (1980).

[35] G. A. Sardanashvily and O. Zakharov, Gauge gravitation theory, Singapore, Singapore: World Scientific (1992) 122 p.

[36] S. W. Hawking and G. F. R. Ellis, The Large Scale Structure of Space-Time, Cambr. Univ. Press, Cambridge, 1973. 404 pp.

[37] G. Sardanashvily, Czech. J. Phys. B **33**, 610 (1983).

[38] D. V. Volkov and V. A. Soroka, JETP Lett. **18**, 312 (1973) [Pisma Zh. Eksp. Teor. Fiz. **18**, 529 (1973)].

[39] V. P. Akulov, D. V. Volkov and V. A. Soroka, JETP Lett. **22**, 187 (1975) [Pisma Zh. Eksp. Teor. Fiz. **22**, 396 (1975)].

[40] P. Nath and R. L. Arnowitt, Phys. Lett. **56B**, 177 (1975).

[41] S. W. MacDowell and F. Mansouri, Phys. Rev. Lett. **38**, 739 (1977) Erratum: [Phys. Rev. Lett. **38**, 1376 (1977)].

[42] D. V. Volkov and A. I. Pashnev, Theor. Math. Phys. **44**, 770 (1980) [Teor. Mat. Fiz. **44**, 321 (1980)].

[43] E. Inonu and E. P. Wigner, Proc. Nat. Acad. Sci. **39**, 510 (1953).

[44] D. J. Cirilo-Lombardo, Found. Phys. **37**, 919 (2007) [Found. Phys. **37**, 1149 (2007)] [Found. Phys. **38**, 99 (2008)] [hep-th/0701195].

[45] J. A. de Azcarraga and J. Lukierski, Z. Phys. C **30**, 221 (1986).

[46] D.J. Cirilo-Lombardo and A. Arbuzov, work in progress.

[47] V.I.Ogievetsky, "Infinite-dimensional algebra of general covariance group as the closure of finite-dimensional algebras of conformal and linear groups," Lett. Nuovo Cim.**8**, 988 (1973).

[48] D.V.Volkov and V.P.Akulov, "Is the Neutrino a Goldstone Particle?," Phys. Lett.**46B**, 109 (1973).

[49] S.Capozziello, D.J.Cirilo-Lombardo and M.De Laurentis, "The Affine Structure of Gravitational Theories: Symplectic Groups and Geometry," Int.J. Geom. Meth. Mod.Phys.**11**, no. 10, 1450081 (2014)

[50] A.B.Borisov and V.I.Ogievetsky, "Theory of Dynamical Affine and Conformal Symmetries as Gravity Theory," Theor.Math.Phys.**21**, 1179 (1975) [Teor.Mat. Fiz.**21**, 329 (1974)].

Fermions, hairy blackholes and hairy wormholes in anti-de Sitter spaces*

Mihailo Čubrović†
Center for the Study of Complex Systems,
Institute of Physics Belgrade, Serbia

Abstract

We discuss the existence, properties and construction (analytical and numerical) of hairy black holes with fermionic matter in asymptotically anti-de-Sitter space. The negative cosmological constant makes hairy black holes stable, and the nucleation mechanism can make the formation of hair at the horizon energetically and entropically preferable to conventional black holes. The difficulties intrinsic to fermions at finite density – the Pauli principle and exchange interactions – require some drastic approximations in calculating the stress-energy tensor and geometry. We will consider several methods on the market – Hartree-Fock, WKB, and fluid-mechanical methods, and consider the dual field theories of these constructions. Then we will apply the same methods to the construction of wormholes; fermions are a natural candidate for wormhole source matter as they have a Dirac sea of negative energies, and negative energy-momentum density is the condition for wormhole formation. The field theory interpretation of wormholes is still open but has to do with strongly entangled systems. The paper combines a pedagogical introduction to the basic methods and results (obtained in the last 10+ years) with an account of fresh research results, mainly on the wormhole applications and non-planar black holes.

1. Introduction

AdS black holes are a favorite topic, not only in relation to holography but also in general: AdS space behaves like a potential box, the cosmological constant provides an effective repulsive force at large distances and the existence of a boundary at spatial infinity makes bound states possible. All of

* The author acknowledges funding provided by the Institute of Physics Belgrade through the grant by the Ministry of Education, Science and Technological Development. The author acknowledges the use of the Sci-Hub service. The results described here would never have been possible without the teaching, help and collaboration from Jan Zaanen, Koenraad Schalm, Yan Liu, Ya-Wen Sun, Elena Gubankova, Mariya Medvedyeva, Vladan Djukić and Nicolas Chagnet.

† e-mail address: mcubrovic@gmail.com

this brings about the famous result that hairy black holes are indisputably possible, and well-studied. In full (global) AdS space, one may have small black holes, which barely see the boundary and radiate like in asymptotically flat space, and large black holes, which reach an equilibrium state with the Hawking radiation at given temperature and remain stable forever (eternal AdS black holes). We will focus on the latter, as they can be treated as (semi)classical stationary systems. Clearly, just like the Hawking radiation, matter and gauge fields can likewise equilibrate between the black hole horizon and AdS boundary, possibly forming hair – by definition, it means nonzero density of some field (and possibly nonzero expectation values of other operators, like charge density, spin, etc) at the horizon itself. This in turn means that the geometry changes as opposed to the no-hair case: the hair itself enters the stress-energy tensor, and the outcome is a hairy black hole geometry, where a horizon still exists but with a different metric. At zero temperature, hair tends to remove extremal black holes in favor of zero-area horizons, with zero Bekenstein-Hawking entropy. We will soon discuss several explicit examples of this phenomenon.

The above story acquires an additional dimension thanks primarily to the AdS/CFT correspondence (gauge/gravity duality) [1, 2, 3] – the fact that the bulk gravity physics is equivalent to a quantum field theory in flat space in one dimension less, whose operators act as boundary sources of the AdS (bulk) fields. The actions in AdS (with field Φ) and in CFT (with field O, which acts as a boundary source to Φ) are equal:

$$\begin{aligned} S_{\rm AdS} &= S_{\rm CFT} \\ S_{\rm AdS} &= \int \mathcal{D}\Phi \exp\left(-\int_{\rm AdS} d^{D+1}x\sqrt{-g}\mathcal{L}_{\rm AdS}\left(\Phi,\partial_\mu\Phi\right) + \oint_\partial d^Dx\sqrt{-h}\mathcal{O}\Phi\right) \\ S_{\rm CFT} &= \int \mathcal{D}\mathcal{O}\exp\left(-\int d^Dx\mathcal{L}_{\rm CFT}\left(\mathcal{O}\right)\right), \end{aligned} \tag{1}$$

where we have denoted by ∂ the boundary of the AdS space, $g_{\mu\nu}$ is the AdS metric and $h_{\mu\nu}$ is the induced metric at the boundary. From now on, integrals over the bulk of AdS will be dnoted just by $\int$, understanding that the integral is over the whole space. At this place we do not intend to explain AdS/CFT and its applications in any detail; suffice to say that one can obtain thermodynamic potentials and correlation functions in field theory, which has found important applications in condensed matter theory, quantum chromodynamics and conformal field theory. Interested readers can consult [4, 5, 6] for reviews. In this work we deal with the gravity side of the correspondence – the formation of a hairy black hole with fermionic matter, which corresponds to a finite electron density phase in field theory. We assume the familiarity with the basic notions of AdS space and quantum field theory in curved spacetime, for example at the level of [7] and [8], respectively.

Mathematically, the topic of this review is the solution of the coupled Einstein-Maxwell-Dirac system with the total action $S_{\rm AdS} = S_{\rm bulk} + S_\partial$.

The bulk action reads:

$$\begin{aligned} S_{\text{bulk}} &= S_{\text{E}} + S_{\text{M}} + S_{\text{Dir}} \\ S_{\text{E}} &= \int d^4x\sqrt{-g}\,(R+6) \\ S_{\text{M}} &= -\int d^4x\sqrt{-g}\frac{\hat{F}^2}{4} \\ S_{\text{Dir}} &= -\int d^4x\sqrt{-g}\left(\frac{1}{2}\bar{\Psi}D_\mu e^\mu_a\Gamma^a\Psi + \frac{1}{2}\bar{\Psi}e^\mu_a\Gamma^a\Psi + m\bar{\Psi}\Psi\right). \end{aligned} \tag{2}$$

Here, $\hat{F}_{\mu\nu} = \partial_\mu A_\nu - \partial_\nu A_\mu$ is the electromagnetic (EM) field strength tensor, and the the cosmological constant in AdS_4 is $6/L^2$, where the AdS radius $L = 1$ is set to unity, as we will mainly work on the Poincare patch of AdS space, so all other dimensionful quantities can be expressed in terms of L. The Dirac bispinor Ψ has mass m and charge q, and the covariant derivative

$$D_\mu = e^a_\mu D_a = \partial_\mu - \frac{\imath}{8}\left[\Gamma^a, \Gamma^b\right]\omega^\mu_{ab} - \imath q A_\mu \tag{3}$$

depends on the spin connection ω^μ_{ab} and the gauge field A_μ, and the gamma matrices satisfy the usual relations $\left[\Gamma^a, \Gamma^b\right] = 2\eta^{ab}$, with the Minkowski metric η. We will be using the mostly plus convention. Obviously, $\Psi = 0$ is a solution, and in this case we get a Schwarzschild black hole if the EM field is also zero, or a charged Reissner-Nordstrom (RN) black hole for nonzero field strength. The question is, are there other solutions, with nonzero profile Ψ? Such solutions describe hairy black holes at finite temperature: the horizon is typically still there, but the geometry is changed. At zero temperature, the black hole might disappear. Since AdS space has a boundary, there is also a boundary contribution to the action, as in (1), depending on extrinsic curvature K, boundary cosmological constant λ and the boundary values of the fields:

$$S_\partial = \oint_\partial d^3x\sqrt{-h}\left[K - \lambda - \frac{1}{2}n_\mu A_\nu\hat{F}^{\mu\nu} - \frac{1}{2}\bar{\Psi}\Psi\right]. \tag{4}$$

The classical equations of motion do not depend on the boundary action. However, S_∂ is still important (1) to make sure there is a good action principle, i.e., that the on-shell solutions are indeed minima of the action[1] (2) to regularize any UV divergences (3) to get correct thermodynamics. The last point will be particularly important: one way to see that the hairy black hole and not the bald black hole is the true vacuum will be the fact that the action on the hairy solution is lower.

Solving the system (2) is a problem in quantum field theory at finite density. We work with classical general relativity (GR) and classical EM

[1]Remember that the (bulk) Euler-Lagrange equations are only a necessary condition for the minimum of the action.

field, but *fermions are never classical*; this is the first important lesson. The Pauli principle always introduces nonlocal correlations which show as the exchange interaction. Another way of saying this is that the pressure of a fermionic gas or fluid always includes the quantum contribution which is absent in both classical and bosonic gas; that is the reason that organized matter such as stars, planets, chairs and notebooks has rigidity and does not collapse onto itself. Therefore, even though we do gravity at $\hbar = 0$, the fermions even at leading order need to be tackled quantum-mechanically. This means calculating the *fermionic determinant*:

$$Z_{\text{Dir}} = \int \mathcal{D}\bar{\Psi}\mathcal{D}\Psi e^{-S_{\text{Dir}}} \text{ " = " } \left[\det\left(D_\mu e^\mu_a \Gamma^a + e^\mu_a \Gamma^a D_\mu + m\right)\right]^{1/2}. \quad (5)$$

We have put the equality sign under quotation marks because the determinant is actually the product of the eigenenergies of all the modes (an infinity of them), which is not only badly divergent (that could be regulated) but is also impossible to calculate because of the *fermion sign problem*, the fact that the fermionic modes enter the path integral with a sign that can be plus or minus. This makes the measure in the path integral (5) non-probabilistic and makes it impossible to expand around a classical solution in a controlled way. Fortunately, the AdS metric turns out to simplify the problem enough that it can be tackled in a way which is tractable and, while of course not exact, can be systematically improved in a perturbative way. This is in fact the motivation behind AdS/CFT modelling of strongly correlated electron systems: the fermion sign problem is fatal for strongly coupled field theories in flat space, but in GR with AdS boundary conditions it transforms into a difficult but doable task.

Is the journey worthwhile? In line with the broad scope of the Belgrade Mathematical Physics Meetings, we have anticipated a broad readership of this paper and thus we have decided to give a very general and perhaps rather dry introduction to the topic of fermionic hairy black holes. This necessarily means that we will not touch upon the many interesting applications: AdS/CFT and its applications to quantum chromodynamics and condensed matter physics, the black hole information problem, the critical phenomena in gravitational collapse and the black hole solutions in string theory. We do discuss one special topic that we currently find very interesting: hairy wormholes generated by fermion matter, where many of the methods used for hairy black holes can be successfully applied. The main task of the paper is to provide a tutorial on the basic methodology and calculation techniques, bringing the reader to the point that he can understand and repeat the calculations from the literature and start doing his own. The existing literature is rather heterogenuous and there is no single text to recommend. We will give the references we deem particularly useful throughout the paper, without the pretention of being exhaustive; the choice of references is certainly dictated also by our prejudices and tastes.

Plan of the paper. In Section 2 we first explain the instabilities of AdS space and AdS black holes to a nonzero density profile of fermions, and introduce the basic concepts that will keep appearing throughout the paper: effective potential and the bound states of the fermionic wavefunctions. In Section 3 we first treat the problem in the consistent one-loop (Hartree-Fock) approximation, calculating the determinant (5) by definition, from the individual wavefunctions for different states. We find this job surprisingly difficult – it is still an active research area. But we are able to give a qualitative picture of the outcome and sketch the phase diagram, depending on the chemical potential μ and fermion mass and charge m, q. As we move toward the high-fermion-density corner of the phase diagram, the things simplify. The simplest and "most classical" limit of the problem is the limit of large density. It is a rule of thumb that for fermions, the role of interactions diminishes as the density grows. At high density, WKB approximation works very well. At highest densities, we find semiclassical fluid with an equation of state that takes into account the fermionic pressure, similar in spirit to the Oppenheimer-Volkov equations for neutron stars. In section 4 we apply these methods to a different topic – hairy wormholes instead of black holes. This problem has recently gained notoriety and might carry some important messages for the black hole information problem. The final section sums up the conclusions.

2. Planar AdS black holes and fermion nucleation

In this and the next section we will focus on large planar black holes on the Poincare patch of AdS space. Large black holes can reach equilibrium with the AdS boundary so they do not emit Hawking radiation and can exist eternally. The Poincare patch of AdS_4 space is a coordinate chart with a single boundary on one end and interior on the other end. It does not cover the whole AdS space but is simpler to work with than global AdS and is good enough to desribe the instability at the horizon. The metric of pure AdS space witout a black hole is given by

$$ds^2 = r^2 \left(-dt^2 + d\vec{x}^2\right) + \frac{dr^2}{r^2} = \frac{1}{z^2}\left(-dt^2 + d\vec{x}^2 + dz^2\right) \tag{6}$$

where $r = 1/z$ is the radial coordinate, t is time and $\vec{x} = (x, y)$ are the transverse coordinates. The AdS boundary is at $r = \infty$ ($z = 0$), and the interior is at $r = 0$ ($z = \infty$). From now on we will mainly use the z coordinate; we will always specify explicitly if a different radial coordinate is used. In AdS/CFT, the radial coordinate corresponds to the energy scale in field theory: the near-boundary region encodes for the physics at high energies, in the ultraviolet (UV), and the deep interior, with z large, is the infrared (IR). Even though we do not consider the CFT dual here, we will still adopt the UV/IR terminology.

In the presence of a point electric charge e we get a Reissner-Nordstrom (RN) black hole with the horizon at $z_h = 1$, with charge e, mass M and temperature T:

$$ds^2 = \frac{1}{z^2}\left(-f(z)dt^2 + d\vec{x}^2 + \frac{dz^2}{f(z)}\right), \quad f(z) = 1 - Mz^3 + e^2z^4$$
$$M = z_h^3 + e^2, \quad A = \frac{ez_h}{2\sqrt{\pi}}(1 - z/z_h)dt, \quad T = \frac{3z_h}{4\pi}\left(1 - \frac{e^2}{3z_h^4}\right) \tag{7}$$

For $e = 0$ we get the Schwarzschild AdS black hole, and for $e = \sqrt{3}z_h^2$ the black hole becomes extremal, with temperature $T = 0$. To see this, remember that the black hole temperature is given by $f(z \to z_H) = 4\pi T(z - z_H) + \ldots$, so plugging in f from above we indeed get the correct expression for T. Importantly, the near-horizon region of a black hole is an AdS space [7]. This IR AdS space (near $z = z_h$) has a priori nothing to do with the AdS asymptotics in the UV (near $z = 0$); it is there also for black holes in flat or dS space. At $T = 0$, rescaling $z - e/\sqrt{3} \mapsto 1/6\epsilon\xi$ and expanding in ϵ to lowest order gives the metric

$$ds^2 = \frac{1}{6}(-dt^2 + d\xi^2) + \frac{e^2}{3}d\vec{x}^2, \quad A_t = \frac{1}{\sqrt{6}\xi}. \tag{8}$$

The is $AdS_2 \times R^2$ geometry, a direct product of AdS with a plane. At finite temperature, a similar rescaling can be worked out, yielding again an AdS_2 throat. Since the throat describes the near-horizon region, instabilities of the black hole can be figured out from possible instabilities of this IR AdS space. Once again, this is *not* the whole AdS_4, which is always stable far from the horizon, in the UV (otherwise our whole classical gravity approach crumbles down), it is just a region near the horizon, in IR.

In order to write the equations of motion, we have to choose a basis for the gamma matrices and the form of the Dirac bispinor (remember that only two out of four components are really independent degrees of freedom). A convenient representation is

$$\Gamma^0 = \sigma^1 \otimes \imath\sigma_2, \; \Gamma^1 = \sigma^1 \otimes \sigma_1, \Gamma^2 = \sigma^1 \otimes \sigma^3, \; \Gamma^z = \sigma^3 \otimes \hat{1}. \tag{9}$$

so that the Dirac equation in a spherically symmetric metric defined as $\mathrm{diag}(g_{tt}, g_{ii}, g_{ii}, g_{zz})$ gives two equivalent decoupled pairs of equations. Taking the Dirac bispinor in the form $\Psi = (\psi_1, \chi_1, \imath\chi_2, \imath\psi_2)^T$, the equations for $\psi_{1,2}$ read [9, 10]:[2]

$$\partial_z\psi_{1,2} \pm \hat{m}\psi_{1,2} - \left(\mp\hat{E} + \hat{k}\right)\psi_{2,1} = 0 \tag{10}$$
$$\hat{m} \equiv m\sqrt{g_{zz}}, \hat{\mu} \equiv \sqrt{\frac{g_{zz}}{-g_{tt}}}A_t, \; \hat{E} \equiv q\hat{\mu} + E\sqrt{\frac{g_{zz}}{-g_{tt}}}, \; \hat{k} \equiv \sqrt{\frac{g_{zz}}{g_{ii}}}k. \tag{11}$$

[2]Since only two components of the Dirac bispinor are independent, the system for $\chi_{1,2}$ yields no new information.

We have Fourier-transformed the derivatives over time and transverse spatial dimensions as $\partial_t = -\imath\omega, \partial_x = \imath k_x, \partial_y = \imath k_y$, and we have exploited the spherical symmetry to set $k_x = k, k_y = 0$. The quantities $\hat{E}$, $\hat{k}$, $\hat{\mu}$ can be informally interpreted as "local" values of the energy, momentum and chemical potential, respectively. The "local" values equal E, k, μ at the AdS boundary, grow monotonously toward the horizon and diverge there, a consequence of the infinite redshift seen by a faraway observer. An important idea is to consider the Schrödinger form of the Dirac equation instead, differentiating (10) once with respect to z, decoupling the equations for $\psi_{1,2}$, and elliminating the first derivatives $\psi'_{1,2}$ by introducing the tortoise coordinate s instead of z. The resulting picture is that of a zero-energy Schrödinger equation, of the form $\partial_s^2\psi_{1,2} - V_{\text{eff}}(s)\psi_{1,2} = 0$, in an effective potential $V_{\text{eff}}(s)$.[3] Near the horizon, the potential is constant at leading order [11]:

$$V_{\text{eff}}(s \to -\infty) = \frac{m^2 + 12k^2/\mu^2 - 2q^2}{(q/\sqrt{2} + k)^2} + \dots \tag{12}$$

It is true that the Schrödinger form is only a consequence of the Dirac equation, not equivalent to it: extra conditions must be imposed on the Schrödinger solution to make it satisfy the Dirac equation. But the effective potential is great for qualitative insights and it contains the basic idea of the black hole instability in a very transparent way. The near-horizon potential can contain bound states if it is negative, hence the instability criterion for a fermionic mode with momentum k is that the numerator of (12) is negative. Fermions fill up the potential well starting from $k = 0$ up to some maximum k for which (12) reaches zero. Therefore, the instability first sets in when V_{eff} is negative for $k \to 0$, so we get our first rule-of-thumb prediction: the black hole will be surrounded by a gas of fermions and become hairy when

$$m < q\sqrt{2}. \tag{13}$$

But this is just one end of the potential well; what happens at the other end? Plugging in the pure AdS metric (6) into (12) we get

$$V_{\text{AdS}}(s \to 0) = \frac{m^2 + m + k^2}{(k + \mu)^2}\frac{1}{s^2} + \dots, \tag{14}$$

which is always non-negative, and grows to infinity. This is good – there is never an instability in the far UV, and the fermionic hair can never come arbitrarily close to the AdS boundary. It also means that bound states in the interior will indeed exist whenever (13) is negative. The physical picture is the following: in the presence of EM and gravitational field of the black hole, fermions are pair-created. These pairs are virtual, and

[3]This is a simple exercise that we will do many times; the reader should be able to do the necessary (straightforward) calculations leading to the expression for $s(z)$ and $V_{\text{eff}}(s)$.

they only have a finite probability of becoming long-living if the external potential energy is large enough. In that case, bound states form, and there is a solution of (2) with nonzero fermion density. In the literature, this is sometimes called fermion nucleation. For scalars, similar logic leads to the Breitenlohner-Freedman bound, which puts a constraint on the scalar mass for the stability of the UV (with fermions, as we have seen, UV is always stable), and in IR it similarly gives a criterion for forming hair [5]. We also see from (14) and Fig. 1 that the potential well becomes shallower as k grows, so the bound states only exist up to some maximum $k = k_F$ which is really the Fermi momentum of the bulk Fermi sea.

From (12,14) we can understand the behavior of the effective potential. In Fig. 1, we give the function $V(s)$ in the whole space, from $z = 0$ ($s = 0$), to $z = z_h$ ($s = -\infty$). The fermionic modes fill the potential well until they reach the energy $E = 0$. From (12), higher modes correspond to higher momentum k. The fermionic density is thus given by a sum over these bound states. The easiest case is in fact an extremely deep well: the energy levels are so dense and so numerous that they can be approximated by a continuum; this is called electron star limit. But the most interesting regime is the one with only a few wavefunctions, which really describes the transition to a hairy solution. This is a much harder nut to crack.

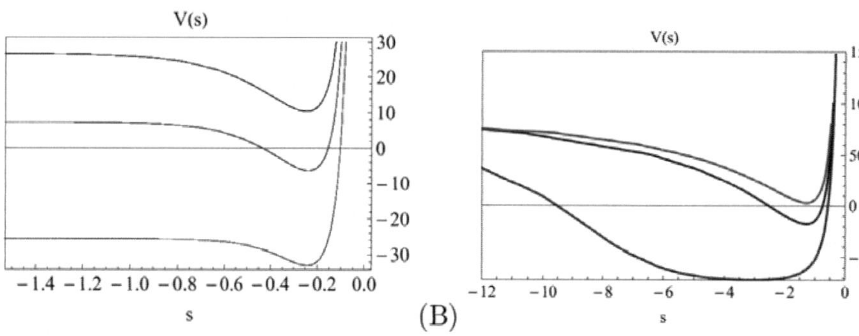

(A) (B)

Figure 1: Effective potential V_{eff}, as a function of the tortoise coordinate $s \equiv \int_0^z dz g_{zz}(z)$, in the RN metric (A), and in the Lifshitz metric (B), for $q = 1$, $m = 0.4$, $\mu = 1$, and three momentum values increasing from violet to blue to red: $k = 1, 2, 3$ (A) and $k = 0, 5, 10$ (B). In both cases, the negative potential well becomes shallower and shallower and eventually disappears as k grows, so we fill the bulk Fermi sea up from $k = 0$ to some maximal $k = k_F$. In the black hole background, the potential is flat for $s \to -\infty$, which corresponds to the AdS_2 near-horizon region and signifies an instability as the bound states extend all the way to the horizon ($s = -\infty$). In the backreacted Lifshitz metric the potential grows for $s \to -\infty$, suggesting that deep IR is stable: the true vacuum is the Lifshitz geometry, not RN. Taken over from [14].

3. Fermionic hair

Now that we have convinced ourselves that hairy solutions, with finite fermion density, have to exist, we need to solve the full system of Einstein, Maxwell and Dirac equations to find them. Clearly, a more general ansatz for the metric than (7) is needed now, and we will write it as

$$ds^2 = -\frac{F(z)G(z)}{z^2}dt^2 + \frac{1}{z^2}d\vec{x}^2 + \frac{1}{F(z)z^2}dz^2, \tag{15}$$

leading to Einstein-Maxwell equations

$$1 - F + zF'/3 - T^{\rm tot}_{tt}FG/3z^2 = 0 \tag{16}$$
$$G' + z(T^{\rm tot}_{tt}/F^2 + T^{\rm tot}_{zz}G) = 0 \tag{17}$$
$$A_t'' - G'/2GA_t' + qn\sqrt{G}/\sqrt{F}z^3 = 0, \tag{18}$$

where $T^{\rm tot}_{\mu\nu}$ is the total stress-energy tensor, both from the electric field (which is easy to find) and from the fermions (which is our big problem). A typical situation in hairy problems is that formulating the physically meaningful boundary conditions is not so easy. Notice the Einstein equations are first-order, so we need one boundary condition for each function (F and G), whereas the Maxwell equation is second-order and requires two boundary conditions. Let us now summarize what boundary behavior we expect on physical grounds.

1. The AdS asymptotics for the metric and gauge field require $F(z \to 0), G(z \to 0) = 1, A_t(z \to 0) = \mu$. So far it's all simple.
2. The main puzzle for the IR geometry is – does the horizon disappear or not? At $T = 0$ we do not expect that the degenerate RN horizon can survive. So we do not expect zeros in F, G but we do expect their derivatives to vanish in order to have a smooth solution (finite derivatives at $z \to \infty$ would likely give divergent curvature). Thus at $T = 0$ we need $F'(z \to \infty) = G'(z \to \infty) = 0$ or, in other words, $F(z \to \infty) = \text{const.} + O(1/z)$ and likewise for G. At finite temperature, general GR arguments suggest there is a horizon at some $z = z_h$ satisfying $F'(z \to z_h) = 4\pi T$.
3. The IR behavior of the gauge field is related to the question: *is all the charge carried by the fermions, or the charge is shared between the fermions and the horizon?* The Gauss-Ostrogradsky theorem for the AdS space, with a UV boundary and either a horizon or a smooth far-away IR takes the form [12]:

$$\oint_\partial d^3x\sqrt{-h}|_{z\to 0} \star \hat{F} = \int d^4x\sqrt{-g}qn + \oint_{\rm IR} d^3x\sqrt{-h_{\rm IR}}|_{z=z_{\rm IR}} \star \hat{F} \tag{19}$$

Here, $\star\hat{F}$ is the coordinate-invariant flux of the 2-form $\hat{F}$, and $h_{\rm IR}$ is the induced metric on the surface normal to the radial direction at

$z_{\rm IR} = z_h$ or $z_{\rm IR} = \infty$, depending on whether there is a horizon or not. In principle, the IR charge might be shared between the horizon and the fermions. However, we will find that in the semiclassical calculation there are no solutions where the charge is shared – any backreaction will always expell all the charge from the IR.

4. The boundary conditions for the Dirac equation present no problems and are pretty standard in AdS space [13]. In the UV, out of the two branches, we want the subleading one, with the motivation to preserve the AdS asymptotics, i.e., to perturb the space as little as possible in the UV. In particular, the near-boundary expansion of (11) gives

$$\psi_1(z \to 0) = \frac{E + \mu q - k}{2m - 1} A_2 z^{5/2-m} + B_1 z^{3/2+m} + \ldots$$
$$\psi_2(z \to 0) = A_2 z^{3/2-m} + \frac{E + \mu q + k}{2m + 1} B_1 z^{5/2+m} + \ldots, \quad (20)$$

so we pick $A_2 = 0$, as the leading contribution for $z \to 0$ comes from the $z^{3/2}$ term. In the IR, the metric determines the boundary conditions: if there is a horizon, we need $\Psi(z = z_h) \to 0$ for stability, if not, then to avoid infinite energy density at large z we require $\partial_z \Psi(z \to \infty) = 0$, for otherwise a nonconstant density profile would give rise to a diverging curvature. The attentive reader should be alarmed: this means two boundary conditions for each component (one in UV and one in IR), but the equations are only first-order. The resolution is that for given momenta, the energy is not arbitrary but fixed by the dispersion relation $E(k)$; thus solving the Dirac equation in an effective potential well introducs energy quantization, as one would expect.

What remains is to find the fermionic stress tensor. Since spinors couple to the spin connection e_a^μ and not directly to the metric, the stress tensor is expressed as

$$T_{\mu\nu} = \left\langle \frac{1}{4} e_{\mu a} \bar{\Psi} \Gamma^a D_\nu \Psi + (\mu \leftrightarrow \nu) \right\rangle, \quad (21)$$

and the expectation value $\langle \ldots \rangle$ reminds us that the fermions are never classical. At zero temperature, the state is pure and can be represented as the sum of (appropriately normalized) radial modes with energies E_ℓ, where ℓ is the radial quantum number, and the energies E_ℓ are all ≤ 0. At finite temperature, the state is mixed and gets a contribution from both positive and negative energies E_ℓ, with thermal weights $w_\ell = \exp\left(-\beta E_\ell\right)/Z$, the partition sum being $Z = \sum_\ell \exp(-\beta E_\ell)$. With this in mind, we can wrote out (21) as

$$T_{tt} = e_{t0} \sum_{\ell=1}^{N} w_\ell \int_0^{k_F} \frac{kdk}{(2\pi)^2} \left(\psi_{1;\ell}^\dagger \psi_{1;\ell} + \psi_{2;\ell}^\dagger \psi_{2;\ell} \right) \left(E_\ell + q A_t \right)$$

$$
\begin{aligned}
T_{ii} &= e_{i1}\sum_{\ell=1}^{N} w_\ell \int_0^{k_F} \frac{kdk}{(2\pi)^2}\left(\psi^\dagger_{1;\ell}\psi_{1;\ell} - \psi^\dagger_{2;\ell}\psi_{2;\ell}\right)k \\
T_{zz} &= e_{z3}\sum_{\ell=1}^{N} w_\ell \int_0^{k_F} \frac{kdk}{(2\pi)^2}\left(\psi^\dagger_{1;\ell}\partial_z\psi_{2;\ell} - \psi^\dagger_{2;\ell}\partial_z\psi_{1;\ell}\right). \qquad (22)
\end{aligned}
$$

For brevity, we write $\psi_{1,2;\ell} \equiv \psi_{1,2}(E\ell, k; z)$. We will consider in detail just the $T = 0$ case, when the weights w_ℓ effectively just pick the ground state and cut off all the others, but we will later discuss the results (without details of the calculations) also at finite T. The spectrum is discrete and gapped in the radial direction, so the integral $\int dE/2\pi$ becomes a sum, however in the transverse directions the system remains gapless, filling the whole (spherical) Fermi sea in the $k-$momentum space, as long as the dispersion relation $E(k) = E_\ell \leq 0$ is satisfied for some ℓ. The highest such k, for which $E_\ell = 0$, is the Fermi momentum k_F, and the possible momenta are $0 \leq k \leq k_F$. It is this continuous quantum number k that makes our life difficult. Here, indeed, our easy path comes to an end, because a self-consistent calculation of the wavefunctions certainly cannot be done in a closed form. Here we must resort to approximations. The number of occupied levels N is a good guide on the kind of approximation one needs to make. One can rephrase it as the ratio Q/q, where Q is the total fermion charge $\int d^4x\sqrt{-g}q\Psi^\dagger\Psi$. The thermodynamic limit, where the number of particles goes to infinity and the charge of an individual fermion to zero so that $N \to \infty, q \to 0, Q = Nq = \text{const.}$, is at one extreme. We expect that the problem approaches the classical regime in this case, and it will turn out to be true. The opposite limit is $Q/q = 1$, with just a single excitation, the hairy black hole at birth. We expect this to be likewise a simple limit, however it will turn out not to be quite true. In-between we dial between the quantum mechanics of $N = 1$ and the classical field theory of $N \to \infty$ [14].

Phase diagram. Before doing that, we can sum up our qualitative knowledge on a phase diagram (Fig. 2). From (12-14), bound states form for small enough m values (panel (A)); if (13) is valid beyond the probe approximation, the borderline is $m = q\sqrt{2}$. Left of this line there is a hairy solution, to the right of it the AdS_2 near-horizon region (and the whole RN black hole) remain. The hairy solutions are best described in different ways depending on the number of filled levels ($N = Q/q$); this is the topic od the rest of this section. One can also plot the situation at finite temperature (panel (B)). The phases remain the same; more precisely, the extremal black hole becomes a finite-temperature black hole, and the hairy solutions also smoothly develop a hairy horizon (thermal horizon with nonzero fermion density $n(z_h)$). What changes is the order of the phase transition: at $T = 0$ it is continuous, and at finite temperature it is discontinuous.

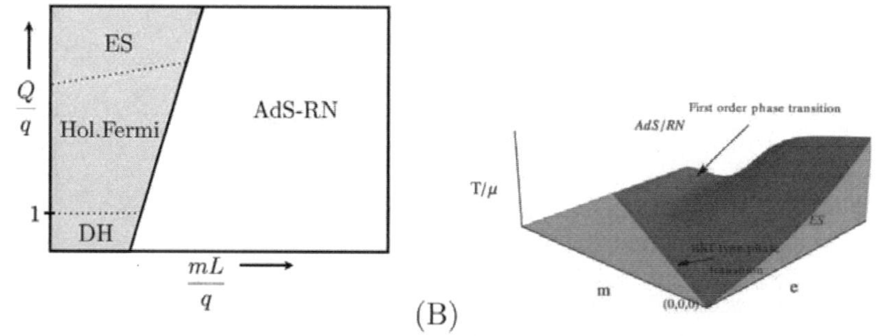

(A) (B)

Figure 2: (A) Phase diagram as a function of the total-to-fermion-charge ratio Q/q (y-axis), and the fermion mass (in units of AdS radius L) over charge ratio mL/q (x-axis). For large masses, the effective potential is positive and the ground state is the bald RN black hole, with quantum critical dual field theory. For smaller masses, hair develops, which corresponds to a Fermi liquid in dual field theory. For $Q \sim q$ (few wavefunctions), the single-wavefunction Dirac hair approximation works; for $Q/q \to \infty$ we approach the semiclassical fluid (electron star) limit; between them there is a smooth crossover with unclear properties, both in AdS and in the holographic dual. Notice different notational conventions for the total charge from the main text (e vs. Q). Taken over from [15]. (B) Adding nonzero temperature as the third axis, we obtain also the thermal phase transitions between the black hole and the hairy solution, which are generically first order, smoothing out to an infinite order (BKT) transition at $T = 0$ – the red line in (B) is the bold black line between the RN and hairy (blue) region in (A).

3.1. Quantum hairy black holes

A controlled approximation is to solve the problem perturbatively, at one-loop order in fermionic fields. This is nothing but the textbook Hartree-Fock (HF) method, but in curved space. Dynamical spacetime makes a big difference: it introduces an additional strongly nonlinear component of the system, making the solution landscape larger and less predictable, and the UV and IR divergences can appear also in the Einstein equations and need explicit regulators. In fact, this is still an open problem – nobody has yet classified the solutions of the Einstein-Maxwell-Dirac system even in the Hartree-Fock approximation, and we do not know what surprises might lurk in this corner of the phase diagram. The HF electrodynamics contains two diagrams, a vacuum bubble that renormalizes the chemical potential as $\hat{\mu}(z) \mapsto \hat{\mu}(z) + \delta\hat{\mu}(z)$ (the Hartree term):

$$\delta\hat{\mu}(z) \equiv q\sum_{\ell=1}^{N}\int\frac{kdk}{2\pi}\left(\psi_1^\dagger(E_\ell,k;z)\psi_1(E_\ell,k;z) + \psi_2^\dagger(E_\ell,k;z)\psi_2(E_\ell,k;z)\right) \tag{23}$$

and the exchange interaction (the Fock term). The explicit z-dependence of the Hartree correction is a gory reminder that the problem is solved in inhomogenous background. This is also the reason why already the Hartree correction is nontrivial: unlike the textbook situation where the shift $\delta\mu$ merely changes the numbers, here it is a radial function $\delta\mu(z)$ and its influence is also qualitative. So far, nobody even tried to do the whole HF calculation, and even just the Hartree term is not easy. We are plagued (1) by the UV divergences introduced by the modes close to $k = k_F$ which, as we have seen, peak most sharply near the boundary and can shatter the AdS space into pieces if not properly renormalized (2) by the IR divergences introduced by the modes with k close to zero, which extend far into large z values and can make the system unstable to forming a naked singularity.

Hard-wall Fermi liquid. The only case which is under good control is the hard-wall model of [12]: the UV divergences are resolved simply by not backreacting on the metric, i.e. solving just the Maxwell-Dirac system in fixed AdS metric (6) even without a black hole, and the IR divergences disappear by cutting off the space at some arbitrary z_0, so that we simply elliminate the IR region. The approximations are rather drastic, but they allow a complete solution. In pure AdS space, the solutions $\psi_{1,2}$ can be found analytically in terms of Bessel functions, the states form discrete and gapped bands, and we only have to solve the Maxwell equation (18). The outcome is given in Fig. 3. Hard wall acts as an infinite potential barrier, so the wavefunctions should die on it, and the condition $\psi_{1,2}(z_0) = 0$ determines the dispersion relation. The wall should not be charged, so in (19) the second term on the right-hand side equals zero, meaning that $A_t'(z_0) = 0$. The picture is that of a Fermi liquid, nicely filling the Fermi sea at momenta $k \leq k_F$ and having long-living quasiparticles. This model is an important starting point for more complicated setups, and has the advantage of being intuitive, but by itself is too simplistic. Indeed, we want to talk about hairy black holes, and here we don't even have one, as it is hidden behind the hard wall!

An attempt to study a simple setup but with a black hole was made in [16]. In this approach, we are limited to a single energy level, $\ell = 1$. This is justified only when the hair is just starting to form, right at the transition point. There is again no backreaction on metric, but the (fixed) metric is now taken to be the RN black hole. This is actually a big jump in difficulty: the wavefunctions oscillate near the horizon at any nonzero energy (Fig. 4(A)), so they can satisfy the IR boundary condition at any energy and momentum (we can always pick the phase so that $\psi'(z_h) = 0$), and the spectrum is continuous as there is no wall to create a gap. This is what forces us to consider the single-mode case: with the gapped hard-wall model we could add a *finite* number of modes, but now there is a *continuum* of them, N going to infinity even for arbitrarily small Q/q. The only way out is to *assume* there one mode only and solve the resulting Dirac-Maxwell system. This setup is convenient for understanding the transition itself, which turns out to be discontinuous (first-order) at finite temperature (Fig. 4(B), and likely infinite-order (Berezinskii-Kosterlitz-Thouless, BKT) at zero temperature, as we shall soon see.

(A) 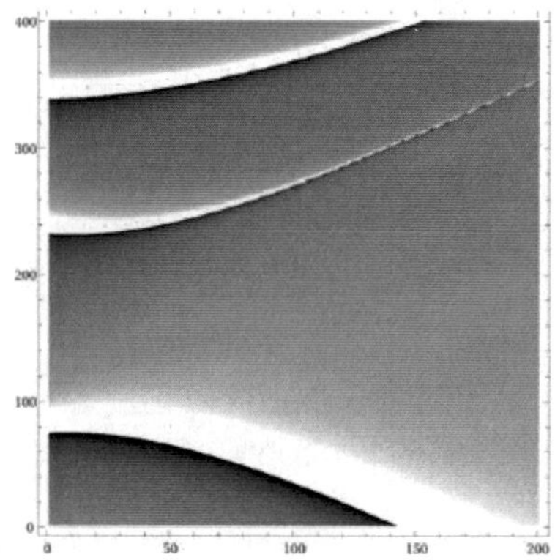 (B)

Figure 3: (A) Dispersion relation $E_\ell(k)$ for the first two electron bands $\ell = 1, 2$ in hard-wall AdS space, for $\mu = 1, m = 1, q = 2$. The first band from bottom is the hole band, not an an electron band – its contribution can be absorbed in the redefinition of the parameters and it does not contribute to hair. The colormap shows the resolvent of the Dirac operator, $(D_z\Gamma^z + \vec{D}\cdot\vec{\Gamma} - m - E)^{-1}$, thus the bright white regions show the places where the resolvent diverges and a discrete bound state is formed. The horizontal axis is the momentum and the vertical axis the energy, both in computational units. (B) Wavefunctions $\psi_{1,2}$ (here for $\ell = 1$ and $k = 1$) are smooth everywhere - what happen exactly at the horizon we do not know in this model, as the space is cut off at $z = 3$.

Quantum electron star. The single-mode approach has taught us a lesson: already at the level of the gauge field only, the changes from the finite fermion density are drastic, and the resulting stress tensor is large at the horizon, so a change of the black hole metric is certainly expected. However, when we try to solve the Einstein equatioons, things become almost intractable. Both UV and IR divergences appear: the former because the currents diverge in continuous space, and the latter because the discrete bands fuse into a continuum in IR. The latter issue is most easily regularized by a hard wall, but a hard wall does not make much sense if we want to backreact on geometry. The regularization of the UV divergences is systematically discussed in [17, 18] and the bottom line is that there is a logarithmic short-distance divergence which can be regularized by point

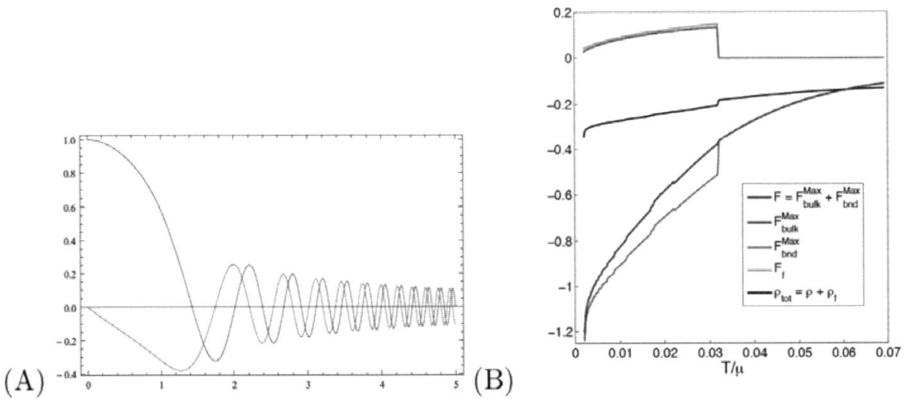

Figure 4: (A) Wavefunctions $\psi_{1,2}$in RN background, for $\ell = 1$ and $k = 1$, always oscillate and approach an essential singularity at the horizon, which indicates an instability: the metric changes and the degenerate horizon disappears. (B) The bulk action (or free energy F, from AdS/CFT correspondence) of the Maxwell (electric field), in blue, consists of the bulk and boundary contribution (dark green and red), the former practically identical to the contribution from fermions. All these are computed from the action (2-4). While the *total* free energy is continuous, it has a cusp, made manifest by the slight jump in density (black), a sign of first-order hair-forming transition.

splitting; in this procedure the cosmological constant becomes renormalized. This is not a drastic change: it will just change the numbers but not qualitative behavior. The IR problem is still unsolved. The approach of [18] is to put the system in global AdS space[4] whose radial slices are spheres, not planes, so the AdS radius provides a regulator. A perhaps more physical approach, motivated by consistent truncations from string theory, is to introduce a non-minimally coupled scalar, i.e., a dilaton that introduces a soft wall and suppresses the IR degrees of freedom in a continuous way, without an abrupt cutoff at some z_0, so the total bulk action is now

$$
\begin{aligned}
S_{\text{bulk}} &= \int d^4x\sqrt{-g}\left[R - V(\Phi) - \frac{1}{2}(\partial\Phi)^2 - \frac{Z(\Phi)}{4}\hat{F}^2\right] - \\
&- \int d^4x\sqrt{-g}\bar{\Psi}\left(\frac{1}{2}D_a\Gamma^a e^{\Phi} + \frac{1}{2}e^{\Phi}D_a\Gamma^a + m\right)\Psi, \qquad (24)
\end{aligned}
$$

where the dilaton potential reproduces the AdS cosmological constant near the boundary, i.e., $\Phi(z \to 0) = 0$ and $V(\Phi \to 0) = 6, Z(\Phi \to 0) = 1$. It is not clear if one can ever remove the IR regulator. That is precisely the

[4]Dual field theory then lives on a sphere instad of a plane.

reason that we regard the dilaton regulator as more physical, since string theory constructions as a rule contain non-minimally coupled scalars, and the action (24) can be obtained by consistent truncation; whereas global AdS is essentially an ad hoc solution, though a very interesting one, with possible applications in AdS/condensed matter duality, where systems that live on surfaces (such as a sphere) appear naturally.

While this is still very much a work in progress,[5] preliminary results suggest that the RN-to-hairy-black-hole transition at zero temperature is an infinite-order (BKT) transition, where all derivatives of S remain smooth (Fig. 5). This is the point where the potential just starts deviating very slightly from the flat IR behavior in Fig. 1(A). At the end of this section we will try to understand this (still conjectural) numerical finding analytically.

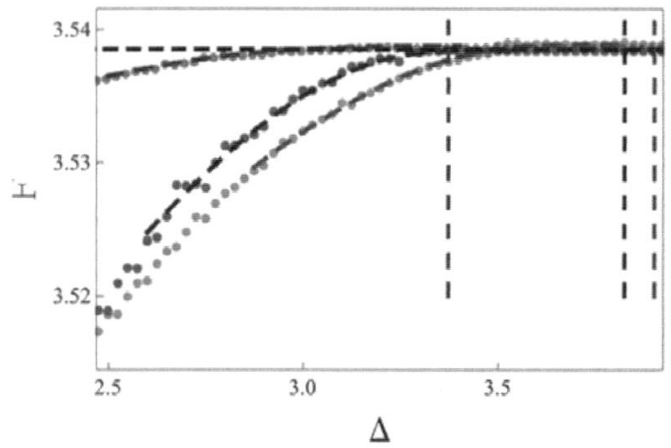

Figure 5: The bulk action (here denoted as free energy F, from AdS/CFT correspondence) as a function of the fermion mass (here denoted as $\Delta = 3/2 + m$) is very well fit by the BKT function $\exp(-c/\sqrt{\Delta_c - \Delta})$. The parameter c is determined by the chemical potential (we plot for three values $mu = 1.0, 1.5, 2.0$ in violet, blue, green). To the right of the transition point the action is independent of m as there is no hair, fermion density is zero, and so nothing depends on the fermion parameters. To the left of the transition point, the fermions form hair of nonzero density. Nobody knows yet how the near-horizon metric changes.

3.2. WKB star and electron star

WKB approach. We have followed the logical chain of reasoning from the point where the hair starts growing, having $Q/q \sim 1$ and deforming the black hole just a slight bit, towards larger and larger hair, eventually reaching the regime $Q/q \gg 1$. But this last regime is the easiest to approach, as the fermions become as close to classical as they can possibly be. A good starting point is the controlled expansion in $\hbar$, where we solve the

[5]With N. Chagnet, V. Djukić and K. Schalm.

Dirac equation in the eikonal approximation or, in other words, the WKB approach [15]. We express the wavefunction as

$$\psi_{1,2} = e^{\imath\theta_\pm}/\sqrt{p}, \quad p \equiv \sqrt{\hat{E}^2 - \hat{m}^2 - \hat{k}^2}, \tag{25}$$

where p has the role of the canonical momentum. The wavefunction is nonzero between the turning points $z_\pm$, determined by the equation $p(z_\pm) = 0$. The explicit from of the phase $\theta_\pm$ as well as higher-order corrections to the phase can be found in [15], but the reader should in fact have no difficulty in deriving them, following the usual WKB procedure (though for the Dirac equation instead of the Schrödinger equation). Now the density and pressure are found by inserting the solution (25) into (22). The procedure can be iterated to obtain self-consistent solutions, but now we solve the whole system including the Einstein equations. It is instructive to plot the total on-shell action (2) as a function of temperature (remember that finite temperature is imposed through the corresponding boundary condition for the metric function F).[6] Fig. 6 plots the dependence $\mathcal{F}(T)$ in the vicinity of the transition value T_c: the derivative $\partial\mathcal{F}/\partial T$ undergoes a jump which is nothing but the entropy $\mathcal{S} \equiv \partial\mathcal{F}/\partial T$. We thus find a *first-order phase transition* at the point when Fermi hair starts forming. Of course, don't forget that the WKB approach is in fact *not* to be trusted very near the transition point: at the transition N changes from 0 to 1, which is far from the regime $N \ll 1$. But the qualitative insight that at finite temperature the system undergoes a non-symmetry-breaking transition is likely robust and we expect to prove it also within the more rigorous fully quantum-mechanical approach of the previous subsection. It is a hairy version of the celebrated Hawking-Page transition [19], and confirms the intuition that the high-tmeperature phase is always a black hole; but now, the low-temperature phase is not simply a gas, but a dense fluid in AdS.

Ploting the density and pressure in Fig. 7(A), one finds that for high values of N they tend to a constant value in deep interior. This motivates the fluid ansatz taken in the electron star limit, now to be considered.

Electron star. Electron star is a charged, AdS version of the neutron stars, described as perfect fluid by the Oppenheimer-Volkov equations. The idea is to *assume* that the fermionic matter is a perfect fluid, and then express the energy density ρ, pressure p and charge density n in terms of *integrals* over energy and momenta (i.e., assume that the bound states are infinitely close, and the gaps between them vanish). The fluid approximation thus becomes exact in the limit of $N \to \infty$, as we expect from a semiclassical approximation. Anticipating the current and stress tensor of the form

$$T_{\mu\nu} = (\rho + p)u_\mu u_\nu + p g_{\mu\nu}, \quad N_\mu = n u_\mu, \tag{26}$$

[6] In AdS/CFT, the bulk on-shell action $\mathcal{S}$ precisely equals the free energy $\mathcal{F}$ of the CFT side. But even without considering the details of the CFT, we can still make use of this interpretation to detect a phase transition in the system.

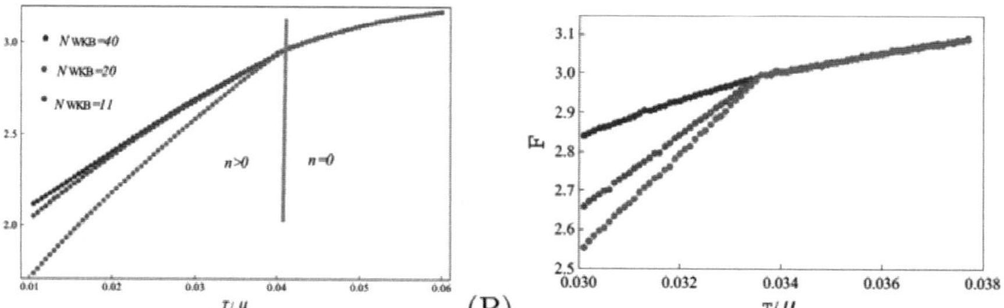

(A) (B)

Figure 6: (A) The on-shell action or free energy as a function of temperature, in the presence of fermions. For low temperatures, the fermion density is finite and the derivative $\partial\mathcal{F}/\partial T$ jumps at $T = T_c$, a sign of first-order transition with the development of the hair. This is in line with the Dirac hair result in the previous figure, and indeed for the lowest number of levels N_{WKB} the transition is the sharpest. In (B) we zoom in into the transition region.

we can write the density starting from (22) and making use the optical theorem to relate it to the imaginary part of the Feynmann propagator G_F. This spells out as

$$\begin{aligned}
\rho &= \int_0^{\hat{E}^2-k^2} \frac{dE}{2\pi}\int_0^{k_F}\frac{d^3k}{(2\pi)^3}\hat{E}\Im\mathrm{Tr}\imath\Gamma^0 G_F(E,k) \\
&= \int_0^{\hat{E}^2-k^2} dE\int\frac{k^2dk}{4\pi^3}\frac{1}{2}\left(1-\tanh\left(\frac{\beta}{2}\hat{E}\right)\right)\mathrm{Tr}(\imath\Gamma^0)^2\delta\left(\hat{E}-\sqrt{k^2+m^2}\right) \\
&= \frac{1}{\pi^2}\int_m^{\hat{\mu}} dE E^2\sqrt{E^2-m^2}. \qquad (27)
\end{aligned}$$

We similarly find the number density n, whereas the pressure need not be computed explicitly: since we work with an isotropic free Fermi fluid, its equation of state has to be $p = \rho - qn\hat{\mu}$. It is here that the approximate nature of the electron star with respect to the WKB star becomes obvious (Fig. 7): in WKB star there is an extra term in the pressure, coming from the nodes of the WKB wavefunction. One can check that the integral in (27) indeed approaches a constant as we go into deep interior. On the other hand, at some z_* when $\hat{\mu}(z_*) = m$ the density falls to zero: the star is a classical object and has a sharp border. So for $0 < z < z_*$ we continue the metric to the RN metric (the metric outside a charged isotropic object).

Since we can express n, ρ, p explicitly, we get a nice system of local ordinary differential equations in F, G, A_t, with all quantum expectation values pulled under the rug. This completes the circle, and brings another universal message: *due to Pauli principle, fermionic operators are never local, except in two extreme cases: when only one state is occupied (so the format of the Slater determinant is* 1×1*, i.e., it contains a single state), or* *when infinitely many states are occupied, so the Slater determinant turns into a classical, continuous probability density.* In Fig. 7 we can see how the WKB solution captures the quantum "tails" near the turning points, which the electron star does not have. It is also instructive to compare this solution to the Oppenheimer-Volkov equations in flat space: in the latter case, $\hat{m} \sim 1/\sqrt{F}$ is always larger than $\hat{\mu} \sim 1/F\sqrt{G}$, unlike in AdS where $\hat{m} \sim 1/z\sqrt{F}$ and for $z > z_*$ it becomes smaller than the local chemical potential, so the integral in (27) has a nonzero range. This is because AdS acts like a potential box that can hold the charged fermions together against electrostatic repulsion. In flat space that does not happen, and we have only neutron stars, not electron stars.

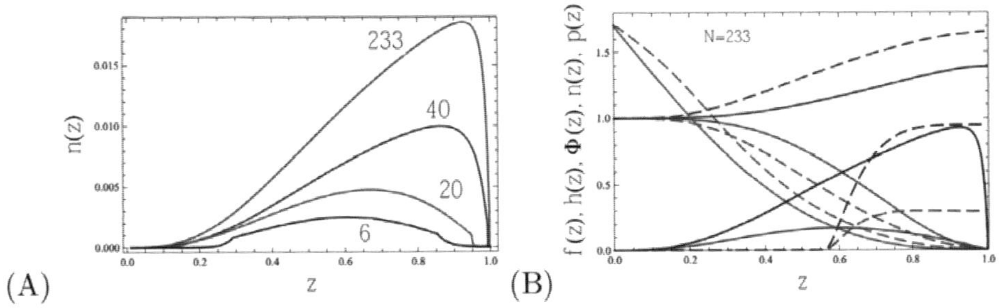

Figure 7: (A) Density of the finite temperature WKB star at various fillings N_{WKB}; besides the classically allowed region, there are also exponentially decaying tails in the classically forbidden region, where $V_{\rm eff} > 0$. (B) In the electron star (fluid limit), there are no such tails and the star has a sharp border. Taken over from [15]. (B) Comparison of the WKB solution (full lines) and the electron star solution (dashed lines) at the same chemical potential, fermion charge and mass. We plot the metric functions f, h (F, G in the main text) in red and violet, the gauge field Φ (A_t in the main text) in green, and density and pressure n, p in blue and dark green. The metric solutions do not differ much, despite the long quantum WKB tails, absent in the electron star.

3.3. Lifshitz metric, BKT transition and the missing pieces

In the framework of the electron star model, the Einstein-Maxwell equations can be solved analytically, thanks to the fact that, in deep IR, $n, \rho, p =$ const. and we can employ a scaling ansatz for the metric. The idea is to match the IR expansion around the scaling solution to the UV expansion around pure AdS. With ansatz of the form $g_{tt} \propto -1/z^\alpha$, $g_{ii} \propto 1/z^\beta$ and $g_{zz} = 1/z^2$ (one metric component we can fix at will as it amounts to picking the gauge for the metric), equations of motion give the IR solution

$$ds^2 = -\frac{1}{z^{2\zeta}}dt^2 + \frac{1}{z^2}d\vec{x}^2 + \frac{1}{z^\zeta}dz^2$$

$$A_t = \frac{1}{z^\zeta}, \quad \mathcal{L} = \frac{2}{z^{2\zeta}} + 6 - \partial_z A_t^2 - nA_t^2 - p_\perp, \tag{28}$$

where in the expression for the total Lagrangian density in the second line, we have inserted in the action (2) the solutions for the metric and the gauge field, as well as the constant (z-independent) solution obtained for ρ in (27) and similarly for n, p. Three important conclusions can be drawn: (1) the IR metric is scale-invariant, with anisotropic scaling of time and space, so that the scaling transformation has the form $t \mapsto \lambda t$, $\vec{x} \mapsto \vec{x}\lambda^{1/\zeta}$ (2) the on-shell Lagrangian density effectively describes a *massive vector field*, with mass squared equal to fermion density n (3) the fermionic contribution to the action equals the pressure. The second point agrees with the known result that Lifshitz black holes are generated by Proca fields [21], and what happens is the Abelian-Higgs mechanism: fermion density acquires a finite expectation value which in turn breaks the $U(1)$ symmetry, giving the photon a mass. The third point is expected within a fluid model, since the action of an ideal Lorentz-invariant (semi)classical fluid equals its pressure [7]. In the fluid limit we can also understand the first-order transition at finite temperature, because it is just a van der Waals-type liquid-gas transition.

We have seen that the thermal transition from RN to a Lifshitz black hole is of first order, and that the $T = 0$ transition is apparently a BKT (infinite order) transition. The latter is not quite clear yet because, as we have emphasized, nobody has yet managed to peek into the deep IR, it remains hidden behind the hard wall. But if we tentatively accept the numerical evidence for the infinite-order transition, can we understand it theoretically? The key lies in understanding how the AdS_2 throat disappears. The conformality-breaking mechanism of [22, 23] gives an idea, though the details are still missing. The crucial moment is that the near-horizon geometry is AdS_2. Right at the horizon ($s \to -\infty$) the potential is approximately constant. In the UV of the AdS_2 throat, which is around some finite value s_0, the potential behaves as $-c/(s - s_0)^2$. This inverse-square potential is known to describe conformal quantum mechanics when $c > -1/4$. For $c = -1/4$ the conformal invariance breaks. discrete states appear and the effective potential is not consistent unless regularized as

$$V_{\text{eff}} = \frac{c}{(s - s_0)^2} - v\delta(s - s_0), \tag{29}$$

and the solution of the effective Schrödinger equation is

$$\psi(r) = c_+(s - s_0)^{\alpha_+} + c_-(s - s_0)^{\alpha_-}, \quad \alpha_\pm = \frac{1}{2} \pm \sqrt{c + \frac{1}{4}}, \tag{30}$$

and the ratio c_+/c_- is given in terms of Bessel functions $J_{1/2}$ and $J_{-1/2}$:

$$\frac{c_+}{c_-} = -\epsilon^{\alpha_- - \alpha_+} \frac{\gamma + \alpha_-}{\gamma + \alpha_+}, \quad \gamma = \sqrt{v} \frac{J_{1/2}(\sqrt{v})}{J_{-1/2}(\sqrt{v})} \tag{31}$$

The solution (30) diverges at $s = s_0$ unless we introduce a cutoff at some distance ϵ from s_0. Imposing the renormalization condition that c_+/c_- remains independent of ϵ, we get the β-function of the renormalization group as (ℓ being the RG scale):

$$\beta \equiv \frac{d\gamma}{d\ell} = (c + 1/4) - (\gamma + 1/2)^2. \tag{32}$$

And we're done: the fixed points of the above flow equation are easily found to be $-\alpha_\mp$. For $\gamma = -\alpha_\mp$ we get the solution for ψ from (30) with $c_\pm = 0$ respectively. The free energy scaling is obtained as $S_{\text{on-shell}} = \mathcal{F} \propto \int d\ell/\beta$, which gives just the form found in Fig. 5. However, the presence of both a hard-wall cutoff in z and the soft-wall dilaton, completely unaccounted for in the above analysis, clearly suggest more work is needed for everything to click together.

4. Wormholes with fermion hair

The lengthy review we have given so far is meant to be self-contained and helpful for those interested in understanding and contributing to the problem of black hole instabilities with fermionic matter. As we have seen, it contains some puzzling questions and is of more than technical interest (after all, the whole field has been active mostly for the last fifteen years or so). But we also want to point out that with the methodological powerhouse of the HF, WKB and fluid methods, one can tackle new problems. A recent issue where fermions at finite density seem very relevant is the search for traversable wormholes.

The motivation for this story lies mainly in the celebrated black hole information paradox: as far as we know, the Hawking radiation is thermalized, meaning that the information content of the matter falling into the black hole is lost. A possible way out or, at least, a way to better understand the issue, is to consider the maximally extended Carter-Penrose diagram of a black hole, which contains two horizons and two spacetimes. If transport between the two were possible, one could imagine that the information is not lost because the matter falling into one horizon is entangled with the matter on the opposite side. This is the idea of the ER=EPR conjecture [24]. In order to build a traversable wormhole, one needs negative that the stress-energy tensor averaged over a geodesic be negative, thus violating the so-called averaged negative energy condition (ANEC) [25, 26]. This will never happen with conventional classical matter. One needs either exotic fields or quantum corrections. Recently however, a few traversable wormholes have been realized with only standard-model matter. The most "conservative" is the setup of [27] which creates negative energy by considering a particle-hole symmetric spectrum of massless fermions in a mangetic monopole field: because of the negative Landau levels, the net energy is negative. The starting point is thus a pair of magnetically charged RN black holes with magnetic charges H and $-H$, with the hope that the negative energy Landau levels will push the averaged stress tensor to large enough absolute values to open up a wormhole. In this way, [27] constructs a quasi-stationary (long-living) wormhole in *assymptotically flat* space. In AdS, negative energy density can easily be constructed by coupling the two boundaries nonlocally: in this way temporary wormholes, opening up for the finite duration of the pertrubation, can be constructed [28], and even eternal wormholes are possible but at the cost of much more exotic boundary CFTs and their couplings [29, 30, 31]. Here we are interested in making a wormhole in a more "down-to-earth" manner, by growing negative-energy fermion levels as in [27]. The task is to make such wormholes more stable, and to see if they survive at higher fermion density rather than just a single wavefunction as in [27]. Here the previously develped methods can help us.

Magnetic electron star. The crucial consequence of the magnetic field is the Landau quantization. The motion along the x-coordinate is quantized into discrete levels, whereas the motion along y is not quantized and introduces degeneracy. The quantization along x-axis makes our life somewhat easier – even without any IR cutoff the ground state wavefunction now has a discrete quantum number, the Landau level m_j. The magnetic field breaks the spherical symmetry of the wavefunctions down to cylindrical, so it is convenient to introduce the polar angles θ, ϕ:

$$ds^2 = -A(z)dt^2 + B(z)dz^2 + C(z)\left(d\theta^2 + \sin^2\theta d\phi^2\right) \qquad (33)$$

and to pick a different gamma matrix basis: $\Gamma^0 = \imath\sigma_1 \otimes \hat{1}$, $\Gamma^1 = \sigma_2 \otimes \hat{1}$, $\Gamma^2 = \sigma_3 \otimes \sigma_1$, $\Gamma^3 = \sigma_3 \otimes \sigma_2$. Separating the variables and representing the wavefunction as

$$\Psi = \sum_{m_j=-j}^{j} \left(\psi_+ (m_j; z), \psi_- (m_j; z)\right) \otimes \left(\eta_1 (m_j; \theta), \eta_3 (m_j; \theta)\right) e^{\imath m_j \phi}, \qquad (34)$$

where j is the total number of Landau levels $j = (H-1)/2$, we get the fully spin-polarized solution ($\eta_2 = 0$) for zero fermion mass:

$$\psi_\pm(m_j; z) = \exp\left(\pm\imath E(m_j) \int_0^z dz' \sqrt{\frac{B(z')}{A(z')}}\right), \qquad (35)$$

$$\eta_1(m_j; \theta) = \frac{e^{\imath H \sin\theta/2}}{\sqrt{\sin\theta}} \left(\tan\frac{\theta}{2}\right)^{m_j}.$$

For nonzero mass, we can perform a Foldy-Wouthuysen transform starting from the above solution. Unlike the massless case considered in [27], the resulting stress-energy tensor will not be traceless, but that is precisely what will guve us extra stability. The reason this is consistent is the Landau quantization: the levels for different m_j are gapped from each other and each Landau level can be treated as a single-particle solution which does not mix with other Landau levels. This results in the stress tensor

$$\langle T_{zz}\rangle = \frac{E_n}{(1+z^2)^2}\left(\sin 2\alpha - \cos 2\alpha\right), \quad \tan\alpha = -m/E(m_j). \tag{36}$$

Fig. 8(A) shows the radial pressure T_{rr} as a function of energy, the outcome being that *positive* stress energy tensor is produced for $0 > E > -m$. In order to avoid this positive contribution, the Landau level spacing has to be large enough, i.e., larger than the mass gap (at zero mass this condition is trivially satisfied, as it simply means that any finite $E(m_j = 1)$ will do; this is the case studied in [27]). The simplest gapping mechanism we can think of is the chemical potential, i.e. an electrostatic field in addition to the magnetostatic one. The black hole thus has to become dyonic, with magnetic charge H and electric charge e. Assuming we have ensured the negativity of (36), we can write it in the form $T_{zz} = -\tau/(1+z^2)^2$, with τ a positive constant. Its magnitude roughly determines the size of the wormhole opening.

(A) 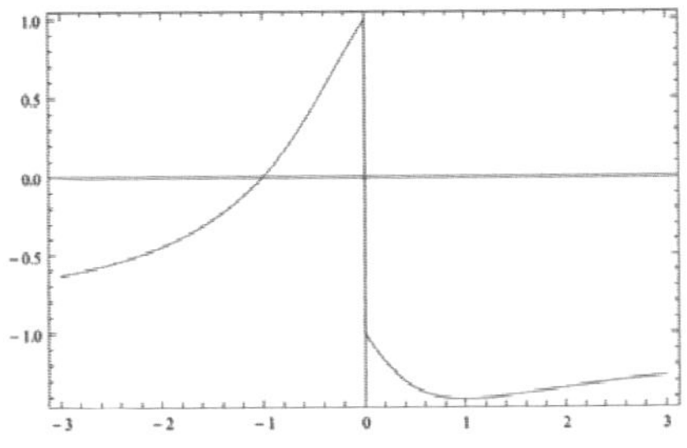

(B) 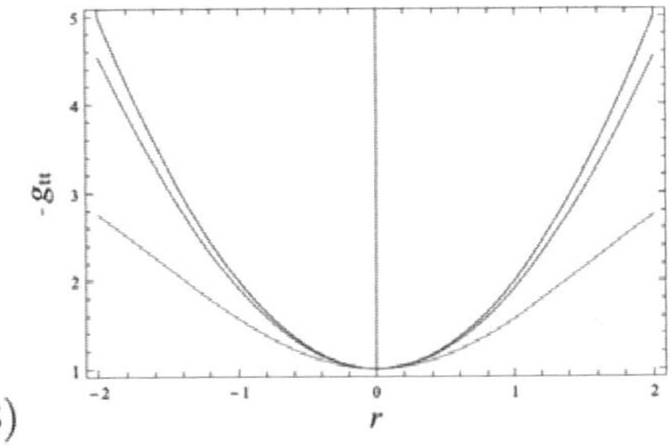

Figure 8: (A) Radial component of the stress-energy tensor $\langle T_{rr}\rangle$ as a function of the (discrete) fermion energy E. Positive contribution only comes when $> E(m_j) > -m$. In order to avoid this range of energies we need a nonzero chemical potential (i.e., electric field, resulting in a dyonic black hole) to stabilize the wormhole with massive fermionic hair. (B) The solution for the metric component g_{tt} in the intermediate region, as a function of the radial coordinate r, for $\tau = 0, 0.05, 0.10$ (black, blue, red). Wormhole solutions (blue, red) are quantitatively very close to the unperturbed black hole (black) but qualitativrly different as there is no zero anymore.

Wormhole solution and matching. Having computed the stress-energy tensor (36), we can solve the Einstein equations. The strategy is again matching the expansions, but now we have three regions: the far region which is asymptotically AdS or even flat (we have mentioned that in the presence of magnetic field discrete bound states can form even in absence of AdS boundary), the intemediate region is a slightly perturbed near-horizon AdS_2 region of our magnetic RN geometry, and the inner region, the wormhole throat that opens up, turns out to be a *global* AdS_2 at leading order, so it has a spherical boundary continuing onto the intermediate regions. The inner, near-global-AdS$_2$ metric in the form (33) at leading order reads

$$A(z) = R_0^2 \left[1 + z^2 - 8\pi\tau \left(z^2 + \left(3z + z^3\right) \arctan z - \log\left(1 + z^2\right)\right)\right]$$
$$B(z) = R_0^4/A(z), \quad C(z) = R_0^2 \left[1 + 8\pi\tau\left(1 + z \arctan z\right)\right]. \tag{37}$$

This solution is to be matched to the intermediate-region solution. Now large z corresponds to the wormhole mouth, i.e., the matching is to be done at large z, where small z is the "center" of the wormhole throat. The solution to match onto is the RN black hole metric:

$$ds^2 = -l^2 f(r)dt^2 + \frac{dr^2}{f(r)} + r^2\left(d\theta^2 + \sin^2\theta d\phi^2\right)$$
$$l = \frac{R_0}{2\pi^2\tau}, \quad R = \frac{r - \sqrt{\pi}\sqrt{e^2 + H^2}}{2\pi^2\tau}. \tag{38}$$

The solution thus exists for any choice of e and H. But for large e (in other words, for a large chemical potential), the density of the hair will increase significantly and we should repeat the WKB star or electron star approach. in AdS this is a simple matter, proving the stability of the configuration even at high densities. The interesting question is, can it work also in assymptotically flat space? In absence of magnetic field, the answer is certainly no – without an AdS boundary, there is nothing to equilibrate the electrostatic repulsion of electrons. But in the presence of magnetic field, one might obtain a stable charged hairy wormhole if the change in the near-horizon geometry is sufficient to effectively decrease the electrostatic energy density. This is the logical immediate task for future work.

We finish this short review of our work in progress on hairy wormholes with a somewhat more ambitious task. The dyonic wormhole model considered here is obviously quite simplistic and artificial. A much more realistic model is to start from a pair of Kerr black holes and see if these can open up a wormhole in a manner analogous to the scenario we have considered. In this case the magnetic field would be generated self-consistently by the (rotating) fermionic hair, removing the need for the magnetic monopole charge. Such an object would come much closer to realistic astrophysical matter.

5. Instead of a conclusion

We have given a crack and practical review of the insights and technologies needed to describe and understand hairy black holes in anti-de Sitter space. The phase diagram in the presence of nonzero fermion density is quite rich, and it involves two deep and universal phenomena. First, the finite-temperature hairy black holes develop through a discontinuous phase transition akin to the Hawking-Page transition (indeed, it is precisely the Hawking-Page transition but at finite density). The standard lore that at high enough temperatures black holes will always form is confirmed. Notice this is true at any fermion mass and charge, and thus at any occupation number, from a single wavefunction to the fluid limit, so the finding is definitely robust. Second, at zero temperature the transition is driven by the fermionic charge and/or chemical potential, i.e., electric charge of the black hole. In this case the black hole vanishes infinitely slowly, in a BKT transition that can be understood as the breaking of the one-dimensional conformal symmetry of the wavefunctions in the effectve inverse-square potential well. This is solely the consequence of the near-horizon physics, independent of the AdS boundary. Similar conformality-breaking infinite-order transitions are known in various backgrounds in string theory. Maybe one could relate the case described here to some consistent top-down model.

As mentioned in the Introduction, we have deliberately left out extensions and applications of the formalism described, for reasons of space and also generality of discussion. The field of applications closest to our experience is the AdS/CFT correspondence. Electrically charged black holes are dual to field theories at finite $U(1)$ density. The transition from a bald black hole to a hairy black hole is thus a transition between two phases at equal chemical potential. How do they differ then? We know that a black hole is dual to the Coulomb (deconfined) phase of some non-Abelian finite-temperature gauge theory [1, 4]; in the simplest setup coming from type IIB string theory, it is the $\mathcal{N} = 4$ supersymmetric $SU(N)$ theory. Coulomb phase means that the $U(1)$ charge is carried by $SU(N)$-gauge-charged operators, in our case fermions ("mesinos") and thus not visible to low-energy probes, since at low energies all operators are likely $SU(N)$-gauge-neutral. The hairy phase describes a dual field theory where the charge is carried by gauge-neutral operators ("baryons") and thus visible to probes such as a photon. This viewpoint was tried and confirmed in [11, 12, 15, 23]. It has realizations in condensed matter systems such as strange metals and heavy fermion materials. In this case, the gauge fields are emergent and arise from the spin-charge separation, and the transition between a black hole and a hairy geometry is a transition between a non-Fermi liquid, where most of the charge is carried by complicated excitatons that are nor directly seen in the spectrum, and a Fermi liquid where the fundamental degrees of freedom are just renormalized electrons. In QCD, this picture describes the phase diagram at intermediate energy scales and finite densities, where a black hole describes quark-gluon plasma, and a hairy solution describes either the color condensate or conventional barionic matter depending on the details of the model. One can learn a lot on AdS/condensed matter and AdS/QCD from [5, 6].

Finally, the search for wormhole solutions and how fermionic hair might stabilize them is likely to become very important in the future, in connection to the quantum information theory and the firewall, ER=EPR and other approaches to the black hole information problem. One can use much of the formalism developed for hairy black holes, but the interpretation is still challanging. It is also unclear how realistic the wormhole proposal is if we work with only conventional, standard model matter, i.e. is it just an important proof of concept or a realistic model?

References

[1] J. Maldacena, *The large N limit of superconformal field theories and super-gravity*, Adv. Math. Theor. Phys. **2**, 231 (1998). [arXiv:hep-th/971120]

[2] S. S. Gubser, I. R. Klebanov and A. M. Polyakov, *Gauge theory correlators from non-critical string theory*, Phys. Lett. B **428**, 105 (1998). [arXiv:hep-th/9802109]

[3] E. Witten, *Anti de Sitter space and holography*, Adv. Math. Theor. Phys. **2**, 253 (1998). [arXiv:hep-th/9802150]

[4] O. Aharony, S. S. Gubser, J. Maldacena, H. Ooguri and Y. Oz, *Large N field theories, string theory and gravity*, Phys. Rep. **323**, 183 (2000). [arXiv:hep-th/9905111]

[5] J. Zaanen, Y.-W. Sun, Y. Liu and K. Schalm, *Holographic duality in condensed matter physics*, Cambridge University Press, 2015.

[6] M. Ammon and J. Erdmenger, *Gauge/gravity duality: foundations and applications*, Cambridge University Press, 2015.

[7] S. W. Hawking and G. F. R. Ellis, *The large-scale structure of space-time*, Cambridge University Press, 2010.

[8] N. D. Birrel and P. C. W. Davies, *Quantum fields in curved space*, Cambridge University Press, 1982.

[9] H. Liu, J. McGreevy, D. Vegh, "Non-Fermi liquids from holography", Phys. Rev. D **83**, 065029 (2011). [arXiv:0903.2477[hep-th]]

[10] M. Čubrović, J. Zaanen and K. Schalm, "String Theory, Quantum Phase Transitions and the Emergent Fermi-Liquid", Science **325**, 439 (2009). [arXiv:0904.1993[hep-th]]

[11] T. Faulkner, H. Liu, J. McGreevy and D. Vegh, *Emergent quantum criticality, Fermi surfaces, and AdS2*, Phys. Rev. D **83**, 125002 (2011). [arXiv:0907.2694 [hep-th]].

[12] S. Sachdev, *A model of a Fermi liquid using gauge-gravity duality*, Phys. Rev. D **84**, 066009 (2011). [arXiv:1107.5321 [hep-th]]

[13] W. Mück and K. S. Viswanathan, *Conformal field theory correlators from classical field theory on anti-de Sitter space: Vector and spinor fields*, Phys. Rev. **D58**, 106006 (1998). [arXiv:hep-th/9805145]

[14] M. Čubrović, Y. Liu, K.'Schalm, Y.-W. Sun and J. Zaanen, *Spectral probes of the holographic Fermi groundstate: Dialing between the electron star and AdS Dirac hair*, Phys. Rev. **D84**, 086002 (2011). [arXiv:1106.1798 [hep-th]]

[15] M. Medvedyeva, E. Gubankova, M. Cubrovic, K. Schalm, J. Zaanen, *Quantum corrected phase diagram of holographic fermions*, JHEP **2013**, 025 (2013). [arXiv:1302.5149[hep-th]].

[16] M. Čubrović, J. Zaanen, K. Schalm, *Constructing the AdS dual of a Fermi liquid: AdS Black holes with Dirac hair*, JHEP **2011**, 17 (2011). [arXiv:1012.5681[hep-th]]

[17] A. Allais, J. McGreevy and X. Josephine Suh, *Quantum electron star*, Phys. Rev. Lett **108**, 231602 (2012). [arXiv:1202.5308[hep-th]]

[18] A. Allais and J. McGreevy, *How to construct a gravitating quantum electron star*, Phys. Rev. **D88**, 066006 (2013). [arXiv:1306.6075[hep-th]]

[19] S. Hawking and D. Page, *Thermodynamics of black holes in anti de Sitter space*, Comm. Math. Phys. **87**, 577 (1977).

[20] S. A. Hartnoll, A. Tavanfar, *Electron stars for holographic metallic criticallity*, Phys. Rev. D **83**, 046003 (2011). [arXiv:1008.2828[hep-th]]

[21] K. Balasubramanian and J. McGreevy, *An analytic Lifshitz black hole*, Phys. Rev. D **80**, 104039 (2009). [arXiv:0909.0263[hep-th]]

[22] D. B. Kaplan, J.-W. Lee, D. T. Son, M. A. Stephanov, *Conformality Lost*, Phys. Rev. D **80**, 125005 (2009). [arXiv:0905.4752[hep-th]].

[23] M. Čubrović, *Confinement/deconfinement transition from symmetry breaking in gauge/gravity duality*, JHEP **2016**, 102 (2016). [arXiv:1605.07849[hep-th]]

[24] J. Maldacena and L. Susskind, *Cool horizons for entangled black holes*, Fortschr. Phys. **61** 781 (2013). [arXiv:1306.0533[hep-th]]

[25] M. S. Morris and K. Thorne, *Wormholes in spacetime and their use for interstellar travel: A tool for teaching general relativity* Am. J. Phys **56**, 395 (1988).

[26] D. Hochberg and M. Visser, *Null energy condition in dynamic wormholes*, Phys. Rev. Lett**81**, 746 (1998).

[27] J. Maldacena, A. Milekhin and F. Popov, *Traversable wormholes in four dimensions*, (2018). [arXiv:1807.04726[hep-th]]

[28] P. Gao, D. L. Jafferis and A. C. Wall, *Traversable wormholes via a double trace deformation*, JHEP **12**, (2017) 151. [arXiv:1608.05687[hep-th]]

[29] Z. Fu, B. Grado-White and D. Marolf, *A perturbative perspective on self-supporting wormholes*, Class. Quant. Grav. **36**, (2019) 045006. [arXiv:1807.07917[hep-th]]

[30] D. Marolf and S. McBride, *Simple perturbatively traversable wormholes from bulk fermions*, (2019). [arXiv:1908.03998[hep-th]]

[31] J. Maldacena and X.-L. Qi, *Eternal traversable wormhole*, (2018). [arXiv:1804.00491[hep-th]]

Courant and Roytenberg bracket and their relation via T-duality *

Ljubica Davidović†
Institute of Physics, Belgrade
Ilija Ivanišević‡
Institute of Physics, Belgrade
Branislav Sazdović§
Institute of Physics, Belgrade

Abstract

We consider the σ-model for closed bosonic string propagating in the coordinate dependent metric and Kalb-Ramond field. Firstly, we consider the generator of both general coordinate and local gauge transformations. The Poisson bracket algebra of these generators is obtained and we see that it gives rise to the Courant bracket. Secondly, we consider generators in a new basis, consisting of the σ derivative of coordinates, as well as the auxiliary currents. Their Poisson bracket algebra gives rise to the Courant bracket, twisted by the Kalb-Ramond field. Finally, we calculate the algebra of the T-dual generator. The Poisson bracket algebra of T-dual generator gives rise to the Roytenberg bracket, equivalent to the bracket obtained by twisting the Courant bracket by the non-commutativity parameter, which is T-dual to the Kalb-Ramond field. We show that the twisted Courant bracket and the Roytenberg bracket are mutually related via T-duality.

1. Bosonic string action

The closed bosonic string is moving in a curved background, associated with the symmetric metric tensor field $G_{\mu\nu} = G_{\nu\mu}$, the antisymmetric Kalb-Ramond tensor field $B_{\mu\nu} = -B_{\nu\mu}$, as well as the scalar dilaton field Φ. If we consider the case of constant dilation field, in a conformal gauge

*This work has been supported in part by The Serbian Ministry of Education, Science and Technological Development, under contract No. 171031.

† e-mail address: ljubica@ipb.ac.rs
‡ e-mail address: ivanisevic@ipb.ac.rs
§ e-mail address: sazdovic@ipb.ac.rs

$g_{\alpha\beta} = e^{2F}\eta_{\alpha\beta}$, the action is given by [1]

$$S = \int_\Sigma d^2\xi \mathcal{L} = \kappa \int_\Sigma d^2\xi \Big[\frac{1}{2}\eta^{\alpha\beta}G_{\mu\nu}(x) + \epsilon^{\alpha\beta}B_{\mu\nu}(x)\Big]\partial_\alpha x^\mu \partial_\beta x^\nu . \tag{1}$$

The integration goes over two-dimensional world-sheet Σ parametrized by $\xi^\alpha(\xi^0 = \tau, \xi^1 = \sigma)$. Coordinates of the D-dimensional space-time are $x^\mu(\xi)$, $\mu = 0, 1, ..., D-1$, $\epsilon^{01} = -1$ and $\kappa = \frac{1}{2\pi\alpha'}$. In a usual way, we derive the expression for the canonical momenta

$$\pi_\mu = \frac{\partial \mathcal{L}}{\partial \dot{x}^\mu} = \kappa G_{\mu\nu}(x)\dot{x}^\nu - 2\kappa B_{\mu\nu}(x)x'^\nu , \tag{2}$$

and obtain the Hamiltonian

$$\mathcal{H}_C = \frac{1}{2\kappa}\pi_\mu(G^{-1})^{\mu\nu}\pi_\nu - 2x'^\mu B_{\mu\nu}(G^{-1})^{\nu\rho}\pi_\rho + \frac{\kappa}{2}x'^\mu G^E_{\mu\nu}x'^\nu , \tag{3}$$

where $G^E_{\mu\nu} = G_{\mu\nu} - 4(BG^{-1}B)_{\mu\nu}$ is the effective metric. It is possible to rewrite the Hamiltonian in a more convenient form

$$\mathcal{H}_C = \frac{1}{4\kappa}(G^{-1})^{\mu\nu}\Big[j_{+\mu}j_{+\nu} + j_{-\mu}j_{-\nu}\Big] . \tag{4}$$

where the currents $j_{\pm\mu}$ are defined by

$$j_{\pm\mu}(x) = j_{0\mu} \pm j_{1\mu} = \pi_\mu + 2\kappa\Pi_{\pm\mu\nu}x'^\nu , \tag{5}$$

and $\Pi_{\pm\mu\nu} = B_{\mu\nu} \pm \frac{1}{2}G_{\mu\nu}$ are composite fields. The $\tau-$component of current j_μ will be marked as an auxiliary current $j_{0\mu} = \pi_\mu + 2\kappa B_{\mu\nu}x'^\nu \equiv i_\mu$. Now, we rewrite the equation (5)

$$j_{\pm\mu} = i_\mu \pm \kappa G_{\mu\nu}x'^\nu . \tag{6}$$

The generalized current is defined by

$$J_C(\xi, \Lambda^C) = \xi^\mu(x)i_\mu + \Lambda^C_\mu(x)\kappa x'^\mu , \tag{7}$$

where ξ^μ and Λ^C_μ are some coordinate dependent parameters. Due to their index structure, the former parameters can be regarded as vector field components and the latter as 1-form components. The suitable language for description of generalized currents is the one of generalized geometry [2]. The generalized vectors are direct sum of elements of tangent and cotangent bundle over some manifold, meaning that the generalized geometry treats vectors and 1-forms on equal footing. Therefore, it is possible to consider the generalized current as the function on generalized vector $\xi \oplus \Lambda^C$.

Besides the algebra of these generalized currents, we are also interested in algebra of T-dual generalized currents. T-duality [1, 3] is inherent to string theory and represents the equivalence of two seemingly different physical theories. The equivalence manifests itself in a way that all quantities in one theory are identified with quantities in its T-dual theory. It was firstly noticed in case of the bosonic closed string with one dimension compactified to a radius R. In that case, mass spectrum is given by [1]

$$M^2 = \frac{K^2}{R^2} + W^2 \frac{R^2}{\alpha'^2}\,, \tag{8}$$

where K is momentum and W winding number. It is obvious that the mass spectrum remains invariant upon simultaneously swapping $K \leftrightarrow W$ and $R \leftrightarrow \frac{\alpha'}{R}$. What can be concluded is that spectrums of two theories that both have one dimension compactified to a circle, where in one case the circle is of small and in the other of large radius, are indistinguishable. The momenta in one theory are winding numbers in its T-dual theory, and vice versa.

We consider the T-duality realized without changing the phase space [4]. Its transformation rules between the canonical variables are given by

$$\pi_\mu \cong \kappa x'^\mu\,. \tag{9}$$

When the above relation is integrated over space parameter σ, we obtain that the T-duality interchanges momenta with the winding numbers. The background fields have their T-dual counterparts as well [5], namely

$$^\star G^{\mu\nu} = (G_E^{-1})^{\mu\nu}, \quad ^\star B^{\mu\nu} = \frac{\kappa}{2}\theta^{\mu\nu}\,, \tag{10}$$

where $\theta^{\mu\nu} = -\frac{2}{\kappa}(G_E^{-1}BG^{-1})^{\mu\nu}$ is the non-commutativity parameter. Two variables are said to be mutually T-dual if they transform one into another when simultaneously both (9) and (10) are applied [4].

Now we follow the same procedure as above for constructing the T-dual generalized currents. Applying (9) and (10) to (5), we obtain the current $l_\pm^\mu$, T-dual to current $j_{\pm\mu}$

$$l_\pm^\mu = k^\mu \pm (G_E^{-1})^{\mu\nu}\pi_\nu, \quad k^\mu = \kappa x'^\mu + \kappa\theta^{\mu\nu}\pi_\nu\,. \tag{11}$$

The Hamiltonian can be expressed in terms of these currents by

$$\mathcal{H}_C = \frac{1}{4\kappa}G^E_{\mu\nu}\left(l_+^\mu l_+^\nu + l_-^\mu l_-^\nu\right). \tag{12}$$

Substituting the expression (11) in (12), we obtain the Hamiltonian in the form (4). The Hamiltonian remains invariant under T-duality.

The generalization of current $l_\pm^\mu$ is given by

$$J_R(\xi_R, \Lambda) = \xi_R^\mu(x)\pi_\mu + \Lambda_\mu(x)k^\mu\,. \tag{13}$$

Instead of auxiliary current i_μ and coordinate $\sigma-$derivative x'^μ, the basis for T-dual generalized currents consist of T-dual auxiliary current k^μ and canonical momenta π_μ. The coefficients ξ^μ_R are arbitrary vector field components, while Λ_μ are arbitrary 1-form components. Next, we will see how these currents are related to the symmetry generators.

2. Symmetry generators algebra

Let us start with the general coordinate transformations. They are generated by

$$\mathcal{G}_{GCT}(\xi) = \int_0^{2\pi} d\sigma \xi^\mu(x)\pi_\mu \,. \tag{14}$$

Action of the general coordinate transformation on background fields is governed by the action of Lie derivative [6]

$$\delta_\xi G_{\mu\nu} = \mathcal{L}_\xi G_{\mu\nu}, \; \delta_\xi B_{\mu\nu} = \mathcal{L}_\xi B_{\mu\nu} \,. \tag{15}$$

The Lie derivative is defined by $\mathcal{L}_\xi = i_\xi d + d i_\xi$, where the interior product i_ξ acts as a contraction with a vector field ξ, while the exterior derivative d extends the concept of the differential of a function to differential forms. The commutator of two Lie derivatives results in another Lie derivative

$$\mathcal{L}_{\xi_1}\mathcal{L}_{\xi_2} - \mathcal{L}_{\xi_2}\mathcal{L}_{\xi_1} = \mathcal{L}_{[\xi_1,\xi_2]_L} \,, \tag{16}$$

where by $[\xi_1, \xi_2]_L$, we marked the Lie bracket between two vector fields ξ_1 and ξ_2. Explicitly, it is given by

$$([\xi_1, \xi_2]_L)^\mu = \xi_1^\nu \partial_\nu \xi_2^\mu - \xi_2^\nu \partial_\nu \xi_1^\mu \,. \tag{17}$$

Calculating the Poisson bracket algebra of general coordinate transformations generators $\mathcal{G}_{GCT}$ (14), we obtain the relation:

$$\{\mathcal{G}_{GCT}(\xi_1), \mathcal{G}_{GCT}(\xi_2)\} = -\mathcal{G}_{GCT}([\xi_1, \xi_2]_L) \,. \tag{18}$$

We note that this algebra gives rise to Lie bracket.

We would like to extend the generator G_{GCT} so that it includes the local gauge transformations as well. They are generated by [6]

$$\mathcal{G}_{LGT}(\Lambda) = \int d\sigma \Lambda_\mu \kappa x'^\mu \,, \tag{19}$$

where Λ_μ are gauge parameters that are 1-form components. The action of local gauge transformations on background fields is given by

$$\delta_\Lambda G_{\mu\nu} = 0, \; \delta_\Lambda B_{\mu\nu} = \partial_\mu \Lambda_\nu - \partial_\nu \Lambda_\mu \,. \tag{20}$$

When gauge parameter is changed so that total derivative of arbitrary function is added $\Lambda_\mu \to \Lambda_\mu + \partial_\mu \lambda$, the action of generator does not change.

Hence, different gauge parameters generate same symmetry and therefore the generator is reducible.

Local gauge transformations are T-dual to general coordinate transformations, due to relation (9). The sum of both generators (14) and (19) results in symmetry generator that is T-dual to itself

$$\mathcal{G}(\xi \oplus \Lambda) = \int d\sigma \Big[\xi^\mu(x)\pi_\mu + \Lambda_\mu \kappa x'^\mu\Big]. \tag{21}$$

Using the language of generalized geometry, we treat the generator (21) as a function of generalized gauge parameter, defined as the direct sum of vector and 1-form parameter $\xi \oplus \Lambda$.

2.1. Courant bracket

The Poisson bracket algebra of the generators (21) is given by

$$\{\mathcal{G}(\xi_1 \oplus \Lambda_1), \mathcal{G}(\xi_2 \oplus \Lambda_2)\} = -\mathcal{G}([\xi_1 \oplus \Lambda_1, \xi_2 \oplus \Lambda_2]_C), \tag{22}$$

where $[\xi_1 \oplus \Lambda_1, \xi_2 \oplus \Lambda_2]_C$ is the Courant bracket [7]. It is considered to be the generalized geometry extension of the Lie bracket. Just like the Lie bracket acts on vectors, Courant bracket acts on generalized vectors. Its full expression is given by

$$[\xi_1 \oplus \Lambda_1, \xi_2 \oplus \Lambda_2]_C = [\xi_1, \xi_2]_L \oplus \Big(\mathcal{L}_{\xi_1}\Lambda_2 - \mathcal{L}_{\xi_2}\Lambda_1 - \frac{1}{2}d(i_{\xi_1}\Lambda_2 - i_{\xi_2}\Lambda_1)\Big). \tag{23}$$

The first term $[\xi_1, \xi_2]_L$ on the right hand side of previous equation gives the vector, while the other terms give 1-form. It has been shown before [8] that the algebra of symmetry transformations gives rise to the Courant bracket.

Courant bracket cannot be a bracket of a Lie algebra, as it does not satisfy the Jacobi identity. However, this does not pose a problem, as the deviation from Jacobi identity is closed form that disappears after the integration and hence correspond to a trivial symmetry. The Jacobiator of Courant bracket is given by [2]

$$[[\xi_1 \oplus \Lambda_1, \xi_2 \oplus \Lambda_2]_C, \xi_3 \oplus \Lambda_3]_C + cyclic = dNij(\xi_1 \oplus \Lambda_1, \xi_2 \oplus \Lambda_2, \xi_3 \oplus \Lambda_3)_C, \tag{24}$$

where the Nijenhuis operator is defined by

$$Nij(\xi_1 \oplus \Lambda_1, \xi_2 \oplus \Lambda_2, \xi_3 \oplus \Lambda_3)_C = \frac{1}{3} < (\xi_1 \oplus \Lambda_1, \xi_2 \oplus \Lambda_2)_C, \xi_3 \oplus \Lambda_3 > +cycl. \tag{25}$$

and $< \xi_1 \oplus \Lambda_1, \xi_2 \oplus \Lambda_2 >= \frac{1}{2}(\xi_1(\Lambda_2) - \xi_2(\Lambda_1))$ is the natural inner product between generalized vectors.

2.2. Twisted Courant bracket

We claimed that it is possible to obtain the generalized currents from the symmetry generators. To demonstrate that, let us now define the new gauge parameter

$$\Lambda^C_\mu = \Lambda_\mu + 2B_{\mu\nu}\xi^\nu \,. \tag{26}$$

This change of parameter can be interpreted as the B-transformation acting on the generalized gauge parameter

$$\mathcal{O}_B\Lambda = \begin{pmatrix} \delta^\mu_\nu & 0 \\ 2B_{\mu\nu} & \delta^\nu_\mu \end{pmatrix} \cdot \begin{pmatrix} \xi^\nu \\ \Lambda_\nu \end{pmatrix} = \begin{pmatrix} \xi^\mu \\ \Lambda^C_\mu \end{pmatrix} \,. \tag{27}$$

It is suitable to rewrite the generator (21) in the new basis

$$\mathcal{G}_C(\xi \oplus \Lambda^C) = \int d\sigma \Big[\xi^\mu i_\mu + \kappa \Lambda^C_\mu x'^\mu \Big] \,. \tag{28}$$

Comparing the expression for the symmetry generator $\mathcal{G}_C$ (28) with the expression for generalized currents (7), we see that this generator is the charge corresponding to the current J_C.

The change of basis results in the appearance of H-flux in the basis algebra

$$\{i_\mu(\sigma), i_\nu(\bar\sigma)\} = -2\kappa B_{\mu\nu\rho} x'^\rho \delta(\sigma - \bar\sigma) \,, \tag{29}$$

where

$$B_{\mu\nu\rho} = (dB)_{\mu\nu\rho} = \partial_\mu B_{\nu\rho} + \partial_\nu B_{\rho\mu} + \partial_\rho B_{\mu\nu} \tag{30}$$

is the H-flux. The Poisson bracket algebra of generators was obtained in [4, 9] and written in the form

$$\{\mathcal{G}_C(\xi_1 \oplus \Lambda^C_1), \mathcal{G}_C(\xi_2 \oplus \Lambda^C_2)\} = -\mathcal{G}_C([\xi_1 \oplus \Lambda^C_1, \xi_2 \oplus \Lambda^C_2]_B) \,. \tag{31}$$

The bracket $[\xi_1 \oplus \Lambda^C_1, \xi_2 \oplus \Lambda^C_2]_B$ is called the Courant bracket, twisted by a 2-form $2B_{\mu\nu}$. This twist is realized by

$$[\mathcal{O}_B(\xi_1 \oplus \Lambda_1), \mathcal{O}_B(\xi_2 \oplus \Lambda_2)]_C - \mathcal{O}_B[\xi_1 \oplus \Lambda_1, \xi_2 \oplus \Lambda_2]_C = H(\xi_1, \xi_2, .) \,. \tag{32}$$

We see that it differs from the Courant bracket (23) by the contraction of H-flux $H = dB$ (30) with two gauge parameters ξ_1 and ξ_2. The full expression for this bracket is given by

$$\begin{aligned} [\xi_1 \oplus \Lambda^C_1, \xi_2 \oplus \Lambda^C_2]_B = [\xi_1, \xi_2]_L \oplus \Big(& \mathcal{L}_{\xi_1}\Lambda^C_2 - \mathcal{L}_{\xi_2}\Lambda^C_1 \\ & - \frac{1}{2} d(i_{\xi_1}\Lambda^C_2 - i_{\xi_2}\Lambda^C_1) + H(\xi_1, \xi_2, .) \Big) \,. \end{aligned} \tag{33}$$

2.3. Roytenberg bracket

Finally, we are going to consider one more transformation of the gauge parameter

$$\xi_R^\mu = \xi^\mu + \kappa\theta^{\mu\nu}\Lambda_\nu\,. \tag{34}$$

This can be written as a generalized gauge parameter transformation, characterized by the action of antisymmetric $\theta^{\mu\nu}$ bi-vector

$$\mathcal{O}_\theta\Lambda = \begin{pmatrix}\delta_\nu^\mu & \kappa\theta^{\mu\nu}\\ 0 & \delta_\mu^\nu\end{pmatrix}\begin{pmatrix}\xi^\nu\\ \Lambda_\nu\end{pmatrix} = \begin{pmatrix}\xi_\mu^R\\ \Lambda_\mu\end{pmatrix}\,. \tag{35}$$

The symmetry generator can be rewritten with new parameters in a new basis

$$\mathcal{G}_\mathcal{R}(\xi_R\oplus\Lambda) = \int d\sigma\Big[\xi_R^\mu\pi_\mu + \Lambda_\mu k^\mu\Big]\,. \tag{36}$$

It is obvious that the generator is the charge corresponding to current J_R (13) and that the T-duality relation $\mathcal{G}_\mathcal{R}(\xi_R\oplus\Lambda)\cong\mathcal{G}_\mathcal{C}(\xi\oplus\Lambda^C)$ holds.

There is a presence of non-geometric fluxes in basis algebra

$$\{k^\mu(\sigma),k^\nu(\bar\sigma)\} = -\kappa Q_\rho{}^{\mu\nu}k^\rho\delta(\sigma-\bar\sigma) - \kappa^2 R^{\mu\nu\rho}\pi_\rho\delta(\sigma-\bar\sigma)\,, \tag{37}$$

where $Q_\rho{}^{\mu\nu} = \partial_\rho\theta^{\mu\nu}$ is Q-flux and $R^{\mu\nu\rho} = \theta^{\mu\sigma}\partial_\sigma\theta^{\nu\rho} + \theta^{\nu\sigma}\partial_\sigma\theta^{\rho\mu} + \theta^{\rho\sigma}\partial_\sigma\theta^{\mu\nu}$ is R-flux. The new generator algebra [4, 10] is given by

$$\{\mathcal{G}_\mathcal{R}(\xi_1^R\oplus\Lambda_1),\,\mathcal{G}_\mathcal{R}(\xi_2^R\oplus\Lambda_2)\} = -\mathcal{G}_\mathcal{R}([\xi_1^R\oplus\Lambda_1,\xi_2^R\oplus\Lambda_2]_R)\,, \tag{38}$$

where the resulting bracket is the Roytenberg bracket [11]. It is a generalization of Courant bracket, obtained by twisting the Courant bracket by some bi-vector. The corresponding bi-vector in our case is the non-commutativity parameter $\kappa\theta^{\mu\nu}$. The Roytenberg bracket differs from the Courant bracket by

$$[\mathcal{O}_\theta(\xi_1\oplus\Lambda_1),\mathcal{O}_\theta(\xi_2\oplus\Lambda_2)]_C - \mathcal{O}_\theta[\xi_1\oplus\Lambda_1,\xi_2\oplus\Lambda_2]_C\,, \tag{39}$$

In its most general form, Roytenberg bracket encompasses all fluxes. In this case, canonical momenta are commutative, meaning that there is no H-flux present in basis algebra. The full expression is

$$\begin{aligned}[\xi_1\oplus\Lambda_1^R,\xi_2\oplus\Lambda_2^R]_R = &\Big([\xi_1,\xi_2]_L - \kappa[\xi_2,\Lambda_1^R\theta]_L + \kappa[\xi_1,\Lambda_2^R\theta]_L\\ &+ \frac{\kappa^2}{2}[\theta,\theta]_S(\Lambda_1^R,\Lambda_2^R,.) - (\mathcal{L}_{\xi_2}\Lambda_1^R - \mathcal{L}_{\xi_1}\Lambda_2^R + \frac{1}{2}d(i_{\xi_1}\Lambda_2^R - i_{\xi_2}\Lambda_1^R))\kappa\theta\Big)\\ &\oplus\Big(\mathcal{L}_{\xi_1}\Lambda_2^R - \mathcal{L}_{\xi_2}\Lambda_1^R - \frac{1}{2}d(i_{\xi_1}b - i_{\xi_2}\Lambda_1^R) - [\Lambda_1^R,\Lambda_2^R]_\theta\Big)\,.\end{aligned} \tag{40}$$

The expression $[\theta,\theta]_S(\Lambda_1^R,\Lambda_2^R,.)$ is the Schouten-Nijenhuis bracket [12] contracted with two 1-forms. It is a generalization of the Lie bracket on the space of multi-vectors. The expression $[\Lambda_1^R, \Lambda_2^R]_\theta$ is the Koszul bracket [13], a generalization of the Lie bracket on the space of differential forms.

We note that 2-form $2B_{\mu\nu}$ and bi-vector $\kappa\theta^{\mu\nu}$ used for twisting the Courant bracket in two cases are mutually T-dual (10). Moreover, both of these brackets appeared in considering charges corresponding to the generalized currents, defined in mutually T-dual bases. Therefore, we can conclude that the T-duality connects twisted Courant and Roytenberg bracket, as long as they are twisted by the mutually T-dual fields.

3. Conclusions

We considered a standard σ-model for closed bosonic string. We wrote Hamiltonian in terms of two types of currents. The components of these currents have been used as bases in which generalized currents were defined. These generalized currents are defined in mutually T-dual bases.

Afterwards, we considered the self T-dual symmetry generator $\mathcal{G}(\xi \oplus \Lambda)$. It takes the generalized gauge parameter $\xi \oplus \Lambda$ as a parameter and generates both general coordinate transformations and local gauge transformations. The Poisson bracket algebra of these generators was calculated and the Courant bracket has been obtained. The Courant bracket appeared in a same way in which Lie bracket was obtained in the case of general coordinate transformations algebra.

Next, we considered the action of B-transformation to the generalized gauge parameter. The symmetry generator written in terms of the resulting gauge parameter is the charge corresponding to the generalized current J_C. The Poisson bracket algebra of these generators gives rise to the Courant bracket twisted by a 2-form $2B$.

In the end, we consider the action of θ-transformation on the generalized gauge parameter. The newly obtained generator is the charge for the generalized current J_R and it gives rise to the Roytenberg bracket. The Roytenberg bracket in general represent the Courant bracket twisted by a bi-vector. In this specific case, the bi-vector is $\kappa\theta^{\mu\nu}$, which is T-dual to the Kalb-Ramond field $B_{\mu\nu}$. Consequently, we conclude that both the twisted Courant bracket and the Roytenberg bracket appear when generalized currents defined in two mutually T-dual bases are considered.

References

[1] K. Becker, M. Becker and J. Schwarz *String Theory and M-Theory: A Modern Introduction* (Cambridge University Press, Cambridge, 2007).

[2] N. Hitchin, *Lectures on generalized geometry* (2010), arXiv:math/1008.0973.

[3] E. Alvarez, L. Alvarez-Gaume and Y. Lozano, *An introduction to T-duality in string theory, Nucl. Phys. Proc. Suppl.* **41** (1995) 1-20.

[4] Lj. Davidović, I. Ivanisević and B. Sazdović, *Courant bracket found out to be T-dual to Roytenberg bracket* arXiv:1903.04832.

[5] T. Buscher, *Path-integral derivation of quantum duality in nonlinear sigma-models, Phys. Lett.* **201** (1988) 466.

[6] Lj. Davidović, B. Sazdović, *The T-dual symmetries of a bosonic string, Eur. Phys. J.* **C 78** (2018) 600.

[7] G. P. L. Courant, *Dirac manifolds, Trans. Amer. Math. Soc.* **319** (1990) 631-661.

[8] B. Zwiebach, *Double Field Theory, T-Duality, and Courant Brackets*(2011) arXiv:1109.1782.

[9] A. Alekseev, T. Strobl, *Current algebras and differential geometry, JHEP* **03** (2005) 035.

[10] N. Halmagyi, *Non-geometric string backgrounds and worldsheet algebras, JHEP* **0807** (2008) 137.

[11] D. Roytenberg, *Quasi-Lie bialgebroids and twisted Poisson manifolds, Letters in Mathematical Physics* **61** (2002) 123.

[12] J. A. de Azcarraga, A. M. Perelomov, J. C. Perez Bueno, *The Schouten-Nijenhuis bracket, cohomology and generalized Poisson structures, J. Phys.* **A29** (1996) 7993-8110.

[13] Y. Kosmann-Schwarzbach, *From Poisson algebras to Gerstenhaber algebras, Annales de l'institut Fourier* **46** (1996) 1243-1274.

T-duality between effective string theories*

Ljubica Davidović[†]
Institute of Physics, Belgrade, Serbia
Branislav Sazdović[‡]
Institute of Physics, Belgrade, Serbia

Abstract

We consider the open bosonic string moving in the constant background. We investigate whether the solving of the constraints, obtained by applying the Dirac procedure to the boundary conditions of the open string, which leads to effective closed string theory, and the T-dualization procedure are commutative. We consider the string with mixed boundary conditions. We start by applying the Dirac procedure to these conditions, which results in two parameter dependent constraints. These constraints are solved and for their solution the effective theory is obtained. On the other hand applying the T-dualization procedure to the initial theory one obtains the T-dual theory. As usual, the form of the theory is such that as if the T-dual boundary conditions are already chosen, so that the T-dual coordinates satisfy exactly the opposite set of the boundary conditions then the corresponding coordinates of the initial string. We apply the Dirac procedure to the T-dual boundary conditions, obtain the parameter dependent constraints and solve them to obtain the T-dual effective theory. We show that the effective theories of the initial and T-dual theory remain T-dual, and find the effective T-duality coordinate transformation laws.

1. Introduction

This article is based on our paper [1]. We consider the open string moving in the constant background fields and its T-dual theory. We choose the mixed boundary conditions, for both initial and T-dual coordinates and solve them using techniques developed in [2, 3, 4]. The initial theory with Neumann boundary conditions for coordinate directions x^a and Dirichlet conditions for the rest of the coordinates x^i is by T-dual coordinate transformation laws equivalent to the T-dual theory with dual coordinates y_a satisfying the Dirichlet and y_i the Neumann boundary conditions.

* This work has been supported by the Serbian Ministry of Education, Science and Technological Development.
† e-mail address: ljubica@ipb.ac.rs
‡ e-mail address: sazdovic@ipb.ac.rs

We treat all boundary conditions as constraints and follow the Dirac procedure. The new constraints are found, first as a Poisson bracket between the hamiltonian and the boundary conditions, and every subsequent as a Poisson bracket between the hamiltonian and the previous constraint. Using the Taylor expansion, we represent this infinite set of constraints we obtain, by only two σ-dependent constraints [5, 6, 7, 8], one for each end-point. Imposing $2\pi-$periodicity, to the variables building the constraints, one observes that the constraints at $\sigma = \pi$ can be expressed in terms of that at $\sigma = 0$, and that in fact solving one pair of constraints one solves the other pair as well.

The σ-dependent constraints are solved by separating the constraints into their even and odd parts under world-sheet parity transformation $(\Omega : \sigma \to -\sigma)$, and separating the variables building them into their even and odd parts. Solving the σ-dependent constraints, one reduces the phase space by half. Halves of the original canonical variables are treated as effective variables: the independent variables and their canonical conjugates. For the solution of the constraints we obtain the effective theories, defined in terms of the effective variables. We examine their characteristics and confirm that the effective theories of two T-dual theories are also T-dual.

2. The initial and T-dual open bosonic string

Let us consider two T-dual bosonic string theories

$$\kappa \int_\Sigma d\xi^2 \partial_+ x^\mu \Pi_{+\mu\nu} \partial_- x^\nu \longleftrightarrow \frac{\kappa^2}{2} \int_\Sigma d\xi^2 \partial_+ y_\mu \Theta_-^{\mu\nu} \partial_- y_\nu, \tag{1}$$

describing the bosonic string and its T-dual moving in constant backgrounds associated with the fields: a metric tensor $G_{\mu\nu}$, a Kalb-Ramond field $B_{\mu\nu}$ and a dilaton field Φ. The integration goes over two-dimensional world-sheet Σ. Σ is usualy parametrized by ξ^α ($\xi^0 = \tau,\ \xi^1 = \sigma$), but here we use the light-cone coordinates given by

$$\xi^\pm = \frac{1}{2}(\tau \pm \sigma), \qquad \partial_\pm = \partial_\tau \pm \partial_\sigma. \tag{2}$$

$x^\mu(\xi)$, $\mu = 0, 1, ..., D-1$ are the coordinates of the D-dimensional space-time, and $y_\mu(\xi)$ are the coordinates of its T-dual. The background field composition is

$$\Pi_{\pm\mu\nu}(x) = B_{\mu\nu}(x) \pm \frac{1}{2} G_{\mu\nu}(x), \tag{3}$$

and the dual background field composition is

$$^\star\Pi_\pm^{\mu\nu} = \frac{\kappa}{2}\Theta_\mp^{\mu\nu} = -(G_E^{-1}\Pi_\mp G^{-1})^{\mu\nu}, \quad (G_E)_{\mu\nu} = (G - 4BG^{-1}B)_{\mu\nu}, \tag{4}$$

where G_E is the effective metric. The T-dual metric is its inverse

$$^\star G^{\mu\nu} = (G_E^{-1})^{\mu\nu}, \tag{5}$$

and a T-dual Kalb-Ramond field is

$$ {}^{\star}B^{\mu\nu} = \frac{\kappa}{2}\theta^{\mu\nu}, \tag{6} $$

where $\theta^{\mu\nu} = -\frac{2}{\kappa}(G_E^{-1}BG^{-1})^{\mu\nu}$ is the noncommutativity parameter.

The coordinates of the initial and the T-dual theory are related by

$$ \partial_\pm x^\mu \cong -\kappa\Theta_\pm^{\mu\nu}\partial_\pm y_\nu, \quad \partial_\pm y_\mu \cong -2\Pi_{\mp\mu\nu}\partial_\pm x^\nu. \tag{7} $$

We consider the block diagonal constant metric and Kalb-Ramond field $G_{\mu\nu} = const$, $B_{\mu\nu} = const$

$$ G_{\mu\nu} = \begin{pmatrix} G_{ab} & 0 \\ 0 & G_{ij} \end{pmatrix}, \quad B_{\mu\nu} = \begin{pmatrix} B_{ab} & 0 \\ 0 & B_{ij} \end{pmatrix}, \tag{8} $$

consequently the composition of the T-dual background fields is also block diagonal $\Theta_\pm^{\mu\nu} = \begin{pmatrix} \Theta_\pm^{ab} & 0 \\ 0 & \Theta_\pm^{ij} \end{pmatrix}$ with components

$$ \begin{aligned} \Theta_\pm^{\mu\nu} &= \\ -\frac{2}{\kappa} & \begin{pmatrix} (G_E{}^{-1})^{ac} & 0 \\ 0 & (G_E{}^{-1})^{ik} \end{pmatrix} \begin{pmatrix} \Pi_{\pm cd} & 0 \\ 0 & \Pi_{\pm kl} \end{pmatrix} \begin{pmatrix} (G^{-1})^{db} & 0 \\ 0 & (G^{-1})^{lj} \end{pmatrix} \\ &= \begin{pmatrix} \theta^{ab} & 0 \\ 0 & \theta^{ij} \end{pmatrix} \mp \frac{1}{\kappa} \begin{pmatrix} (G_E^{-1})^{ab} & 0 \\ 0 & (G_E^{-1})^{ij} \end{pmatrix}. \end{aligned} \tag{9} $$

where $(G_E)_{ab} = G_{ab} - 4B_{ac}(G^{-1})^{cd}B_{db}$, $(G_E)_{ij} = G_{ij} - 4B_{ik}(G^{-1})^{kl}B_{lj}$ and $\theta^{ab} = -\frac{2}{\kappa}(G_E^{-1})^{ac}\, B_{cd}(G^{-1})^{db}$, $\theta^{ij} = -\frac{2}{\kappa}(G_E^{-1})^{ik}\, B_{kl}(G^{-1})^{lj}$.

2.1. Boundary conditions

We will consider the boundary conditions of the initial theory

$$ \gamma_\mu^{(0)}\delta x^\mu\Big|_0^\pi = 0, \quad \gamma_\mu^{(0)} = \kappa\big(\Pi_{+\mu\nu}\partial_- x^\nu + \Pi_{-\mu\nu}\partial_+ x^\nu\big). \tag{10} $$

For each of the space-time coordinates of the open bosonic string one can fulfill the boundary conditions (10) by choosing either the Neumann or the Dirichlet boundary condition. Let us choose the Neumann condition for coordinates x^a, $a = 0, 1, \ldots, p$ and the Dirichlet condition for coordinates x^i, $i = p+1, \ldots, D-1$, which read

$$ \begin{aligned} &\textit{Neumann}: \quad \gamma_a^{(0)}\Big|_{\partial\Sigma} = 0, \quad {}_N\gamma_a^0 \equiv \gamma_a^{(0)} = \kappa\big(\Pi_{+ab}\partial_- x^b + \Pi_{-ab}\partial_+ x^b\big), \\ &\textit{Dirichlet}: \quad \kappa\dot{x}^i\Big|_{\partial\Sigma} = 0, \quad {}_D\gamma_0^i \equiv \kappa\dot{x}^i. \end{aligned} \tag{11} $$

The boundary conditions of the T-dual string are

$$^\star\gamma^{(0)\mu}\,\delta y_\mu\Big|_0^\pi = 0, \quad ^\star\gamma^{(0)\mu} = \frac{\kappa^2}{2}\Big[\Theta_-^{\mu\nu}\partial_- y_\nu + \Theta_+^{\mu\nu}\partial_+ y_\nu\Big]. \tag{12}$$

The T-dual theory is equivalent to a open string theory (1) with chosen boundary conditions (11), if the T-dual boundary conditions are fulfilled in a Neumann way for coordinates y_i and in a Dirichlet way for y_a

$$\begin{aligned} &\textit{Neumann}: \ ^\star\gamma^{(0)i}\Big|_{\partial\Sigma} = 0, \quad {}_N^\star\gamma_0^i \equiv {}^\star\gamma^{(0)i} = \frac{\kappa^2}{2}\Big[\Theta_-^{ij}\partial_- y_j + \Theta_+^{ij}\partial_+ y_j\Big], \\ &\textit{Dirichlet}: \ \kappa\dot{y}_a\Big|_{\partial\Sigma} = 0, \quad {}_D^\star\gamma_a^0 \equiv \kappa\dot{y}_a. \end{aligned} \tag{13}$$

This is because of the T-duality transformation law (7), which gives

$$-\kappa\dot{x}^\mu \cong {}^\star\gamma^{(0)\mu}(y), \quad \gamma_\mu^{(0)}(x) \cong -\kappa\dot{y}_\mu, \tag{14}$$

and consequently

$$\begin{aligned} {}_D\gamma_0^i &\equiv \kappa\dot{x}^i \cong -{}^\star\gamma^{(0)i} \equiv -{}_N^\star\gamma_0^i, \\ {}_N\gamma_a^0 &\equiv \gamma_a^{(0)} \cong -\kappa\dot{y}_a = -{}_D^\star\gamma_a^0. \end{aligned} \tag{15}$$

So, T-dualization interchanges the Dirichlet and Neumann boundary conditions.

3. Canonical form of the boundary conditions and their consistency by Dirac procedure

We will treat the T-dual boundary conditions (11) and (13), of theories (1) as constraints and we will apply the Dirac consistency procedure, following [2, 3]. This will be done in a canonical description, because the conditions can be expressed in terms of the currents building the energy-momentum tensors, and consequently the hamiltonians. So, to calculate new constraints one only uses the algebra of these currents.

The canonical hamiltonian densities of theories (1) are

$$\mathcal{H}_C = T_- - T_+ = \frac{1}{4\kappa}(G^{-1})^{\mu\nu}\Big[j_{+\mu}j_{+\nu} + j_{-\mu}j_{-\nu}\Big], \tag{16}$$

and

$$^\star\mathcal{H}_C = {}^\star T_- - {}^\star T_+ = \frac{1}{4\kappa}({}^\star G^{-1})_{\mu\nu}\Big[{}^\star j_+^{\mu}\,{}^\star j_+^{\nu} + {}^\star j_-^{\mu}\,{}^\star j_-^{\nu}\Big], \tag{17}$$

where $T_\pm = \mp\frac{1}{4\kappa}(G^{-1})^{\mu\nu}j_{\pm\mu}j_{\pm\nu}$ and ${}^\star T_\pm = \mp\frac{1}{4\kappa}({}^\star G^{-1})_{\mu\nu}\,{}^\star j_\pm^{\mu}\,{}^\star j_\pm^{\nu}$ are the energy-momentum tensors. The currents are given by

$$j_{\pm\mu} = \pi_\mu + 2\kappa\Pi_{\pm\mu\nu}x'^\nu, \quad {}^\star j_\pm^\mu = {}^\star\pi^\mu + 2\kappa\,{}^\star\Pi_\pm^{\mu\nu}y'_\nu, \tag{18}$$

where the momenta are $\pi_\mu = -2\kappa B_{\mu\nu}x'^\nu + \kappa G_{\mu\nu}\dot{x}^\nu$ and ${}^\star\pi^\mu = -\kappa^2\theta^{\mu\nu}y'_\nu + \kappa(G_E^{-1})^{\mu\nu}\dot{y}_\nu = -2\kappa\,{}^\star B^{\mu\nu}y'_\nu + \kappa\,{}^\star G^{\mu\nu}\dot{y}_\nu$. Using the last relations the currents can be rewritten in terms of coordinates as

$$j_{\pm\mu} = \kappa G_{\mu\nu}\partial_\pm x^\nu, \quad {}^\star j_\pm^\mu = \kappa(G_E^{-1})^{\mu\nu}\partial_\pm y_\nu. \tag{19}$$

The algebra of the initial currents [9] in a constant background is given by

$$\begin{aligned} \{j_{\pm\mu}(\sigma), j_{\pm\nu}(\bar{\sigma})\} &= \pm 2\kappa\, G_{\mu\nu}\,\delta'(\sigma-\bar{\sigma}), \\ \{j_{\pm\mu}(\sigma), j_{\mp\nu}(\bar{\sigma})\} &= 0. \end{aligned} \tag{20}$$

The initial Neumann (N) and Dirichlet (D) boundary conditions (11) are in a canonical form given by

$$\begin{aligned} {}_N\gamma_a^0 &= \Pi_{+ab}(G^{-1})^{bc}j_{-c} + \Pi_{-ab}(G^{-1})^{bc}j_{+c}, \\ {}_D\gamma_0^i &= \kappa\dot{x}^i = \frac{1}{2}(G^{-1})^{ij}(j_{+j}+j_{-j}). \end{aligned} \tag{21}$$

The T-dual conditions in a canonical form are

$$\begin{aligned} {}^\star_N\gamma_0^i &= {}^\star\Pi_{+ij}({}^\star G^{-1})^{jk}\,{}^\star j_{-k} + {}^\star\Pi_{-ij}({}^\star G^{-1})^{jk}\,{}^\star j_{+k}, \\ {}^\star_D\gamma_a^0 &= \frac{1}{2}({}^\star G^{-1})_{ab}\Big({}^\star j_+^b + {}^\star j_-^b\Big). \end{aligned} \tag{22}$$

Although the Neumann boundary conditions transform into Dirichlet conditions of the T-dual theory by (15), and Dirichlet's to Neumann, one can observe that the form of Neumann and Dirichlet conditions has not changed in T-dual theories.

Following the Dirac procedure, one can impose consistency to these constraints. The additional constraints are defined for every $n \geq 1$ by

$${}_N\gamma_a^n = \{H_C, {}_N\gamma_a^{n-1}\}, \quad {}_D\gamma_n^i = \{H_C, {}_D\gamma_{n-1}^i\}, \tag{23}$$

with $H_C = \int d\sigma\mathcal{H}_C$ being the canonical hamiltonian of the initial theory, and similarly for the T-dual theory. All the constraints can be gathered into only two σ dependent constraints. They are obtained by multiplying every constraint by an appropriate degree of the world-sheet space parameter σ and adding the terms together

$$\Gamma_a^N(\sigma) = \sum_{n\geq 0}\frac{\sigma^n}{n!}\,{}_N\gamma_a^n\Big|_{\sigma=0}, \quad \Gamma_D^i(\sigma) = \sum_{n\geq 0}\frac{\sigma^n}{n!}\,{}_D\gamma_n^i\Big|_{\sigma=0}. \tag{24}$$

We obtain the explicit form of the sigma dependent constraints of the initial theory

$$\begin{aligned} \Gamma_a^N(\sigma) &= \Pi_{+ab}(G^{-1})^{bc}j_{-c}(\sigma) + \Pi_{-ab}(G^{-1})^{bc}j_{+c}(-\sigma), \\ \Gamma_D^i(\sigma) &= \frac{1}{2}(G^{-1})^{ij}\Big[j_{+j}(-\sigma) + j_{-j}(\sigma)\Big], \end{aligned} \tag{25}$$

and of its T-dual

$$\begin{aligned} {}^\star\Gamma^i_N(\sigma) &= \frac{\kappa}{2}\Big[\Theta^{ij}_- G^E_{jk}\,{}^\star j^k_-(\sigma) + \Theta^{ij}_+ G^E_{jk}\,{}^\star j^k_+(-\sigma)\Big], \\ {}^\star\Gamma^D_a(\sigma) &= \frac{1}{2} G^E_{ab}\big[{}^\star j^b_+(-\sigma) + {}^\star j^b_-(\sigma)\big]. \end{aligned} \tag{26}$$

One can check that the σ-dependent constraints are of the second class and one therefore can solve them.

Obviously, the parameter dependent constraints are given in terms of currents depending on either σ or $-\sigma$. So, it is useful to express the canonical variables into their even and odd parts, with respect to σ. For the initial coordinates one has

$$x^\mu = q^\mu + \bar{q}^\mu, \quad q^\mu = \sum_{n\geq 0} \frac{\sigma^{2n}}{(2n)!} x^{(2n)\mu}\Big|_{\sigma=0}, \quad \bar{q}^\mu = \sum_{n\geq 0} \frac{\sigma^{2n+1}}{(2n+1)!} x^{(2n+1)\mu}\Big|_{\sigma=0}, \tag{27}$$

and for the momenta one has

$$\pi_\mu = p_\mu + \bar{p}_\mu, \quad p_\mu = \sum_{n\geq 0} \frac{\sigma^{2n}}{(2n)!} \pi^{(2n)}_\mu\Big|_{\sigma=0}, \quad \bar{p}_\mu = \sum_{n\geq 0} \frac{\sigma^{2n+1}}{(2n+1)!} \pi^{(2n+1)}_\mu\Big|_{\sigma=0}. \tag{28}$$

Similarly, we separate the T-dual variables into the odd and even parts

$$\begin{aligned} y_\mu &= k_\mu + \bar{k}_\mu, \\ {}^\star\pi^\mu &= {}^\star p^\mu + {}^\star\bar{p}^\mu. \end{aligned} \tag{29}$$

After this separations the constraints become

$$\begin{aligned} \Gamma^N_a(\sigma) &= 2(BG^{-1})_a{}^b p_b + \bar{p}_a - \kappa(G_E)_{ab}\bar{q}'^b, \\ \Gamma^i_D(\sigma) &= (G^{-1})^{ij}\Big[p_j - \kappa G_{jk} q'^k + 2\kappa B_{jk}\bar{q}'^k\Big], \end{aligned} \tag{30}$$

and

$$\begin{aligned} {}^\star\Gamma^i_N(\sigma) &= -2(G^{-1}B)^i{}_j\,{}^\star p^j + {}^\star\bar{p}^i - \kappa(G^{-1})^{ij}\bar{k}'_j, \\ {}^\star\Gamma^D_a(\sigma) &= G^E_{ab}\Big[{}^\star p^b + \kappa^2\theta^{bc}\bar{k}'_c - \kappa(G^{-1}_E)^{bc}k'_c\Big]. \end{aligned} \tag{31}$$

This form of constraints is solvable and has a solution

$$\begin{aligned} \bar{p}_a &= 0, \qquad \bar{q}'^a = -\theta^{ab}p_b, \\ q'^i &= 0, \qquad p_i = -2\kappa B_{ij}\bar{q}'^j, \end{aligned} \tag{32}$$

and in T-dual theory

$$\begin{aligned} {}^\star\bar{p}^i &= 0, \qquad \bar{k}'_i = -\frac{2}{\kappa}B_{ij}\,{}^\star p^j, \\ {}^\star p^a &= -\kappa^2\theta^{ab}\bar{k}'_b, \qquad k'_a = 0. \end{aligned} \tag{33}$$

Solving the constraints has reduced the phase space by half. The σ-derivative of coordinates and the momenta for the solution (32) are

$$x'^\mu = \begin{cases} q'^a - \theta^{ab} p_b, & \mu{=}a, \\ \bar{q}'^i, & \mu{=}i, \end{cases} \tag{34}$$

and

$$\pi_\mu = \begin{cases} p_a, & \mu{=}a, \\ \bar{p}_i - 2\kappa B_{ij} \bar{q}'^j, & \mu{=}i. \end{cases} \tag{35}$$

The T-dual coordinate σ-derivative and the T-dual momenta for the solution (33) are

$$y'_\mu = \begin{cases} \bar{k}'_a, & \mu{=}a, \\ k'_i - \frac{2}{\kappa} B_{ij} {}^\star p^j, & \mu{=}i, \end{cases} \tag{36}$$

and

$$^\star\pi^\mu = \begin{cases} ^\star\bar{p}^a - \kappa^2 \theta^{ab} \bar{k}'_b, & \mu{=}a, \\ ^\star p^i, & \mu{=}i. \end{cases} \tag{37}$$

When one considers the constraints at the other string end-point $\sigma = \pi$, one multiplies every constraint with the appropriate power of $\sigma - \pi$ and sums the products into ${}_\pi\Gamma^N_a(\sigma)$ and ${}_\pi\Gamma^i_D(\sigma)$. The sigma dependent constraints at π and 0 differ by the following exchange

$$j_{+a}(2\pi - \sigma) \longleftrightarrow j_{+a}(-\sigma), \quad j_{+i}(2\pi - \sigma) \longleftrightarrow j_{+i}(-\sigma). \tag{38}$$

It follows that the extension of a domain [2] of the variables building the currents, i.e. original coordinates and momenta and demand their 2π-periodicity $x^\mu(\sigma{+}2\pi) = x^\mu(\sigma)$, and $\pi_\mu(\sigma{+}2\pi) = \pi_\mu(\sigma)$ makes the solutions (32) and (33) the universal solution for both $\sigma-$dependent constraints at $\sigma = 0, \pi$.

4. Effective theories for the solution of boundary conditions

Next, we will substitute the solution of the constraints into the canonical hamiltonians, to obtain the effective hamiltonians. Using the equations of motion for momenta, we will find the corresponding effective lagrangians. Effective theories describe the closed effective strings.

$$
\begin{array}{ccc}
\kappa\int d\xi^2\partial_+x^\mu\Pi_{+\mu\nu}\partial_-x^\nu & \overset{T}{\longleftrightarrow} & \frac{\kappa^2}{2}\int d\xi^2\partial_+y_\mu\Theta_-^{\mu\nu}\partial_-y_\nu \\
\Big\downarrow {\scriptstyle \Gamma=0} & & \Big\downarrow {\scriptstyle {}^\star\Gamma=0} \\
? & & ?
\end{array}
$$

The solution of the initial boundary conditions is such that the effective hamiltonian will be given in terms of odd $\bar q^i, \bar p_i$ in Dirichlet directions and even q^a, p_a in Neumann directions. It reads

$$\mathcal{H}^{eff} = \mathcal{H}_N^{eff}(q^a, p_a) + \mathcal{H}_D^{eff}(\bar q^i, \bar p_i), \tag{39}$$

where

$$
\begin{aligned}
\mathcal{H}_N^{eff}(q^a, p_a) &= \frac{\kappa}{2}\, q'^a G^E_{ab} q'^b + \frac{1}{2\kappa} p_a (G_E^{-1})^{ab} p_b, \\
\mathcal{H}_D^{eff}(\bar q^i, \bar p_i) &= \frac{\kappa}{2}\, \bar q'^i G_{ij} \bar q'^j + \frac{1}{2\kappa} \bar p_i (G^{-1})^{ij} \bar p_j.
\end{aligned} \tag{40}
$$

Simmilarlly, the effective T-dual hamiltonian is given in terms of odd $\bar k_a, {}^\star\bar p^a$ in Dirichlet directions and even $k_i, {}^\star p^i$ in Neumann directions. The corresponding effective lagrangians will consequently depend on both even and odd coordinate parts $q^a, \bar q^i$ and $k_i, \bar k_a$. The effective T-dual hamiltonian is

$${}^\star\mathcal{H}^{eff} = {}^\star\mathcal{H}_D^{eff}(\bar k_a, {}^\star\bar p^a) + {}^\star\mathcal{H}_N^{eff}(k_i, {}^\star p^i), \tag{41}$$

where

$$
\begin{aligned}
{}^\star\mathcal{H}_D^{eff}(\bar k_a, {}^\star\bar p^a) &= \frac{\kappa}{2}\, \bar k'_a (G_E^{-1})^{ab} \bar k'_b + \frac{1}{2\kappa} {}^\star\bar p^a (G_E)_{ab} {}^\star\bar p^b, \\
{}^\star\mathcal{H}_N^{eff}(k_i, {}^\star p^i) &= \frac{\kappa}{2}\, k'_i (G^{-1})^{ij} k'_j + \frac{1}{2\kappa} {}^\star p^i G_{ij} {}^\star p^j.
\end{aligned} \tag{42}
$$

The effective Lagrangians of the above effective hamiltonians are

$$\mathcal{L}^{eff} = \Big[\pi_\mu \dot x^\mu\Big]\Big|_{\Gamma_\mu=0} - \mathcal{H}^{eff}, \quad {}^\star\mathcal{L}^{eff} = \Big[{}^\star\pi^\mu \dot y_\mu\Big]\Big|_{{}^\star\Gamma_\mu=0} - {}^\star\mathcal{H}^{eff}. \tag{43}$$

The effective lagrangians can be separated into

$$\mathcal{L}^{eff} = \mathcal{L}_N(q,p) + \mathcal{L}_D(\bar q, \bar p), \quad {}^\star\mathcal{L}^{eff} = {}^\star\mathcal{L}_D(\bar k, {}^\star\bar p) + {}^\star\mathcal{L}_N(k, {}^\star p), \tag{44}$$

with

$$
\begin{aligned}
&\mathcal{L}_N(q,p) = p_a \dot q^a - \mathcal{H}_N^{eff}(q^a, p_a), \quad \mathcal{L}_D(\bar q, \bar p) = \bar p_i \dot{\bar q}^i - \mathcal{H}_D^{eff}(\bar q^i, \bar p_i), \\
&{}^\star\mathcal{L}_D(\bar k, {}^\star\bar p) = {}^\star\bar p^a \dot{\bar k}_a - {}^\star\mathcal{H}_D^{eff}(\bar k_a, {}^\star\bar p^a), \quad {}^\star\mathcal{L}_N(k, {}^\star p) = {}^\star p_i \dot k_i - {}^\star\mathcal{H}_N^{eff}(k_i, {}^\star p^i).
\end{aligned} \tag{45}
$$

The explicit forms of the effective lagrangians are found by eliminating the momenta from (45), using the equations of motion for them

$$p_a = \kappa G^E_{ab}\dot{q}^b, \quad \bar{p}_i = \kappa G_{ij}\dot{\bar{q}}^j, \tag{46}$$

and

$$^\star\bar{p}^a = \kappa (G_E^{-1})^{ab}\dot{\bar{k}}_b, \quad ^\star p^i = \kappa (G^{-1})^{ij}\dot{k}_j. \tag{47}$$

For these equations the σ-derivatives of the initial and T-dual coordinates, given by (34) and (36), become

$$x'^\mu = \begin{cases} q'^a + 2(G^{-1}B)^a{}_b\,\dot{q}^b, & \mu=a, \\ \bar{q}'^i, & \mu=i, \end{cases} \tag{48}$$

and

$$y'_\mu = \begin{cases} \bar{k}'_a, & \mu=a, \\ k'_i - 2(BG^{-1})_i{}^j\,\dot{k}_j, & \mu=i. \end{cases} \tag{49}$$

In order to find the expression for the initial and the T-dual coordinate we need to introduce a double coordinate $\tilde{q}^a$ of the even part of the initial coordinate q^a

$$\dot{\tilde{q}}^a = q'^a, \quad \tilde{q}'^a = \dot{q}^a, \tag{50}$$

and a double coordinate $\tilde{k}_i$ of the even part of the T-dual coordinate k_i

$$\dot{\tilde{k}}_i = k'_i, \quad \tilde{k}'_i = \dot{k}_i. \tag{51}$$

The coordinates become

$$x^\mu = \begin{cases} q^a + 2(G^{-1}B)^a{}_b\,\tilde{q}^b, & \mu=a, \\ \bar{q}^i, & \mu=i, \end{cases} \tag{52}$$

and

$$y_\mu = \begin{cases} \bar{k}_a, & \mu=a, \\ k_i - 2(BG^{-1})_i{}^j\,\tilde{k}_j, & \mu=i. \end{cases} \tag{53}$$

So, after elimination of the momenta the effective lagrangians become

$$\mathcal{L}^{eff} = \mathcal{L}_N(q) + \mathcal{L}_D(\bar{q}), \quad ^\star\mathcal{L}^{eff} = {}^\star\mathcal{L}_D(\bar{k}) + {}^\star\mathcal{L}_N(k), \tag{54}$$

where the lagrangians (45) reduced to

$$\mathcal{L}_N(q) = \frac{\kappa}{2} G^E_{ab} \eta^{\alpha\beta} \partial_\alpha q^a \partial_\beta q^b, \qquad \mathcal{L}_D(\bar{q}) = \frac{\kappa}{2} G_{ij} \eta^{\alpha\beta} \partial_\alpha \bar{q}^i \partial_\beta \bar{q}^j,$$
$$^\star\mathcal{L}_D(\bar{k}) = \frac{\kappa}{2} (G_E^{-1})^{ab} \eta^{\alpha\beta} \partial_\alpha \bar{k}_a \partial_\beta \bar{k}_b, \qquad ^\star\mathcal{L}_N(k) = \frac{\kappa}{2} (G^{-1})^{ij} \eta^{\alpha\beta} \partial_\alpha k_i \partial_\beta k_j. \tag{55}$$

5. T-duality between effective theories

Introducing the effective coordinates and T-dual effective coordinates

$$Q^\mu = \begin{bmatrix} q^a \\ \bar{q}^i \end{bmatrix}, \quad K_\mu = \begin{bmatrix} \bar{k}_a \\ k_i \end{bmatrix}, \tag{56}$$

and their corresponding canonically conjugated momenta

$$P_\mu = \begin{bmatrix} p_a \\ \bar{p}_i \end{bmatrix}, \quad ^\star P^\mu = \begin{bmatrix} ^\star\bar{p}^a \\ ^\star p^i \end{bmatrix}, \tag{57}$$

one can rewrite the effective hamiltonians (39) and (41) as

$$\mathcal{H}^{eff} = \frac{\kappa}{2} Q'^\mu G^{eff}_{\mu\nu} Q'^\nu + \frac{1}{2\kappa} P_\mu (G^{-1}_{eff})^{\mu\nu} P_\nu,$$
$$^\star\mathcal{H}^{eff} = \frac{\kappa}{2} K'_\mu {}^\star G^{\mu\nu}_{eff} K'_\nu + \frac{1}{2\kappa} {}^\star P^\mu ({}^\star G^{-1}_{eff})_{\mu\nu} {}^\star P^\nu, \tag{58}$$

where

$$G^{eff}_{\mu\nu} = \begin{pmatrix} G^E_{ab} & 0 \\ 0 & G_{ij} \end{pmatrix}, \qquad ^\star G^{\mu\nu}_{eff} = \begin{pmatrix} (G_E^{-1})^{ab} & 0 \\ 0 & (G^{-1})^{ij} \end{pmatrix}. \tag{59}$$

The T-duality relations

$$\kappa x'^\mu \cong {}^\star\pi^\mu, \quad \kappa y'_\mu \cong \pi_\mu, \tag{60}$$

for the solution of the boundary conditions and separating the odd and even parts reduce to

$$\kappa q'^a \cong {}^\star\bar{p}^a, \quad \kappa \bar{k}'_a \cong p_a,$$
$$\kappa \bar{q}'^i \cong {}^\star p^i, \quad \kappa k'_i \cong \bar{p}_i, \tag{61}$$

which gives

$$\kappa Q'^\mu \cong {}^\star P^\mu, \quad \kappa K'_\mu \cong P_\mu. \tag{62}$$

Additionally, the background fields (59) are T-dual as expected, because by T-duality the metric should transform to the inverse of the effective metric.

In absence of the effective Kalb-Ramond field this means the T-dual metric should be inverse to the initial metric, what is just the case

$$(G^{eff}_{\mu\nu})^{-1} = \begin{pmatrix} G^E_{ab} & 0 \\ 0 & G_{ij} \end{pmatrix}^{-1} = \begin{pmatrix} (G_E^{-1})^{ab} & 0 \\ 0 & (G^{-1})^{ij} \end{pmatrix} = {}^\star G^{\mu\nu}_{eff}. \quad (63)$$

Using (62) and (63) we can conclude that the effective hamiltonians (58) are T-dual to each other.

The corresponding lagrangians (54) are given by

$$\mathcal{L}^{eff} = \dot{Q}^\mu P_\mu - \mathcal{H}^{eff}(Q, P), \quad {}^\star\mathcal{L}^{eff} = \dot{K}_\mu {}^\star P^\mu - {}^\star\mathcal{H}^{eff}(K, {}^\star P), \quad (64)$$

which for the equations of motion for momenta (46) and (47)

$$P_\mu = \kappa G^{eff}_{\mu\nu} \dot{Q}^\nu, \quad {}^\star P_\mu = \kappa\, {}^\star G^{eff}_{\mu\nu} \dot{K}_\nu, \quad (65)$$

become

$$\mathcal{L}^{eff} = \frac{\kappa}{2} \eta^{\alpha\beta}\, \partial_\alpha Q^\mu\, G^{eff}_{\mu\nu}\, \partial_\beta Q^\nu, \quad {}^\star\mathcal{L}^{eff} = \frac{\kappa}{2} \eta^{\alpha\beta}\, \partial_\alpha K_\mu\, {}^\star G^{\mu\nu}_{eff}\, \partial_\beta K_\nu. \quad (66)$$

Combining (62) with (65) one obtains

$$Q'^\mu \cong {}^\star G^{\mu\nu}_{eff} \dot{K}_\nu, \quad K'_\mu \cong G^{eff}_{\mu\nu} \dot{Q}^\nu. \quad (67)$$

Therefore, the effective and T-dual effective variables Q^μ and K_μ are connected by

$$\partial_\pm K_\mu \cong \pm G^{eff}_{\mu\nu} \partial_\pm Q^\nu. \quad (68)$$

This is the T-dual effective coordinate transformation law. Using it together with (63), one can conclude that the effective lagrangians (66) are T-dual. This law is in agreement with the T-dual coordinate transformation law (7), for $B_{\mu\nu} = 0$

$$\partial_\pm y_\mu \cong \pm G_{\mu\nu} \partial_\pm x^\nu, \quad (69)$$

keeping in mind that the metric is replaced by the effective metric $G_{\mu\nu} \to G^{eff}_{\mu\nu}$.

6. Conclusion

In the present paper we showed that the T-dualization procedure and the solving of the mixed boundary conditions, treated as constraints in the Dirac consistency procedure, do commute. The results are summarized in the following diagram

$$
\begin{array}{ccc}
\kappa \int d\xi^2 \partial_+ x^\mu \Pi_{+\mu\nu} \partial_- x^\nu & \overset{T}{\longleftrightarrow} & \frac{\kappa^2}{2} \int d\xi^2 \partial_+ y_\mu \Theta_-^{\mu\nu} \partial_- y_\nu \\
\Big\downarrow{\scriptstyle \Gamma=0} & & \Big\downarrow{\scriptstyle {}^\star\Gamma=0} \\
\frac{\kappa}{2} \int d\xi^2 \partial_+ Q^\mu G^{eff}_{\mu\nu} \partial_- Q^\nu & \overset{T}{\longleftrightarrow} & \frac{\kappa}{2} \int d\xi^2 \partial_+ K_\mu {}^\star G^{\mu\nu}_{eff} \partial_- K_\nu .
\end{array}
$$

We started by considering the string described by the open string sigma model (left up). The string is moving in the constant metric $G_{\mu\nu}$ and a constant Kalb-Ramond field $B_{\mu\nu}$. We chose the Neumann boundary conditions for some directions x^a and the Dirichlet boundary conditions for all other directions x^i. We treated these conditions as canonical constraints, and applied the Dirac procedure. The boundary conditions where rewritten in a canonical form in terms of currents building the energy-momentum tensor components. By Dirac procedure the new constraints are found commuting the hamiltonian with the known constraints. The canonical form of constraints allowed us a simple calculation of the exact form of the infinitely many constraints. From these constraints we formed two σ-dependent constraints, for every string endpoint, by multiplying every obtained constraint with the appropriate power of σ for constraints in $\sigma = 0$ and power of $\pi - \sigma$ for constraints in $\sigma = \pi$, and adding these terms together into Taylor expansions. The constraints at $\sigma = 0$ and $\sigma = \pi$ were found to be equivalent by imposing 2π-periodicity condition for the canonical variables x^μ and π_μ.

The σ-dependent constraints are of the second class. To solve them we introduced even and odd parts of the initial canonical variables. We found the solution and expressed the σ-derivative of the initial coordinate x'^μ and the initial momentum π_μ in terms of even parts q^a, p_a of x^μ, π_μ in Neumann directions and of their odd parts $\bar{q}^i, \bar{p}_i$ in Dirichlet directions, see (34) and (35.) For the solution of constraints $\Gamma = 0$, the theory reduced to the effective theory (left down). We obtained the effective hamiltonian (39). For the equations of motion for momenta, we obtained the corresponding effective lagrangian (54).

We also considered the T-dual (right up) of the initial theory. We applied the Dirac procedure to the mixed boundary conditions of the T-dual theory. The constraints where solved, which reduced the phase space to $\bar{k}_a, {}^\star\bar{p}^a$ in D-sector and $k_i, {}^\star p^i$ in N-sector. For the solution of T-dual constraints ${}^\star\Gamma = 0$, we obtained the T-dual effective hamiltonian (41), as well as the corresponding T-dual effective lagrangian (54) (right down).

The canonically conjugated effective variables are now pairs q^a, p_a and $\bar{q}^i, \bar{p}_i$ for the initial and $k_i, {}^\star p^i$ and $\bar{k}_a, {}^\star\bar{p}^a$ for the T-dual effective theory. The effective variables in both effective theories satisfy the modified Poisson brackets. This is different, in comparison to the choice of the Neumann boundary conditions for all directions [2, 3, 4]. In that case, solving the constraints leads to full elimination of odd variables. For constant initial background fields, considered in this paper the effective fields are constant. But, the nongeometricity obtained earlier in [2, 3, 4] can still be seen. It appears in a fact that coordinates of the initial and T-dual theories, can not be expressed without an introduction of double coordinates, see (52) and (53).

The obtained effective theories, defined in terms of the effective variables, where compared using the T-dualization procedure. It was confirmed that the corresponding background fields (the effective metrics $G^{eff}_{\mu\nu}$ and ${}^\star G^{\mu\nu}_{eff}$ (59)) are T-dual to each other. Also, the effective variables of the initial effective theory are confirmed to be T-dual to the T-dual effective variables of the T-dual effective theory, by obtaining the T-duality law connecting them. This law was an appropriate reduction of the standard T-duality coordinate transformation law. Therefore, we showed the T-duality of the reduced bosonic string theories.

References

[1] Lj. Davidović and B. Sazdović, *Eur. Phys. J.* **C 79** (2019) 770.

[2] Lj. Davidović and B. Sazdović, *Phys. Rev.* **D 83** (2011) 066014.

[3] Lj. Davidović and B. Sazdović, *JHEP* **08** (2011) 112.

[4] Lj. Davidović and B. Sazdović, *Eur. Phys. J.* **C 72** (2012) 2199.

[5] B. Sazdović, *Eur. Phys. J.* **C44** (2005) 599.

[6] B. Nikolić and B. Sazdović, *Phys. Rev.* **D74** (2006) 045024.

[7] B. Nikolić and B. Sazdović, *Phys. Rev.* **D75** (2007) 085011.

[8] B. Nikolić and B. Sazdović, *Adv. Theor. Math. Phys.* **14** (2010) 1.

[9] Lj. Davidović and B. Sazdović, *Eur. Phys. J.* **C 78** (2018) 600.

Nonassociative differential geometry and gravity*

Marija Dimitrijević Ćirić†
University of Belgrade, Faculty of Physics
Studentski trg 12-16, Belgrade, Serbia

Abstract

In this short contribution we introduce a nonassocaitive deformation of differential geometry and General Relativity. The nonassociativity is based on the string theory nongeometric R-flux. We use the twist formalism to consistently deform the algebra of infinitesimal diffeomorphisms into the quasi Hopf algebra of (deformed) infnitesimal diffeomorphisms and introduce the NA deformation of differential geometry. In particualr, we define the Levi-Civita connection, curvature tensor and torsion. The space-time quantities (curvature, torsion) are obtained by the zero momenum leaf projection to the space-time. The vacuum Einstein equation in space-time, expanded up to first order in the deformation parameter $\kappa\hbar$ is obtained.

1. Introduction

In the context of string theory, it is expected that the closed string sector provides a framework for a quantum theory of gravity. Namely, the massless bosonic modes of the closed string sector contain gravitational degrees of freedom such as the metric, the B-field, and the dilaton. In particular, in locally non-geometric backgrounds one expects to find a low-energy limit of closed string theory which is described by an effective nonassociative theory of gravity on spacetime.

Attempts to formulate a consistent effective gravity theory in the spacetime, starting from the nonasocaitive phase space of closed strings were done in [1, 2, 3]. There the construction is done using the twist approach. The twist approach provides a well defined way to introduce the noncommutative/nonassocaitive differential geometry and the notions of connections and curvature. The essential step [2, 3] is the projection from the phase space to the spacetime via the zero momentum leaf. In this short contribution we explain how the metric aspects of nonassociative differential

* This work has been supported by the Action MP1405 "Quantum Structure of Spacetime" from the European Cooperation in Science and Technology (COST) and by Project ON171031 of the Serbian Ministry of Education, Science and Technological Development.

† e-mail address: dmarija@ipb.ac.rs

geometry and vacuum Einstein equations in a model based on a locally non-geometric flux R are developed. The contribution is based on work done in [2, 3].

After a short overview of nonassociativity and noncommutativity in physics, in next section we develop the nonassocaitive differential geometry based on a cochain twist. This we use in Section 3 to construct a nonassociative theory of gravity on spacetime. Explicit expressions for the torsion, curvature, Ricci tensor and Levi-Civita connection in nonassociative Riemannian geometry on phase space are obtained. Using the projection to the zero momentum leaf, we construct the R-flux corrections to the Ricci tensor on spacetime, and comment on the potential implications of these corrections.

2. Nonassociative differential geometry

First ideas of a space-time with noncommuting (NC) coordinates appeared in the early days of quantum mechanics. In the 1930 Heisenberg proposed to introduce noncommuting coordinte operators in order to regularize the divergent electron self-energy. In the 1947 Snyder construced the first model of a NC space-time [4]. The renormalization theory, developed in the 1950es sucesfully solved the problem of divergences in quantum filed theory, so the idea of noncommutativity was not developed furhter. However, in the 1990es new results from string theory and searches for quantum gravity and quantum space-time renewed interest in noncommutative geometry. Lot of work has been done in formulating quantum field theory and gravity in a NC space-time, see reviews [5].

In the similar way, first ideas on nonassociativity of coordinates in physics go back to the early days of quantum mechanics. Jordan formulated a version of quantum mechanics [6] whith a new composition between herimitean observables

$$A \circ B = \frac{1}{2}((A+B)^2 - A^2 - B^2).$$

This composition gives a hermitean observable again, but it is nonassocaitive. Nambu further developed a modification of classical mechanics by introducing a Nambu-Poisson bracket $\{f, g, h\}$ instead of the usual Poisson bracket [7]. Nambu-Poisson bracket fulfils the fundamental identity instead of the usual Jacobi identity. The idea of nonassociativity in physics was then forgoten until recenty. Namley, it was discovered that symmetries of closed string field theory close a strong homotopy Lie-algebra, L_∞ algebra. These algebras can be seen as a generalization of the usual Lie algebra that do not fulfil the Jacobi identity, but higher homotopy relations instead. Nonassociative $\star$-products were develeoped from (closed) string theory in locally non-geometric backgrounds [8]. We will work with a model of a nonasocaitive phase space that originates from closed strings moving in a non-geometric background defined by the constant R-flux.

The phase space $\mathcal{M}$ has coordinates $x^A = (x^\mu, \tilde{x}_\mu = p_\mu)$, $\partial_A = \big(\partial_\mu, \tilde{\partial}^\mu = \frac{\partial}{\partial p_\mu}\big)$ with $A = 1, \dots 2d$. We introduce a deformation by a cochain twist $\mathcal{F}$

$$\mathcal{F} = \exp\Big(-\tfrac{\mathrm{i}\hbar}{2} (\partial_\mu \otimes \tilde{\partial}^\mu - \tilde{\partial}^\mu \otimes \partial_\mu) - \tfrac{\mathrm{i}\kappa}{2} R^{\mu\nu\rho} (p_\nu\, \partial_\rho \otimes \partial_\mu - \partial_\mu \otimes p_\nu\, \partial_\rho)\Big) , \quad (1)$$

with $R^{\mu\nu\rho}$ totally antisymmetric and constant, and $\kappa := \frac{\ell_s^3}{6\hbar}$. This twist fails to fulfill the 2-cocycle condition

$$\Phi\, (\mathcal{F} \otimes 1)\, (\Delta \otimes \mathrm{id})\mathcal{F} = (1 \otimes \mathcal{F})\, (\mathrm{id} \otimes \Delta)\mathcal{F} \ . \quad (2)$$

The associator Φ is given by

$$\Phi = \exp\big(\, \hbar\kappa\, R^{\mu\nu\rho}\, \partial_\mu \otimes \partial_\nu \otimes \partial_\rho\, \big) =: \phi_1 \otimes \phi_2 \otimes \phi_3 = 1 \otimes 1 \otimes 1 + O(\hbar\kappa). \quad (3)$$

In the following text we will use the notation: $\mathcal{F} = \mathrm{f}^\alpha \otimes \mathrm{f}_\alpha$, $\mathcal{F}^{-1} = \bar{\mathrm{f}}^\alpha \otimes \bar{\mathrm{f}}_\alpha$, $\Phi^{-1} =: \bar{\phi}_1 \otimes \bar{\phi}_2 \otimes \bar{\phi}_3$. The R-matrix encodes the braiding and it is defined by

$$\mathcal{R} = \mathcal{F}^{-2} =: \mathrm{R}^{\,\alpha} \otimes \mathrm{R}_{\,\alpha}, \quad (4)$$

The inverse of the R-matrix is then

$$\mathcal{R}^{-1} = \mathcal{F}^2 =: \overline{\mathrm{R}}^{\,\alpha} \otimes \overline{\mathrm{R}}_{\,\alpha}.$$

The phase space $\mathcal{M}$ is invariant under the action of infinitesimal diffeomorphism. The Hopf algebra of infinitesimal diffeomorphisms $U\mathsf{Vec}(\mathcal{M})$ is given by:

$$\begin{aligned} &[u, v] = (u^B \partial_B v^A - v^B \partial_B u^A)\partial_A, \\ &\Delta(u) = 1 \otimes u + u \otimes 1, \\ &\epsilon(u) = 0, S(u) = -u. \end{aligned}$$

The twist (1) deforms this Hopf algebra into a quasi-Hopf algebra of infinitesimal diffeomorphisms $U\mathsf{Vec}^{\mathcal{F}}(\mathcal{M})$. The deformation is such that the algebra structure does not change, the coproduct is deformed

$$\Delta^{\mathcal{F}}\xi = \mathcal{F}\,\Delta\,\mathcal{F}^{-1},$$

while the counit and the antipod do not change: $\epsilon^{\mathcal{F}} = \epsilon, S^{\mathcal{F}} = S$.

On the basis vector fields the twist acts as

$$\begin{aligned} \Delta_{\mathcal{F}}(\partial_\mu) &= 1 \otimes \partial_\mu + \partial_\mu \otimes 1 \ , \\ \Delta_{\mathcal{F}}(\tilde{\partial}^\mu) &= 1 \otimes \tilde{\partial}^\mu + \tilde{\partial}^\mu \otimes 1 + \mathrm{i}\,\kappa\, R^{\mu\nu\rho}\, \partial_\nu \otimes \partial_\rho \ . \end{aligned}$$

We use the covariance under the infinitesimal diffeomorphisms as a guiding principle to define a NA phase space. Namley, we know that the differential geometry on $\mathcal{M}$ is covatiant under $U\mathsf{Vec}(\mathcal{M})$. Then we demand that the NA differential geometry on $\mathcal{M}$ should be covariant under the action of twist-deformed insintesimal diffeomorphism $U\mathsf{Vec}^{\mathcal{F}}(\mathcal{M})$. This means that for any $U\mathsf{Vec}(\mathcal{M})$-module algebra $\mathcal{A}$ (functions, forms, tensors) and for $a, b \in \mathcal{A}$, $u \in \mathsf{Vec}(\mathcal{M})$

$$u(ab) = u(a)b + au(b),$$

where the action on product is defined by the Leibniz rule, that is the coproducts of vectors u. The twist (1) deforms this into: $U\mathsf{Vec}(\mathcal{M}) \to U\mathsf{Vec}^{\mathcal{F}}(\mathcal{M})$ and $\mathcal{A} \to \mathcal{A}_\star$ with

$$ab \to a \star b = \bar{\mathrm{f}}^{\,\alpha}(a) \cdot \bar{\mathrm{f}}_{\,\alpha}(b).$$

Then $\mathcal{A}_\star$ is a $U\mathsf{Vec}^{\mathcal{F}}(\mathcal{M})$-module algebra

$$\xi(a \star b) = \xi_{(1)}(a) \star \xi_{(2)}(b),$$

for $\xi \in U\mathsf{Vec}^{\mathcal{F}}(\mathcal{M})$ and the action is via the twisted coproduct $\Delta^{\mathcal{F}}\xi = \xi_{(1)} \otimes \xi_{(2)}$.

The new composition is noncommutative

$$a \star b = \bar{\mathrm{f}}^{\,\alpha}(a) \cdot \bar{\mathrm{f}}_{\,\alpha}(b) = \bar{\mathrm{R}}^{\alpha}(b) \star \bar{\mathrm{R}}_{\alpha}(a) =: {}^{\alpha}b \star {}_{\alpha}a$$

and it is also nonassociative:

$$(a \star b) \star c = {}^{\phi_1}a \star ({}^{\phi_2}b \star {}^{\phi_3}c).$$

The noncommutativity is controled by the inverse of the R-matrix, while the nonassociativity is controled by the associator Φ.

In particular, the algebra of functions $C^\infty(\mathcal{M})$ is deformed to $C^\infty(\mathcal{M})_\star$

$$\begin{aligned} f \star g &= \bar{\mathrm{f}}^{\,\alpha}(f) \cdot \bar{\mathrm{f}}_{\,\alpha}(g) \\ &= f \cdot g + \tfrac{\mathrm{i}\hbar}{2}\left(\partial_\mu f \cdot \tilde{\partial}^\mu g - \tilde{\partial}^\mu f \cdot \partial_\mu g\right) + \mathrm{i}\,\kappa\, R^{\mu\nu\rho}\, p_\nu\, \partial_\rho f \cdot \partial_\mu g + \cdots, \end{aligned} \tag{5}$$

For the special case of phase space coordinates we get

$$\begin{aligned} [x^\mu \stackrel{\star}{,} x^\nu] &= 2\,\mathrm{i}\,\kappa\, R^{\mu\nu\rho}\, p_\rho, \\ [p_\mu \stackrel{\star}{,} p_\nu] &= 0, \quad [x^\mu \stackrel{\star}{,} x^\nu] = \mathrm{i}\,\hbar\,\delta^\mu{}_\nu, \\ [x^\mu \stackrel{\star}{,} x^\nu \stackrel{\star}{,} x^\rho] &= \ell_s^3\, R^{\mu\nu\rho}. \end{aligned} \tag{6}$$

The exterior algebra of differential forms $\Omega^\sharp(\mathcal{M})$ is deformed to $\Omega^\sharp(\mathcal{M})_\star$ with

$$\begin{aligned} \omega \wedge_\star \eta &= \bar{\mathrm{f}}^{\,\alpha}(\omega) \wedge \bar{\mathrm{f}}_{\,\alpha}(\eta), \\ f \star \mathrm{d}x^A &= \mathrm{d}x^C \star \left(\delta^A{}_C\, f - \mathrm{i}\,\kappa\, \mathscr{R}^{AB}{}_C\, \partial_B f\right), \end{aligned} \tag{7}$$

with non-vanishing components $\mathscr{R}^{x^\mu, x^\nu}{}_{\tilde{x}_\rho} = R^{\mu\nu\rho}$. Especcially, for the basis 1-forms

$$\begin{aligned} (\mathrm{d}x^A \wedge_\star \mathrm{d}x^B) \wedge_\star \mathrm{d}x^C &= {}^{\phi_1}(\mathrm{d}x^A) \wedge_\star \left({}^{\phi_2}(\mathrm{d}x^B) \wedge_\star {}^{\phi_3}(\mathrm{d}x^C)\right) \\ &= \mathrm{d}x^A \wedge_\star (\mathrm{d}x^B \wedge_\star \mathrm{d}x^C) = \mathrm{d}x^A \wedge \mathrm{d}x^B \wedge \mathrm{d}x^C. \end{aligned}$$

Exterior derivative d is undeformed. It fulfills $\mathrm{d}^2 = 0$ and the undeformed Leibniz rule

$$\mathrm{d}(\omega \wedge_\star \eta) = \mathrm{d}\omega \wedge_\star \eta + (-1)^{|\omega|}\, \omega \wedge_\star \mathrm{d}\eta. \tag{8}$$

Duality or the $\star$-pairing is defined as

$$\langle\, \omega\, ,\, u\, \rangle_\star = \big\langle\, \bar{\mathrm{f}}^{\,\alpha}(\omega)\, ,\, \bar{\mathrm{f}}_{\,\alpha}(u)\, \big\rangle\, . \tag{9}$$

Finally, the Lie-derivative can consistently be deformed to a $\star$-Lie drivative as

$$\begin{aligned}
&\mathcal{L}^\star_u(T) = \mathcal{L}_{\bar{\mathrm{f}}^{\,\alpha}(u)}(\bar{\mathrm{f}}_{\,\alpha}(T)), &(10)\\
&\mathcal{L}^\star_u(\omega \wedge_\star \eta) = \mathcal{L}^\star_{\bar\phi_1 u}(^{\bar\phi_2}\omega) \wedge_\star {}^{\bar\phi_3}\eta + {}^{\alpha}(^{\bar\phi_1\,\bar\varphi_1}\omega) \wedge_\star \mathcal{L}^\star_{\alpha(^{\bar\phi_2\,\bar\varphi_2}u)}(^{\bar\phi_3\,\bar\varphi_3}\eta),\\
&[\mathcal{L}^\star_u, \mathcal{L}^\star_v]_\bullet = [\bar{\mathrm{f}}^{\,\alpha}\mathcal{L}^\star_u, \bar{\mathrm{f}}_{\,\alpha}\mathcal{L}^\star_v] = \mathcal{L}^\star_{[u,v]_\star},
\end{aligned}$$

with $[u,v]_\star = \big[\bar{\mathrm{f}}^{\,\alpha}(u), \bar{\mathrm{f}}_{\,\alpha}(v)\big]$ and

$$\big[u,[v,z]_\star\big]_\star = \big[[^{\bar\phi_1}u, {}^{\bar\phi_2}v]_\star, {}^{\bar\phi_3}z\big]_\star + \big[{}^{\alpha}(^{\bar\phi_1\,\bar\varphi_1}v), [_{\alpha}(^{\bar\phi_2\,\bar\varphi_2}u), {}^{\bar\phi_3\,\bar\varphi_3}z]_\star\big]_\star .$$

It is well known that the usual Lie derivative $\mathcal{L}_u$ generates a one parameter family of diffeomorphisms. However, a relation of $\mathcal{L}^\star_u$ with diffeomorphism symmetry in space-time still needs to be understood fully [9, 10].

After formulating the basic notions of the NA differential geometry, we now define a $\star$-connection by

$$\begin{aligned}
\nabla^\star \,:\, \mathsf{Vec}_\star &\longrightarrow \mathsf{Vec}_\star \otimes_\star \Omega^1_\star\\
u &\longmapsto \nabla^\star u\, , &(11)\\
\nabla^\star(u \star f) &= (^{\bar\phi_1}\nabla^\star(^{\bar\phi_2}u)) \star {}^{\bar\phi_3}f + u \otimes_\star \mathrm{d}f. &(12)
\end{aligned}$$

A connection defined in this way satisfies the right Leibniz rule, for $u \in \mathsf{Vec}_\star$ and $f \in A_\star$. In particular

$$\begin{aligned}
&\nabla^\star\partial_A =: \partial_B \otimes_\star \Gamma^B_A =: \partial_B \otimes_\star (\Gamma^B_{AC} \star \mathrm{d}x^C)\, . &(13)\\
&\mathrm{d}_{\nabla^\star}(\partial_A \otimes_\star \omega^A) = \partial_A \otimes_\star (\mathrm{d}\omega^A + \Gamma^A_B \wedge_\star \omega^B),
\end{aligned}$$

for $\omega^A \in \Omega^\sharp_\star$.

Once we defined the $\star$-connection, the torsion and the curvature tensors can be defined straightforwardly. The torsion we define as

$$\begin{aligned}
&\mathsf{T}^\star := \mathrm{d}_{\nabla^\star}(\partial_A \otimes_\star \mathrm{d}x^A) \,:\, \mathsf{Vec}_\star \otimes_\star \mathsf{Vec}_\star \to \mathsf{Vec}_\star,\\
&\mathsf{T}^\star(\partial_A, \partial_B) = \partial_C \star (\Gamma^C_{AB} - \Gamma^C_{BA}) =: \partial_C \star \mathsf{T}^C{}_{AB}.
\end{aligned}$$

In the coordinate basis, the torsion-free condition reduces to $\Gamma^C_{AB} = \Gamma^C_{BA}$.

The curvature tensor we define as

$$\begin{aligned}
&\mathsf{R}^\star := \mathrm{d}_{\nabla^\star} \bullet \mathrm{d}_{\nabla^\star} \,:\, \mathsf{Vec}_\star \longrightarrow \mathsf{Vec}_\star \otimes_\star \Omega^2_\star,\\
&\mathsf{R}^\star(\partial_A) = \partial_C \otimes_\star (\mathrm{d}\Gamma^C_A + \Gamma^C_B \wedge_\star \Gamma^B_A) = \partial_C \otimes_\star \mathsf{R}^C_A,
\end{aligned}$$

The Ricci tensor is a contraction of the curvature tensor

$$\begin{aligned}\mathsf{Ric}^\star(u,v) &:= -\langle\, \mathsf{R}^\star(u,v,\partial_A)\,,\, \mathrm{d}x^A \,\rangle_\star \\ \mathsf{Ric}^\star &= \mathsf{Ric}_{AD}\star(\mathrm{d}x^D\otimes_\star \mathrm{d}x^A).\end{aligned} \tag{14}$$

Commponents of the Ricci tensor in the coordinate basis can be calculated from $\mathsf{Ric}_{BC} := \mathsf{Ric}^\star(\partial_B,\partial_C)$ and they are given by

$$\begin{aligned}\mathsf{Ric}_{BC} &= \partial_A\Gamma^A_{BC} - \partial_C\Gamma^A_{BA} + \Gamma^A_{B'A}\star\Gamma^{B'}_{BC} - \Gamma^A_{B'C}\star\Gamma^{B'}_{BA} \\ &+ \mathrm{i}\,\kappa\,\Gamma^A_{B'E}\star\big(\mathscr{R}^{EG}{}_A\,(\partial_G\Gamma^{B'}_{BC}) - \mathscr{R}^{EG}{}_C\,(\partial_G\Gamma^{B'}_{BA})\big) \\ &+ \mathrm{i}\,\kappa\,\mathscr{R}^{EG}{}_A\,\partial_G\partial_C\Gamma^A_{BE} - \mathrm{i}\,\kappa\,\mathscr{R}^{EG}{}_A\,\partial_G\big(\Gamma^A_{B'E}\star\Gamma^{B'}_{BC} - \Gamma^A_{B'C}\star\Gamma^{B'}_{BE}\big) \\ &+\kappa^2\,\mathscr{R}^{AF}{}_D\,\big(\mathscr{R}^{EG}{}_A\,\partial_F(\Gamma^D_{B'E}\star\partial_G\Gamma^{B'}_{BC}) - \mathscr{R}^{EG}{}_C\,\partial_F(\Gamma^D_{B'E}\star\partial_G\Gamma^{B'}_{BA})\big)\ .\end{aligned} \tag{15}$$

Unfortunately, a scalar curvature cannot be defined along these lines. That is, it cannot be seen as a map and the inverse metric tensor needed for the definiton. Due to nonassociativity it is not straightforward to define the inverse of the metric tensor. Namely

$$G^{MN}\star G_{NP} = \delta^P_M, \ \ \text{but}\ \ (G^{MN}\star G_{NP})\star f \neq G^{MN}\star(G_{NP}\star f).$$

This problem we hope to solve in our future work.

3. Nonassociative deformation of General Relativity

The connection of General Relativity $\Gamma^{\mathsf{LC}\,\rho}_{\mu\nu}$ is a Levi-Civita connection: it is torssion-free and metric compatible $\nabla_\alpha g_{\mu\nu} = 0$. In this section we generalize this conditions to the NA phase space. First we define a metric tensor $\mathsf{g}^\star \in \Omega^1_\star\otimes_\star\Omega^1_\star$ and then we demand

$$^\star\nabla \mathsf{g}^\star = 0. \tag{16}$$

In addition, the connection is torsion free $\Gamma^C_{AB} = \Gamma^C_{BA}$. Expanding the above condition up to first order in $\hbar\kappa$ we find

$$\begin{aligned}\Gamma^{S(0,0)}_{AD} &= \Gamma^{\mathsf{LC}\,S}_{AD} = \tfrac{1}{2}\,\mathsf{g}^{SQ}\,(\partial_D\mathsf{g}_{AQ} + \partial_A\mathsf{g}_{DQ} - \partial_Q\mathsf{g}_{AD})\ , \\ \Gamma^{S(0,1)}_{AD} &= -\frac{\mathrm{i}\,\hbar}{2}\,\mathsf{g}^{SP}\,\big((\partial_\mu\mathsf{g}_{PQ})\,\tilde\partial^\mu\Gamma^{\mathsf{LC}\,Q}_{AD} - (\tilde\partial^\mu\mathsf{g}_{PQ})\,\partial_\mu\Gamma^{\mathsf{LC}\,Q}_{AD}\big)\ , \\ \Gamma^{S(1,0)}_{AD} &= \mathrm{i}\,\kappa\,R^{\alpha\beta\gamma}\Big(\tilde{\mathsf{g}}^S{}_\gamma\,\mathsf{g}_{\beta N}\,(\partial_\alpha\Gamma^{\mathsf{LC}\,N}_{AD}) - \mathsf{g}^{SM}\,p_\beta\,(\partial_\gamma\mathsf{g}_{MN})\,\partial_\alpha\Gamma^{\mathsf{LC}\,N}_{AD}\Big)\ , \\ \Gamma^{S(1,1)}_{AD} &= \frac{\hbar\,\kappa}{2}\,R^{\alpha\beta\gamma}\Big[\ \text{long expression}\ + (\partial_\alpha\mathsf{g}^{SQ})\,(\partial_\beta\mathsf{g}_{QP})\,\partial_\gamma\Gamma^{\mathsf{LC}\,P}_{AD}\Big].\end{aligned} \tag{17}$$

Here we labeled $\tilde{\mathsf{g}}^S{}_\gamma = \mathsf{g}^{SM}\,\delta_{M,\tilde{x}_\gamma}$. Notice that $\Gamma^{S(0,1)}_{AD}$ and $\Gamma^{S(1,0)}_{AD}$ are purely imaginary, while $\Gamma^{S(1,1)}_{AD}$ is real. For g_{MN} that does not depend on the momenta p_μ, only the last term in $\Gamma^{S(1,1)}_{AD}$ remains.

The phase space vacuum Einstein equations are given by

$$\mathsf{Ric}_{BC} = 0 \,. \tag{18}$$

Using (14) and (17) we obtain vacuum Einstein equations in phase space expanded up to first order in $\hbar\kappa$. However, we are interested in the deformation of General Relativity in space-time. To extract the space-time vacuum Einstein equations from (18) we develop a method of projection which we describe in the following.

We start from objects in the space-time M, for example the space-time metric tensor $g = g_{\mu\nu} \mathrm{d}x^\mu \otimes \mathrm{d}x^\nu$. Then we lift these objects to the phase space $\mathcal{M}$ foliated with leaves of constant momenta. Note that each leave is diffeomorphic to M. We do all calculation using the NA differential geometry tools developed in the previous section. Final results (in phase space) we project to the space-time, using again a leaf of constant momenta. In particular, we will use the zero-momentum leaf to perform the projection. Our procedure is ilustrated by the diagram

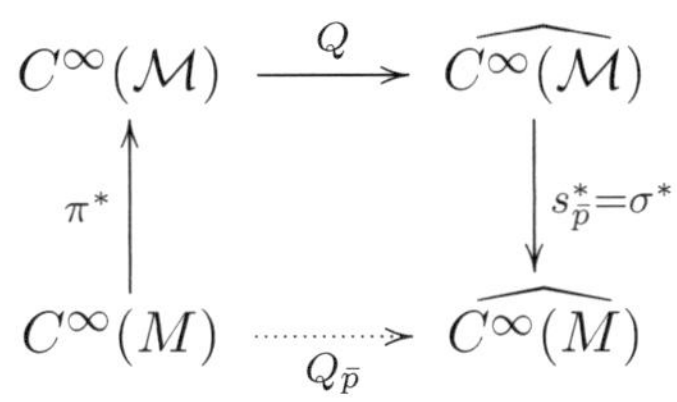

Applying this procedure to the metric tensor $g = g_{\mu\nu} \mathrm{d}x^\mu \otimes \mathrm{d}x^\nu$ gives the phase space metric tensor $\hat{\mathsf{g}}_{MN}\, \mathrm{d}x^M \otimes \mathrm{d}x^N$ with

$$\big(\hat{\mathsf{g}}_{MN}(x)\big) = \begin{pmatrix} \mathsf{g}_{\mu\nu}(x) & 0 \\ 0 & \mathsf{h}^{\mu\nu}(x) \end{pmatrix} \,. \tag{19}$$

To have a non-singular metric we had to introduce an additional nondegenerate bilinear $\mathsf{h}(x)^{\mu\nu}\, \mathrm{d}\tilde{x}_\mu \otimes \mathrm{d}\tilde{x}_\nu$. Its choice is arbitrary, we pick up the simplest and the most natural choice $\mathsf{h}(x)^{\mu\nu} = \eta^{\mu\nu}$.

Now we can do all calculations in phase space, using the NA differential geometry. In particular, we calculate Ric_{BC} in terms of g_{AB}, (14), (17). In the end we project the result to space-time using the zero section $x \mapsto \sigma(x) = (x, 0)$. In particular, the projection of Ricci tensoris given by

$$\begin{aligned} \mathsf{Ric} \;\; &\to \;\; \mathsf{Ric}^{\star\circ} = \mathsf{Ric}^\circ_{\mu\nu}\, \mathrm{d}x^\mu \otimes \mathrm{d}x^\nu, \\ \mathsf{Ric}^\circ_{\mu\nu}(x) \;\; &= \;\; \sigma^*(\mathsf{Ric}_{\mu\nu})(x,p) = \mathsf{Ric}_{\mu\nu}(x,0). \end{aligned}$$

Components of the lifted metric $\hat{\mathsf{g}}_{MN}\, \mathrm{d}x^M \otimes \mathrm{d}x^N = \mathsf{g}_{MN} \star (\mathrm{d}x^M \otimes_\star \mathrm{d}x^N)$, expanded up to first order in $\hbar\kappa$ are given by

$$\mathsf{g}_{MN}(x) = \begin{pmatrix} \mathsf{g}_{\mu\nu}(x) & \frac{\mathrm{i}\,\kappa}{2}\, R^{\sigma\nu\alpha}\, \partial_\sigma \mathsf{g}_{\mu\alpha} \\ \frac{\mathrm{i}\,\kappa}{2}\, R^{\sigma\mu\alpha}\, \partial_\sigma \mathsf{g}_{\alpha\nu} & \eta^{\mu\nu}(x) \end{pmatrix} \,. \tag{20}$$

Finally, components of the Ricci tensor in space-time, expanded up to first order in $\hbar\kappa$ follow as:

$$\begin{aligned}\mathsf{Ric}^{\circ}_{\mu\nu} = \mathsf{Ric}^{\mathsf{LC}}_{\mu\nu} + \tfrac{\ell_s^3}{12}\, R^{\alpha\beta\gamma} \Big(&\partial_\rho\big(\partial_\alpha \mathsf{g}^{\rho\sigma}\,(\partial_\beta \mathsf{g}_{\sigma\tau})\,\partial_\gamma \Gamma^{\mathsf{LC}\,\tau}_{\mu\nu}\big) \\ &-\partial_\nu\big(\partial_\alpha \mathsf{g}^{\rho\sigma}\,(\partial_\beta \mathsf{g}_{\sigma\tau})\,\partial_\gamma \Gamma^{\mathsf{LC}\,\tau}_{\mu\rho}\big) \\ &+\partial_\gamma \mathsf{g}_{\tau\omega}\big(\partial_\alpha(\mathsf{g}^{\sigma\tau}\,\Gamma^{\mathsf{LC}\,\rho}_{\sigma\nu})\,\partial_\beta\Gamma^{\mathsf{LC}\,\omega}_{\mu\rho} - \partial_\alpha(\mathsf{g}^{\sigma\tau}\,\Gamma^{\mathsf{LC}\,\rho}_{\sigma\rho})\,\partial_\beta\Gamma^{\mathsf{LC}\,\omega}_{\mu\nu} \\ &+(\Gamma^{\mathsf{LC}\,\sigma}_{\mu\rho}\,\partial_\alpha \mathsf{g}^{\rho\tau} - \partial_\alpha\Gamma^{\mathsf{LC}\,\sigma}_{\mu\rho}\,\mathsf{g}^{\rho\tau})\,\partial_\beta\Gamma^{\mathsf{LC}\,\omega}_{\sigma\nu} \\ &-(\Gamma^{\mathsf{LC}\,\sigma}_{\mu\nu}\,\partial_\alpha \mathsf{g}^{\rho\tau} - \partial_\alpha\Gamma^{\mathsf{LC}\,\sigma}_{\mu\nu}\,\mathsf{g}^{\rho\tau})\,\partial_\beta\Gamma^{\mathsf{LC}\,\omega}_{\sigma\rho}\big)\Big). \end{aligned} \tag{21}$$

The vacuum Einstein equations in space-time are given by

$$\mathsf{Ric}^{\circ}_{\mu\nu} = 0. \tag{22}$$

4. Conclusions

In this short contribution, we described how the R-flux (via NA differential geometry) generates non-trivial dynamical consequences on spacetime. The first order corrections are are independent of $\hbar$ (they are first order in $\kappa\hbar = \frac{\ell_s^3}{6}$) and real-valued. To obtain the results in space-time, we used the projection to the zero momentum leaf. Note that pulling back to a leaf of constant momentum $p = p^\circ$ (generally) gives a non-vanishing imaginary contribution $\mathsf{Ric}^{(1,0)}_{\mu\nu}\big|_{p=p^\circ}$ to the spacetime Ricci tensor. Also, n-triproducts calculated on the zero momentum leaf [2] coincide with those proposed in [8].

We took the simplest choice of $\mathsf{h}(x)^{\mu\nu} = \eta^{\mu\nu}$. This can be changed to a more general metric. In relation with Born geometry discussed in [11] we can say that in our model nonassociativity does not generates curved momentum space. It might generate a more genral deformations with $\mathsf{h}(x)^{\mu\nu} \neq \eta^{\mu\nu}$. This has to be investigated further.

There are lots of projects to be discussed in our future work. Some of them are: phenomenological consequences of the R-flux induced corrections to GR solutions, construction of scalar curvature, adding matter fields, full Einstein equations. The twisted diffeomorphism symmetry has to be understood better, in particular its relation with the L_∞ structure.

References

[1] R. Blumenhagen and M. Fuchs, *Towards a theory of nonassociative gravity*, JHEP 1607 (2016) 019.

[2] P. Aschieri and R. Szabo, *Triproducts, nonassociative star products and geometry of R-flux string compactifications*, J.Phys.Conf.Ser. 634 (2015) no.1, 012004.

[3] P. Aschieri, M. Dimitrijević Ćirić and R. Szabo, *Nonassociative differential geometry and gravity with non-geometric fluxes*, JHEP 1802 (2018) 036.

[4] H. S. Snyder, *Quantized space-time*, Phys. Rev. **71** (38), 1947.

[5] P. Aschieri, M. Dimitrijević, P. Kulish, F. Lizzi, and J. Wess, *Noncommutative Spacetimes*, Springer Verlag, 2009.

R. J. Szabo, *Quantum Gravity, Field Theory and Signatures of Noncommutative Spacetime*, Gen. Rel. Grav. I (42), 1-29 (2010).

R. J. Szabo, *An Introduction to Nonassociative Physics*, PoS CORFU2018 (2019) 100.

[6] P. Jordan, *Über Eine Klasse Nichtassociativer Hyperkomplexer Algebren*, Nachr. Ges. Wiss. Göttingen, 569-575 (1932).

[7] Y. Nambu, *Generalized Hamiltonian dynamics*, Phys. Rev. D **7**, 2405-2412 (1973).

[8] D. Lüst, *T-duality and closed string noncommutative (doubled) geometry*, JHEP **1012** (2010) 084,

R. Blumenhagen and E. Plauschinn, *Nonassociative gravity in string theory?*, J. Phys. A **44** (2011) 015401.

[9] M. Bojowald, S. Brahma, U. Buyukcam, M. Ronco, *Extending general covariance: Moyal-type noncommutative manifolds*, Phys. Rev. D **98** 026031 (2018).

[10] M. Dimitrijević Ćirić, G. Giotopoulos, V. Radovanović and R. J. Szabo, in preparation.

[11] L. Freidel, R. G. Leigh, Dj. Minić, *Born Reciprocity in String Theory and the Nature of Spacetime*, Phys. Lett. B **730** (2014) 302-306.

Causality in nonlocal gravity*

Stefano Giaccari†
Department of Sciences, Holon Institute of Technology (HIT),
52 Golomb St., Holon 5810201, Israel
Leonardo Modesto‡
Department of Physics, Southern University of Science and
Technology, Shenzhen 518055, China

Abstract

The definition of quantum gravity is hindered by the difficulty of reconciling the requirements of renormalizability and unitarity. Recently it has also been pointed out that some higher-derivative which may appear in modifications of Einstein-Hilbert gravity terms are associated with a violation of causality at scales larger than the Planck length [1]. This can be avoided by adding an infinite tower of massive higher spin particles, which are in fact expected in a weakly coupled string theory. On the other hand this argument poses a crucial test for the causality of quantum gravity theories containing only conventional particles with spin not greater than 2.

We review several results about a class of weakly nonlocal purely gravitational (or coupled to matter) theories that are compatible with perturbative unitarity and finiteness at quantum level. In particular,we argue the requirement of causality can be satisfied avoiding some higher derivative terms.This result is justified computing Shapiro's time delay in terms of tree-level scattering amplitudes for nonlocal gravity models with and without matter. We also show how generic nonlocal gravity theories consistent with causality can be obtained by a field redefinition from standard local theories.

1. Introduction

The major challenge of quantum gravity lies in the difficulty of reconciling renormalizability and perturbative unitarity. In fact, the Einstein-Hilbert action is not power counting renormalizable, but, if we introduce the infinite number of counterterms generated by the renormalization procedure, it is

* The research of S.G. was supported by the Israel Science Foundation (ISF), grant No. 244/17. S.G. is also thankful to the University of Zagreb and the Croatian Science Foundation for support under the project No. 8946 during the first part of this research.

† e-mail address: stefanog@hit.ac.il

‡ e-mail address: lmodesto@sustech.edu.cn

perturbatively unitary. On the other hand, it is possible to build higher-derivatives theories of quantum gravity that are renormalizable with finitely many counterterms, but are non-unitary. This is the case for example of the celebrated Stelle theory [2]. In the literature a number of possible solutions to this puzzle have been discussed. Among them in recent years nonlocal models received a great deal of attention both for their capability to produce an interesting phenomenology and for their remarkable properties upon quantization. In particular, a class of weakly nonlocal gravitational theories has been proven to be super-renormalizable (or finite) and perturbatively unitary[3, 4, 5, 6, 7, 8]. These theories have also been studied in connection to cosmological backgrounds and black hole solutions [9, 10, 11, 12, 13].

However the main concern about nonlocal theories is surely causality, whose investigation has only recently been addressed in a systematic way. In a classical theory the problem is closely related to a mathematically sound formulation of the initial value problem whereas from the quantum perspective nonlocality implies a new formulation of the Bogoliubov causality condition for local interactions. Remarkably, for quasi-local interactions where the nonlocality shows up only at scales smaller than the nonlocality scale ℓ_Λ, non causal effects remain confined within the scale ℓ_Λ [14, 15]. This has also led to the idea that for asymptotically free nonlocal theories, this violation of microcausality may actually be undetectable [16].

In this note, we consider another notion of causality introduced by Gao and Wald [17], according to which it is impossible to send signals faster than what is allowed by the asymptotic causal structure of the spacetime. Its violation has been recently discussed by Camanho, Edelstein, Maldacena, and Zhiboedov [1], in particular in connection to Shapiro's time delay, which is one of the classical tests of general relativity (GR). Light propagating near a compact object should suffer a time delay compared with the same propagation in flat spacetime. Therefore, if we get a negative time delay, or actually a time advancement, we have a causality violation. Exploiting the relation between the Shapiro time delay and the scattering amplitudes for gravitating particles in the eikonal approximation, the authors of [1] have proven that these causality problems are produced only by the form of the on-shell three-point functions of the theory. Therefore the most general higher derivative gravity theory giving rise to a causality violation is at most cubic in the Riemann tensor. In particular their analysis applies to a high energy scattering process in which gravity is still weakly coupled. This can be achieved if the impact parameter b can be chosen such that

$$\ell_{\rm P} \ll b \ll \ell_\Lambda \,, \tag{1}$$

where $\ell_{\rm P}$ the Planck scale, and ℓ_Λ the non locality scale. In such cases the loss of causality can be evaded by adding massive higher spin particles with spin $J > 2$ and mass $m^2 \sim \ell_\Lambda^{-2}$.

In the following we want to argue that the terms responsible for this causality violation do not need to show up in a nonlocal theory of quantum gravity of the kind studied in [3, 4, 5, 6, 7, 8]. So these theories turn out to be consistent even without the introduction of an infinite tower of massive higher spin fields. First, in section 2. we review some essential features of the class of theories under consideration, in particular their renormalization properties and perturbative unitarity. Then, in section 3. we review results about scattering amplitudes in weakly nonlocal theories pointing out the crucial role played by a theorem relating tree-level amplitudes in theories related by field redefinitions [22, 23]. In section 4. we finally report about the results of [24], where the problem of causality has been addressed.

2. Weakly nonlocal gravity

We investigate the class of theories defined by the action

$$S_{\rm g} = \frac{2}{\kappa_D^2} \int d^D x \sqrt{-g}\, [R + G_{\mu\nu}\gamma(\Box) R^{\mu\nu} + V(\mathcal{R})] \,, \tag{2}$$

where $\kappa_D^2 = 32\pi G$. Given the non-locality scale $\sigma \equiv \ell_\Lambda^2$, the form factor $\gamma(\Box)$ is defined by

$$\gamma(\Box) = \frac{e^{H(\sigma\Box)} - 1}{\Box} \,, \tag{3}$$

where the function $\exp H(z)$ is asymptotically polynomial in a conical region C around the real axis, namely

$$|\exp H(z)| \to |z|^{\gamma+\mathrm{N}+1} \quad \text{for} \quad |z| \to +\infty, \tag{4}$$

with N an integer defined in terms of the spacetime dimension D so that $2N+4 = D$ (if D is even) or $2N+4 = D+1$ (if D is odd). This condition is necessary to avoid the appearance of nonlocal counterterms in the UV regime. An example due to Tomboulis [4] is

$$H_T(z) = \frac{1}{2}\left[\Gamma\left(0, p(z)^2\right) + \gamma_E + \log\left(p(z)^2\right)\right] , \tag{5}$$

where $p(z)$ is a polynomial of degree γ+N+1, $\Gamma(a,z)$ the incomplete Gamma function and γ_E the Euler-Mascheroni constant. The local potential $V(\mathcal{R})$ is at least cubic in the curvature, namely $V \sim O(\mathcal{R}^3)$, but quadratic in the Ricci tensor, and is taken to contain at most $2\gamma + 2N + 4$ derivatives.

These choices are motivated by inspection of quantum divergence in the UV regime. In fact the graviton propagator scales as $k^{-(2\gamma+2N+4)}$ and the vertices contain terms whose leading behavior is just the inverse. This determines the upper bound on the superficial degree of divergence of any graph G, $\omega(G) \equiv DL + (V - I)(2\gamma + 2N + 4)$. We find in a spacetime of even or odd dimension respectively:

$$\omega(G)_{even} = D_{even} - 2\gamma(L-1), \quad \omega(G)_{odd} = D_{odd} - (2\gamma+1)(L-1) \,. \tag{6}$$

Thus, if $\gamma > D_{even}/2$ or $\gamma > (D_{odd} - 1)/2$, only 1-loop divergences survive. Therefore, the theory is super-renormalizable and only a finite number of operators of mass dimension up to M^D has to be included in the action for renormalization in even dimensions. In odd dimensions, due to dimensional reasons, there are no divergences at one loop and the theory is automatically finite. The freedom in the choice of the potential $V(\mathcal{R})$ can be used to "kill" the one-loop divergences in even dimensions. For example, in 4 dimensions it is possible to prove that the two quartic killer operators

$$s_1 \, R^2 \, \Box^{\gamma-2} R^2 + s_2 \, R_{\mu\nu} R^{\mu\nu} \, \Box^{\gamma-2} R_{\rho\sigma} R^{\rho\sigma} \tag{7}$$

give contributions to the beta functions of the couplings for R^2 and $R^2_{\mu\nu}$ which are linear in their front coefficients s_1 and s_2 so that finiteness can be achieved by choosing them so that $\beta_{R^2} = \beta_{R^2_{\mu\nu}} = 0$. The crucial point for the following is that the killer terms should be in general at least quadratic in the Ricci tensor.

One can easily find from the kinetic term the two-point function in the harmonic gauge ($\partial^\mu h_{\mu\nu} = 0$)

$$\mathcal{O}^{-1} \sim \frac{1}{k^2 e^{H(k^2/\Lambda^2)}} \left(P^{(2)} - \frac{P^{(0)}}{D-2} \right) , \tag{8}$$

where $P^{(0)}$ and $P^{(2)}$ are the usual spin 2 and spin 0 projectors. Therefore perturbative unitarity together with the absence of gauge invariant poles other than the graviton pole requires that $\exp H(z)$ be real and positive on the real axis and without zeros on the whole complex plane $|z| < +\infty$. This choice however implies a subtlety related to the fact that amplitudes are well-defined as integrals along certain loop integration contours and in a certain regime of external momenta, which is typically the Euclidean one. The vertices we have defined in order to achieve UV finiteness must decrease sufficiently fast along some directions in the complex plane, namely the ones corresponding to Euclidean momenta. However for the non-polynomial entire functions this necessarily implies a fast growth in other directions in the complex plane, thus generally preventing the usual Wick rotation. This could generically point at a violation of perturbative unitarity. However, the theories considered in this note turn out to be unitary at perturbative level to all perturbative orders in the loop expansion as rigorously and extensively proved in [18] and more recently in [19, 20, 21] The proof is based on an analytic continuation of the external particles' energies from imaginary to real values. It turns out that the Landau singularities, and the discontinuities of the amplitudes are the same of a local theory at any perturbative order in the loop expansion. This is a consequence of the classical spectrum of the theory that is the same of the local theory. Therefore, the Cutkosky cutting rules are the same of the local theory. Finally, there is no contribution of cut diagrams corresponding to anomalous thresholds to the imaginary part of the scattering amplitudes as proved in [19, 21]. Indeed, if a diagram is cut in less then two or more then two parts the contribution to the discontinuities vanishes as a consequence of the energy momentum conservation.

3. Scattering amplitudes in higher derivative gravity theories

We here review some results about scattering amplitudes in higher derivative gravity theories, in particular four-graviton tree-level ones in the case when all higher derivative terms are at least quadratic in the Ricci tensor. For the sake of clarity, we first consider an action

$$\begin{aligned} S_g &= -2\kappa_D^{-2}\int d^Dx\,\sqrt{-g}\Big(R + R\gamma_0 R + R_{\mu\nu}\gamma_2 R^{\mu\nu} + \\ &\quad (R_{\mu\nu\rho\sigma}\gamma_4 R^{\mu\nu\rho\sigma} - 4R_{\mu\nu}\gamma_4 R^{\mu\nu} + R\gamma_4 R)\Big)\,, \end{aligned} \tag{9}$$

where γ_0, γ_2, γ_4 are generic functions of $\sigma\Box$. As observed in [22], in such cases the computation can be addressed by standard Feynman diagram techniques due to a number of simplifications. First of all, we note the last term is the famous Gauss-Bonnet density (GB), which is topological in four dimensions whereas for generic higher dimensions it gives rise to vertices only. Furthermore, as the process involves only three-graviton vertices with two gravitons on-shell and one off-shell and a four-graviton vertex with all external legs on-shell, we can make full use of the linearized vacuum equation of motion for the physical field $h_{\mu\nu}$ in the harmonic gauge, i.e. $\Box h_{\mu\nu} = 0$. Actually, we can choose polarizations satisfying the conditions $\partial^\mu h_{\mu\nu} = h^\mu_\mu = 0$ all along the computation, which greatly simplifies the algebra. In fact, these conditions imply that all the scalar operators are vanishing on-shell at linear order in $h_{\mu\nu}$, including the scalar curvature $R^{(1)}$ and the root of metric determinant $\sqrt{-g}^{(1)}$. One can further show that $R^{(1)}_{\mu\nu} = 0$ due to the linearized EOM. We can express all the amplitudes in terms of the Mandelstam variables $s = 4E^2$, $t = -2E^2(1-\cos\theta)$ and $u = -2E^2(1+\cos\theta)$, with E the energy and θ the scattering angle in the center-of-mass reference frame.

In the case where γ_0, γ_2, γ_4 are constants, for gravitons with positive helicity, one finds for $D = 4, 5, 6$

$$\mathcal{A}^{D=4}(++,++) = i\frac{1}{\kappa_4^2}\frac{4E^2}{\sin^2\theta}\,, \tag{10}$$

$$\begin{aligned} &\mathcal{A}^{D=5}(++,++) = \\ &-i\frac{2}{\kappa_5^2}\left\{\frac{16E^6\gamma_4^2\left[1+8E^2(3\gamma_0+\gamma_2)\right]}{(1-4E^2\gamma_2)\left[3+4E^2(16\gamma_0+5\gamma_2)\right]} - 2E^2\frac{1}{\sin^2\theta}\right\}, \end{aligned} \tag{11}$$

$$\mathcal{A}^{D=6}(++,++) = \\ -i\frac{2}{\kappa_6^2}\left\{\frac{8E^6\gamma_4^2\left[1+8E^2(3\gamma_0+\gamma_2)\right]}{(1-4E^2\gamma_2)\left[1+2E^2(10\gamma_0+3\gamma_2)\right]} - 2E^2\frac{1}{\sin^2\theta}\right\}. \quad (12)$$

Remarkably, in 4 dimensions, where γ_4 cannot enter the amplitude because of the Gauss-Bonnet theorem, the result coincides with the one expected in Einstein theory by dimensional analysis and symmetry arguments . In particular no term scaling as E^4 in the UV shows up as it would be natural to expect in a four derivative theory. This is the result of non-trivial cancellations between the massive poles in the propagator and the three-graviton vertices and between the contact and exchange diagrams. This result can be also understood as the one consistent with the natural expectation in the limit where the Einstein term can be dropped out only leaving the scaleless quadratic terms. They would be expected to naively give amplitudes $\sim E^4$, but this cannot happen because the graviton field is dimensionless and there is no other scale. So the amplitude is actually expected to vanish. In $D > 4$ γ_4 enters the amplitude, but only quadratically whereas the expected linear contribution is absent due to a cancellation between the contact diagram with a vertex from the Gauss-Bonnet term and the exchange diagrams with two different vertices (one from GB, the second one from standard terms R, R^2 or $R^2_{\mu\nu}$). Whereas in the ultraviolet regime the amplitude scales as E^4, in the infrared one finds arbitrary powers of E^2 associated with the massive poles in the propagators which cannot cancel with the three-graviton vertices of the Gauss-Bonnet density.

For weakly nonlocal gravity the amplitude can be also performed straightforwardly if $\gamma_4(\Box) = 0$. Infact, as on-shell $\mathbf{R} \sim O(h^2)$ and $\mathbf{Ric} \sim O(h^2)$ (whereas $\mathbf{Riem} \sim O(h)$), the form factors are spectators in the expansion in the number of gravitons and many results for the Stelle gravity apply to the general nonlocal theory. For the three exchange diagrams and the contact one, we find

$$\mathcal{A}_s(++,++) = \\ -2\kappa_4^{-2}\left(-\frac{9}{8}\frac{t(s+t)}{s}+\frac{9}{32}\gamma_2(s)\left(s^2+(s+2t)^2\right)+\frac{9}{8}s^2\gamma_0(s)\right), \quad (13)$$

$$\mathcal{A}_t(++,++) = -2\kappa_4^{-2}\left(-\frac{1}{8}\frac{\left(s^3-5s^2t-st^2+t^3\right)(s+t)^2}{s^3t}\right. \\ \left.+\frac{1}{16}\gamma_2(t)\frac{\left(2s^4-10s^3t-s^2t^2+4st^3+t^4\right)(s+t)^2}{s^4}+\frac{1}{8}\gamma_0(t)\frac{t^2(s+t)^4}{s^4}\right), \quad (14)$$

$$\mathcal{A}_u(++,++) = -2\kappa_4^{-2}\left(-\frac{1}{8}\frac{\left(s^3-5s^2u-su^2+u^3\right)(s+u)^2}{s^3u}+\right.$$

$$\left.\frac{1}{16}\gamma_2(u)\frac{\left(2s^4-10s^3u-s^2u^2+4su^3+u^4\right)(s+u)^2}{s^4}+\frac{1}{8}\gamma_0(u)\frac{u^2(s+u)^4}{s^4}\right), \tag{15}$$

$$\mathcal{A}_{\rm contact}(++,++) =$$

$$-2\kappa_4^{-2}\left(-\frac{1}{4}\frac{s^4+s^3t-2st^3-t^4}{s^3}-\frac{9}{32}\gamma_2(s)\left(s^2+(s+2t)^2\right)-\frac{9}{8}s^2\gamma_0(s)\right.$$

$$-\frac{1}{16}\gamma_2(t)\frac{\left(2s^4-10s^3t-s^2t^2+4st^3+t^4\right)(s+t)^2}{s^4}-\frac{1}{8}\gamma_0(t)\frac{t^2(s+t)^4}{s^4}-$$

$$\left.\frac{1}{16}\gamma_2(u)\frac{\left(2s^4-10s^3u-s^2u^2+4su^3+u^4\right)(s+u)^2}{s^4}-\frac{1}{8}\gamma_0(u)\frac{u^2(s+u)^4}{s^4}\right), \tag{16}$$

where the possible poles associated with γ_0 and γ_2 cancel separetely in each channel. Once again, the amplitude

$$\mathcal{A}(++,++) = \mathcal{A}_s(++,++)+\mathcal{A}_t(++,++)+$$

$$\mathcal{A}_u(++,++)+\mathcal{A}_{\rm contact}(++,++) = \mathcal{A}(++,++)_{\rm EH}\,, \tag{17}$$

coincides with the one in Einstein-Hilbert theory.

These results, which may look somewhat surprising, find actually a very natural explanation in terms of a field redefinition theorem [22, 23] that allows to map nonlocal field theories to local ones at tree-level.

In particular, let us consider two general weakly nonlocal actions, namely $S'(g,\Phi_a)$ and $S(g',\Phi'_a)$, respectively defined in terms of the fields g,Φ_a and g',Φ'_a, where g is the metric and Φ_a a set of matter or gauge fields, and such that

$$S'(g,\Phi_a) = S(g,\Phi_a)+E_i^g(g,\Phi_a)F_{ij}^g(g,\Phi_a)E_j^g(g,\Phi_a)$$

$$+E_a^\Phi(g,\Phi_c)F_{ab}^\Phi(g,\Phi_c)E_b^\Phi(g,\Phi_c)\,, \tag{18}$$

where F^g and F^Φ can contain derivative operators or weakly nonlocal operators of the covariant $\Box$ operator, and

$$E_i^g = \frac{\delta S}{\delta g_i}\,,\quad E_a^\Phi = \frac{\delta S}{\delta \Phi_a} \tag{19}$$

are the EOM of the theory with action $S(g, \Phi_a)$. The statement of the the theorem is that there exists a field redefinition

$$g'_i = g_i + \Delta^g_{ij} E^g_j \,, \qquad \Delta^g_{ij} = \Delta^g_{ji},$$
$$\Phi'_a = \Phi_a + \Delta^\Phi_{ab} E^\Phi_b \,, \qquad \Delta^\Phi_{ab} = \Delta^\Phi_{ba} \,, \tag{20}$$

such that, perturbatively in $F^{g,\Phi}$, but to all orders in powers of $F^{g,\Phi}$, we have the equivalence

$$S'(g, \Phi) = S(g', \Phi') \,. \tag{21}$$

The indices i, a on fields we encode all Lorentz, group indices, and the spacetime dependence of the fields. Δ^g_{ij} (Δ^Φ_{ab}) could be a weakly nonlocal or quasi-polynomial operator acting linearly on the EOM E^g_j (E^Φ_a), and they are defined perturbatively in powers of the operators $F^{g,\Phi}$, namely

$$\Delta^g_{ij} = F^g_{ij} + \ldots \quad \text{or} \quad \Delta^\Phi_{ab} = F^\Phi_{ab} + \ldots \,. \tag{22}$$

The claim above can be straightforwardly checked at the first order in the Taylor expansion for the functional $S(g', \Phi'_a)$

$$S(g', \Phi'_a) = S(g + \delta g, \Phi + \delta\Phi) \approx S(g) + \frac{\delta S}{\delta g_i}\delta g_i + \frac{\delta S}{\delta \Phi_a}\delta\Phi_a$$
$$= S(g) + E^g_i \,\delta g_i + E^\Phi_a \,\delta\Phi_a \,, \tag{23}$$

which is consistent with the equivalence (21) if we assume the field redefinitions (20) with the chosen coefficients (22). The theorem states the equivalence of the two theories only perturbatively in $F^{g,\Phi}$, so that the two theories do not need to be equivalent in all aspects. For example $S'(g, \Phi)$ can have additional poles in the spectrum compared with $S(g', \Phi')$ and also the quantum behaviors of the the theories can be completely different. However, the theorem applies to all the n-points tree-level functions whose external legs are on the mass-shell shared by the two theories, and this explains the results found by direct computation for theories that are quadratic in both the Ricci and scalar curvature and lack a term quadratic in the Riemann tensor.

4. Shapiro's time delay

The results of the previous section can be nicely translated in the language of Shapiro's time delay. This can in fact be recovered from the scattering amplitudes in the so-called eikonal approximation, which resums a particular set of diagrams (horizontal ladders) in the deflectionless limit $t/s << 1$. s is large compared to the inverse of the nonlocality scale ℓ_Λ^{-2}, but still well below the Planck scale so that the theories we are considering are still weakly coupled. Under favorable circumstances the amplitude exponentiates in the impact parameter space [25, 26]

$$iA_{\text{eik}} = 2s \int d^{D-2}\vec{b}\, e^{-i\vec{q}\cdot\vec{b}} \left[e^{i\delta(b,s)} - 1\right], \tag{24}$$

where the phase is given by

$$\delta(b,s) = \frac{1}{2s} \int \frac{d^{D-2}\vec{q}}{(2\pi)^{D-2}} e^{i\vec{q}\cdot\vec{b}} A_{\text{tree}}(s, -\vec{q}^2)\,. \tag{25}$$

Shapiro's time delay is then given by

$$\Delta t = 2\partial_E \delta(E, b)\,. \tag{26}$$

where E is the energy of the probe-particle.

In particular, for the action (9), with $\gamma_0 = \gamma_2 = 0$ and $\gamma_4 = \lambda_{GB}$ a constant, one finds for the four graviton amplitude in the Regge limit [27]

$$A_t = A_{t\text{EH}} + A_{t\text{GB}} \approx -\frac{8\pi G s^2}{t}(\epsilon_1 \cdot \epsilon_3)(\epsilon_2 \cdot \epsilon_4)$$

$$+\frac{\kappa_D^2 \lambda_{\text{GB}} s^2}{t} \left(k_2^\mu k_4^\nu \epsilon_{2\nu}^\rho \epsilon_{4\rho\mu} \epsilon_1 \cdot \epsilon_3 + k_1^\mu k_3^\nu \epsilon_{1\nu}^\rho \epsilon_{3\rho\mu} \epsilon_2 \cdot \epsilon_4\right),$$

where $\kappa_D^2 = 32\pi G$, the momenta of the four gravitons k_1, k_2, k_3, k_4 are all incoming, i.e. $\sum_{i=1}^4 k_i = 0$, and ϵ_i $(i = 1, \ldots, 4)$ are the polarizations of the gravitons. Choosing the metric $ds^2 = -dudv + \sum_{i=1}^{D-2} (dx^i)^2$, we can evaluate this amplitude in the following momentum configuration

$$\begin{aligned} k_{1\mu} &= \left(k_u, \frac{\vec{q}^{\,2}}{16k_u}, \frac{\vec{q}}{2}\right), \quad k_{3\mu} = -\left(k_u, \frac{\vec{q}^{\,2}}{16k_u}, -\frac{\vec{q}}{2}\right) \\ k_{2\mu} &= \left(\frac{\vec{q}^{\,2}}{16k_v}, k_v, -\frac{\vec{q}}{2}\right), \quad k_{4\mu} = -\left(\frac{\vec{q}^{\,2}}{16k_v}, k_v, \frac{\vec{q}}{2}\right) \\ s &\simeq 4k_u k_v, \quad t \simeq -(\vec{q})^2, \end{aligned} \tag{27}$$

where we just kept the leading order in the t/s expansion, assuming $t/s \gg 1$. We also take the polarizations $\epsilon^{\mu\nu} = \epsilon^\mu \epsilon^\nu$, given by

$$\epsilon_1^\mu = \left(-\frac{\vec{q}\cdot\vec{e}_1}{2k_u}, 0, \vec{e}_1\right), \quad \epsilon_3^\mu = \left(\frac{\vec{q}\cdot\vec{e}_3}{2k_u}, 0, \vec{e}_3\right) \tag{28}$$

$$\epsilon_2^\mu = \left(0, \frac{\vec{q}\cdot\vec{e}_2}{2k_v}, \vec{e}_2\right), \quad \epsilon_4^\mu = \left(0, -\frac{\vec{q}\cdot\vec{e}_4}{2k_v}, \vec{e}_4\right). \tag{29}$$

Choosing $e_1 = e_3$ and $e_2 = e_4$ we can compute the phase (25) for the Einstein-Hilbert term

$$\delta_{\rm g}(b,s) = \frac{\Gamma\left(\frac{D-4}{2}\right)}{\pi^{\frac{D-4}{2}}}\frac{Gs}{b^{D-4}}(e_1\cdot e_3)(e_2\cdot e_4)\,, \tag{30}$$

and for the Gauss-Bonnet term

$$\begin{aligned}\delta_{\rm GB}(b,s) &= 4\lambda_{\rm GB}\left(\left(e_1^{ij}e_1^{ij}\right)e_2^{ij}e_2^{ik} + \left(e_2^{ij}e_2^{ij}\right)e_1^{ij}e_1^{ik}\right)\partial_{b_i}\partial_{b_j}\frac{\Gamma\left(\frac{D-4}{2}\right)}{\pi^{\frac{D-4}{2}}}\frac{Gs}{b^{D-4}} \\ &= -4\lambda_{\rm GB}\frac{\Gamma\left(\frac{D-4}{2}\right)}{\pi^{\frac{D-4}{2}}}\frac{Gs}{b^{D-2}}\left[2(e_1\cdot e_1)(e_2\cdot e_2)\right. \\ &\quad \left.-(D-2)(n\cdot e_1)^2-(D-2)(n\cdot e_2)^2\right]\,, \end{aligned} \tag{31}$$

where $\vec{n} \equiv \vec{b}/b$. The total contribution to the phase is given by the sum of (30) and (31), namely

$$\delta_{\rm g-GB}(b,s) = \delta_{\rm g}(b,s) + \delta_{\rm GB}(b,s). \tag{32}$$

Finally, the Shapiro's time delay is

$$\begin{aligned}\Delta t_{\rm g-GB} &= \frac{\Gamma\left(\frac{D-4}{2}\right)}{\pi^{\frac{D-4}{2}}}\frac{16EG}{b^{D-4}}(e_1\cdot e_1)(e_2\cdot e_2) \\ &\left[1+\frac{4\lambda_{\rm GB}(D-2)(D-4)}{b^2}\left(\frac{(n\cdot e_1)^2}{e_1\cdot e_1}+\frac{(n\cdot e_2)^2}{e_2\cdot e_2}-\frac{2}{D-2}\right)\right]. \end{aligned} \tag{33}$$

We can see that if the impact factor b^2 becomes small, $b^2 < \lambda_{\rm GB}$, the third term in (33) can be bigger then the first two, depending on the sign of $\lambda_{\rm GB}$ and the polarizations. Therefore, we can have a time advance and causality is violated.

On the other hand, we saw in the previous sections that for theories that are quadratic in both the Ricci and scalar curvature and lack a term quadratic in the Riemann tensor, the tree-level amplitude exactly coincides with the Einstein-Hilbert one. The corresponding time delay is

$$\Delta t_{\rm g} = \frac{\Gamma\left(\frac{D-4}{2}\right)}{\pi^{\frac{D-4}{2}}}\frac{16EG}{b^{D-4}}(e_1\cdot e_3)(e_2\cdot e_4)\,, \tag{34}$$

and of course no time advancement is possible.

Actually, *any nonlocal theory that is tree-level equivalent by the field redefinition theorem to a causal local one is causal too. In other words, given a causal (possibly local) theory, the theorem provides an algorithm for constructing a full class of higher derivative (even non-local) causal theories.*

An explicit example of a nonlocal theory involving gravity, one gauge field, and a scalar field is the one given by the action

$$\mathcal{L} = \frac{2}{\kappa_D^2}\left[R + \left(G_{\mu\nu} - \frac{\kappa_D^2}{2}(T^A_{\mu\nu} + T^\phi_{\rho\sigma})\right) F_{\rm g}^{\mu\nu,\rho\sigma} \left(G_{\rho\sigma} - \frac{\kappa_D^2}{2}(T^A_{\rho\sigma} + T^\phi_{\rho\sigma})\right)\right]$$
$$-\frac{1}{4}F_{\mu\nu}F^{\mu\nu} + \nabla_\mu F^{\mu\nu}\, F^{\rm A}\, \nabla_\rho F^\rho{}_\nu$$
$$+\frac{1}{2}\phi(\Box - m^2)\phi + \phi(\Box - m^2)\, F^\phi\, (\Box - m^2)\phi\,, \qquad (35)$$

where the analytic functions of the d'Alembertian operator $F^{\rm g}$, F^A, F^ϕ and the second rank tensors $T^A_{\mu\nu}$, $T^\phi_{\mu\nu}$ are defined as follows,

$$F_{\rm g}^{\mu\nu,\rho\sigma} \equiv \left(g^{\mu\rho}g^{\nu\sigma} - \frac{1}{2}g^{\mu\nu}g^{\rho\sigma}\right)\left(\frac{e^{H_{\rm g}(\Box)} - 1}{\Box}\right),$$
$$F^{\rm A} \equiv \frac{1}{2}\left(\frac{e^{H_A(\Box)} - 1}{\Box}\right),$$
$$F^\phi \equiv \frac{1}{2}\left(\frac{e^{H_\phi(\Box - m^2)} - 1}{\Box - m^2}\right),$$
$$T^A_{\mu\nu} \equiv F_{\mu\sigma}F^\sigma_\nu - \frac{1}{4}F_{\mu\nu}F^{\mu\nu},$$
$$T^\phi_{\mu\nu} \equiv \partial_\mu\phi\partial_\nu\phi - \frac{1}{2}g_{\mu\nu}(\partial_\lambda\phi\partial^\lambda\phi + m^2\phi^2)\,. \qquad (36)$$

$H_{\rm g}$, H_A and H_ϕ are form factors suitably chosen so that the theory is unitary and finite at quantum level in odd dimension (in particular in $D = 5$). In particular the theory (35) has the same spectrum of the equivalent local theory, namely the graviton, the photon, and the real scalar field (see [5, 7, 8] for more details) and by the field redefinition theorem all the tree-level n-point functions for the theory (35) are identical to the ones in local Einstein-Hilbert gravity coupled to the local Maxwell field, and a local scalar field. It is straightforward to prove that the theory above satisfies the field redefinition theorem, namely it is equivalent to Einstein's gravity minimally coupled to the electromagnetic field and scalar matter. Therefore, all the tree-level n-point functions for the theory (35) are identical to the ones that one can compute in local Einstein-Hilbert gravity couple to the local Maxwell field, and a local scalar field. In particular we can consider the elastic scattering of gravitons on massive scalars, and the photon-graviton scattering, whose amplitudes read

$$A(h,\phi;h,\phi)_{2;2} = 8\pi G\,\frac{(m^4 - su)^2}{t(s + m^2)(u + m^2)}\,,$$

$$A(h,\phi;h,\phi)_{-2;2} = 8\pi G\,\frac{m^4 t}{(s+m^2)(u+m^2)}\,, \tag{37}$$

$$A(h,A;h,A)_{1,2;1,2} = -8\pi G\,\frac{u^2}{t}\,,$$

$$A(h,A;h,A)_{1,-2;1,-2} = -8\pi G\,\frac{s^2}{t}\,, \tag{38}$$

plus the ones one can obtain by parity conjugation.

Taking the massless limit of (37), the helicity flip amplitude vanishes, while

$$A(h,\phi;h,\phi)_{2;2} = 8\pi G\,\frac{su}{t}\,. \tag{39}$$

All the above amplitudes in the eikonal limit $s \ll t$ simplify to:

$$-8\pi G\,\frac{s^2}{t}\,, \tag{40}$$

and the time delay is the same we have computed for the four graviton amplitudes. Therefore, the nonlocal theory (35) is causal as well as the local Einstein-Maxwell-scalar theory.

One could wonder whether causality can still be preserved in a theory where the nonlocality explicitly shows up in the amplitudes, i.e. in cases where the field redefinition theorem cannot be applied. An indication that this is actually possible comes from the theory whose action consists of (2) and the minimally coupled ordinary two-derivatives scalar matter,

$$S = S_{\rm g} + \int d^D x \sqrt{-g}\left(-\frac{1}{2}g^{\mu\nu}\partial_\mu\phi\partial_\nu\phi - \frac{1}{2}m^2\phi^2\right). \tag{41}$$

Using the graviton propagator (8), the tree-level gravitational scattering amplitude for 2-scalars in 2-scalars can be easily obtained (we here assume $m = 0$):

$$A_s = -8\pi G\frac{ut}{s}e^{-H(s)}\,, \quad A_t = -8\pi G\frac{s(s+t)}{t}e^{-H(t)}\,,$$

$$A_u = -8\pi G\frac{st}{u}e^{-H(u)}\,. \tag{42}$$

In the Regge limit $t \ll s$the leading contribution comes from the amplitude in the t-channel, namely

$$A_t(s,t) \approx -8\pi G\frac{s^2}{t}e^{-H(t)}\,. \tag{43}$$

For $D > 4$ there are no issues related to infrared divergences and we can now compute the phase (25) in $D = 5$,

$$\begin{aligned} \delta(b,s) &= \frac{1}{2s}\int \frac{d^3\vec{q}}{(2\pi)^3} e^{i\vec{q}\cdot\vec{b}} A_t(s,-\vec{q}^2) \\ &= 4\pi G s \int \frac{d^3\vec{q}}{(2\pi)^3} e^{i\vec{q}\cdot\vec{b}} \frac{e^{-H(-\vec{q}^2)}}{\vec{q}^2} \\ &= \frac{2Gs}{\pi}\int dq \frac{\sin(bq)}{bq}\, e^{-H(-q^2)} , \end{aligned} \tag{44}$$

where $q = |\vec{q}|$. In particular, for the form factor

$$e^{-\sigma\Box} , \tag{45}$$

which emerges naturally in string field theory [29, 30, 31, 32], one finds the analytic result

$$\delta(b,s)_{\rm SFT} = Gs\,\frac{{\rm Erf}(b/2\ell_\Lambda)}{b} , \tag{46}$$

which reduces to the one in Einstein's theory for $b \gg \ell_\Lambda$, namely

$$\delta(b,s)_{\rm SFT} \quad \to \quad \delta_{\rm EH}(b,s) = \frac{Gs}{b} . \tag{47}$$

The corresponding time delays are

$$\Delta t_{\rm SFT} = \frac{16EG}{\pi}\,\pi\frac{{\rm Erf}(b/2\ell_\Lambda)}{b} , \tag{48}$$

$$\Delta t_{\rm EH} = \frac{16EG}{\pi}\frac{\pi}{b} . \tag{49}$$

In Fig.1 we plot $\Delta t_{\rm T}$ for the form factor (5), which has been obtained numerically, together with $\Delta t_{\rm SFT}$ and $\Delta t_{\rm EH}$. Very similar results can be obtained for different values of α and γ

From the analytical results as well as from the plots, it is clear that the Shapiro's time delay never becomes negative and the causality condition is satisfied up to and beyond the non locality scale ℓ_Λ.

5. Conclusions

Weakly nonlocal theories are an interesting arena where such crucial ideas about quantum gravity as ultraviolet finiteness, perturbative unitarity and causality can be tested in a very straightforward way thanks to the powerful formalism of quantum field theory. In particular we have given evidence that in a lot of cases we can extend Einstein-Hilbert gravity without violating

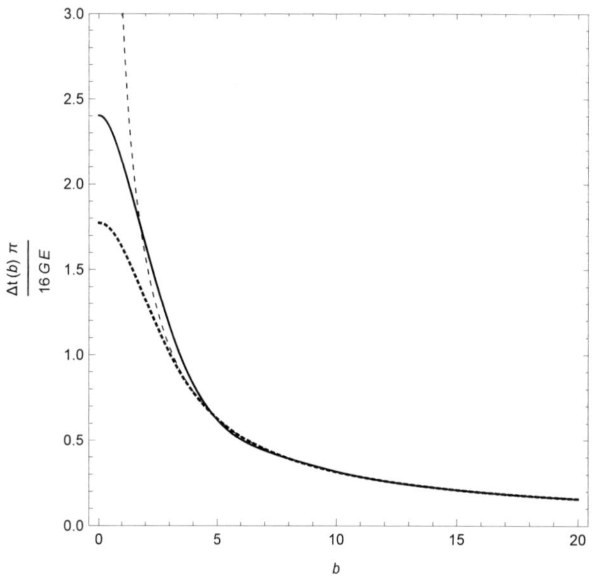

Figure 1: From top to bottom the lines represent respectively the following Shapiro's delays: Δt_{EH}, Δt_{T} (for $\gamma = 3$), and Δt_{SFT}. We also assumed $\ell_\Lambda = 1$.

the notion of causality related to Shapiro's time delay and discussed in [1]. Contrary to what happens in weakly coupled string theory, causality is not achieved by the introduction of an infinite tower of massive higher spin fields, but by avoiding the higher-derivative terms which could cause a Shapiro time advance. This has been proven to be possible in several cases. In particular a field redefinition theorem allows to construct a wide class of nonlocal theory for matter coupled to gravity compatible with causality. As a particular applications of the theorem, we have discussed the Einstein-Maxwell-Scalar nonlocal field theory, which can be proven to be causal, unitary, and finite in the ultraviolet. Other examples discussed in [24] are the $N = 1$ nonlocal supergravity [28] and Lee-Wick gravity [33, 34, 35, 36, 37]. In general, causality represents a valuable guide principle to understand what kind of higher derivative terms can show up in a consistent quantum gravity theory.

References

[1] X. O. Camanho, J. D. Edelstein, J. Maldacena and A. Zhiboedov, "Causality Constraints on Corrections to the Graviton Three-Point Coupling," JHEP **1602**, 020 (2016) doi:10.1007/JHEP02(2016)020 [arXiv:1407.5597 [hep-th]].

[2] K. S. Stelle, "Renormalization of Higher Derivative Quantum Gravity," Phys. Rev. D **16**, 953 (1977).

[3] Y. V. Kuz'min, "The Convergent Nonlocal Gravitation. (in Russian)," Sov. J. Nucl. Phys. **50**, 1011 (1989) [Yad. Fiz. **50**, 1630 (1989)].

[4] E. T. Tomboulis, "Renormalization and unitarity in higher derivative and nonlocal gravity theories," Mod. Phys. Lett. A **30**, 1540005 (2015). E. T. Tomboulis, hep-th/9702146.

[5] L. Modesto, "Super-renormalizable Quantum Gravity," Phys. Rev. D **86**, 044005 (2012) [arXiv:1107.2403 [hep-th]].

[6] T. Biswas, A. Conroy, A. S. Koshelev and A. Mazumdar, "Generalized ghost-free quadratic curvature gravity," Class. Quant. Grav. **31**, 015022 (2014) Erratum: [Class. Quant. Grav. **31**, 159501 (2014)] [arXiv:1308.2319 [hep-th]].

[7] L. Modesto and L. Rachwal, "Super-renormalizable and finite gravitational theories," Nucl. Phys. B **889**, 228 (2014) [arXiv:1407.8036 [hep-th]].

[8] L. Modesto and L. Rachwal, "Universally finite gravitational and gauge theories," Nucl. Phys. B **900**, 147 (2015) [arXiv:1503.00261 [hep-th]].

[9] Y. D. Li, L. Modesto and L. Rachwal, "Exact solutions and spacetime singularities in nonlocal gravity," JHEP **1512**, 173 (2015) [arXiv:1506.08619 [hep-th]].

[10] L. Modesto and L. Rachwal, "Finite Conformal Quantum Gravity and Nonsingular Spacetimes," arXiv:1605.04173 [hep-th].

[11] A. S. Koshelev, L. Modesto, L. Rachwal and A. A. Starobinsky, "Occurrence of exact R^2 inflation in non-local UV-complete gravity," JHEP **1611**, 067 (2016) [arXiv:1604.03127 [hep-th]].

[12] A. S. Koshelev, K. Sravan Kumar and A. A. Starobinsky, "R^2 inflation to probe non-perturbative quantum gravity," arXiv:1711.08864 [hep-th].

[13] I. Dimitrijevic, B. Dragovich, A. S. Koshelev, Z. Rakic and J. Stankovic, "Cosmological Solutions of a Nonlocal Square Root Gravity," Phys. Lett. B **797**, 134848 (2019) doi:10.1016/j.physletb.2019.134848 [arXiv:1906.07560 [gr-qc]].

[14] E. T. Tomboulis, "Nonlocal and quasilocal field theories," Phys. Rev. D **92**, no. 12, 125037 (2015) doi:10.1103/PhysRevD.92.125037 [arXiv:1507.00981 [hep-th]].

[15] G. Calcagni, L. Modesto and G. Nardelli, "Initial conditions and degrees of freedom of non-local gravity," JHEP **1805**, 087 (2018) Erratum: [JHEP **1905**, 095 (2019)] doi:10.1007/JHEP05(2018)087, 10.1007/JHEP05(2019)095 [arXiv:1803.00561 [hep-th]].

[16] F. Briscese and L. Modesto, "Unattainability of the Trans-Planckian regime in Nonlocal Quantum Gravity," arXiv:1912.01878 [hep-th].

[17] S. Gao and R. M. Wald, "Theorems on gravitational time delay and related issues," Class. Quant. Grav. **17**, 4999 (2000) doi:10.1088/0264-9381/17/24/305 [gr-qc/0007021].

[18] R. Pius and A. Sen, "Cutkosky rules for superstring field theory," JHEP **1610**, 024 (2016) Erratum: [JHEP **1809**, 122 (2018)] doi:10.1007/JHEP09(2018)122, 10.1007/JHEP10(2016)024 [arXiv:1604.01783 [hep-th]].

[19] F. Briscese and L. Modesto, "Cutkosky rules and perturbative unitarity in Euclidean nonlocal quantum field theories," arXiv:1803.08827 [gr-qc].

[20] P. Chin and E. T. Tomboulis, "Nonlocal vertices and analyticity: Landau equations and general Cutkosky rule," JHEP **1806**, 014 (2018) doi:10.1007/JHEP06(2018)014 [arXiv:1803.08899 [hep-th]].

[21] R. Pius and A. Sen, "Unitarity of the Box Diagram," JHEP **1811**, 094 (2018) doi:10.1007/JHEP11(2018)094 [arXiv:1805.00984 [hep-th]].

[22] P. Donà, S. Giaccari, L. Modesto, L. Rachwal and Y. Zhu, "Scattering amplitudes in super-renormalizable gravity," JHEP **1508**, 038 (2015) doi:10.1007/JHEP08(2015)038 [arXiv:1506.04589 [hep-th]].

[23] D. Anselmi and M. Halat, "Renormalizable acausal theories of classical gravity coupled with interacting quantum fields," Class. Quant. Grav. **24**, 1927 (2007) doi:10.1088/0264-9381/24/8/003 [hep-th/0611131].

[24] S. Giaccari and L. Modesto, "Causality in Nonlocal Gravity," arXiv:1803.08748 [hep-th].

[25] D. N. Kabat and M. Ortiz, "Eikonal quantum gravity and Planckian scattering," Nucl. Phys. B **388**, 570 (1992) [hep-th/9203082].

[26] M. Ciafaloni and D. Colferai, "Rescattering corrections and self-consistent metric in Planckian scattering," JHEP **1410**, 85 (2014) [arXiv:1406.6540 [hep-th]].

[27] B. Bellazzini, C. Cheung and G. N. Remmen, "Quantum Gravity Constraints from Unitarity and Analyticity," Phys. Rev. D **93**, no. 6, 064076 (2016) [arXiv:1509.00851 [hep-th]].

[28] S. Giaccari and L. Modesto, "Nonlocal supergravity," Phys. Rev. D **96**, no. 6, 066021 (2017) [arXiv:1605.03906 [hep-th]].

[29] V. A. Kostelecky and S. Samuel, "Collective Physics in the Closed Bosonic String," Phys. Rev. D **42**, 1289 (1990). doi:10.1103/PhysRevD.42.1289

[30] G. Calcagni and G. Nardelli, "String theory as a diffusing system," JHEP **1002**, 093 (2010) doi:10.1007/JHEP02(2010)093 [arXiv:0910.2160 [hep-th]].

[31] G. Calcagni and L. Modesto, "Nonlocality in string theory," J. Phys. A **47**, no. 35, 355402 (2014) [arXiv:1310.4957 [hep-th]].

[32] G. Calcagni and L. Modesto, "Nonlocal quantum gravity and M-theory," Phys. Rev. D **91**, no. 12, 124059 (2015) [arXiv:1404.2137 [hep-th]].

[33] L. Modesto and I. L. Shapiro, "Superrenormalizable quantum gravity with complex ghosts," Phys. Lett. B **755**, 279 (2016) [arXiv:1512.07600 [hep-th]].

[34] L. Modesto, "Super-renormalizable or finite Lee-Wick quantum gravity," Nucl. Phys. B **909**, 584 (2016) [arXiv:1602.02421 [hep-th]].

[35] D. Anselmi and M. Piva, "A new formulation of Lee-Wick quantum field theory," JHEP **1706**, 066 (2017) doi:10.1007/JHEP06(2017)066 [arXiv:1703.04584 [hep-th]].

[36] D. Anselmi and M. Piva, "Perturbative unitarity of Lee-Wick quantum field theory," Phys. Rev. D **96**, no. 4, 045009 (2017) doi:10.1103/PhysRevD.96.045009 [arXiv:1703.05563 [hep-th]].

[37] D. Anselmi, "Fakeons And Lee-Wick Models," JHEP **1802**, 141 (2018) doi:10.1007/JHEP02(2018)141 [arXiv:1801.00915 [hep-th]].

Hyperspherical three-body variables applied to Sakumichi and Suganuma's lattice QCD data*

James Leech†
University of St. Andrews, The Old Burgh School,
Abbey Walk, St Andrews, Fife KY16 9LB, Scotland

Milovan Šuvakov‡
Institute of Physics Belgrade, University of Belgrade,
Pregrevica 118, Zemun, P.O.Box 57, 11080 Beograd, Serbia

* on sabbatical leave of absence at
Department of Health Sciences Research, Mayo Clinic,
200 First Street SW, Rochester, MN 55905, USA

V. Dmitrašinović§
Institute of Physics Belgrade, University of Belgrade,
Pregrevica 118, Zemun, P.O.Box 57, 11080 Beograd, Serbia

April 17, 2020

Abstract

We analyse the 2015 lattice QCD calculations of the three-quark potential by Sakumichi and Suganuma [1] using hyperspherical three-body variables, following Refs. [2, 3]. We show that the triangle shape-dependence of Sakumichi & Suganuma's three-quark confinement potential evaluated at $\beta = 5.8$ differs from the one evaluated at $\beta = 6.0$ by 2%. This difference corresponds precisely to the difference between the Y- and Δ-string potentials in the isosceles and right-angled triangle geometries, thus setting an upper and a lower bound on the mean value.

1. Introduction

In spite of decades-long efforts, [1, 4, 5, 6], the functional form of the three-heavy-quark potential in lattice QCD is still not well known. Unlike

---

* We thank Naoyuki Sakumichi for providing the unpublished data. The work of M. Š and V.D. was supported by the Serbian Ministry of Science and Technological Development under Grants No. OI 171037 and III 41011.

† e-mail address: jamesleechpublic@gmail.com

‡ e-mail address: suki@ipb.ac.rs

§ e-mail address: dmitrasin@ipb.ac.rs

the confining part of the two-body potential, which depends only on one variable - the distance between two bodies (quark and antiquark) - the three-body potential depends on three (scalar) variables.

These three variables may be defined in different ways, perhaps the simplest one being the set of three two-body separations. The hyperspherical coordinates, see [2, 3] and references therein, consist of a (dimensional) hyperradius $R = \sqrt{\frac{1}{3}\sum_{i<j}^{3}(\mathbf{r}_i - \mathbf{r}_j)^2} = \sqrt{\sum_{i}^{3}(\mathbf{R}_{\text{g.b.}} - \mathbf{r}_i)^2}$, which is proportional to the root-mean-square distance of the three particles from their geometrical barycenter $\mathbf{R}_{\text{g.b.}} = \frac{1}{3}\sum_{i}^{3}\mathbf{r}_i$ [1], and scales linearly with λ: $R \to \lambda R$ and thus measures the size of the system; and of two dimensionless (shape) variables, that may be expressed as hyperangles, or some functions thereof. The scaling transformation affects only the hyperradius, whereas the permutation symmetry affects only the two "shape space" variables/hyperangles.

Thus far, most of analyses of lattice 3-body potentials had been in terms of (only) one scalar variable, chosen according to the tastes and prejudices of the authors, usually as the Δ- or Y-string lengths, l_Δ, l_Y, respectively. As both l_Δ and l_Y are linearly proportional to the hyperradius R, both parametrizations effectively "measure" the linear hyperradial dependence of the confining 3-body potential, averaged over some set of shape-space points. Thus, agreement, or disagreement of such a fit with the lattice data must depend on the choice of shape-space points.

The Δ- and Y-string lengths, l_Δ, l_Y, differ little in most of the shape space, the relative difference being the largest near and at "two-body collision points", where the Coulomb interaction also peaks. Thus, an effective disambiguation between Δ- and Y-string potentials demands either a very precise lattice measurement in regions where the difference is small, or a very precise subtraction of the Coulomb term in regions where the difference is large. In either case, the task is difficult. Thus, the final results may depend on the initial choice of geometrical configurations on the lattice, in addition to the usual statistical and systematic errors/uncertainties. These geometrical configurations have been generally subject of entirely arbitrary choice(s) of various authors.

We have analysed Koma & Koma's and Takahashi et al.'s lattice results [4, 5] in terms of three hyperspherical coordinates for the first time, in Ref. [2, 3]. Whereas the old Takahashi et al. data was insufficient to produce a continuous functional shape-space dependence, Koma & Koma's [4] results yielded two continuous, generally smooth functional dependences along two orthogonal lines in the shape space. There are several important differences between Komas' and Sakumichi's calculations: [4] had a 24^4 lattice at $\beta = 6.0$, with 221 three-quark geometries and only one gauge configuration; [1] had a $16^3 \times 32$ lattice at $\beta = 5.8$ with 101 three-quark

[1] which equals the physical center-of-mass $\mathbf{R}_{\text{g.b.}} = \mathbf{R}_{\text{CM}}$ when all three masses are equal

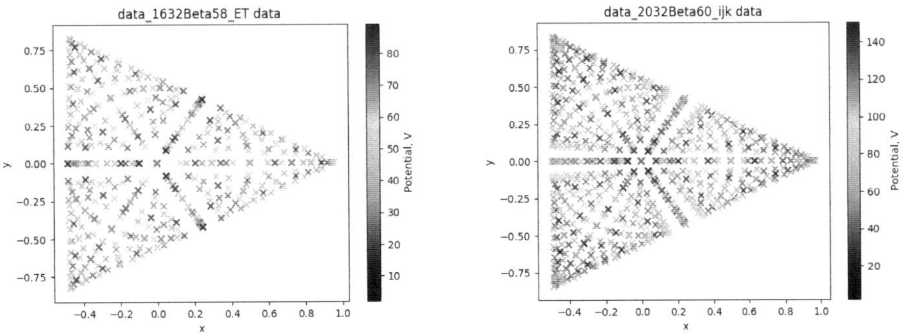

Figure 1: Distribution of Sakumichi & Suganuma's 3q configurations at $\beta = 5.8$ (l.h.s.) and $\beta = 6.0$ (r.h.s.).

geometries, and a $20^3 \times 32$ lattice at $\beta = 6.0$ and 211 three-quark geometries, with 1000 and 2000 gauge configurations, respectively.

2. Analysis of lattice data

In this report we subject Sakumichi & Suganuma's lattice data [1] to the same kind of analysis as Koma & Koma's [4] in Refs. [2, 3]. Fig. 1 depicts all configurations of the three-body system in Refs. [1, 4] including their permutations. The three lines that cross the origin represent isosceles triangles. Three lines orthogonal to them represent the right-angled triangles - one such line is at $y = -0.5$. One can see here that these lines are the only two sets of geometric configurations that are common to both Koma and Suganuma data. Therefore, we shall use them both.

We assume that the total three-quark potential V_{3q} has the form

$$V_{3q}(\alpha, \phi, R) = -\frac{A(\alpha, \phi)}{R} + B(\alpha, \phi)R + C, \tag{1}$$

henceforth referred to as the Coulomb + linear potential Ansatz. The first term represents the sum of QCD Coulomb pairwise interactions, which is dominant at small values of the hyper-radius R. The second term represents the confinement potential, which is linear in R and dominant at large values of hyper-radius R, and the third term - C - is a constant. Here $A(\phi, \alpha)$ is assumed to be the (standard) sum of of pair-wise Coulomb terms, and $B(\phi, \alpha)$ is the unknown hyper-angular dependence of the linearly rising confining potential. Our goal is to determine $B(\phi, \alpha) \simeq V/R$ using the lattice data and the well-known hyper-angular and hyper-radial dependences of the two-body Coulomb term, as explained in Refs. [2, 3].

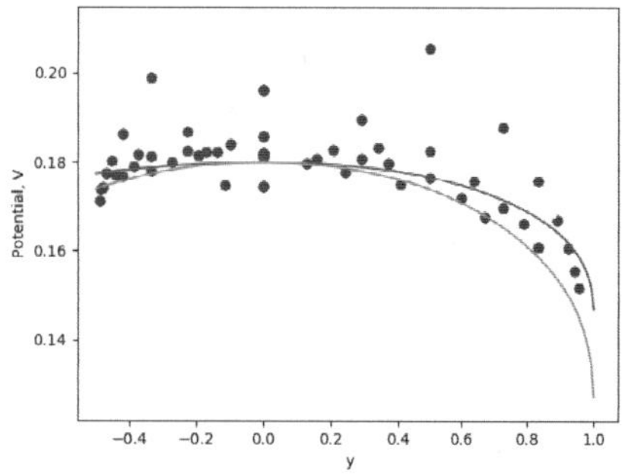

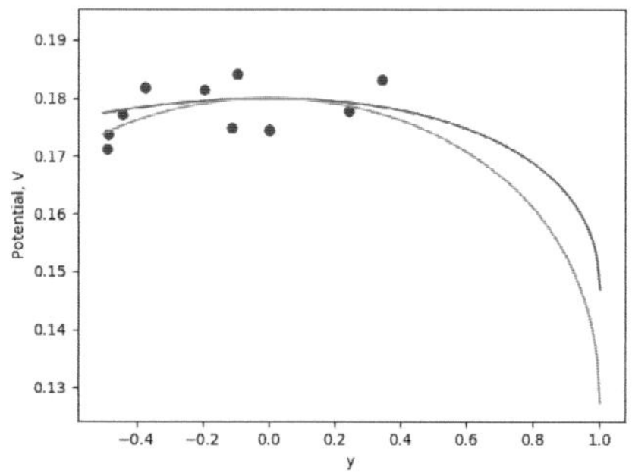

Figure 2: Extracted values of Sakumichi & Suganuma's V/R for isosceles triangles ($y = 0$), at $\beta = 5.8$: a) all (l.h.s); b) with hyper-radius filter $R \geq 8.0$ (r.h.s). The blue line represents the Δ-string, the yellow line is the Y-string.

2.1. The $\beta = 5.8$ case

It can be seen in Fig. 2.a that for the isosceles triangle configurations in the [1] $\beta = 5.8$ data set, all of the $B(x)$ values form a scattered set of points, some of which, though not all, lie between the Δ and Y-string potentials' functional forms. As one imposes the hyperradial filter $R \geq 8.0$, the scatter reduces, as does the number of points, Fig. 2.b, but no single, smooth curve emerges. The corresponding graphs for the right-angled triangles are shown in Fig. 3. Note that the hyperradial filter $R \geq 7.0$ eliminates many of the low-lying points and that almost all of the remaining points fall onto the (upper, blue) Δ-string prediction. The only discrepancy are the multiple points at the isosceles right-angled configuration.

2.2. The $\beta = 6.0$ case

Similarly: a) (again) the hyperradial filter $R \geq 8.0$ eliminates the many widely scattered low-R points in Fig. 4, and almost all of the remaining points fall onto the (lower, yellow) Y-string prediction in Fig. 5; b) the resulting y dependence of $V(y)/R$ for right-angled triangles in Fig. 4 is practically equivalent to the Y-string (the discrepancies are several multiple points near the isosceles right-angled configuration), and differs markedly from that at $\beta = 5.8$ in Fig. 2. This difference is (almost entirely) reducible to a constant shift.

3. Conclusions

We see that the results of our analysis are beset with several ambiguities:

1) Sakumichi & Suganuma [1] have calculated the potential at two different values of the coupling constant $\beta = 5.8$ and $\beta = 6.0$, but in two

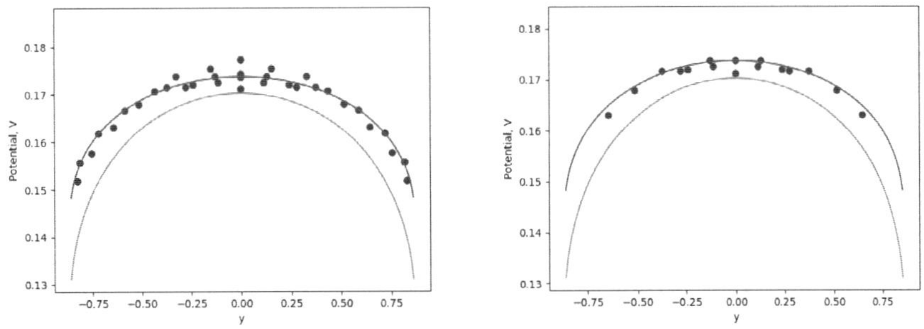

Figure 3: Extracted values of V/R for: a) all right triangles ($x = 0.5$) (l.h.s.) b) right triangles ($x = 0.5$) with hyper-radius filter $R \geq 7.0$, at $\beta = 5.8$, (r.h.s.). The blue line represents the Δ-string, yellow line is the Y-string.

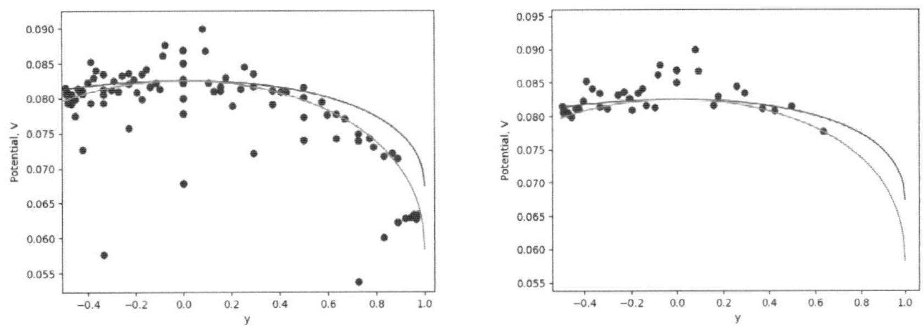

Figure 4: Extracted values of V/R for: isosceles triangles ($y = 0$), at $\beta = 6.0$. a) unrestricted data (l.h.s.); b) with hyper-radius filter $R \geq 9.0$. (r.h.s.).

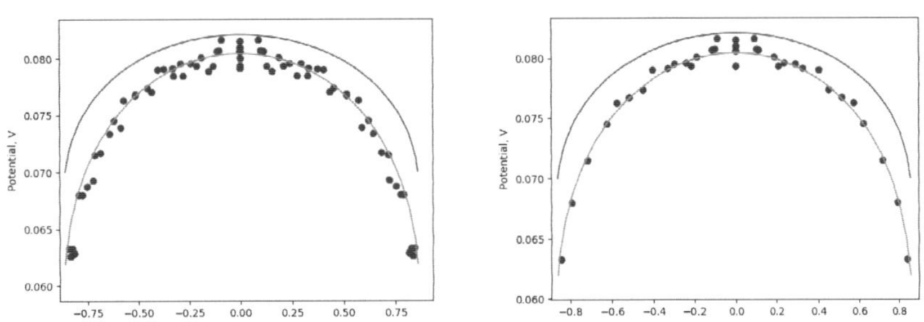

Figure 5: Extracted values of V/R for: a) all right triangles ($x = 0.5$) (l.h.s.) b) right triangles ($y = 0$) with hyper-radius larger than 8.0, at $\beta = 6.0$, (r.h.s.).

somewhat different sets of geometric and gauge configurations. These two calculations ought to agree with each other after rescaling (renormalization), which they do not (in their entirety), see Figs. 3, 5 and Figs. 2 and 4.

2) Sakumichi & Suganuma's [1] geometric configurations differ somewhat from Koma & Koma's [4], even in the small region of overlap. Namely, none of Sakumichi's right-angled triangles are exactly right-angled: there is always a small discrepancy, which leaves open the question of the correct normalization of the confining potential V/R in Figs. 3 and 5.

The shape-dependent part of the confining potential V/R must/can not change under scaling/renormalization, [7], therefore the $\beta = 5.8$ and $\beta = 6.0$ functional dependences must be identical, which they are not. The difference between the two potentials is consistent with an overall (constant) shift, however, which might be reconcilable with statistical and/or systematic errors.

References

[1] N. Sakumichi and H. Suganuma, Phys. Rev. D **92**, no. 3, 034511 (2015),

[2] James Leech, Milovan Šuvakov, and V. Dmitrašinović, "Hyperspherical Three-Body Variables Applied to Lattice QCD Data", p. 253 - 258, Proceedings of the "Ninth Mathematical Physics Meeting", eds. Branko Dragovich, Igor Salom and Marko Vojinović, Institute of Physics, Belgrade, Serbia, (2017).

[3] James Leech, Milovan Šuvakov, and V. Dmitrašinović, "Application of Hyperspherical Three-Body Variables to Lattice QCD Data", Proceedings of the International Workshop "Excited QCD 2018", Mt. Kopaonik, March 2018, Act. Phys. Pol. Suppl. **B 11**, 435 (2018).

[4] Y. Koma and M. Koma, Phys. Rev. D **95**, no. 9, 094513 (2017).

[5] T. T. Takahashi, H. Suganuma, Y. Nemoto and H. Matsufuru, Phys. Rev. D **65**, 114509 (2002)

[6] C. Alexandrou, P. de Forcrand and A. Tsapalis, "Static three-quark SU(3) and four-quark SU(4) potentials," Phys. Rev. D **65**, 054503 (2002).

[7] Jan Smit, "Introduction to Quantum Fields on a Lattice – 'a robust mate"' Cambridge University Press, Cambridge UK, (2003).

XYZ Gaudin model with boundary terms*

Nenad Manojlović[†]
Departamento de Matemática, Faculdade de Ciências e Tecnologia
Universidade do Algarve, Campus de Gambelas,
PT-8005-139 Faro, Portugal
and
Grupo de Física Matemática da Universidade de Lisboa
Departamento de Matemática, Faculdade de Ciências
Campo Grande, Edifício C6, 1749-016 Lisboa, Portugal

Igor Salom[‡]
Institute of Physics, University of Belgrade
Pregrevica 118, 11080 Belgrade, Serbia

Nuno Cirilo António[§]
Centro de Análise Funcional e Aplicações
Instituto Superior Técnico, Universidade de Lisboa
Av. Rovisco Pais, 1049-001 Lisboa, Portugal

Abstract

We study the elliptic Gaudin model as a quasi-classical limit of the XYZ Heisenberg spin chain with the most general K-matrix. In particular, we give the generating function of the Gaudin Hamiltonians with boundary terms.

1. Introduction

There is extensive literature on the subject of Gaudin algebras, discussing its various generalizations in many diferent context [1, 2, 3, 4, 5, 6, 7, 8, 9, 10, 11, 12, 13, 14, 15, 16, 17, 18, 19, 20, 21, 22, 23, 24, 25, 26, 27, 28, 29, 30, 31, 32, 33, 34, 35, 36, 37, 38, 39, 40, 41, 42, 43]. Here we derive the generating function of the Gaudin Hamiltonians with boundary terms in the elliptic case following Sklyanin's approach in the periodic case [4], as we have done for the rational and trigonometric Gaudin model [44, 45, 43, 46, 47]. In the section 2. we review the XYZ Heisenberg spin chain. Our starting point is

* I.S. was supported in part by the Serbian Ministry of Education, Science and Technological Development under grant number ON 171031.

† e-mail address: nmanoj@ualg.pt

‡ e-mail address: isalom@ipb.ac.rs

§ e-mail address: nantonio@math.ist.utl.pt

the corresponding R-matrix [48, 49] and its most relevant properties. The relevant Lax operator as well as the definition of the Sklyanin algebra [50] is our next step. Then we study the properties of the central element of the RLL-relations, the so-called quantum determinant of the Lax operator, and the representation theory of the Sklyanin algebra which follows from the fusion procedure [50, 51, 52]. Sklyanin showed how to introduce non-periodic boundary conditions which are compatible with the integrability of the bulk system[27]. Following his approach we study the general solution of the relevant Reflection Equation [53, 54, 55] which at the same time defines the corresponding solution of the so-called dual Reflection Equation. The Reflection Equation Algebra provides the suitable algebraic framework for studying the non-periodic XYZ Heisenberg spin chain. This algebra is generated by the entries of the so-called Sklyanin monodromy matrix. As Sklyanin showed [27], the trace of the product of this monodromy matrix with the solution of the dual Reflection Equation yields the transfer matrix of the chain, i.e. the generator of an Abelian subalgebra which is relevant for the dynamics of the system. Moreover, the Sklyanin monodromy matrix admits a central element of the Reflection Equation Algebra. In the section 3. we study the non-periodic elliptic Gaudin model as the quasi-classical limit of the XYZ Heisenberg spin chain. Following Sklyanin's approach in the rational case [4] we show how the linear combination of the quantum determinant of the elliptic monodromy matrix together with its trace yields the generation function of the elliptic Gaudin model, in the periodic case. To study the non-periodic elliptic Gaudin model we follow the same approach we have used successfully in the rational [44] as well as in the trigonometric case [47]. Namely, we combine the classical r-matrix together with the general solution of the classical Reflection Equation into the so-called non-unitary classical r-matrix which is the key notion in the study of the non-periodic Gaudin model. The relevant Lax operator can then be defined in somewhat natural way [44, 45, 46, 47]. This Lax operator and the non-unitary classical r-matrix determine the linear bracket which is antisymmetric and satisfies the Jacobi identity, defining the dynamical symmetry algebra which is of utmost importance for the solvability of the system. The generating function of the Gaudin Hamiltonian is then introduced in a similar way like for the other non-periodic Gaudin models [44, 45, 46, 47]. Finally, in the appendix A we summarise some fundamental identities satisfied by the Jacobi elliptic functions used in the text.

2. XYZ Heisenberg spin chain

The R-matrix relevant for the XYZ Heisenberg spin chain is given by [48, 49]

$$R(\lambda, \eta, \kappa) = \mathbb{1} + \sum_{\alpha=1}^{3} W_\alpha(\lambda, \eta, \kappa)\ \sigma^\alpha \otimes \sigma^\alpha, \tag{1}$$

where

$$\begin{aligned} W_1(\lambda,\eta,\kappa) &= \frac{\operatorname{cn}(\lambda+\eta,\kappa)\ \operatorname{sn}(\eta,\kappa)}{\operatorname{sn}(\lambda+\eta,\kappa)\ \operatorname{cn}(\eta,\kappa)}, \\ W_2(\lambda,\eta,\kappa) &= \frac{\operatorname{dn}(\lambda+\eta,\kappa)\ \operatorname{sn}(\eta,\kappa)}{\operatorname{sn}(\lambda+\eta,\kappa)\ \operatorname{dn}(\eta,\kappa)}, \\ W_3(\lambda,\eta,\kappa) &= \frac{\operatorname{sn}(\eta,\kappa)}{\operatorname{sn}(\lambda+\eta,\kappa)}, \end{aligned} \tag{2}$$

the functions $\operatorname{sn}(\lambda,\kappa)$, $\operatorname{cn}(\lambda,\kappa)$, and $\operatorname{dn}(\lambda,\kappa)$ are the usual Jacobi elliptic functions, λ is a spectral parameter, η is a quasi-classical parameter, κ is the modulus and σ^α, $\alpha = 1,2,3$, are the Pauli matrices

$$\sigma^\alpha = \begin{pmatrix} \delta_{\alpha 3} & \delta_{\alpha 1} - \imath\delta_{\alpha 2} \\ \delta_{\alpha 1} + \imath\delta_{\alpha 2} & -\delta_{\alpha 3} \end{pmatrix}. \tag{3}$$

The R-matrix (1) satisfies the Yang-Baxter equation

$$R_{12}(\lambda-\mu)R_{13}(\lambda)R_{23}(\mu) = R_{23}(\mu)R_{13}(\lambda)R_{12}(\lambda-\mu)\ . \tag{4}$$

In the present case Yang-Baxter equation reduces to the following matrix equation

$$\begin{aligned} \sum_{\alpha,\beta,\gamma=1}^{3} \epsilon_{\alpha\beta\gamma}\,&(W_\beta(\lambda-\mu)W_\gamma(\lambda) - W_\alpha(\lambda-\mu)W_\gamma(\mu) + W_\alpha(\lambda)W_\beta(\mu) \\ &-W_\gamma(\lambda-\mu)W_\beta(\lambda)W_\alpha(\mu))\,\sigma^\alpha\otimes\sigma^\beta\otimes\sigma^\gamma = 0, \end{aligned} \tag{5}$$

or to the six scalar equations

$$\begin{aligned} &W_2(\lambda-\mu)W_3(\lambda) - W_1(\lambda-\mu)W_3(\mu) + W_1(\lambda)W_2(\mu) - W_3(\lambda-\mu)W_2(\lambda)W_1(\mu) = 0, \\ &W_3(\lambda-\mu)W_1(\lambda) - W_2(\lambda-\mu)W_1(\mu) + W_2(\lambda)W_3(\mu) - W_1(\lambda-\mu)W_3(\lambda)W_2(\mu) = 0, \\ &W_1(\lambda-\mu)W_2(\lambda) - W_3(\lambda-\mu)W_2(\mu) + W_3(\lambda)W_1(\mu) - W_2(\lambda-\mu)W_1(\lambda)W_3(\mu) = 0, \\ &W_3(\lambda-\mu)W_2(\lambda) - W_1(\lambda-\mu)W_2(\mu) + W_1(\lambda)W_3(\mu) - W_2(\lambda-\mu)W_3(\lambda)W_1(\mu) = 0, \\ &W_1(\lambda-\mu)W_3(\lambda) - W_2(\lambda-\mu)W_3(\mu) + W_2(\lambda)W_1(\mu) - W_3(\lambda-\mu)W_1(\lambda)W_2(\mu) = 0, \\ &W_2(\lambda-\mu)W_1(\lambda) - W_3(\lambda-\mu)W_1(\mu) + W_3(\lambda)W_2(\mu) - W_1(\lambda-\mu)W_2(\lambda)W_3(\mu) = 0, \end{aligned} \tag{6}$$

which are consequences of the definition (2) of the functions $W_\alpha(\lambda) \equiv W_\alpha(\lambda,\eta,\kappa)$, $\alpha = 1,2,3$, the identities (92) and the addition theorems of the Jacobi elliptic functions (93) - (95).

In the following, we study some important properties of the R-matrix (1). Evidently, this R-matrix has the parity invariance

$$R_{21}(\lambda) = R_{12}(\lambda), \tag{7}$$

temporal invariance

$$R^t_{12}(\lambda) = R_{12}(\lambda), \tag{8}$$

and it is regular at $\lambda = 0$,

$$R_{12}(0) = 2\mathcal{P}, \tag{9}$$

here $\mathcal{P}$ is the permutation matrix in $\mathbb{C}^2 \otimes \mathbb{C}^2$.

With the aim of checking the unitarity property we calculate the product

$$R(\lambda)R(-\lambda) = \left(1 + \sum_{\alpha=1}^{3} W_\alpha(\lambda)W_\alpha(-\lambda)\right)\mathbb{1} + \sum_{\alpha=1}^{3} \left(W_\alpha(\lambda) + W_\alpha(-\lambda)\right)\sigma^\alpha \otimes \sigma^\alpha$$
$$- \sum_{\alpha,\beta,\gamma,\rho=1}^{3} \epsilon_{\alpha\beta\gamma}\, \epsilon_{\alpha\beta\rho}\, W_\alpha(\lambda)W_\beta(-\lambda)\, \sigma^\gamma \otimes \sigma^\rho. \tag{10}$$

A straightforward calculation shows that

$$\begin{aligned} W_1(\lambda) + W_1(-\lambda) - W_2(\lambda)W_3(-\lambda) - W_3(\lambda)W_2(-\lambda) &= 0, \\ W_2(\lambda) + W_2(-\lambda) - W_3(\lambda)W_1(-\lambda) - W_1(\lambda)W_3(-\lambda) &= 0, \\ W_3(\lambda) + W_3(-\lambda) - W_1(\lambda)W_2(-\lambda) - W_2(\lambda)W_1(-\lambda) &= 0. \end{aligned} \tag{11}$$

Therefore the R-matrix (1) has the unitarity property

$$R(\lambda)R(-\lambda) = \rho(\lambda, \eta, \kappa)\; \mathbb{1} \tag{12}$$

where the function $\rho(\lambda, \eta, \kappa)$ is given by

$$\rho(\lambda, \eta, \kappa) = \left(1 + \sum_{\alpha=1}^{3} W_\alpha(\lambda)W_\alpha(-\lambda)\right) = 4\frac{\mathrm{sn}^2(\eta, \kappa)}{\mathrm{sn}^2(2\eta, \kappa)}\; \frac{\mathrm{sn}^2(\lambda, \kappa) - \mathrm{sn}^2(2\eta, \kappa)}{\mathrm{sn}^2(\lambda, \kappa) - \mathrm{sn}^2(\eta, \kappa)}. \tag{13}$$

The R-matrix (1) has the crossing symmetry

$$R(\lambda) = \left(\sigma^2 \otimes \mathbb{1}\right) R^{t_1}(-\lambda - 2\eta) \left(\sigma^2 \otimes \mathbb{1}\right), \tag{14}$$

were t_1 denotes the transpositions in the first space of the tensor product $\mathbb{C}^2 \otimes \mathbb{C}^2$. Consequently, the crossing unitarity property holds

$$R^{t_1}(\lambda)R^{t_1}(\lambda - 4\eta) = \rho(\lambda + 2\eta, \eta, \kappa)\; \mathbb{1}. \tag{15}$$

In what follows we study an inhomogeneous XYZ spin chain with N sites, with the local space V_m, that is the $2s + 1$ dimensional spin s representation space of the Sklyanin algebra and inhomogeneous parameter α_j.

$$\mathcal{H} = \bigotimes_{m=1}^{N} V_m.$$

Following [50] we introduce the Lax operator

$$\mathbb{L}_{0q}(\lambda) = \mathbb{1} \otimes S^0 + \sum_{\alpha=1}^{3} W_\alpha(\lambda, \eta, \kappa)\, \sigma^\alpha \otimes S^\alpha, \tag{16}$$
$$= \begin{pmatrix} S^0 + W_3(\lambda)S^3 & W_1(\lambda)S^1 - \imath W_2(\lambda)S^2 \\ W_1(\lambda)S^1 + \imath W_2(\lambda)S^2 & S^0 - W_3(\lambda)S^3 \end{pmatrix},$$

were S^0, S^1, S^2, S^3 are the generators of the Sklyanin algebra $U_{\tau,\eta}(sl(2))$ [50]. The generators of the Sklyanin algebra satisfy the following relations

$$\begin{aligned} \left[S^1, S^2\right] &= \imath \left(S^0 S^3 + S^3 S^0\right), \\ \left[S^2, S^3\right] &= \imath \left(S^0 S^1 + S^1 S^0\right), \\ \left[S^3, S^1\right] &= \imath \left(S^0 S^2 + S^2 S^0\right), \\ \left[S^0, S^1\right] &= \imath\, J_{23} \left(S^2 S^3 + S^3 S^2\right), \\ \left[S^0, S^2\right] &= \imath\, J_{31} \left(S^3 S^1 + S^1 S^3\right), \\ \left[S^0, S^3\right] &= \imath\, J_{12} \left(S^1 S^2 + S^2 S^1\right), \end{aligned} \tag{17}$$

where

$$J_{23} = \frac{W_2(\lambda-\mu)W_3(\lambda)W_2(\mu) - W_3(\lambda-\mu)W_2(\lambda)W_3(\mu)}{W_1(\lambda) - W_1(\lambda-\mu)W_1(\mu)}, \tag{18}$$

$$J_{23} = \frac{W_3(\lambda-\mu)W_2(\mu)W_3(\lambda) - W_2(\lambda-\mu)W_3(\mu)W_2(\lambda)}{W_1(\mu) - W_1(\lambda-\mu)W_1(\lambda)}, \tag{19}$$

$$J_{31} = \frac{W_3(\lambda-\mu)W_1(\lambda)W_3(\mu) - W_1(\lambda-\mu)W_3(\lambda)W_1(\mu)}{W_2(\lambda) - W_2(\lambda-\mu)W_2(\mu)}, \tag{20}$$

$$J_{31} = \frac{W_1(\lambda-\mu)W_3(\mu)W_1(\lambda) - W_3(\lambda-\mu)W_1(\mu)W_3(\lambda)}{W_2(\mu) - W_2(\lambda-\mu)W_2(\lambda)}, \tag{21}$$

$$J_{12} = \frac{W_1(\lambda-\mu)W_2(\lambda)W_1(\mu) - W_2(\lambda-\mu)W_1(\lambda)W_2(\mu)}{W_3(\lambda) - W_3(\lambda-\mu)W_3(\mu)}, \tag{22}$$

$$J_{12} = \frac{W_2(\lambda-\mu)W_1(\mu)W_2(\lambda) - W_1(\lambda-\mu)W_2(\mu)W_1(\lambda)}{W_3(\mu) - W_3(\lambda-\mu)W_3(\lambda)}. \tag{23}$$

Actually, the quantities J_{12}, J_{23} and J_{31} do not depend on the spectral parameter and are given by [50, 51]

$$J_{12} = \frac{W_1^2(\lambda) - W_2^2(\lambda)}{W_3^2(\lambda) - 1} = (1-\kappa^2)\, \frac{\mathrm{sn}^2(\eta,\kappa)}{\mathrm{cn}^2(\eta,\kappa)\mathrm{dn}^2(\eta,\kappa)}, \tag{24}$$

$$J_{23} = \frac{W_2^2(\lambda) - W_3^2(\lambda)}{W_1^2(\lambda) - 1} = \kappa^2\, \frac{\mathrm{sn}^2(\eta,\kappa)\mathrm{cn}^2(\eta,\kappa)}{\mathrm{dn}^2(\eta,\kappa)}, \tag{25}$$

$$J_{31} = \frac{W_3^2(\lambda) - W_1^2(\lambda)}{W_2^2(\lambda) - 1} = -\frac{\mathrm{sn}^2(\eta,\kappa)\mathrm{dn}^2(\eta,\kappa)}{\mathrm{cn}^2(\eta,\kappa)}. \tag{26}$$

A straightforward calculation shows that

$$J_{12} + J_{23} + J_{31} + J_{12}J_{23}J_{31} = 0. \tag{27}$$

Therefore

$$J_{\alpha\beta} = -\frac{J_\alpha - J_\beta}{J_\gamma}, \tag{28}$$

with

$$J_1 : J_2 : J_3 = \frac{\mathrm{cn}(2\eta,\kappa)}{\mathrm{cn}^2(\eta,\kappa)} : \frac{\mathrm{dn}(2\eta,\kappa)}{\mathrm{dn}^2(\eta,\kappa)} : 1. \tag{29}$$

Also, it is important to notice that [51]

$$\kappa = \frac{J_{12}+1}{J_{12}+J_{23}}\sqrt{-J_{23}J_{31}}, \qquad \mathrm{sn}^2(\eta,\kappa) = \frac{J_{12}+J_{23}}{J_{12}+1}. \tag{30}$$

The quadratic Casimir elements of the Sklyanin algebra are given by

$$C_0 = (S^0)^2 + \sum_{\alpha=1}^{3}(S^\alpha)^2, \tag{31}$$

$$C_2 = \sum_{\alpha=1}^{3} J_\alpha (S^\alpha)^2. \tag{32}$$

Following [50], it maybe useful to introduce another Casimir element of the Sklyanin algebra

$$C_1 = C_0 - C_2 = (S^0)^2 + \sum_{\alpha=1}^{3}(1 - J_\alpha)\,(S^\alpha)^2. \tag{33}$$

Evidently the two dimensional representation of the Sklyanin algebra is given by the R-matrix (1). Therefore in this case

$$S^0 = \mathbb{1} \quad \text{and} \quad S^\alpha = \sigma^\alpha. \tag{34}$$

In this representation, evidently,

$$C_0 = 4\mathbb{1} \quad \text{and} \quad C_2 = (J_1 + J_2 + J_3)\,\mathbb{1}. \tag{35}$$

The other irreducible representations of the Sklyanin algebra are constructed by the so-called fusion procedure [52, 49, 51, 50]. In particular, the three dimensional representation can be obtained from

$$\begin{aligned} R_{1(23)}(\lambda) &= P_{23}^{+} R_{1,23}(\lambda) P_{23}^{+} \\ &= P_{23}^{+} R_{12}(\lambda+\eta) R_{13}(\lambda-\eta) P_{23}^{+}, \end{aligned} \tag{36}$$

where

$$P_{23}^{+} = \mathbb{1} - P_{23}^{-} = \mathbb{1} - \frac{1}{4}\, R_{23}(-2\eta)\,. \tag{37}$$

It follows that, with a suitable choice of bases, the generators of the Sklyanin algebra are represented by the following set of matrices [50]

$$S^0 = \begin{pmatrix} J_3 & 0 & J_1 - J_2 \\ 0 & J_1 + J_2 - J_3 & 0 \\ J_1 - J_2 & 0 & J_3 \end{pmatrix}, \tag{38}$$

$$S^1 = \sqrt{2J_2J_3}\begin{pmatrix} 0 & 1 & 0 \\ 1 & 0 & 1 \\ 0 & 1 & 0 \end{pmatrix}, \tag{39}$$

$$S^2 = \sqrt{2J_3J_1}\begin{pmatrix} 0 & -\imath & 0 \\ \imath & 0 & -\imath \\ 0 & \imath & 0 \end{pmatrix}, \tag{40}$$

$$S^3 = 2\sqrt{J_1J_2}\begin{pmatrix} 1 & 0 & 0 \\ 0 & 0 & 0 \\ 0 & 0 & -1 \end{pmatrix}. \tag{41}$$

Straightforward substitution of formulae (38) - (41) into (31) and (32) yields

$$C_0 = (J_1 + J_2 + J_3)^2\, \mathbb{1}, \qquad C_2 = (8J_1J_2J_3)\, \mathbb{1}. \tag{42}$$

The commutation relations of the generators of the Sklyanin algebra (17) guarantee the RLL-relations for the XYZ Lax operator (16)

$$R_{12}(\lambda - \mu)\mathbb{L}_{1q}(\lambda)\mathbb{L}_{2q}(\mu) = \mathbb{L}_{2q}(\mu)\mathbb{L}_{1q}(\lambda)R_{12}(\lambda - \mu). \tag{43}$$

The RLL-relations admit a central element

$$\mathbb{D}\left[\mathbb{L}(\lambda)\right] = \mathrm{tr}_{00'} P_{00'}^{-} \mathbb{L}_{0q}(\lambda - \eta)\mathbb{L}_{0'q}(\lambda + \eta), \tag{44}$$

where

$$P_{00'}^{-} = \frac{\mathbb{1} - \mathcal{P}_{00'}}{2} = \frac{1}{4}\, R_{00'}(-2\eta)\,. \tag{45}$$

A straightforward calculation shows that

$$\Big[\mathbb{D}\left[\mathbb{L}(\mu)\right],\, \mathbb{L}(\nu)\Big] = 0\,. \tag{46}$$

Substituting (16) into (44) yields

$$\mathbb{D}\left[\mathbb{L}(\lambda)\right] = (S^0)^2 - \sum_{\alpha=1}^{3} W_\alpha(\lambda - \eta)W_\alpha(\lambda + \eta)(S^\alpha)^2. \tag{47}$$

In particular, in the case of the R-matrix, its quantum determinant is given by

$$\mathbb{D}\left[R(\lambda)\right] = \left(1 - \sum_{\alpha=1}^{3} W_\alpha(\lambda-\eta)W_\alpha(\lambda+\eta)\right)\mathbb{1} = \rho(\lambda+\eta,\eta,\kappa)\mathbb{1}\ . \tag{48}$$

In general, the central $\mathbb{D}\left[\mathbb{L}(\lambda)\right]$ can be expressed in terms of the Casimir elements of the Sklyanin algebra using (28), (19), (21) and (23),

$$\mathbb{D}\left[\mathbb{L}(\lambda)\right] = C_0 - \frac{1 + W_3(\lambda-\eta)W_3(\lambda+\eta)}{J_3}C_2\ . \tag{49}$$

The quantum determinant $\mathbb{D}\left[\mathbb{L}(\lambda)\right]$ can be written in terms of C_0 and C_1 as follows

$$\begin{aligned}\mathbb{D}\left[\mathbb{L}(\lambda)\right] = &\left(1 - \frac{1 + W_3(\lambda-\eta)W_3(\lambda+\eta)}{J_3}\right)C_0 \\ &+ \left(\frac{1 + W_3(\lambda-\eta)W_3(\lambda+\eta)}{J_3}\right)C_1.\end{aligned} \tag{50}$$

The Lax operator (16) satisfies the following identity

$$\mathbb{L}_{0q}(\lambda)\,\sigma_0^2\mathbb{L}_{0q}^{t_0}(\lambda-2\eta)\sigma_0^2 = \mathbb{1}\otimes\mathbb{D}\left[\mathbb{L}(\lambda-\eta)\right], \tag{51}$$

were t_0 denotes the transpositions in the auxiliary space $\mathbb{C}^2$.

The so-called monodromy matrix

$$T(\lambda) = \mathbb{L}_{0N}(\lambda-\alpha_N)\cdots\mathbb{L}_{01}(\lambda-\alpha_1) \tag{52}$$

is used to describe the system. Notice that $T(\lambda)$ is a two-by-two matrix in the auxiliary space $V_0 = \mathbb{C}^2$, whose entries are operators acting in $\mathcal{H}$

$$T(\lambda) = \begin{pmatrix} A(\lambda) & B(\lambda) \\ C(\lambda) & D(\lambda) \end{pmatrix}. \tag{53}$$

From RLL-relations it follows that the monodromy matrix satisfies the RTT-relations

$$R_{00'}(\lambda-\mu)T_0(\lambda)T_{0'}(\mu) = T_{0'}(\mu)T_0(\lambda)R_{00'}(\lambda-\mu). \tag{54}$$

The RTT-relations define the commutation relations for the entries of the monodromy matrix.

In the periodic case the modified Algebraic Bethe Ansatz [48, 56] yields the spectrum of the spin-s XYZ Heisenberg Hamiltonian

$$H = -\frac{1}{2}\sum_{m=1}^{N}\left(J_1 S_m^1 S_{m+1}^1 + J_2 S_m^2 S_{m+1}^2 + J_3 S_m^3 S_{m+1}^3\right). \tag{55}$$

A way to introduce non-periodic boundary conditions which are compatible with the integrability of the bulk system, was developed in [27].

The compatibility condition between the bulk and the boundary of the system takes the form of the so-called Reflection Equation. It is written in the following form for the left reflection matrix acting on the space $V_1 = \mathbb{C}^2$ at the first site, $K^-(u) \in \mathrm{End}(\mathbb{C}^2)$

$$R_{12}(u-v)K_1^-(u)R_{21}(u+v)K_2^-(v) = K_2^-(v)R_{12}(u+v)K_1^-(u)R_{21}(u-v). \tag{56}$$

The general solution of the reflection equation above, for the R-matrix (16), can be written as follows [53, 54, 55]

$$K^-(u) = \begin{pmatrix} \mathrm{sn}(u+a) - d\,\mathrm{sn}(u-a) & b\,\mathrm{sn}(2u)\dfrac{c(1-\tau\,\mathrm{sn}^2(u)) + 1 + \tau\,\mathrm{sn}^2(u)}{1-\tau^2\,\mathrm{sn}^2(u)\,\mathrm{sn}^2(a)} \\ b\,\mathrm{sn}(2u)\dfrac{c(1-\tau\,\mathrm{sn}^2(u)) - 1 - \tau\,\mathrm{sn}^2(u)}{1-\tau^2\,\mathrm{sn}^2(u)\,\mathrm{sn}^2(a)} & -\mathrm{sn}(u-a) + d\,\mathrm{sn}(u+a) \end{pmatrix}, \tag{57}$$

here a, b, c, d are arbitrary constants.

Due to the properties of the Yang R-matrix the dual reflection equation can be presented in the following form

$$R_{12}(v-u)K_1^+(u)R_{21}(-u-v-2\omega)K_2^+(v) = K_2^+(v)R_{12}(-u-v-2\omega)K_1^+(u)R_{21}(v-u).$$

One can then verify that the mapping

$$K^+(u) = K^-(-u-\omega)$$

is a bijection between solutions of the reflection equation and the dual reflection equation. After substitution of into the dual reflection equation one gets the reflection equation with shifted arguments.

We use Sklyanin approach to integrable spin chains with non-periodic boundary conditions [27]. The Sklyanin monodromy matrix $\mathcal{T}(\lambda)$ is

$$\mathcal{T}_0(\lambda) = T_0(\lambda)K_0^-(\lambda)\widetilde{T}_0(\lambda). \tag{58}$$

The monodromy matrix $\widetilde{T}_0(\lambda)$ is such that its RTT-relations (54) can be recast as follows

$$\widetilde{T}_{0'}(\mu)R_{00'}(\lambda+\mu)T_0(\lambda) = T_0(\lambda)R_{00'}(\lambda+\mu)\widetilde{T}_{0'}(\mu), \tag{59}$$

$$\widetilde{T}_0(\lambda)\widetilde{T}_{0'}(\mu)R_{00'}(\mu-\lambda) = R_{00'}(\mu-\lambda)\widetilde{T}_{0'}(\mu)\widetilde{T}_0(\lambda). \tag{60}$$

Then, by construction, the exchange relations of the monodromy matrix $\mathcal{T}(\lambda)$ are

$$R_{00'}(\lambda-\mu)\mathcal{T}_0(\lambda)R_{0'0}(\lambda+\mu)\mathcal{T}_{0'}(\mu) = \mathcal{T}_{0'}(\mu)R_{00'}(\lambda+\mu)\mathcal{T}_0(\lambda)R_{0'0}(\lambda-\mu). \tag{61}$$

The Reflection Equation Algebra admits a central element, the so-called Sklyanin determinant [27],

$$\Delta\left[\mathcal{T}(\lambda)\right] = \mathrm{tr}_{00'} P^{-}_{00'} \mathcal{T}_0(\lambda - \eta/2) R_{00'}(2\lambda) \mathcal{T}_{0'}(\lambda + \eta/2). \tag{62}$$

The open chain transfer matrix is given by the trace of the monodromy $\mathcal{T}(\lambda)$ over the auxiliary space V_0 with an extra reflection matrix $K^+(\lambda)$,

$$t(\lambda) = \mathrm{tr}_0 \left(K^+(\lambda)\mathcal{T}(\lambda)\right). \tag{63}$$

The reflection matrix $K^+(\lambda)$ is the corresponding solution of the dual reflection equation.

The commutativity of the transfer matrix for different values of the spectral parameter

$$[t(\lambda), t(\mu)] = 0, \tag{64}$$

is guaranteed by the dual reflection equation and the exchange relations of the monodromy matrix $\mathcal{T}(\lambda)$.

In the spin-$\frac{1}{2}$ case, in the homogeneous limit, this transfer matrix yields the Hamiltonian with the boundary terms of the form:

$$H = \sum_{m=1}^{N-1} H_{m,m+1} + \left(A_- \sigma_1^z + B_- \sigma_1^+ + C_- \sigma_1^-\right) + \left(A_+ \sigma_N^z + B_+ \sigma_N^+ + C_+ \sigma_N^-\right), \tag{65}$$

where the first term corresponds to the expression 55 (here without the periodic boundary condition). The separation of variables method [57, 58] was used to obtain the spectrum of the above transfer matrix by S. Faldella and G. Niccoli [59].

3. Generalized XYZ Gaudin model

The key observation is that the R-matrix (1) admits the following expansion [16]

$$R(\lambda, \eta) = \mathbb{1} + 2\eta r(\lambda) + \mathcal{O}(\eta^2), \tag{66}$$

where the classical r-matrix is given by

$$r(\lambda) = \sum_{\alpha=1}^{3} w_\alpha(\lambda)\, \sigma^\alpha \otimes \sigma^\alpha, \tag{67}$$

where

$$w_1(\lambda) = \frac{\mathrm{cn}(\lambda, \kappa)}{\mathrm{sn}(\lambda, \kappa)}, \quad w_2(\lambda) = \frac{\mathrm{dn}(\lambda, \kappa)}{\mathrm{sn}(\lambda, \kappa)}, \quad w_3(\lambda) = \frac{1}{\mathrm{sn}(\lambda, \kappa)}, \tag{68}$$

the functions $\mathrm{sn}(\lambda,\kappa)$, $\mathrm{cn}(\lambda,\kappa)$, and $\mathrm{dn}(\lambda,\kappa)$ are the usual Jacobi elliptic functions and σ^α, $\alpha = 1,2,3$, are the Pauli matrices (3). Evidently, this classical r-matrix has the parity invariance property

$$r_{21}(\lambda) = r_{12}(\lambda), \tag{69}$$

and, due to the fact that $w_\alpha(\lambda)$ (68) are odd function of λ cf. (91), it also has the unitarity property

$$r_{21}(-\lambda) = -r_{12}(\lambda). \tag{70}$$

Notice that in the case of the r-matrix (67) the classical Yang-Baxter equation

$$[r_{13}(\lambda), r_{23}(\mu)] + [r_{12}(\lambda-\mu), r_{13}(\lambda) + r_{23}(\mu)] = 0, \tag{71}$$

reduces to the following three identities

$$\begin{aligned} w_1(\lambda)\, w_2(\mu) &= -w_2(\lambda-\mu)\, w_3(\lambda) + w_1(\lambda-\mu)\, w_3(\mu), \\ w_3(\lambda)\, w_1(\mu) &= -w_1(\lambda-\mu)\, w_2(\lambda) + w_3(\lambda-\mu)\, w_2(\mu), \\ w_2(\lambda)\, w_3(\mu) &= -w_3(\lambda-\mu)\, w_1(\lambda) + w_2(\lambda-\mu)\, w_1(\mu), \end{aligned} \tag{72}$$

which are consequences of the definition (68) of the functions $w_\alpha(\lambda)$, $\alpha = 1,2,3$, the identities (92) and the addition theorems of the Jacobi elliptic functions (93) - (95).

The Lax operator of the chain (16) admits the following expansion

$$\mathbb{L}_{0q}(\lambda,\eta) = \mathbb{1} + 2\eta\, \ell_{0q}(\lambda) + \mathcal{O}(\eta^2), \tag{73}$$

where

$$\ell_{0q}(\lambda) = \sum_{\alpha=1}^{3} w_\alpha(\lambda)\, \sigma_0^\alpha \otimes S^\alpha. \tag{74}$$

Therefore the expansion of the monodromy matrix (52) reads

$$T_0(\lambda,\eta) = \mathbb{1} + 2\eta\, L_0(\lambda) + \eta^2\, T_0^{(2)}(\lambda) + \mathcal{O}(\eta^3), \tag{75}$$

where the Gaudin Lax operator is given by

$$L_0(\lambda) = \sum_{m=1}^{N} \ell_{0n}(\lambda - \alpha_m). \tag{76}$$

The RTT-relations (54) imply the so-called Sklyanin linear bracket for the Gaudin Lax operator

$$[L_0(\lambda), L_{0'}(\mu)] = [r_{00'}(\lambda-\mu), L_0(\lambda) + L_{0'}(\mu)], \tag{77}$$

with the above classical r-matrix (67).

It can be shown that the transfer matrix of the chain and the quantum determinant of the monodromy matrix admit the following expansions

$$\begin{aligned} t(\lambda,\eta) &= 1 + \eta^2 \ \mathrm{tr}_0 T_0^{(2)}(\lambda) + \mathcal{O}(\eta^3), \\ \mathbb{D}\left[T_0(\lambda,\eta)\right] &= 1 + \eta^2 \left(\mathrm{tr}_0 T_0^{(2)}(\lambda) + 4\ \mathrm{tr}_0 L_0^2(\lambda)\right) + \mathcal{O}(\eta^3). \end{aligned} \tag{78}$$

Thus the generating function $\tau(\lambda)$ of the Gaudin Hamiltonians in the elliptic case can be obtain as a difference

$$\mathbb{D}\left[T_0(\lambda,\eta)\right] - t(\lambda,\eta) = 4\eta\ \mathrm{tr}_0 L_0^2(\lambda) + \mathcal{O}(\eta^3), \tag{79}$$

with, as expected,

$$\tau(\lambda) = \mathrm{tr}_0 L_0^2(\lambda). \tag{80}$$

Evidently, $\tau(\lambda)$ commute for different values of the spectral parameter,

$$[\tau(\lambda), \tau(\mu)] = 0. \tag{81}$$

As in the rational and the trigonometric case, the expansion into partial fractions yields the corresponding Gaudin Hamiltonians

$$\tau(\lambda) = \sum_{m=1}^{N} \wp(\lambda - \alpha_m)\ s_m(s_m+1) + \sum_{m=1}^{N} \zeta(\lambda - \alpha_m)\ H_m + H_0, \tag{82}$$

where $\wp$ is the Weierstrass P function, ζ is the zeta function, H_0 is λ independent and

$$H_m = 2 \sum_{m \neq n} \sum_{\alpha=1}^{3} w_\alpha(\alpha_n - \alpha_m)\ S_n^\alpha S_m^\beta. \tag{83}$$

Sklyanin and Takebe [16, 17] obtained the spectrum of the generating function both by the modified Algebraic Bethe Ansatz and by the separation of variables method.

In order to define the Gaudin model with boundary terms we consider the following non-unitary classical r-matrix [44, 47]

$$r_{00'}^K(\lambda,\mu) = r_{00'}(\lambda - \mu) - K_{0'}(\nu) r_{00'}(\lambda+\mu) K_{0'}^{-1}(\mu), \tag{84}$$

where

$$K_0(\lambda) \equiv K_0^-(\lambda). \tag{85}$$

It is straightforward to check that this r-matrix satisfies the classical Yang-Baxter equation [44, 47]

$$[r_{32}^K(\lambda_3,\lambda_2), r_{13}^K(\lambda_1,\lambda_3)] + [r_{12}^K(\lambda_1,\lambda_2), r_{13}^K(\lambda_1,\lambda_3) + r_{23}^K(\lambda_2,\lambda_3)] = 0. \tag{86}$$

The corresponding Lax operator is given by

$$\mathcal{L}_0(\lambda) = \sum_{m=1}^{N} \left(\ell_{0m}(\lambda - \alpha_m) + K_0(\lambda)\ell_{0m}(\lambda + \alpha_m)K_0^{-1}(\lambda)\right). \tag{87}$$

Evidently, it satisfies the following linear bracket relations

$$[\mathcal{L}_0(\lambda), \mathcal{L}_{0'}(\mu)] = \left[r^K_{00'}(\lambda,\mu), \mathcal{L}_0(\lambda)\right] - \left[r^K_{0'0}(\mu,\lambda), \mathcal{L}_{0'}(\mu)\right]. \tag{88}$$

By definition this linear bracket is obviously anti-symmetric. It obeys the Jacobi identity since the r-matrix $r^K_{00'}(\lambda,\mu)$ satisfies the classical Yang-Baxter equation.

The generating function $\tau(\lambda)$ of the Gaudin Hamiltonians with boundary terms is given by

$$\tau(\lambda) = \mathrm{tr}_0\, \mathcal{L}_0^2(\lambda). \tag{89}$$

The generating function for different values of the spectral parameter obviously commute,

$$[\tau(\lambda), \tau(\mu)] = 0. \tag{90}$$

Thus, defining the family of the Gaudin Hamiltonians in involution.

In this way, we have achieved the main objective of this work. Our next step will be to solve explicitly the model using either the modified algebraic Bethe ansatz [48], which was successfully used in the periodic case [16], or the separation of variables method [57, 58], used to solve the periodic elliptic Gaudin model [17] as well as the non-periodic elliptic chain [59].

A Jacobi elliptic functions

In this appendix we briefly review some basic identities which follow from the definition of the Jacobi elliptic functions. In particular,

$$\mathrm{sn}(-\lambda,\kappa) = -\mathrm{sn}(\lambda,\kappa), \quad \mathrm{cn}(-\lambda,\kappa) = \mathrm{cn}(\lambda,\kappa), \quad \mathrm{dn}(-\lambda,\kappa) = \mathrm{dn}(\lambda,\kappa), \tag{91}$$

$$\mathrm{sn}^2(\lambda,\kappa) + \mathrm{cn}^2(\lambda,\kappa) = 1, \quad \mathrm{dn}^2(\lambda,\kappa) + \kappa^2\mathrm{sn}^2(\lambda,\kappa) = 1. \tag{92}$$

The addition theorems for the Jacobi elliptic functions read

$$\mathrm{sn}(\lambda+\mu,\kappa) = \frac{\mathrm{sn}(\lambda,\kappa)\mathrm{cn}(\mu,\kappa)\mathrm{dn}(\mu,\kappa) + \mathrm{cn}(\lambda,\kappa)\mathrm{dn}(\lambda,\kappa)\mathrm{sn}(\mu,\kappa)}{1-\kappa^2\mathrm{sn}^2(\lambda,\kappa)\mathrm{sn}^2(\mu,\kappa)}, \tag{93}$$

$$\mathrm{cn}(\lambda+\mu,\kappa) = \frac{\mathrm{cn}(\lambda,\kappa)\mathrm{cn}(\mu,\kappa) - \mathrm{sn}(\lambda,\kappa)\mathrm{dn}(\lambda,\kappa)\mathrm{sn}(\mu,\kappa)\mathrm{dn}(\mu,\kappa)}{1-\kappa^2\mathrm{sn}^2(\lambda,\kappa)\mathrm{sn}^2(\mu,\kappa)}, \tag{94}$$

$$\mathrm{dn}(\lambda+\mu,\kappa) = \frac{\mathrm{dn}(\lambda,\kappa)\mathrm{dn}(\mu,\kappa) - \kappa^2\mathrm{sn}(\lambda,\kappa)\mathrm{cn}(\lambda,\kappa)\mathrm{sn}(\mu,\kappa)\mathrm{cn}(\mu,\kappa)}{1-\kappa^2\mathrm{sn}^2(\lambda,\kappa)\mathrm{sn}^2(\mu,\kappa)}. \tag{95}$$

Some other important identities follow from the formulae above

$$\mathrm{sn}(\lambda+\mu,\kappa)\,\mathrm{sn}(\lambda-\mu,\kappa)=\frac{\mathrm{sn}^2(\lambda,\kappa)-\mathrm{sn}^2(\mu,\kappa)}{1-\kappa^2\mathrm{sn}^2(\lambda,\kappa)\mathrm{sn}^2(\mu,\kappa)}, \tag{96}$$

$$\mathrm{cn}(\lambda+\mu,\kappa)\,\mathrm{cn}(\lambda-\mu,\kappa)=\frac{\mathrm{cn}^2(\mu,\kappa)-\mathrm{sn}^2(\lambda,\kappa)\,\mathrm{dn}^2(\mu,\kappa)}{1-\kappa^2\mathrm{sn}^2(\lambda,\kappa)\mathrm{sn}^2(\mu,\kappa)}, \tag{97}$$

$$\mathrm{dn}(\lambda+\mu,\kappa)\,\mathrm{dn}(\lambda-\mu,\kappa)=\frac{\mathrm{dn}^2(\mu,\kappa)-\kappa^2\mathrm{sn}^2(\lambda,\kappa)\,\mathrm{cn}^2(\mu,\kappa)}{1-\kappa^2\mathrm{sn}^2(\lambda,\kappa)\mathrm{sn}^2(\mu,\kappa)}. \tag{98}$$

References

[1] M. Gaudin, *Diagonalisation d'une classe d'hamiltoniens de spin*, J. Physique 37 (1976) 1087–1098.

[2] M. Gaudin, *La fonction d'onde de Bethe*, chapter 13 Masson, Paris, 1983.

[3] M. Gaudin, *The Bethe Wavefunction*, Cambridge University Press, 2014.

[4] E. K. Sklyanin, *Separation of variables in the Gaudin model*, Zap. Nauchn. Sem. Leningrad. Otdel. Mat. Inst. Steklov. (LOMI) 164 (1987) 151–169; translation in J. Soviet Math. 47 (1989) 2473–2488.

[5] A. A. Belavin and V. G. Drinfeld. *Solutions of the classical Yang-Baxter equation for simple Lie algebras* (in Russian), Funktsional. Anal. i Prilozhen. 16 (1982), no. 3, 1–29; translation in Funct. Anal. Appl. 16 (1982) no. 3, 159-180.

[6] M. A. Semenov-Tian-Shansky, *Quantum and classical integrable systems*, in Integrability of Nonlinear Systems, Lecture Notes in Physics Volume 495 (1997) 314-377.

[7] B. Jurčo, *Classical Yang-Baxter equations and quantum integrable systems*, J. Math. Phys. Volume 30 (1989) 1289–1293.

[8] B. Jurčo, *Classical Yang-Baxter equations and quantum integrable systems (Gaudin models)*, in Quantum groups (Clausthal, 1989), Lecture Notes in Phys. Volume 370 (1990) 219–227.

[9] B. Feigin, E. Frenkel, and N. Reshetikhin, *Gaudin model, Bethe ansatz and correlation functions at the critical level*, Commun. Math. Phys. 166 (1994) 27–62.

[10] N. Reshetikhin and A. Varchenko, *Quasiclassical asymptotics of solutions to the KZ equations*, in Geometry, Topology and Physics, Conf. Proc. Lecture Notes Geom., pages 293–273. Internat. Press, Cambridge, MA, 1995.

[11] F. Wagner and A. J. Macfarlane, *Solvable Gaudin models for higher rank symplectic algebras. Quantum groups and integrable systems (Prague, 2000)* Czechoslovak J. Phys. 50 (2000) 1371–1377.

[12] T. Brzezinski and A. J. Macfarlane, *On integrable models related to the osp(1,2) Gaudin algebra*, J. Math. Phys. 35 (1994), no. 7, 3261–3272.

[13] P. P. Kulish and N. Manojlović, *Bethe vectors of the* $osp(1|2)$ *Gaudin model*, Letters in Mathematical Physics **Vol. 55** (2001) 77-95.

[14] P. P. Kulish and N. Manojlović, *Creation operators and Bethe vectors of the* $osp(1|2)$ *Gaudin model*, J. Math. Phys. 42 no. 10 (2001) 4757–4778.

[15] P. P. Kulish and N. Manojlović, *Trigonometric* $osp(1|2)$ *Gaudin model*, J. Math.Phys. 44 no. 2 (2003) 676–700.

[16] E. K. Sklyanin and T. Takebe, *Algebraic Bethe ansatz for the XYZ Gaudin model*, Physics Letters A 219 (1996) 217–225.

[17] E. K. Sklyanin and T. Takebe, *Separation of Variables in the Elliptic Gaudin Model*, Commun. Math. Phys. 204 (1999) 17– 38.

[18] K. Hikami, P. P. Kulish and M. Wadati, *Integrable Spin Systems with Long-Range Interaction*, Chaos, Solitons & Fractals Vol. 2 No. 5 (1992) 543–550.

[19] K. Hikami, P. P. Kulish and M. Wadati, *Construction of Integrable Spin Systems with Long-Range Interaction*, J. Phys. Soc. Japan Vol. 61 No. 9 (1992) 3071–3076.

[20] K. Hikami, *Gaudin magnet with boundary and generalized Knizhnik-Zamolodchikov equation*, J. Phys. A Math. Gen. **28** (1995) 4997–5007.

[21] W. L. Yang, R. Sasaki and Y. Z. Zhang, *$\mathbb{Z}_n$ elliptic Gaudin model with open boundaries*, JHEP 09 (2004) 046.

[22] W. L. Yang, R. Sasaki and Y. Z. Zhang, *A_{n-1} Gaudin model with open boundaries*, Nuclear Physics B 729 (2005) 594–610.

[23] K. Hao, W.-L. Yang, H. Fan, S. Y. Liu, K. Wu, Z. Y. Yang and Y. Z. Zhang, *Determinant representations for scalar products of the XXZ Gaudin model with general boundary terms*, Nuclear Physics **B 862** (2012) 835–849.

[24] A. Lima-Santos, *The $sl(2|1)^{(2)}$ Gaudin magnet with diagonal boundary terms*, J. Stat. Mech. (2009) P07025.

[25] E. K. Sklyanin, *Boundary conditions for integrable equations*, (Russian) Funktsional. Anal. i Prilozhen. 21 (1987) 86–87; translation in Functional Analysis and Its Applications Volume 21, Issue 2 (1987) 164–166.

[26] E. K. Sklyanin, *Boundary conditions for integrable systems*, in the Proceedings of the VIIIth international congress on mathematical physics (Marseille, 1986), World Sci. Publishing, Singapore, (1987) 402–408.

[27] E. K. Sklyanin, *Boundary conditions for integrable quantum systems*, J. Phys. A: Math. Gen. **21** (1988) 2375–2389.

[28] T. Skrypnyk, *Non-skew-symmetric classical r-matrix, algebraic Bethe ansatz, and Bardeen-Cooper-Schrieffer-type integrable systems*, J. Math. Phys. 50 (2009) 033540, 28 pages.

[29] T. Skrypnyk, *"Z2-graded" Gaudin models and analytical Bethe ansatz*, Nuclear Physics B 870 (2013), no. 3, 495–529.

[30] N. Cirilo António, N. Manojlović and Z. Nagy, *Trigonometric $s\ell(2)$ Gaudin model with boundary terms*, Reviews in Mathematical Physics **Vol. 25** No. 10 (2013) 1343004 (14 pages).

[31] J. Cao, H. Lin, K. Shi and Y. Wang, *Exact solutions and elementary excitations in the XXZ spin chain with unparallel boundary fields*, Nucl. Phys. B 663 (2003) 487–519.

[32] R. I. Nepomechie, *Bethe ansatz solution of the open XXZ chain with nondiagonal boundary terms*, J. Phys. A 37 (2004), no. 2, 433–440.

[33] D. Arnaudon, A. Doikou, L. Frappat, E. Ragoucy and N. Crampé, *Analytical Bethe ansatz in $gl(N)$ spin chains*, Czechoslovak J. Phys. 56 (2006), no. 2, 141–148.

[34] C. S. Melo, G. A. P. Ribeiro and M. J. Martins, *Bethe ansatz for the $XXX-S$ chain with non-diagonal open boundaries*, Nuclear Phys. B 711, no. 3 (2005) 565–603.

[35] J. Cao, W.-L. Yang, K. Shi and Y. Wang, *Off-diagonal Bethe ansatz solution of the XXX spin chain with arbitrary boundary conditions*, Nuclear Physics B 875 (2013) 152–165.

[36] J. Cao, W.-L. Yang, K. Shi and Y. Wang, *Off-diagonal Bethe ansatz solutions of the anisotropic spin-1/2 chains with arbitrary boundary fields*, Nuclear Physics B 877 (2013) 152–175.

[37] E. Ragoucy, *Coordinate Bethe ansätze for non-diagonal boundaries*, Rev. Math. Phys. 25 (2013), no. 10, 1343007.

[38] S. Belliard, N. Crampé and E. Ragoucy, *Algebraic Bethe ansatz for open XXX model with triangular boundary matrices*, Lett. Math. Phys. 103 No. 5 (2013) 493–506.

[39] S. Belliard and N. Crampé *Heisenberg XXX model with general boundaries: eigenvectors from algebraic Bethe ansatz*, SIGMA Symmetry Integrability Geom. Methods Appl. 9 (2013), Paper 072, 12 pp.

[40] R. A. Pimenta and A. Lima-Santos, *Algebraic Bethe ansatz for the six vertex model with upper triangular K-matrices*, J. Phys. A 46 No. 45 (2013) 455002, 13 pp.

[41] N. Cirilo António, N. Manojlović and I. Salom, *Algebraic Bethe ansatz for the XXX chain with triangular boundaries and Gaudin model*, Nuclear Physics B 889 (2014) 87-108.

[42] K. Hao, J. Cao, T. Yang and W.-L. Yang, *Exact solution of the XXX Gaudin model with the generic open boundaries*, `arXiv:1408.3012`.

[43] N. Manojlović and I. Salom, *Algebraic Bethe ansatz for the XXZ Heisenberg spin chain with triangular boundaries and the corresponding Gaudin model*, Nuclear Physics **B 923** (2017) 73-106; `arXiv:1705.02235`.

[44] N. Cirilo António, N. Manojlović, E. Ragoucy and I. Salom, *Algebraic Bethe ansatz for the $s\ell(2)$ Gaudin model with boundary*, Nuclear Physics **B 893** (2015) 305-331; `arXiv:1412.1396`.

[45] I. Salom and N. Manojlović, *Creation operators of the non-periodic $sl(2)$ Gaudin model*, Proceedings of the 8th Mathematical Physics meeting: Summer School and Conference on Modern Mathematical Physics, 24 - 31 August 2014, Belgrade, Serbia, SFIN **XXVIII** Series A: Conferences No. A1, ISBN: 978-86-82441-43-4, (2015) 149–155.

[46] I. Salom, N. Manojlović and N. Cirilo António, *Generalized $s\ell(2)$ Gaudin algebra and corresponding Knizhnik-Zamolodchikov equation*, Nuclear Physics **B 939** (2019) 358-371.

[47] N. Manojlović and I. Salom, *Algebraic Bethe ansatz for the trigonometric $s\ell(2)$ Gaudin model with triangular boundary*, Symmetry 12 (2020) 352; `arXiv:1709.06419`.

[48] L. A. Takhtajan and L. D. Faddeev, *The quantum method for the inverse problem and the XYZ Heisenberg model*, (in Russian) Uspekhi Mat. Nauk **34** No. 5 (1979) 13–63; translation in Russian Math. Surveys **34** No.5 (1979) 11–68.

[49] P. P. Kulish and E. K. Sklyanin, *Quantum spectral transform method. Recent developments*, Lect. Notes Phys. **151** (1982), 61–119.

[50] E. K. Sklyanin, *Some algebraic structures connected with the Yang-Baxter equation*, Funkts. Analiz i ego Prilozh. **16** No. 4 (1982) 27–34; translation in Functional Analysis and Applications **17** (1983) 263–270.

[51] E. K. Sklyanin, *Some algebraic structures connected with the Yang-Baxter equation. Representations of quantum algebras*, Funkts. Analiz i ego Prilozh. **17** No. 4 (1983) 34–48; translation in Functional Analysis and Applications **17** (1984) 273–284.

[52] P. P. Kulish, N. Yu. Reshetikhin and E. K. Sklyanin, *Yang-Baxter equation and representation theory I*, Letters in Mathematical Physics **5** (1981) 393–403.

[53] H. J. de Vega and A. Gonzalez-Ruiz, *Boundary K matrices for the XYZ, XXZ and XXX spin chains*, J. Phys. A **27** (1994) 6129–6137.

[54] T. Inami and H. Konno, *Integrable XYZ spin chain with boundaries*, J. Phys. A **27** (1994) L913-L918.

[55] Y. Komori and K. Hikami, *Elliptic K-matrix associated with Belavin's symmetric R-matrix*, Nucl. Phys. B **494** (1997) 687–701.

[56] T. Takebe, *Generalized Bethe ansatz with the general spin representations of the Sklyanin algebra*, J. Phys. A: Math. Gen. 25 (1992) 1071-1083.

[57] E. K. Sklyanin, *Separation of variables - new trends*, Prog. Theor. Phys. Suppl. **118** (1995) 35–60.

[58] J. M. Maillet and G. Niccoli, *On Separation of Variables for Reflection Algebras*, J. Stat. Mech. **1909**, no. 9, (2019) 094020.

[59] S. Faldella and G. Niccoli, *SOV approach for integrable quantum models associated with general representations on spin-1/2 chains of the 8-vertex reflection algebra*, J. Phys. A **47** (2014) 115202.

Numerical study of classical motions of two equal-mass opposite-charge ions in a Paul trap*

Srdjan Marjanović†
Institute of Physics Belgrade, University of Belgrade,
Pregrevica 118, 11080 Beograd, Serbia
† present address:
Microsoft Development Center Serbia,
Španskih boraca 3, Beograd 11070, Serbia

V. Dmitrašinović‡
Institute of Physics Belgrade, University of Belgrade,
Pregrevica 118, Zemun, P.O.Box 57, 11080 Beograd, Serbia

April 22, 2020

Abstract

We conduct a numerical test of nonlinear stability, following Danby's definition and Nauenberg's implementation, of two oppositely charged, equal-mass ions in a Paul trap operated at a single frequency, as a function of initial conditions. We find a small, but non-negligible domain of stability in the space of trap parameters, which opens the door to experimental searches.

Introduction

Numerous new few-body systems bound by the Coulomb potential *in vacuo* have recently been discovered in numerical searches, see Refs. [1, 2] and references therein. No such system can be detected in a terrestrial experiment, due to the presence of Earth's gravitational field, so their detection requires a confinement device, i.e., an ion trap, such as the Paul one, which, in turn, significantly impacts on the dynamics of the few-body system.

Paul traps have long been known to confine charged particles, ranging from atomic ions to macroscopic charged objects. The case of two, or more, oppositely charged, yet equal-mass ions has only recently been studied, however: Refs. [3, 4, 5] showed that it is possible to simultaneously confine

* This work was supported by the Serbian Ministry of Science and Technological Development under grant numbers OI 171037 and III 41011.

† e-mail address: gmsrki@gmail.com

‡ e-mail address: dmitrasin@ipb.ac.rs

opposite charge ions with markedly different charge-to-mass ratios in a Paul trap operated at two frequencies. It is the equal-mass limit of the two-body opposite-charge case in a single-frequency Paul trap that interests us here.

The question is if and when two equal-mass opposite-charge macroscopic particles' classical periodic orbit is stable in a Paul trap? It is not difficult to show analytically in the adiabatic limit that an electrically charged bound state of two opposite-charge, equal mass particles remains confined, though its center-of-mass may move around the trap, in a Paul trap operated at a single frequency. The case of electrically neutral bound states must be studied numerically, however, as they may, but need not become deconfined by the influence of the trap's field, depending on the precise values of the trap parameters and initial conditions of the pair.

In this note we present the results of a numerical investigation of the stability domain of two equal-mass, and equal, but opposite-sign charged macroscopic bodies periodic orbit in a Paul trap operated at a single frequency. We establish a nonlinear stability domain in the space of control parameters using the Danby-Nauenberg method [6, 7, 8].

Preliminaries

The Keplerian elliptic periodic orbit is *marginally* linearly stable *in vacuo*, but not so in a Paul trap: the trap's time-variable electric field makes this system non-Hamiltonian. Therefore, only a numerical analysis can determine its stability. The nonlinear stability domain of solutions to the two-body problem was obtained by a numerical integration of the equations of motion, and then measuring whether, for a small initial deviation from the initial conditions, the orbit remains confined after some, sufficiently large number n of periods. We found that $n \simeq 250 - 500$ was sufficient for a test of nonlinear stability, in agreement with similar conclusions by Nauenberg [8] in the case of Newtonian three-body systems [1].

The primary goal of this numerical simulation is to establish if and when the motion of the pair's center of mass (CM) is confined. Zero total charge of a pair suggests decoupling from the electric field and thus deconfinement, so we investigate that case as the critical one.

The simulations were implemented using the (standard) fifth-order Runge-Kutta algorithm. A Coulomb pair moving on a Keplerian elliptic orbit *in vacuo* has an intrinsic frequency f (inversely proportional to the period T) associated with its size, or energy. The applied electric field in the Paul trap, on the other hand has its own (independent) frequency f_0. The ratio of these two frequencies (both of which are scale-dependent) represents one

[1] Nauenberg [8] showed that "this nonlinear stability domain looks surprisingly similar to the stability domain of the Mathieu equation with an additional nonlinear restoring or damping term" and that "a linear curve fits the lower part of the domain with a slope that is ... in good agreement with the numerical linear stability results of Danby [6] and Bennett [7]".

dimension in the space of scale-invariant parameters. The scale independent properties, such as the stability of the motion, can only depend on this ratio.

We have investigated the parameter space of initial conditions where the frequency f of the free-space Keplerian orbit of the charged pair is comparable to the frequency f_0 of the applied field, i.e., when the system is close to resonance. Thus we study the most interesting case of strong coupling between Coulomb interaction and the "driving force" of the Paul trap field.

Results

We have kept the mass of individual particles fixed at 1 AMU and the charge of the particles at $\pm 1e$. Both particles were initially placed on the x-axis, with equal velocities parallel to the y-axis, but in opposite directions (positive particle has $v_{y0} > 0$), thus ensuring that the particles were confined to $x - y$ plane. The x coordinates and velocities of particles in any given simulation were chosen to correspond with the orbit (defined by the Coulomb interaction, without any external fields) that has a predefined nominal rotation frequency f.

The longest-lived orbits were observed starting from the center of the trap. We scanned a region of initial conditions with nominal rotation frequencies f ranging from $0.6 f_0$ to $2.2 f_0$, with $f_0 = 2$ MHz being the frequency of the applied field. By fixing this specific value of the driving force frequency, we have also set the length scale (in millimeters).

The second parameter over which the initial conditions were scanned is the displacement of the particle pair from the center of the trap. The displacement is equal to the starting x-coordinate of the particle center of mass. The total length of time in each simulation is $125 \mu s$, which represents 250 periods of the applied field. Then, we displaced the particles from the center of the trap and measured how much longer the bound motion of the charged pair was observed. We have calculated, as a measure of stability, the number of cycles (rotations in the center of mass frame) completed during one simulation run, see Fig. 1. We performed 2400 simulations in total.

It should be immediately apparent in Fig. 1 that for displacement values larger than $1.2 mm$, the trapping field tears the pair apart and the particles continue on as individually trapped charges. For lower values of displacement, however, we observe a non-trivial structure with three distinct regions.

1) The first, the "least stable" (dark green to blue) region with 150-200 completed rotations centered on $f/f_0 = 1$ extends the furthest, out to $1.2 mm$ of displacement. Note that, although this is not directly visible in this graph, we have checked that all of the trajectories in this region ended breaking up before the simulation completed (after roughly $80 \mu s$).

2) In the second (light green) region, up to $0.5 mm$ of displacement and around $f/f_0 = 1$, the particles actually remained bound by the Coulomb

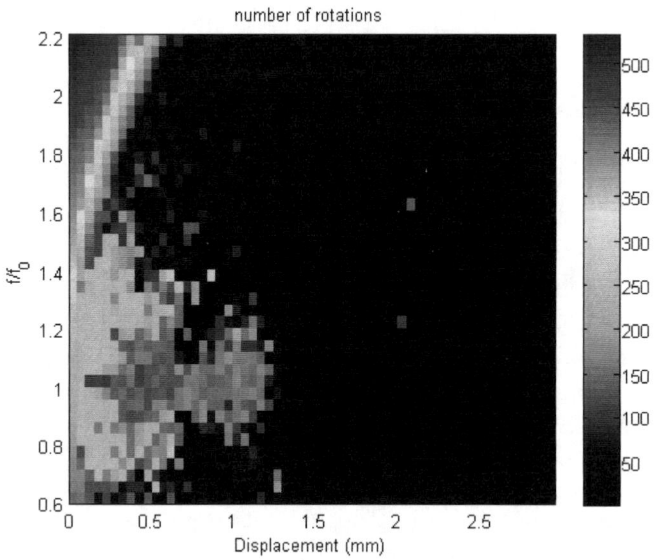

Figure 1: (color on-line) The number of completed periods/rotations (see color code on the right-hand side) in the CM reference frame as a function of the initial displacement from the center of the trap (in mm, on the x-axis) and of the frequency ratio f/f_0 (on the y axis). The conditions in the upper left-hand corner of the graph (dark red) allowed for two times higher rotation count when driven by two times higher frequency, with the region of acceptable displacement growing with the frequency of the applied field.

interaction for the duration of the simulation completing their nominal 250 rotations.

3) Finally, the third (yellow to red) region is that of higher frequency ratio $f/f_0 \geq 1.5$ (the upper left-hand-side corner) that has a significantly higher count of completed orbits n. Given that the nominal frequency f is higher than f_0 in this region, an increase of the number of completed orbits would be expected on this account alone, but, more interestingly, the region of stability expands with the increasing frequency ratio f/f_0, thus indicating that pairs with rapidly rotating orbits "feel" the effects of the Paul trap field to a lesser extent than the slower rotating ones. It is in this region that one should try and conduct experiments.

Conclusions

In this paper we have investigated the nonlinear stability of electrically neutral equal-mass two-body periodic orbits in a Paul trap. As the Paul trap's electric field depends on time, and is generally not spherically symmetric, the angular momentum and energy are not conserved, thus eliminating two integrals of motion. Consequently, a test of stability of periodic solutions to equations of motion had to be purely numerical.

Our numerical stability calculations show that a small, yet clearly outlined domain exists in the space of parameters of a Paul trap, where a Kepler-like motion of pair of opposite-charged ions is nonlinearly stable and where observation of such a periodic Coulomb 2-body orbit is likely.

The present study is meant to invite an experimental observation of such motion, which should be a precursor to similar investigations of periodic systems with three, or more charged particles, such as those in Refs. [1, 2], moving in a Paul trap.

Acknowledgments

This work was financed by the Serbian Ministry of Science and Technological Development under grant numbers OI 171037 and III 41011. All numerical work was done on the Zefram cluster, Laboratory for gaseous electronics, Center for non-equilibrium processes, at the Institute of Physics, Belgrade.

References

[1] M. Šindik, A. Sugita, M. Šuvakov, and V. Dmitrašinović, Phys. Rev. **E 98**, 060101(R) (2018); arXiv 1812.08654, [physics.comp-ph].

[2] Marco Fenucci, and Ángel Jorba, Commun. Nonlin. Sci. Num. Simul. **83**, 105105 (2020).

[3] D. Trypogeorgos, and C. Foot, Phys. Rev. **A 94**, 023609 (2016),

[4] G. Geyer and R. Blümel, Journal of Undergraduate Research in Physics **25**, 1-6 (2012).

[5] Nathan Leefer, Kai Krimmel, William Bertsche, Dmitry Budker, Joel Fajans, Ron Folman, Hartmut Häffner, Ferdinand Schmidt-Kaler, Hyperfine Interactions, **238**, 12 (2017).

[6] J. M. A. Danby, Astron. Jour. **69**, 165 (1964).

[7] A. Bennett, Icarus, **4**, 177 (1965).

[8] Michael Nauenberg, Astron. Jour. **124**, 2332-2338 (2002).

Piecewise flat metrics and quantum gravity*

Aleksandar Miković†
Departamento de Matemática and COPELABS,
Universidade Lusófona, Av. do Campo Grande, 376, Lisbon, Portugal

Abstract

We introduce a physical piecewise linear metric associated to a Regge triangulation of a smooth 4-manifold. We describe the basic properties of the corresponding geometry in the cases of the Euclidean and the Minkowski signature. In the Minkowski case, we describe the Regge action and how to define the corresponding path integral for the casual triangulations. We also discus the Regge path integral for a triangulation associated to the Friedman-Lemaitre-Robertson-Walker cosmological model and briefly study the corresponding wavefunctions, namely the Hartle-Hawking and the Vilenkin wavefunction.

1. Introduction

The idea of using manifold triangulations and the corresponding edge lengths to define the path integral for general relativity (GR) was introduced by Tulio Regge and his colaborators in the 1960's, and it was further developed by other researchers, see [1] for a review and references. However, finding a smooth-manifold limit of the Regge path integral turned out to be an elusive task, and Regge's approach was abandoned by most of the researchers, although in the past 20 years there has been a revival of the quantum Regge calculus in the form of the spin foam models [2] and in the form of the casual dynamical triangulation models [3].

The appearence of the superstring theory in the 1980's as a theory of quantum gravity [4], introduced the idea that the finitenness of a quantum gravity (QG) theory can be achieved by changing the mathematical structure which describes the spacetime. In the superstring theory case, this was realised by using a loop manifold instead of a manifold. In the 1990's appared the approach of non-commutative geometry [5], where the spacetime manifold was replaced by a non-commutative manifold. Each of these approaches has its advantages and disadvantages, and the main problem of the superstring theory is the aparent absence in Nature of the supersymmetry, while in the case of the non-commutative geometry approach, it

* This work has been supported by GFMUL.

† e-mail address: amikovic@ulusofona.pt

is a non-geometric nature of the non-commutative manifolds, so that it is difficult to see what is their physical meaning.

Recently a new mathematical structure for the spacetime was proposed, and this is the piecewise linear (PL) manifold structure associated to a smooth manifold triangulation [6, 7]. This is essentially the Regge formulation of GR, but with the assumption that the basic structure of the spacetime is a PL manifold, while a smooth manifold is an approximation valid when the number of space-time cells (4-simplexes) is large and their size is small. This idea is analogous to a situation in hydrodynamics, where a fluid, which is a large collection of molecules, can be approximated as a continuous medium at the scales much larger than the intermolecular distance.

In this paper we are going to introduce a PL metric associated to a triangulation of a smooth manifold. Note that for a manifold triangulation there is a Cayley-Menger metric [10, 11]. However, the CM metric is not dimensionless so that it cannot be a physical metric. The physical PL metric can be introduced as a rescaled CM metric and we will describe the basic properties of the physical PL metric, especially in the case of the Minkowski signature, since in the existing literature, the PL geometry is predominantly discussed in the Euclidean case, see for example [11]. We are also going to define the path integral (PI) in the Minkowski signature case for casual triangulations, since the existing definition [6] applies only to triangulations where all of the edge lengths are spacelike. We will then discus the path integral for triangulations associated to the Friedman-Lemaitre-Robertson-Walker (FLRW) cosmological model, since it is a physical toy model where some key properties of PLQG can be studied.

In section 2 we will describe the Regge calculus, which was originally formulated for the Riemannian manifolds, and we will introduce a physical PL metric. We will also explain what are the problems of the associated path integrals. In section 3 we will formulate the Regge calculus for a pseudo-Riemannian manifold with a Minkowski signature and define the corresponding physical PL metric, while in section 4 we will define a Regge path integral for a causal triangulation. In section 5 we will define and discus the HH wavefunction for a triangulation corresponding to the FLRW cosmological model, while in section 6 we will do the same for the Vilenkin wavefunction. In section 7 we will present our conclusions.

2. Regge calculus

The Regge discretization of GR [9], amounts to replacing the smooth spacetime manifold M with a simplicial complex $T(M)$ which corresponds to a triangulation of M, while the metric on $T(M)$ is determined by the set of the edge lengths

$$\{L_\epsilon > 0 \,|\, \epsilon \in T(M)\}\,. \tag{1}$$

Given the edge lengths (1), one would like to define a metric on the PL manifold $T(M)$ such that the PL metric on each 4-simplex σ of $T(M)$ is flat and of the euclidean signature, i.e. $(+,+,+,+)$.

This can be done by using the Cayley-Menger metric [11]

$$G_{\mu\nu}(\sigma) = L^2_{0\mu} + L^2_{0\nu} - L^2_{\mu\nu}\,, \tag{2}$$

where the five vertices of σ are labeled as $0,1,2,3,4$ and $\mu,\nu = 1,2,3,4$. Although the CM metric is flat in a four-simplex, it is not dimensionless and hence it is not diffeomorphic to $g_{\mu\nu} = \delta_{\mu\nu}$. This can be remedied by defining a new PL metric

$$g_{\mu\nu}(\sigma) = \frac{G_{\mu\nu}(\sigma)}{(\mathrm{det}G(\sigma))^{1/4}}\,, \tag{3}$$

which is a rescaled CM metric such that $\det g_{\mu\nu} = 1$, provided $\det G_{\mu\nu} > 0$.

Insuring the euclidean signature of a PL metric requires the following restrictions

$$\det G(\sigma) > 0\,, \tag{4}$$
$$\det G(\tau) > 0\,, \tag{5}$$
$$\det G(\Delta) > 0\,, \tag{6}$$

for every 4-simplex σ, every tetrahedron τ and every triangle Δ of $T(M)$. The last inequality is equivalent to the triangular inequalities for the edge lengths of a triangle. These inequalities permit us to define the volumes of n-simplexes via the Cayley-Menger determinants [10]

$$\det G(\sigma_n) = 2^n (n!)^2 V^2(\sigma_n)\,, \quad n = 2,3,4\,. \tag{7}$$

Note that for an arbitrary assignment of L_ϵ, the volumes V_n can be positive, zero or imaginary. Just taking the strict triangular inequalities will insure the positivity of the triangle areas, but then some of the higher volumes can be zero or negative. Hence all of the three inequalities must be imposed.

The Einstein-Hilbert (EH) action on M is given by

$$S_{EH} = \int_M \sqrt{\mathrm{det}g}\, R(g)\, d^4x\,, \tag{8}$$

where $R(g)$ is the scalar curvature associated to a metric g. On $T(M)$ the EH action becomes the Regge action

$$S_R(L) = \sum_{\Delta\in T(M)} A_\Delta(L)\,\delta_\Delta(L)\,, \tag{9}$$

where $A_\Delta = V(\Delta)$ and the deficit angle δ_Δ is given by

$$\delta_\Delta = 2\pi - \sum_{\sigma\supset\Delta} \theta_\Delta^{(\sigma)}\,. \tag{10}$$

A dihedral angle $\theta_\Delta^{(\sigma)}$ is defined as the angle between the 4-vector normals associated to the two tetrahedrons that share the triangle Δ, and it is given by

$$\sin\theta_\Delta^{(\sigma)} = \frac{4}{3}\frac{A_\Delta V_\sigma}{V_\tau V_{\tau'}} \,. \tag{11}$$

Given the Regge action (9), the corresponding Euclidean path integral can be written as

$$Z_E = \int_D \prod_{\epsilon=1}^{N_1} dL_\epsilon \, \mu(L) \, e^{-S_R(L)/l_P^2} \,, \tag{12}$$

where D is a subspace of $(\mathbf{R}_+)^{N_1}$ consistent with the triangular inequalities. One can introduce the path-integral measure $\mu(L)$ as

$$\mu(L) = \prod_{\epsilon=1}^{N_1} (L_\epsilon)^\alpha \,, \tag{13}$$

where α is a constant, and usually $\alpha = 1$.

The immediate problem with the choices (12) and (13) is that the finiteness of Z_E is not guaranteed because $S_R(L)$ is not bounded from bellow and the measure (13) does not fall off sufficiently quickly for large L_ϵ and negative α. A simple way to remedy this is to complexify the Euclidean path integral via

$$Z_{EC} = \int_D \prod_{\epsilon=1}^{N_1} dL_\epsilon \, \mu(L) \, e^{iS_R(L)/l_P^2} \,. \tag{14}$$

However, the problem with (14) is that it is not clear how to relate it to a path integral for the Minkowski signature metrics, i.e. how to do the Wick rotation.

3. Minkowski PL metric

The problems with the euclidean path integrals (12) and (14) can be avoided by using the Minkowski signature metric from the very beginning. In order to formulate the Regge action in the Minkowski signature case, we need to discuss certain aspects which are absent in the euclidean case.

The novelty in the Minkowski case is that L_ϵ^2 can be positive or negative, so that $L_\epsilon \in \mathbf{R}_+$ or $L_\epsilon \in i\,\mathbf{R}_+$. Consequently we have to indicate in $T(M)$ which edges are space-like (S) and which edges are time-like (T). We will not use the light-like edges ($L_\epsilon^2 = 0$). Although one can triangulate a pseudo-Riemannian manifold such that all the edges are spacelike, it is much simpler and more natural to use the triangulations where we have both the spacelike and the timelike edges.

The CM metric is now given by the same expression as in the euclidean case (2), while the physical PL metric is given by

$$g_{\mu\nu}(\sigma) = \frac{G_{\mu\nu}(\sigma)}{|\det G(\sigma)|^{1/4}} \quad , \tag{15}$$

where the modulus of the determinant accounts for the fact that now $\det G(\sigma) < 0$.

In order to insure the Minkowski signature of the PL metric we need to impose

$$\det G(\sigma) < 0 \quad , \tag{16}$$

for any four-simplex σ in $T(M)$. This is analogous to the first restriction in the euclidean case (4). However, there is no need for the analogs of the second (5) and the third restriction (6), since the signatures of $\det G(\tau)$ and $\det G(\Delta)$ are not fixed in the Minkowski case. Namely, $\det G(\tau) > 0$ if τ belongs to a euclidean hyper-plane of $g_{\mu\nu}(\sigma)$, while $\det G(\tau) < 0$ if τ belongs to a Minkowski hyper-plane. Also $\det G(\Delta) > 0$ if Δ belongs to a euclidean plane while $\det G(\Delta) < 0$ if Δ belongs to a Minkowski plane.

The volumes of n-simplexes can be defined as

$$(V_n)^2 = \frac{|\det G_n|}{2^n (n!)^2} > 0 \,, \quad n = 2, 3, 4 \,, \tag{17}$$

so that $V_n > 0$. Note that in the $n = 1$ case we should distinguish between the labels $L_\epsilon \in \mathbf{C}$ and their volumes $|L_\epsilon| > 0$. We will also use an equivalent labeling $L_\epsilon \to |L_\epsilon|$ with an indication S or T for an edge ϵ.

Given that the edge lengths can take real or imaginary values in a Minkowski space, this implies that the angles between the vectors can be real or complex. Let us consider the angles in a Minkowski plane. Such angles can be defined as

$$\cos\alpha = \frac{\vec{u}\cdot\vec{v}}{||\vec{u}||\,||\vec{v}||}\,, \quad \sin\alpha = \sqrt{1-\cos^2\alpha}\,, \quad \alpha \in \mathbf{C}\,, \tag{18}$$

where $\vec{u} = (u_1, u_0)$, $\vec{u}\cdot\vec{v} = u_1 v_1 - u_0 v_0$ and $||\vec{u}|| = \sqrt{\vec{u}\cdot\vec{u}}$.

Consider two spacelike vectors $\vec{u} = (1, 0)$ and $\vec{v} = (\cosh a, \sinh a)$, $a \in \mathbf{R}$. Since $||\vec{u}|| = ||\vec{v}|| = 1$ then

$$\cos\alpha = \cosh a\,, \quad \sin\alpha = i\sinh a \Rightarrow \alpha = i\,a\,. \tag{19}$$

In the case of a spacelike vector $\vec{u} = (1, 0)$ and a timelike vector $\vec{v} = (\sinh a, \cosh a)$, we have $||\vec{u}|| = 1$ and $||\vec{v}|| = i$, so that

$$\cos\alpha = -i\sinh a\,, \quad \sin\alpha = \cosh a \Rightarrow \alpha = \frac{\pi}{2} - i\,a\,. \tag{20}$$

And if we have two timelike vectors $\vec{u} = (0,1)$ and $\vec{v} = (\sinh a, \cosh a)$, then

$$\cos\alpha = \cosh a\,, \quad \sin\alpha = i\sinh a \Rightarrow \alpha = i\,a\,. \tag{21}$$

The definition (18) then implies that the sum of the angles between two intersecting lines in a Minkowski plane is 2π.

In order to define the dihedral angles in the Minkowski case we will introduce

$$(v_n)^2 = \frac{\det G_n}{2^n (n!)^2}\,, \quad n = 2,3,4\,, \tag{22}$$

so that $v_n = V_n$ for $\det G_n > 0$ or $v_n = i\,V_n$ for $\det G_n < 0$. In the $n = 1$ case we have $v_\epsilon = L_\epsilon$ for a spacelike edge or $v_\epsilon = iL_\epsilon$ for a timelike edge where $L_\epsilon > 0$. Then the angle between two edges in a triangle is given by

$$\sin\alpha_\pi^{(\Delta)} = \frac{2\,v_\Delta}{v_\epsilon\, v_{\epsilon'}}\,, \tag{23}$$

where π the common point (known as the hinge).

The dihedral angle between two triangles sharing an edge in a tetrahedron is given by

$$\sin\phi_\epsilon^{(\tau)} = \frac{3}{2}\frac{v_\epsilon\, v_\tau}{v_\Delta\, v_{\Delta'}}\,, \tag{24}$$

while the dihedral angle between two tetrahedrons sharing a triangle in a four-simplex is given by

$$\sin\theta_\Delta^{(\sigma)} = \frac{4}{3}\frac{v_\Delta\, v_\sigma}{v_\tau\, v_{\tau'}} \quad . \tag{25}$$

The formulas (23),(24) and (25) are generalizations the corresponding euclidean formulas such that $V_n \to v_n$, and the novelty in the Minkowski case is that $\sin\theta$ is not restricted to the interval $[-1,1]$, but $\sin\theta \in \mathbf{R}$ or $\sin\theta \in i\mathbf{R}$. This also means that the Minkowski dihedral angles can take the complex values.

In the case of a dihedral angle $\theta_\Delta^{(\sigma)}$ there are two possibilities. If the triangle Δ is in a Minkowski (ST) plane, then θ will be an angle in an orthogonal Euclidean (SS) plane, so that $\sin\theta = \sin a$. If Δ is in an SS plane, then θ will be in an orthogonal ST plane, so that $\sin\theta = \cosh a$ or $\sin\theta = i\sinh a$.

The deficit angle will then take the following values:

$$\delta_\Delta = 2\pi - \sum_{\sigma\supset\Delta} \theta_\Delta^{(\sigma)} \in \mathbf{R}\,, \tag{26}$$

when Δ is an ST triangle, while

$$\delta_\Delta = 2\pi - \sum_{\sigma\supset\Delta} \theta_\Delta^{(\sigma)} \in \frac{\pi}{2}\mathbf{Z} + i\mathbf{R}\,, \tag{27}$$

when Δ is an SS triangle. Note that for an SS triangle the triangle inequalities are valid, while for an ST triangle they do not apply.

The appearance of the complex values for the deficit angles in the Minkowski signature case raises the question of how to generalize the euclidean Regge action such that the new action is real. A proposal for a lorentzian Regge action was given in [3]

$$\tilde{S}_R = \sum_{\Delta \in SS} A_\Delta \frac{1}{i} \delta_\Delta + \sum_{\Delta \in ST} A_\Delta \, \delta_\Delta \,. \tag{28}$$

However, the problem with this definition is that a priori $\tilde{S}_R \in \mathbf{R} + i\frac{\pi}{2}\mathbf{Z}$, so that one has to verify for a given triangulation that $Im\, \tilde{S}_R = 0$.

In order to avoid this difficulty, we will take

$$\tilde{S}_R = Re \left(\sum_{\Delta \in SS} A_\Delta \frac{1}{i} \delta_\Delta \right) + \sum_{\Delta \in ST} A_\Delta \, \delta_\Delta \,. \tag{29}$$

This definition can be justified by the fact that the authors of [3] have verified that $Im\, \tilde{S}_R = 0$ for a special class of triangulations, which are physically relevant, and they are called the casual triangulations.

4. GR path integral

In the case of the Minkowski signature PL metrics, the definition of the path integral requires some extra assumptions. Let $M = \Sigma \times [0, n]$, $n \in \mathbf{N}$, be a spacetime manifold with two boundaries diffeomorphic to a 3-manifold Σ. We will use a time-ordered triangulation, which is also known as a causal triangulation [3]

$$T(M) = \cup_{k=0}^{n-1} \tilde{T}_k \left(\Sigma \times [k, k+1] \right) \,, \tag{30}$$

where $\tilde{T}_k$ is a triangulation of a slab $\Sigma \times [k, k+1]$ such that

$$\partial \tilde{T}_k = T_k(\Sigma) \cup T_{k+1}(\Sigma)$$

and T_k are triangulations of Σ. We then choose $v_\epsilon = L_\epsilon$ for $\epsilon \in T_k(\Sigma)$ and $v_\epsilon = iL_\epsilon$ for $\epsilon \in \tilde{T}_k \setminus (T_k \cup T_{k+1})$.

The corresponding path integral can be defined as

$$Z_R = \int_D \prod_{\epsilon=1}^{N_1} dL_\epsilon \, \mu(L) \, e^{i\tilde{S}_R(L)/l_P^2} \,, \tag{31}$$

where D is a region of $(\mathbf{R}_+)^{N_1}$ consistent with a PL Minkowski geometry[1]. The path-integral measure μ must be chosen such that it makes Z_R convergent. Furthermore, if we want that a quantum effective action $\Gamma(L)$ which

[1]The easiest way to determine D is to embed $T(M)$ into a 5-dimensional Minkowski space.

can be associated to (31), becomes $\tilde{S}_R(L)$ in the classical limit ($L_\epsilon \gg l_P$), then the measure μ has to obey

$$\ln \mu(\lambda L_1, \cdots, \lambda L_N) \approx O(\lambda^a)\,, \quad a \geq 2\,, \tag{32}$$

for $\lambda \to +\infty$, see [6, 8]. A choice for μ can be made such that it is consistent with (32) and it insures the diffeomorphism invariance of the smooth-spacetime effective action [6]. It is given by

$$\mu(L) = \exp\left(-V_4(L)/L_0^4\right)\,, \tag{33}$$

where V_4 is the volume of $T(M)$ and L_0 is a new parameter in the theory. A set of possible values for L_0 can be fixed by requiring that the effective cosmological constant coincides with the observed value, see [8, 7].

5. Hartle-Hawking wavefunction

In canonical quantization of GR the spacetime M must have the "slab" topology $\Sigma \times I$, where the interval I is a subset of $\mathbf{R}$, and the spacetime metric is given by

$$ds^2 = -(N^2 - n^i n_i)dt^2 + 2n_i\, dt dx^i + h_{ij}\, dx^i dx^j\,, \tag{34}$$

where (x^i, t) are coordinates on $\Sigma \times I$, N is the lapse and n^i is the shift vector, while h_{ij} is a metric on Σ. The canonical analysis of the AH action for the metric (34) gives that the canonical variables are h_{ij} and its canonically conjugate momenta p_h^{ij}, which are constrained by the diffeomorphism constraints $D_i(p_h, h)$ and the Hamiltonian constraint $W(p_h, h)$. The canonical quantization then gives that a wavefunction $\Psi(h)$ has to be invariant under the 3-diffeomorphisms of Σ and $\Psi(h)$ has to obey the Wheeler-de Witt (WdW) equation

$$\hat{W}(\hat{p}_h, \hat{h})\Psi(h) = 0\,, \tag{35}$$

where $\hat{W}$ is an operator obtained by substituting the variables in the function $W(p_h, h)$ with the operators $\hat{p_h}^{ij}$ and $\hat{h}_{ij}$.

Solving the WdW equation (35) in general case is notoriously difficult. However, Hartle and Hawking have proposed a way to construct a solution which describes an initial wavefunction of the universe [12]. It is given by a path integral

$$\Psi_0(h) = \int \mathcal{D}g\, e^{-\int_M d^4x \sqrt{g}(R(g)+\lambda)/l_P^2}\,, \tag{36}$$

where the spacetime M has the topology of a cup such that $\partial M = \Sigma$ and the metrics g have the euclidian signature such that $g|_{\partial M} = h$.

Even the Hartle-Hawking (HH) wavefunction $\Psi_0(h)$ can be calculated only in some special cases. These are the minisuperspace models where the spacetime metric has a finite number of the degrees of freedom (DOF). For example, the Friedmann-Lemaitre-Robertson-Walker (FLRW) metric is given by

$$ds^2 = -N^2(t)\,dt^2 + a^2(t)\,(dx^2 + dy^2 + dz^2)\,, \tag{37}$$

and there are two DOF, the scaling factor a and the laps N.

Consequently

$$\Psi_0(a) = \int_J dN \int \mathcal{D}a \exp\left(-\int_I dt\, L_E(a,\dot{a},N)/l_P^2\right)\,, \tag{38}$$

in the $N(t) =$ constant gauge where $J, I \subseteq \mathbf{R}$ and L_E is the euclidean metric extension of the Lagrangian $L(a,\dot{a},N)$ corresponding to the metric (37).

For a general minisuperspace model the path integral (38) can be calculated only approximately by using the stationary phase approximation. Also, in order to obtain a solution of the WdW equation the interval J has to be promoted into a contour in the complex plane, see [13, 14]. Even in the Lorentzian version of (38), one has to extend J into a complex plane contour [15].

The PLQG formulation also offers a possibility to calculate the HH wavefunction, since one can mimic the minisuperspace models by using simple triangulations where many of the edge lengths are the same. Also there is an advantage that the spacetime geometry is transparent in the PL case, so that all the domains of integration are uniquely determined and there is no need for complex domains of integration since the convergence is achieved through the PI measure (33).

For example, let us consider the case when $M = S^4$ (a 4-sphere) such that $\partial M = \Sigma = S^3$ (a three-sphere). We will then consider a triangulation $T(S^4) = \sigma_6$ (a 4-dimensional simplicial complex based on six points), which we embed in $\mathbf{R}^5$, with $T(S^3) = \sigma_5$ (a 3-dimensional subcomplex of σ_6 based on 5 points), such that $L_\epsilon = l > 0$ for $\epsilon \in \sigma_5$, $L_\epsilon = s > 0$ for $\epsilon \in \sigma_6 \setminus \sigma_5$. Consequently

$$S_R(l,s) = \frac{5\sqrt{3}}{2}\, l^2\, \delta_1(l,s) + \frac{5}{2}\, l \sqrt{s^2 - \frac{l^2}{4}}\, \delta_2(l,s)\,, \tag{39}$$

where

$$\delta_1 = 2\pi - 2\alpha\,, \quad \delta_2 = 2\pi - 3\beta\,, \tag{40}$$

and

$$\sin\alpha = \frac{\sqrt{s^2 - \frac{3l^2}{8}}}{\sqrt{s^2 - \frac{l^2}{3}}}\,, \quad \sin\beta = \frac{2\sqrt{2}\sqrt{s^2 - \frac{3l^2}{8}}\sqrt{s^2 - \frac{l^2}{4}}}{3(s^2 - \frac{l^2}{3})}\,. \tag{41}$$

The HH path integral is then given by

$$\Psi_0(l) = \int_{l_1}^{\infty} ds\, \mu(l,s)\, \exp\left(-S_R(l,s)/l_P^2\right)\,, \tag{42}$$

where $l_1 = \sqrt{\frac{3}{8}}\, l$ and

$$\mu(l,s) = \exp\left(-\frac{5\sqrt{2}\, l^3}{48\, L_0^4} \sqrt{s^2 - \frac{3l^2}{8}} \right) . \tag{43}$$

The integral (42) is convergent becasue when $s \to +\infty$ we have $\alpha \to \pi/2$, $\beta \to \arcsin(2\sqrt{2}/3)$ so that

$$\mu\, e^{-S_R/l_P^2} \approx e^{-\frac{5\sqrt{3}\pi}{4}\, l^2/l_P^2} \exp\left(-(\lambda_0\, l^2 + \delta) ls/l_P^2 \right) , \tag{44}$$

where $\lambda_0 = 5\sqrt{2}\, l_P^2/(48 L_0^4)$ and $\delta = 2\pi - 3\arcsin(2\sqrt{2}/3)$.

The HH integral is also convergent for the trivial measure $\mu = 1$, which corresponds to $\lambda_0 = 0$. However, when a cosmological constant (CC) term is included, the HH integral will be divergent for a sufficiently negative λ, since then

$$\mu\, e^{-S_R/l_P^2} \approx e^{-\frac{5\sqrt{3}\pi}{4}\, l^2/l_P^2} \exp\left(-[(\lambda_0 + \lambda) l^2 + \delta] ls/l_P^2 \right) , \tag{45}$$

and $(\lambda_0 + \lambda) l^2 + \delta < 0$. This illustrates the earlier observation that a Euclidean GR path integral is not always convergent.

A natural question arises in relation to the PL HH wavefunction (42), and that is whether it satisfies a WdW equation. Note that the scale factor and the lapse are given by

$$a = l/l_0 \,, \quad N = s/t_0 \quad , \tag{46}$$

where l_0 is a unit of length and t_0 is a unit of time. One can then ask is there a WdW operator $\hat{W}$ such that $\hat{W}\Psi_0(a) = 0$? More precisely, are there some constants α, β and γ such that

$$\hat{W}\Psi_0 = \alpha \frac{1}{a}\frac{d^2\Psi}{da^2} + \beta \frac{d}{da}\left(\frac{1}{a}\frac{d\Psi}{da}\right) + \gamma \frac{d^2}{da^2}\left(\frac{\Psi_0}{a}\right) + \lambda\, a^3 \Psi_0 = 0 \,. \tag{47}$$

The equation (47) does not necessarilly holds in PLQG, becase the WdW equation corresponds to a smooth manifold M, while we have a PL manifold $T(M)$. However, when $N_1 \to \infty$, i.e. in the smooth-manifold limit, we expect that

$$\hat{W}_{T(M)} \to \hat{W}_M \,, \tag{48}$$

where $\hat{W}_{T(M)}$ is an operator such that $\hat{W}_{T(M)}\Psi_0(a) = 0$.

The bosonic matter can be coupled to gravity by using the euclidean PL metric (3). In the case of a scalar field ϕ we have

$$S_m = \sum_{\sigma \in T(M)} V_\sigma\, \mathcal{L}_\sigma \,, \tag{49}$$

where

$$\mathcal{L}_\sigma = -\frac{1}{2} g^{\mu\nu}(\sigma)\, D_\mu\phi\, D_\nu\phi \,-\, U(\phi_0) \tag{50}$$

and $U(\phi)$ is the scalar field potential. The metric $g^{\mu\nu}(\sigma)$ is given by the inverse matrix of (3), while

$$D_\mu \phi = \frac{\phi_\mu - \phi_0}{L_{0\mu}}\,, \tag{51}$$

where $\phi_\mu = \phi(\xi_\mu)$ and $\phi_0 = \phi(\xi_0)$, while ξ_0 and ξ_μ are the verticies of σ.

In the case of the triangulation corresponding to the Regge action (39), we have the following HH wavefuction

$$\Psi(l,f) = \int_{l_1}^{\infty} ds \int_{-\infty}^{\infty} d\varphi\, \mu(l,s) \exp\left(-\frac{1}{l_P^2}\left[S_R(l,s) - S_m(l,s,\varphi,f)\right]\right)\,, \tag{52}$$

where $\phi(\xi) = f$ for $\xi \in \sigma_5$ and $\phi(\xi) = \varphi$ for $\xi \in \sigma_6 \setminus \sigma_5$. Note that in the euclidean case the matter action is $-S_m$.

A necessary but not sufficient condition for the convergence of (52) is that the integral

$$\int_{-\infty}^{\infty} d\varphi \exp\left(\frac{1}{l_P^2} S_m(l,s,\varphi,f)\right)\,, \tag{53}$$

is convergent. For the polynomial potentials $U(\phi)$ this requires that $U(\phi) > 0$ for $\phi \to \pm\infty$.

The fermionic matter can be coupled by using the PL tetrads $e_\mu^a(\sigma)$, defined by

$$e_\mu^a(\sigma)\, e_\nu^b(\sigma)\, \eta_{ab} = g_{\mu\nu}(\sigma)\,, \tag{54}$$

and the PL spin connection $\omega_\mu^{ab}(\epsilon^*)$, where ϵ^* is the dual edge connecting the centers of two adjacent 4-simplexes, see [16].

6. Vilenkin wavefunction

We saw in the previous section that the HH path integral for the PL version of the FLRW minisuperspace model was not always convergent. This is a manifestation of a more general problem that a Euclidean path integral is not always convergent, because, beside the sign of the cosmological constant, the sign of the scalar curvature and the interval of its values affect the convergence of the path integral. One can remedy this situation by using the Minkowski signature metrics, but as we saw in section 4, the definition of a Minkowski path integral requires the spacetime topology of a slab or a cylinder, which is different from the cup topology which was used to define the HH path integral. However, one can simulate the cup topology by collapsing the cylinder to a cone, which is essentially the Vilenkin proposal for the wavefunction of the universe, see [17].

Note that the path integral (31) is a function of the initial edge lengths $L_\epsilon = l_\epsilon$ on $T_0(\Sigma)$ and the final edge lengths $L_\epsilon = l'_\epsilon$ on $T_n(\Sigma)$. This function is known as the propagator, which we denote as $G(l,l')$. We denote as $G(h,h')$ the smooth-manifold version of $G(l,l')$, where h is the metric on the final boundary Σ and h' is the metric on the initial boundary Σ. The smooth propagator satisfies a non-homogeneous WdW equation

$$\hat{W}(\hat{p}_h,\hat{h})\, G(h(x),h'(y)) = \prod_{y\in\Sigma} \delta(h(x)-h'(y))\,. \tag{55}$$

The Vilenkin proposal for the wavefunction of the universe is to take $h'=0$, so that $G(h,0)$ satisfies the WdW equation for $h\neq 0$, see [17]. In the PL case, the analog of the Vilenkin wavefunction will be the propagator $G(l,0)$, and it is easy to see that it will be identical to the Lorentzian version of the HH wavefunction for the conical triangulation from the section 5.

Let us consider 6 points in $\mathbf{R}^5$ such that

$$v_{jk}^2 = l^2\,,\quad v_{j5}^2 = 3l^2/8 - t^2 \equiv l_1^2 - t^2\,,\quad t\in[0,\infty)\,, \tag{56}$$

where $j,k = 1,2,...,4$ and $j\neq k$. This corresponds to an embedding of σ_6 into $\mathbf{R}^5$ with coordinates (x,y,z,w,t) such that σ_5 is embedded into the spatial hyperplane $t=0$ with the point $(0,0,0,0,0)$ corresponding to the center of the 3-sphere which contains the spatial points $(x_j,y_j,z_j,w_j,0)$ such that $d_{ij}=l$.

Note that

$$v_{j5} = \begin{cases} s & t\le l_1 \\ is & t > l_1, \end{cases} \tag{57}$$

where

$$s = \sqrt{|l_1^2 - t^2|} \tag{58}$$

and $l_1 = \sqrt{3l/8}$ is the radius of the 3-sphere.

We will then define the lorentzian HH integral as

$$\Psi_0(l) = \int_0^{l_1} ds\,\mu(s,l)\, e^{i(\tilde{S}_R(s,l)+\lambda V_4)/l_P^2} + \int_0^{\infty} ds\,\mu(s,l)\, e^{i(\tilde{S}_R(s,l)+\lambda V_4)/l_P^2}\,, \tag{59}$$

where

$$S_R(l,s) = \frac{5\sqrt{3}}{2}\, l^2\, \delta_1(l,s) + \frac{5}{2}\, l\sqrt{\left|s^2 - \frac{l^2}{4}\right|}\, \delta_2(l,s) \tag{60}$$

and $\tilde{S}_R$ is given by (29), while

$$\delta_1 = 2\pi - 2\alpha\,,\quad \delta_2 = 2\pi - 3\beta\,, \tag{61}$$

and

$$V_4(l,s) = \frac{5\sqrt{2}}{48}\, l^3 \sqrt{\left|s^2 - \frac{3}{8}\, l^2\right|}\,. \tag{62}$$

The expressions for α and β will be given by the definition (25) and by the embedding (56). In order to see what is $\tilde{S}_R$, it is helpful to write the dihedral angles in terms of the t variable

$$\sin\alpha = \begin{cases} \frac{it}{\sqrt{l_3^2-t^2}} = i\sinh a\,, & t < l_3 \\ \frac{t}{\sqrt{t^2-l_3^2}} = \cosh a\,, & t > l_3 \end{cases} \tag{63}$$

and

$$\sin\beta = \begin{cases} \frac{2\sqrt{2}}{3}\frac{it\sqrt{l_2^2-t^2}}{l_3^2-t^2} = i\sinh b\,, & t < l_3 \\ \frac{2\sqrt{2}}{3}\frac{t\sqrt{t^2-l_2^2}}{t^2-l_3^2} = \sin b\,, & t > l_3 \end{cases}\,, \tag{64}$$

where $l_2 = l/\sqrt{8}$ and $l_3 = l/\sqrt{24}$.

Note that these expressions can be obtained from the Euclidean expressions (41) by performing an analytic continuation $t \to it$ (a Wick rotation).

The convergence of the lorentzian HH integrals (59) reduces to the convergence of the second integral. This integral is convergent due to the large-s asymptotics

$$\mu\, e^{iS_R} \approx C\, e^{-\lambda_0 l^3 s/l_P^2}\, e^{i(\lambda l^2+\delta)ls/l_P^2}\,, \tag{65}$$

and due to the asymptotics for $t \to l_3^{\pm 0}$

$$a \approx -\frac{1}{2}\ln|t-l_3|\,, \quad b \approx -\ln|t-l_3|\,. \tag{66}$$

7. Conclusions

Note that the Vilenkin PL wavefunction (59) is defined for any value of the cosmological constant λ, while the Hartle-Hawking PL wavefunction (38) is defined only for

$$\lambda > -\lambda_0 - \frac{\delta}{l^2}\,. \tag{67}$$

This reflects the fact that the Lorentzian path integral (31) has better convergence properties than the Euclidean path integral (12). Also, the Wilenkin PL wavefunction is the same as the lorentzian version of the HH PL wavefunction, a fact which indicates that this may be true in the smooth limit approximation.

Note that the propagator $G(h,h')$ is not the same as the Schrodinger equation propagator $K(\tilde{h},T;\tilde{h}',T')$, where K satisfies the deparametrized form of the WdW equation

$$\left[i\hbar\frac{\partial}{\partial T} - \hat{H}(p_{\tilde{h}},\tilde{h},T)\right] K(\tilde{h},T;\tilde{h}',T') = 0\,. \tag{68}$$

Namely, one can perform a deparametrization of the Hamiltonian constraint $W(p_h, h)$, which amounts to performing a canonical transformation $(p_h, h) \to (p_{\tilde{h}}, P_T; \tilde{h}, T)$ such that

$$W(p_h, h) = 0 \Leftrightarrow P_T + H(p_{\tilde{h}}, \tilde{h}, T) = 0 . \quad (69)$$

Consequently, the canonical quantization with respect to the new canonical variables will give the Schrodinger equation (68) in the gauge $\partial_i T = 0$, i.e. a gauge choice where T does not depend on the spatial coordinates x^i [2].

For example, in the case of the relativistic particle, when the WdW equation becomes the KG equation, we have for the propagator

$$G(\vec{x}, t) = \int_{\mathbf{R}^4} d\omega\, d^3\vec{k}\, \frac{e^{-i\omega t + i\vec{k}\cdot\vec{x}}}{\omega^2 - \vec{k}^2 - \omega_0^2} , \quad (70)$$

where $\omega = p_0/\hbar$, $\vec{k} = \vec{p}/\hbar$ and $\omega_0 = m/\hbar$, while for the Schrodinger propagator one obtains

$$K(\vec{x}, t) = \int_{\mathbf{R}^3} d^3\vec{k}\, e^{-it\sqrt{\vec{k}^2 + \omega_0^2} + i\vec{k}\cdot\vec{x}} . \quad (71)$$

In the PL context, the Schrodinger propagator becomes $K(\tilde{l}_\epsilon, T_k; \tilde{l}'_\epsilon, T_{k'})$, where $T_k = kl_0$ is an PL analogue of the time variable T and $k, k' \in \mathbf{N}$. The choice of the time variable can be implemented through the following restrictions

$$V(\Sigma_k) = f(k)\, l_0^3 , \quad k = 0, 1, 2, ..., n , \quad (72)$$

where $f(k)$ is a given function. The restrictions (72) have to be imposed in the path integral (31) in order to obtain the Schrodinger propagator K.

Note that any solution of a deparametrized WdW equation can be written as

$$\Phi(\tilde{h}, T) = \int \mathcal{D}\tilde{h}'\, K(\tilde{h}, T; \tilde{h}', 0)\, \Phi_0(\tilde{h}') , \quad (73)$$

where $\Phi_0(\tilde{h}')$ is the initial wavefunction of the universe (WFU). The functional $\Phi_0(\tilde{h})$ is arbitrary, in contrast to the HH or the Vilenkin vawefunctions, which should satisfy the WdW equation $\hat{W}\Psi(h) = 0$. It looks like that there is a greater freedom in choosing a WFU in the Schrodinger framework than in the WdW framework, so that this question deserves a further study.

The study of the HH path integral with matter (52) and its lorentzian version are obvious further steps in the PL approach to the problem of the WFU.

[2] Existence of such a gauge choice is called the problem of time in canonical GR, see [18].

Another interesting problem for a further study would be a determination of the smooth limits of the PL Hartle-Hawking and the Vilenkin wavefunctions which can be done by using a conical spacetime triangulation where the spatial simplicial complex σ_5 is replaced by σ_n and then let $n \to \infty$.

References

[1] T. Regge and R.M. Williams, J. Math. Phys. 41 (2000) 3964

[2] A. Perez, Living Rev. Relativ. 16 (2013) 3

[3] J. Ambjorn, A. Goerlich, J. Jurkiewicz and R. Loll, Phys. Reports 519 (2012) 127

[4] M.B. Green, J.H. Schwarz and E. Witten, Superstring Theory, Cambridge University Press, 1987

[5] A. Connes, Noncommutative Geometry, Academic Press, 1994;
J. Madore, An Introduction to Noncommutative Differential Geometry and its Physical Applications, Cambridge University Press, 1999

[6] A. Miković, Adv. Theor. Math. Phys. 21 (2017) 631

[7] A. Miković and M. Vojinović, SFIN 31 (2018) 267

[8] A. Miković and M. Vojinović, EPL 110 (2015) no.4, 40008

[9] T. Regge, Nuovo Cimento 19 (1961) 558

[10] A. Cayley, Cambridge Mathematical Journal, 2 (1841) 267;
K. Menger, Mathematische Annalen, 100 (1928) 75.

[11] H.W. Hamber, Gen. Rel. Grav. 41 (2009) 817-876

[12] J.B. Hartle and S.W. Hawking, Phys. Rev. D 28 (1983) 2960

[13] J.J. Halliwell and J. Louko, Phys. Rev. D 39 (1989) 2206

[14] J.J. Haliwell and J.B. Hartle, Phys. Rev. D 41 (1990) 1815

[15] J. Feldbrugge, J.-L. Lehners and N. Turok, Phys. Rev. D 95 (2017) 103508

[16] G. Gionti, Class. Quant. Grav. 22 (2005) 4217

[17] A. Vilenkin and M. Yamada, Phys. Rev. D 99 (2019) 066010

[18] C.J. Isham, NATO Sci. Ser. C 409 (1993) 157

From quantum foundations of quantum field theory, string theory and quantum gravity to dark matter and dark energy*

Đorđe Minić†
Department of Physics, Virginia Tech, Blacksburg, VA 24061, U.S.A.

ABSTRACT

We review our recent work on quantum foundations of quantum mechanics, quantum field theory and quantum gravity (formulated as metastring theory) and various implications for the problems of dark matter and dark energy. The first point concerns the new understanding of quantum theory via the concept of quantum (modular) spacetime endowed with manifest non-locality that is consistent with causality. This view implies the consistency of the fundamental length and Lorentz symmetry, based on the principle of relative (observer dependent) locality. The geometry of such quantum spacetime is encoded in the new concept of Born geometry. This in turn leads to a novel understanding of quantum field theory in a manifestly bi-local representation endowed with metaparticle quanta. A fully dynamical quantum spacetime, with a dynamical Born geometry, leads to quantum gravity (a quantum theory of dynamical geometry of quantum non-locality) in the guise of metastring theory. This generic formulation of string theory implies a radiatively stable positive cosmological constant (viewed as a curvature of the dual spacetime) as a model of dark energy in the observed classical spacetime, as well as metaparticle quanta (the zero modes of the metastring) as the natural quanta of dark matter in this approach.

1. Introduction and overview

In this talk I review recent work [1, 2, 3, 4, 5, 6, 7, 8, 9, 10] on quantum foundations of quantum mechanics, quantum field theory and quantum gravity

*DM is grateful to L. Freidel, R. G. Leigh, J. Kowalski-Glikman, P. Berglund, T. Hübsch, D. Edmonds, T. Takeuchi, D. C. Dai, D. Stojkovic, J. H. Simonetti, M. Kavic and V. Jejjala, for recent illuminating collaborations and insightful discussions. DM thanks the organizers of the international conference of the Polish Society on Relativity, the Bangkok workshop on High Energy Theory, the Multimessenger Universe conference at Penn State's Institute for Gravitation and the Cosmos, the 10th Mathematical Physics Meeting in Belgrade, Serbia, the Miami winter conferences, and the conference on quantum gravity phenomenology in Granada, Spain, for kind hospitality and invitations to present this talk. The research of DM is supported in part by the US Department of Energy (under grant DE-SC0020262) and the Julian Schwinger Foundation.

† e-mail address: dminic@vt.edu

(in the form of metastring theory) as well as unique implications for the problems of dark matter and dark energy. The starting point addresses the new understanding of quantum theory using the concept of quantum, or modular, spacetime endowed with manifest non-locality that is consistent with causality. This view implies the consistency of the fundamental length and Lorentz symmetry, based on the principle of observer based, or relative, locality. The geometry of such quantum spacetime is encoded in the new concept of Born geometry. This leads to a new understanding of quantum field theory in a manifestly bi-local representation endowed with metaparticle quanta. A fully dynamical quantum spacetime (a dynamical Born geometry) leads to a theory of quantum gravity in the form of metastring theory (a robust quantum theory whose geometry is the Born geometry of quantum non-locality). This generic formulation of string theory implies a radiatively stable positive cosmological constant (dark energy) [11] in the observed classical spacetime and metaparticle quanta (the zero modes of the metastring) representing the natural quanta of dark matter [12] (correlated to dark energy and visible matter).

The logic of our story is very similar to the path that leads from the Minkowski geometry of special relativity via relativistic non-gravitational field theory to a dynamical spacetime of general relativity. In our case we start with a hidden geometry in quantum theory (Born geometry) and proceed to its dynamical implementation in quantum gravity (formulated as a metastring theory) with implications for quantum field theory (formulated in a way that takes into account the hidden Born geometry) with implications for the observed world: metaparticles as dark matter quanta, and dark energy emerging from the geometry of the dual spacetime. In some sense this story is a sharpening of the modern approaches to non-perturbative quantum physics [13], using a simple but crucial insight about a completeness of quantum kinematics of discretized physical systems [14].

In particular, in [5] we have shown that any quantum theory is endowed with a generic quantum polarization associated with modular spacetime [4]. The generic polarization manifestly realizes quantum non-locality (associated with quantum superposition principle) that is consistent with causality, and reveals a novel underlying geometry structure, that we call Born geometry [1, 2], which unifies symplectic, orthogonal and conformal geometries. Born geometry turns out to be fundamental in a particular quantum theory that consistently propagates in this geometry - this turns out to be string theory formulated in a generalized-geometric and intrinsically non-commutative, doubled, form (that we call metastring theory) [3, 6, 7, 8]. The zero modes of the metastring define a new concept, called metaparticle, that explicitly realizes the geometry of modular spacetime [9], and that could be considered as an explicit prediction of the modular representation of quantum theory. The metaparticle [10] is associated with a modular generalization of quantum fields which can be viewed as low energy remnants of metastring fields.

Note that from this new viewpoint [1, 2, 3, 4, 5, 6, 7, 8, 9, 10] quantum gravity is essentially defined as "gravitization of the quantum", that is, as a theory of a dynamical Born geometry (for more on Born geometry, see [15]). As such it incorporates the concept of Born reciprocity [16] as a covariant implementation of T-duality, the fundamental relation between short and long distance physics in string theory, as well as the new idea of relative (or observed dependent) locality [17].

In the first part of this talk we describe the hidden quantum spacetime geometry underlying the generic representation of quantum theory, including quantum field theory (which renders the generic modular representation manifestly non-local) and then in the second part we find that the same (and, in general, dynamical) geometric structure underlies metastring theory, a manifestly T-duality covariant formulation of string theory, viewed as a consistent theory of quantum gravity and matter. Thus quantum gravity "gravitizes" the quantum spacetime geometry. Finally, in the last part we discuss a robust effective description of such a theory of quantum gravity at long distance that leads to a non-commutative (yet covariant) effective field theory, with metaparticle quanta, as implied by an intrinsic non-commutativity of closed string theory, and we discuss some of its natural consequences for the problems of dark energy and dark matter.

2. Quantum theory from quantum spacetime

The fundamental reason for the existence of modular polarizations in quantum theory can be seen as follows [5]: If one imagine that a quantum system is formulated on a lattice (as assumed in the modern (Wilsonian) non-perturbative approaches [13]), then a theorem due to Zak [14] states that a complete set of quantum numbers needed to describe any quantum system would require both quantum numbers associated with the lattice and its inverse. This is easy to see by realizing that non-commuting Hermitian operators, such as coordinates and momenta, $[q, p] = i\hbar$, when exponentiated, together with the appropriate lattice spacing a and its inverse $\frac{2\pi\hbar}{a}$, commute, that is, $[\exp(\frac{i}{\hbar}q\frac{2\pi\hbar}{a}), \exp(\frac{i}{\hbar}pa)] = 0$. Such unitary observables were labeled as "modular" by Aharonov [18], and have also appeared in other discussions of the fundamental issues in quantum theory [19]. Note that these variables are purely quantum, in the sense that their formal $\hbar \to 0$ limit is singular. Also, even their commutators are zero, the associated Poisson brackets are non-zero, as these are unitary (phase) variables. Finally, the classical limit is defined by starting with a modular formulation and defining an appropriate "extensification" [5]. Thus, in principle, there are many consistent classical limits (as suggested by the consistent history approach to quantum theory [20]). As we shall in what follows these purely quantum variables will appear in the context of quantum field theory as well.

So we are instructed to look at a complete set of unitary operators (as opposed to a complete set of Hermitian operators). Let us look at the simplest example of the q and p operators. The commuting subalgebra of the original non-commuting $[q, p] = i\hbar$ algebra, can be completely described by self-dual lattices (endowed with the natural symplectic form (ω) coming from the commutator bracket). These in turn represent a discretization of a phase space (in general fully covariant) defined by q and p and when lifted to the original non-commutative algebra, require extra data associated with the lift that is described by a doubly orthogonal ($O(d, d)$, where d denotes the spacetime dimension) metric η (a symmetric counterpart of the anti-symmetric ω, associated with $Sp(2d)$ transformations). Finally, in order to define the vacuum state on this self-dual lattice, we need a conformal structure $O(2, 2(d-1))$ [5]. This triplet of structures define what we call Born geometry [1, 2] associated with the modular representation of quantum theory [5], which naturally captures quantum non-locality that is consistent with causality, given the quantum nature of the unitary operators and the fact that the triple intersection of $Sp(2d)$, $O(d, d)$ and $O(2, 2(d-1))$, gives the Lorentz group [6].

We now formalize these insights about the hidden quantum spacetime geometry of quantization [5], which will, perhaps surprisingly, take us all the way to quantum gravity in the guise of an intrinsically non-commutative formulation of string theory. We start with the Heisenberg (or Weyl-Heisenberg) group, which is generated, on the level of the corresponding algebra, by the familiar position $\hat{q}^a$ and momentum $\hat{p}_b$ operators:

$$[\hat{q}^a, \hat{p}_b] = i\hbar\delta^a_b. \tag{1}$$

It will be convenient to introduce a length scale λ and a momentum scale ϵ, with $\lambda\epsilon = \hbar$. Then, let us introduce the following notation $\hat{x}^a \equiv \hat{q}^a/\lambda$, $\hat{\tilde{x}}_a \equiv \hat{p}_a/\epsilon$,, with $[\hat{x}^a, \hat{\tilde{x}}_b] = i\delta^a_b$. Even more compactly let us suggestively write

$$\mathbb{X}^A \equiv (x^a, \tilde{x}_a)^T, \quad [\hat{\mathbb{X}}^a, \hat{\mathbb{X}}^b] = i\omega^{AB}, \tag{2}$$

with $\frac{1}{2}\omega_{AB}dX^A dX^B = \frac{1}{\hbar}dp_a \wedge dq^a$, where $\omega_{AB} = -\omega_{BA}$ is the canonical symplectic form on phase space $\mathcal{P}$. The Heisenberg group $H_{\mathcal{P}}$ is generated by Weyl operators [21] ($\mathbb{K}$ stands for the pair $(\tilde{k}, k)$ and $\omega(\mathbb{K}, \mathbb{K}') = k \cdot \tilde{k}' - \tilde{k} \cdot k'$)

$$W_{\mathbb{K}} \equiv e^{2\pi i\omega(\mathbb{K}, \mathbb{X})}. \tag{3}$$

These form a central extension of the translation algebra

$$W_{\mathbb{K}} W_{\mathbb{K}'} = e^{2\pi i\omega(\mathbb{K}, \mathbb{K}')} W_{\mathbb{K}+\mathbb{K}'}. \tag{4}$$

The projection $\pi : H_{\mathcal{P}} \to \mathcal{P}$ (where $\pi : W_{\mathbb{K}} \to \mathbb{K}$) defines a line bundle over $\mathcal{P}$ (in principle a covariant phase space of quantum probes). In this formulation, states are sections of degree one

$$W_{\mathbb{K}'}\Phi(\mathbb{K}) = e^{2\pi i\omega(\mathbb{K}, \mathbb{K}')}\Phi(\mathbb{K} + \mathbb{K}'). \tag{5}$$

In this language, geometric quantization means to take a Lagrangian $L \in \mathcal{P}$, so that states descend to square integrable functions on L.

A Lagrangian submanifold L is a maximally isotropic subspace L with $\omega|_L = 0$, and thus $\{\partial/\partial q^a\} \in T\mathcal{P}$ defines a Lagrangian submanifold, or "space". (Indeed, $\omega(\partial/\partial q^a, \partial/\partial q^b) = 0$.) This can be understood as a classical characterization of space (and in the covariant context, of spacetime), as a "slice" of phase space. How about a purely quantum characterization of space? We claim that quantum theory reveals a new notion of quantum space (and, more covariantly, a new notion of quantum spacetime).

Note that for space-like separations the operators of a local quantum field theory commute. Thus in order to understand the meaning of quantum spacetime (quantum Lagrangian), we need to look at a maximally commuting subalgebra of the Heisenberg algebra and the representation that diagonalizes it. Thus, borrowing from notions of non-commutative algebra and non-commutative geometry [22] (such as the theorem of Gelfand-Naimark [23]), we can say that a Lagrangian submanifold is a maximally commutative subgroup of the Heisenberg group. If we accept this notion of a Lagrangian, then the quantum regime is very different from the classical regime. In particular the vanishing Poisson bracket $\{f(q), g(p)\}$ requires either f or g to be constant. However, the vanishing commutator $[f(\hat{q}), g(\hat{p})] = 0$ requires only that the functions be commensurately periodic

$$e^{i\alpha\hat{p}}e^{i\beta\hat{q}} = e^{i\hbar\alpha\beta}e^{i\beta\hat{q}}e^{i\alpha\hat{p}}, \quad \alpha\beta = 2\pi/\hbar. \tag{6}$$

What is interesting here is that similar considerations led Aharonov to introduce *modular variables* to describe purely quantum phenomena, such as interference [24].

2.1. Modular variables

Modular variables are described in great detail in the very insightful book by Aharonov and Rohrlich [24], where one can find detailed bibliography on this subject[1]. The fundamental question posed there was as follows: how does one capture interference effects (due to the fundamental linearity of quantum theory) in terms of Heisenberg operators? For example, what are the quantum observables that can measure the relative phase responsible for interference in a double-slit experiment? No polynomial functions of the operators $\hat{q}$ and $\hat{p}$ can detect such phases, but operators that translate in space, such as $e^{iR\hat{p}/\hbar}$, do. Thus the modular variables denoted $[\hat{q}]$ and $[\hat{p}]$, which are defined modulo a length scale R (the slit spacing being a natural choice), play a central role, where

$$[p]_{2\pi\hbar/R} = p \mod (2\pi\hbar/R), \qquad [q]_R = q \mod (R). \tag{7}$$

The shift operator $e^{iR\hat{p}/\hbar} = e^{iR[\hat{p}]/\hbar}$ shifts the position of a particle state (say an electron in the double-slit experiment) by a distance R and is a function

[1] See also [19, 18].

of the modular momenta[2]. These modular variables (the main examples being the Aharonov-Bohm and Aharonov-Casher phases [24]) satisfy non-local operator equations of motion. For example, given the Hamiltonian, $\hat{H} = \hat{p}^2/2m + V(\hat{q})$, the Heisenberg equation of motion for the shift operator is,

$$e^{-iR\hat{p}/\hbar} \frac{d}{dt} e^{iR\hat{p}/\hbar} = -\frac{iR}{\hbar} \left(\frac{V(\hat{q}+R) - V(\hat{q})}{R} \right). \tag{8}$$

Modular variables are fundamentally non-local in a non-classical sense, since we see here that their evolution depends on the value of the potential at distinct locations. Remarkably, thanks to the uncertainty principle, this dynamical non-locality does not lead to a violation of causality [24]. One of the characteristic features of these variables is that they do not have classical analogues; indeed, the limit $\hbar \to 0$ of $[p]_{h/R}$ is ill-defined. Also modular variables capture entanglement of continuous q, p variables.

Note that modular variables are, in general, covariant and, also, contextual[3]. In other words, they carry specific experimental information, such as the length R between the two-slits. However, in the context of quantum gravity such scales are automatically built in, and the contextuality is in principle removed. Also, the fundamental dynamical equations for modular variables are non-local in quantum gravity because of the presence of the fundamental length.

When exponentiated (i.e. when understood as particular Weyl operators), the modular variables naturally commute. In other words, given $[x^a, \tilde{x}_b] = \frac{i}{2\pi}\delta^a_b$, the following commutator of modular operators vanishes [5]

$$[e^{2\pi i x}, e^{2\pi i \tilde{x}}] = 0. \tag{9}$$

Thus a quantum algebra of modular variables possesses more commutative directions than a classical Poisson algebra, because the Poisson bracket of modular variables does not vanish, $\{e^{2\pi i x}, e^{2\pi i \tilde{x}}\} \neq 0$.

Here we make a historical note [26]: The above non-local equations of motion were essentially written by Max Born, in the very first paper which used the phrase "Quantum Mechanics" in its title, in 1924, one year before the Heisenberg breakthrough paper. Actually, Heisenberg crucially used Born's prescription of replacing classical equations by the corresponding difference equations, in order to derive what we now call the canonical commutation relations (properly written by Born and Jordan) from the Bohr-Sommerfeld quantization conditions.[4]

[2] There is a very close similarity here to the Galois quantum theory discussed in [25].

[3] Aharonov and collaborators have pushed the logic associated with modular variables to argue for a new kind of weak measurements of such non-local variables that capture the superposition principle of quantum theory. Similarly, Aharonov and collaborators argue for a time symmetric formulation of quantum theory [24].

[4] Note that the above discrete operatorial (Born-Heisenberg-Jordan) equations of motion for modular variables can be generalized to other quantum systems [27]. Also, the

2.2. Modular spacetime and Born geometry of quantum theory

Returning to the subject of quantum Lagrangians, notice that the quantum Lagrangian is analogous to a Brillouin cell in condensed matter physics. The volume and shape of the cell are given by λ and ϵ (i.e. $\hbar$ and G_N (α')) The uncertainty principle is implemented in a subtle way: we can specify a point in modular cell, but if so, we can't say *which* cell we are in.

This means that there is a more general notion of quantization [5], beyond that of geometric quantization. Instead of selecting a classical polarization L (the arguments of the wave function, or the arguments of a local quantum field) we can choose a *modular polarization*. In terms of the Heisenberg group all that is happening is that in order to have a commutative algebra, we need only

$$\omega(\mathbb{K}, \mathbb{K}') \in 2Z, \quad W_{\mathbb{K}} W_{\mathbb{K}'} = e^{2\pi i \omega(\mathbb{K}, \mathbb{K}')} W_{\mathbb{K}+\mathbb{K}'} = W_{\mathbb{K}'} W_{\mathbb{K}'}. \tag{10}$$

This defines a lattice Λ in phase space $\mathcal{P}$. Finally, we specify a "lift" of the lattice from the phase space $\mathcal{P}$ to the Heisenberg group $H_{\mathcal{P}}$.

Maximally commuting subgroups $\hat{\Lambda}$ of the Heisenberg group correspond to lattices that are integral and self-dual with respect to ω [29]. Given W_λ where $\lambda \in \Lambda$ there is a lift to $\hat{\Lambda}$ which defines "modular polarization"

$$U_\lambda = \alpha(\lambda) W_\lambda, \tag{11}$$

where $\alpha(\lambda)$ satisfies the co-cycle condition

$$\alpha(\lambda)\alpha(\mu) e^{\pi i \omega(\lambda,\mu)} = \alpha(\lambda + \mu), \quad \lambda, \alpha \in \Lambda. \tag{12}$$

One can parametrize a solution to the co-cycle condition by introducing a symmetric bilinear from η and setting (with $\eta(\mathbb{K}, \mathbb{K}') = k \cdot \tilde{k}' + \tilde{k} \cdot k'$,)

$$\alpha_\eta(\lambda) \equiv e^{i\frac{\pi}{2}\eta(\lambda,\lambda)}. \tag{13}$$

Finally, when we choose a classical Lagrangian L, there is a special state that we associate with the vacuum: it is translation invariant (which in our context can be interpreted as "empty space"). In modular quantization there is no such translation invariant state (because of the lattice structure). The best we can do is to choose a state that minimizes an "energy", which requires the introduction of another symmetric bilinear form, that we call, again suggestively, H. This means, first, that we are looking for operators such that

$$[\hat{\mathbb{P}}_A, \Phi] = \frac{i}{2\pi}\partial_A \Phi, \qquad \Phi(\hat{\mathbb{X}} + \lambda) = \Phi(\hat{\mathbb{X}}), \tag{14}$$

where the modular observables $\Phi(\hat{\mathbb{X}} + \lambda) = \Phi(\mathbb{X})$ are generated by the lattice observables U_λ with $\lambda \in \Lambda$. Translation invariance would be the

appearance of the "covariant hidden spacetime $\tilde{x}$" bears some similarity to the Koopman-von-Neumann theory [28].

condition $\hat{\mathbb{P}}|0\rangle = 0$. Since this is not possible, the next natural choice is to minimize the translational energy. Therefore we pick a positive definite metric H_{AB} on $\mathcal{P}$, and we define [5]

$$\hat{E}_H \equiv H^{AB}\hat{\mathbb{P}}_A\hat{\mathbb{P}}_B, \tag{15}$$

and demand that $|0\rangle_H$ be the ground state of $\hat{E}_H$. This is indeed the most natural choice and it shows that we cannot fully disentangle kinematics (i.e., the definition of translation generators) from dynamics. In the Schrödinger case, since the translation generators commute, the vacuum state $\hat{E}|0\rangle = 0$ is also the translation invariant state and it carries no memory of the metric H needed to define the energy. In our context, due to the non-commutativity of translations, the operators $\hat{E}_H$ and $\hat{E}_{H'}$ do not commute. As a result the vacuum state depends on H, in other words $|0\rangle_H \neq |0\rangle_{H'}$, and it also possesses a non-vanishing zero point energy.

Thus, modular quantization involves the introduction of three quadratic forms (ω, η, H), i.e. what we call *Born geometry* [1, 2], which underlies the geometry of modular variables.

As we will see, in the context of metastring theory, a choice of polarization is a choice of a spacetime within $\mathcal{P}$ but the most general choice is a *modular polarization* that we have discussed above. From the foundational quantum viewpoint Born geometry (ω, η, H) arises as a parametrization of such quantizations, which results in a notion of quantum spacetime, that we call *modular spacetime*. In particular, a one dimensional (1d) modular line is a two dimensional (2d) torus that is compact and not-simply connected. Finally, large spacetimes of canonical general relativity (and its extensions, like string theory) result as a "many-body" phenomenon, through a process of tensoring (entanglement) of unit modular cells, that we refer to as "extensification" [5].

In particular, the symplectic structure ω found in

$$ds^2_\omega = \frac{1}{2}\omega_{AB}d\mathbb{X}^A d\mathbb{X}^B = \frac{1}{\hbar}dp_a \wedge dq^a,$$

is encoded in the canonical Heisenberg commutator between q^a and p_a. The generalized, quantum, metric H comes from the Born rule in quantum theory $ds^2_H = H_{AB}d\mathbb{X}^A d\mathbb{X}^B = \frac{1}{\hbar}(\frac{dq_a dq^a}{G_N} + G_N dp_a dp^a)$. For weak gravity, this metric reduces to the spacetime metric (where spacetime can be viewed as a slice of phase space). Due to gravity's extreme weakness, we only see spacetime metric at low energies. (The ratio ϵ/λ defines a tension; if this is identified with c^3/G_N, it is enormous, $\sim 10^{32} kg/sec$.) Therefore, in this formulation the usual dynamical spacetime metric is the low energy leftover of the quantum metric. Finally, the polarization (or locality metric) η encodes the distinction between spacetime-like and energy-momentum-like aspects of phase space (and in this sense it defines an analog of the "causal" structure in phase space) $ds^2_\eta = \eta_{AB}d\mathbb{X}^A d\mathbb{X}^B = \frac{2}{\hbar}dp_a dq^a$. This new metric captures the essence of relative locality - when η is constant we have absolute locality. Curving η also means "gravitizing the quantum". In general all three elements of Born geometry, ω, η and H are dynamical and curved in metastring theory, as we will discuss in what follows.

Also, we have that the Lorentz group (in D spacetime dimensions) lies at the intersection of the symplectic, neutral and doubly orthogonal groups [5],

$$\mathrm{O}(1, d-1) = \mathrm{Sp}(2d) \cap \mathrm{O}(d, d) \cap \mathrm{O}(2, 2(d-1)), \tag{16}$$

which sheds new light on the origin of quantum theory through compatibility of the causal (Lorentz) structure and non-locality captured by the discreteness of quantum spacetime. This also captures the role of relative (observer-dependent) locality [17] needed to resolve the apparent contradiction between discreteness of quantum spacetime and Lorentz symmetry.

Let us end this discussion of quantum mechanics by a few comments regarding the Stone-von Neumann theorem [30] which asserts that all representations of the Heisenberg group are unitarily equivalent. Normally, we think of this as a choice of basis in phase space (a choice of polarization or classical Lagrangian), and all such choices are related by Fourier transform. Similarly, one can pass from a classical polarization (such as the Schrödinger representation) to a modular polarization via the Zak transform [14] (see section 3). Note that, there is a connection on the line bundle over phase space that has unit flux through a modular cell. (This is very similar to Integer Quantum Hall effect.) A modular wave function is quasi-periodic

$$\Psi(x + a, \tilde{x}) = e^{2i\pi a\tilde{x}}\Psi(x, \tilde{x}), \quad \Psi(x, \tilde{x} + \tilde{a}) = \Psi(x, \tilde{x}). \tag{17}$$

The quasi-periods correspond to the tails of an Aharonov-Bohm [31] potential attached to a unit flux. In particular, vacuum states must have at least one zero in a cell, which leads to theta functions (the Zak transforms of Gaussians). Note that from the point of modular polarization, the familiar Schrödinger polarization is just a singular limit[5].

3. Quantum field theory and quantum spacetime

Now we make some general comments about quantum field theory in the modular form, following the general modular formulation of any quantum theory. (Later we will discuss how such a formulation of quantum field theory comes emerges from quantum gravity at large distances.) This new modular polarization of quantum field theory reveals new structures and sheds new light on both the short distance (UV) and long distance (IR) physics of quantum fields, and the new definition of the continuum limit of quantum field theories which is self-dual with respect to the UV and

[5] This discussion can be also extended to a path integral formulation in modular polarization as done by Yigit Yargic in his Perimeter Scholars International master thesis work under Laurent Freidel.

IR properties (resembling some crucial properties of non-commutative field theories [32]. In particular, the modular representation of quantum field theory introduces dual "electric" and "magnetic" variables, which are in general, non-commuting. This general message extends our results in the context of the 2d conformal field theory formulation of string theory in which the explicit non-commutativity of such "electric" and "magnetic" variables has been explicitly demonstrated [7, 8].

The general modular representation can be defined in terms of the Zak transform of a Schrödinger representation (given in terms of wave functions). Given a square normalizable wave function $\psi(x)$ (where $x \equiv q/a$ and $\tilde{x} \equiv p/b$, and where $ab = 2\pi\hbar$) belonging to a Hilbert space, one defines the modular representation as the following lattice Fourier transform (or Zak transform)

$$\psi_a(x, \tilde{x}) \equiv \sqrt{a} \sum_n e^{-2\pi i n \tilde{x}} \psi(a(x+n)). \tag{18}$$

Note that if $\psi(x)$ is a Gaussian, its Zak transform, the modular $\psi_a(x, \tilde{x})$ is given by the doubly-period theta function associated with the lattice. (The inverse Zak transform

$$\Phi(x+n) \equiv \frac{1}{\sqrt{a}} \int_0^1 d\tilde{x} e^{2\pi i n \tilde{x}} \Phi_a(a^{-1}x, \tilde{x}), \tag{19}$$

illustrates that the usual Schrödinger representation is really singular, and thus not generic.)

Now, if one second quantizes $\psi(x)$ one naturally ends up with a quantum field operator $\phi(x)$. Similarly, the second quantization of the modular $\psi_a(x, \tilde{x})$ would lead to a modular quantum field operator $\phi(x, \tilde{x})$

$$\phi(x) \to \phi(x, \tilde{x}). \tag{20}$$

Note that the usual wave functional approach to quantum field theory (defined in terms of functionals $\Psi[\phi(x)]$ should be now defined in terms of wave functionals of modular fields $\Psi[\phi(x, \tilde{x})]$, where the fields ϕ and their duals $\tilde{\phi}$ in general do not commute, and thus the wave functional can be still chosen to be a functional of ϕ in a very particular polarization. However, now we have more freedom in the general modular polarization. Thus the dual momenta p and $\tilde{p}$ (to x and $\tilde{x}$ respectively) lead, via the canonical minimal coupling prescription, not only to the usual fields ϕ but also to their duals $\tilde{\phi}$ (see section 5). This procedure defines the modular polarization of quantum field theory in terms of the functional Zak transform of the original wave functional

$$\Psi[\phi(x)] \to \Psi[\phi(x, \tilde{x}), \tilde{\phi}(x, \tilde{x})]. \tag{21}$$

For example, the Gaussian wave functionals with non-trivial momentum kernels (such as the ones found in the context of non-trivial interacting theories like 2+1 and 3+1 dimensional compact QED [33] as well as 2+1 and 3+1 dimensional Yang Mills theory [34] would be mapped into functional theta functions.

Once again, we can understand this new view of continuum quantum field theory as a sharpening of the canonical picture of a quantum field theory formulated on a lattice (as assumed in the modern, Wilsonian, non-perturbative approaches [13]), and a theorem due to Zak [14] which states that a complete set of quantum numbers needed to describe any quantum system would require both quantum numbers associated with the lattice and its inverse. Thus, in general, before we take a naive continuum limit, we need to work with the full set of quantuum spacetime data, represented by the lattice and its dual (spacetime and its dual) as well as fields and dual fields.

On the level of the path integral formulation (the open path integral defines the above wave functional in terms of boundary data), one would have to work with an explicit formulation of quantum field theory in terms of ("electric") quantum fields $\phi(x,\tilde{x})$ and their ("magnetic") duals $\tilde{\phi}(x,\tilde{x})$

$$\int \text{``}D\phi(x,\tilde{x})D\tilde{\phi}(x,\tilde{x})\text{''} e^{\frac{i}{\hbar}S_{chiral}[\phi(x,\tilde{x}),\tilde{\phi}(x,\tilde{x})]}. \tag{22}$$

There is no overcounting here, because the action will turn out to be chiral. So even though one has formally doubled the number of variables, the chiral nature of the theory, keeps the total number of degrees of freedom the same as in the purely electric formulation. Similarly the measure of the path integral takes into account the chiral nature of this formally doubled formulation of quantum field theory. That is why the doubled measure is written under the quotation marks. By formally introducing the doublet $\mathbb{Z} = (\phi(x,\tilde{x}), \tilde{\phi}(x,\tilde{x}))$, we can write this doubled path integral as

$$\int (D\mathbb{Z})_{chiral} e^{\frac{i}{\hbar}S_{chiral}[\mathbb{Z}]}. \tag{23}$$

As usual, this formulation has to be properly regulated, but now, with two cut-offs, with the continuum limit defined in a symmetric, self-dual way with respect to the double RG flows thus resembling non-commutative field theory [32], albeit with full spacetime covariance. This new insight on quantum field theory should be important not only in the context of quantum non-locality in quantum field theory, but also in the realms of strong coupling and deep infrared.

4. Quantum gravity and dynamical quantum spacetime

The unexpected outcome of our research on the foundations of quantum mechanics and quantum field theory is that this fundamental quantum geometry of quantum theory can be realized in the context of metastring theory, where this quantum geometry is "gravitized" (i.e. dynamical). At the classical level, metastring theory [1, 2, 3, 4, 5, 6, 7, 8, 9, 10] can be thought of as a formulation of string theory in which the target space is doubled in such a way that T-duality acts linearly on the coordinates. This doubling means that momentum and winding modes appear on an equal footing. We refer to the target space as a phase space since the metastring action requires the presence of a background symplectic form ω. The metastring formulation also requires the presence of geometrical structures that generalize to phase space the spacetime metric and the B-field (where the B-field originates from the symplectic structure ω). In fact, in the metastring we have not one but *two* notions of a metric. The first metric η is a neutral metric that defines a bi-Lagrangian structure and allows to define the classical spacetime as a Lagrangian sub-manifold[6] — more precisely, the classical spacetime is defined as a null subspace for η which is also Lagrangian for ω. The second metric H is a metric of signature $(2, 2(D-1))$ that encodes the geometry along the classical spacetime (of dimension D) as well as the transverse energy-momentum space geometry. In this formulation, T-duality exchanges the Lagrangian sub-manifold with its image under $J = \eta^{-1}H$. Classical metastring theory is defined by the following action [3] which realizes the above comments about quantum field theory in the modular polarization for the special case of a two-dimensional world-sheet quantum field theory[7]

$$\hat{S} = \frac{1}{4\pi}\int_\Sigma d^2\sigma\Big(\partial_\tau \mathbb{X}^A(\eta_{AB} + \omega_{AB})(\mathbb{X})\partial_\sigma \mathbb{X}^B - \partial_\sigma \mathbb{X}^A H_{AB}(\mathbb{X})\partial_\sigma \mathbb{X}^B\Big), \quad (24)$$

where $\mathbb{X}^A$ are dimensionless coordinates on phase space and the fields η, H, ω are all dynamical (i.e., in general dependent on $\mathbb{X}$) phase space fields. In the context of a flat metastring we have constant η_{AB}, H_{AB} and ω_{AB}

$$\eta_{AB} \equiv \begin{pmatrix} 0 & \delta \\ \delta^T & 0 \end{pmatrix}, \quad H_{AB} \equiv \begin{pmatrix} h & 0 \\ 0 & h^{-1} \end{pmatrix}, \quad \omega_{AB} = \begin{pmatrix} 0 & \delta \\ -\delta^T & 0 \end{pmatrix}, \quad (25)$$

where δ^μ_ν is the d-dimensional identity matrix and $h_{\mu\nu}$ is the d-dimensional Lorentzian metric, T denoting transpose.

In view of our general comments regarding the modular polarization (quantum spacetime polarization) in quantum field theory, the metastring sheds new light on some old questions regarding the continuum limit of string theory [36] as well as the Wilsonian approach to string theory [37]. In the metastring formulation [3] it is convenient, as suggested by the

[6]We remind the reader that in symplectic geometry, a Lagrangian subspace is a half-dimensional submanifold of phase space upon which the symplectic form pulls back to zero. In simple terms, a Lagrangian submanifold might be the subspace coordinatized by the q's within the phase space coordinatized by q's and p's.

[7]See also [35].

double field formalism [38], to introduce dimensionless coordinates $\mathbb{X}^A \equiv (X^\mu/\lambda, P_\mu/\varepsilon)^T$ on phase space[8], or equivalently, $\mathbb{X}^A \equiv (x^a, \tilde{x}_a)^T$, where λ and ϵ represent the fundamental spacetime and energy-momentum scales. Here $\hbar = \lambda\epsilon$ and $\alpha' = \frac{\lambda}{\epsilon}$. Given a pair (H, η) it is natural to consider the operator $J \equiv \eta^{-1}H$. The consistency of string theory requires J to be a chiral structure, that is, a real structure ($J^2 = 1$) compatible with η, implying that J is an $O(D, D)$ transformation (realizing generalized T-duality in target space). These three structures, the symplectic $Sp(2D)$ ω, the $O(D, D)$ η and the $SO(2, 2(D-1))$ H, define the new concept of Born geometry [1, 2, 3, 4, 5] (see also [15]) which unifies the complex geometry of quantum theory with the metrical geometry of general relativity and the symplectic geometry of canonical Hamiltonian dynamics [40]. Note that in the phase space formulation the local phase space coordinates $\mathbb{X}$ are *quasiperiodic* $\mathbb{X}^A(\sigma + 2\pi) = \mathbb{X}^A(\sigma) + \Delta^A$, where Δ^A is the corresponding quasiperiod (which either vanishes for the canonical Polyakov string or is given by the winding number in the usual treatment of T-duality on compact spaces).

The worldsheet formulation of the metastring is chiral. Thus, even though the fields are doubled the central charges (left and right) are $c_L = c_R = D$ and we still have $D = 26$ for criticality. The metastring is not manifestly invariant under the worldsheet Lorentz transformations and it contains monodromies $\mathbb{X}^A(\sigma + 2\pi) = \mathbb{X}^A(\sigma) + \Delta^A$. The usual Polyakov string can be obtained by integrating out the dual $\tilde{X}$, for constant η and H backgrounds, and by supposing that the monodromies are in the kernel of $(\eta - \omega)$. T-duality is implemented in target space by the action of the chiral J operator ($J \equiv \eta^{-1}H$, $J^2 = 1$): $\mathbb{X} \to J(\mathbb{X})$.

The target space of the metastring is not spacetime, but, to first order, a chiral phase space $\mathcal{P}$ equipped by the symplectic structure ω, and the bilagrangian structure, and in particular, the polarization metric η which relates to the symplectic connection of the Fedosov deformation quantization [41] and thus leads to the star product of deformation quantization, and finally, the quantum H metric which relates to the complex structure in the context of geometric quantization [42], leading to the concept of Hilbert spaces. This classical Born geometry implements the ideas of Born duality in string theory [1, 2].

The classical equations of motion of the metastring $\partial_\tau \mathbb{X}^A - (J\partial_\sigma \mathbb{X})^A = 0$, implies the relation between momenta and monodromies $2\pi P = J(\Delta)$. There is *soldering* between worldsheet null coordinates $\sigma^\pm \equiv \sigma \pm \tau$ and the chiral target space structure $\partial_\pm \mathbb{X}^A - (P_\pm \mathbb{X})^A = 0$, where the chiral projector is defined as $2P_\pm = (1 \pm J)$. This allows us to liberate the left geometry from the right geometry (which is reminiscent of twistor theory). The careful analysis of the metastring action [3] shows that its symplectic form is $\Omega = \frac{1}{4\pi}\int \delta\mathbb{X}^A \eta_{AB} \nabla_\sigma \delta\mathbb{X}^B$, where ∇ is the generalized Fedosov connection

[8] See also [39] as an early example of a phase space formulation of string theory (and thus of relative, observer-dependent, locality in string theory).

found in the Fedosov deformation quantization approach [41].

Also, the operator product expansion of the metastring vertex operators $V_k = \epsilon_k e^{i\mathbb{K}\mathbb{X}}$, (i.e. modular variables) lead to the restriction of $\mathbb{K}$ on a double Lorentzian integral lattice Γ, that by modular invariance, must be self-dual. These exist in $D = 2mod(8)$, and are unique. Criticality gives a very unique lattice $\Gamma = \Pi_{1,25} \times \Pi_{1,25}$. This fact, in turn, leads to the large symmetry structure found by Borcherds in the study of the monstrous moonshine [43] [9].

As already noted, the metastring is chiral. This requires the introduction of a preferred worldsheet time coordinate which is fundamentally Lorentzian [3][10]. How can this be consistent with modular invariance? The answer is given by employing the Giddings-Wolpert-Krichever-Novikov construction [46]: given a Riemann surface, provided a choice of local coordinates around punctures is labeled by one scalar, there exists a unique Abelian differentional e with imaginary periods. The real part of this Abelian differential is the modular invariant time $\tau = Re(e)$. The zeros of e represent interaction points where the worldsheet Lorentzian cones double. Cutting the Riemann surface along the real trajectory of e we obtain a string decomposition of the surface. The Nakamura graphs [47] encode this decomposition and give a very effective cell decomposition of moduli space. Thus Nakamura graphs are the natural Feynman diagrams for closed strings [48][11].

4.1. Non-commutativity in quantum gravity/metastring theory

The metastring formulation points to an unexpected fundamental non-commutativity of closed string theory, that we address in what follows. It is well established [50] that the structure of the zero mode algebra of the compactified closed string depends on a lattice of momenta $(\Lambda, 2\eta)$ which is integral and self-dual with respect to a neutral metric: a so-called Narain lattice [51]. In [1, 2, 3, 4, 5, 6, 7, 8] we have refined this structure and we have shown that in fact the kinematical structure of the string zero modes depends on a *para-hermitian lattice*: a triple (Λ, η, ω), where Λ is a subgroup of $\mathbb{R}^{2d}$ that describes the lattice of wave-covectors $\lambda\mathbb{K}$, with λ the string length, η is a neutral metric, a symmetric bilinear form of signature (d, d), and ω is an invertible two-form. This structure needs to satisfy two compatibility conditions: first, the lattice Λ is assumed to be integral with respect to the para-hermitian structure, i.e., $(\eta \pm \omega)(\lambda\mathbb{K}, \lambda\mathbb{K}') \in \mathbb{Z}$,

[9]For a string theory related discussion, see [44].

[10]See also Witten's treatment of Feynman's $i\epsilon$ prescription in string theory, which requires doubling of Xs and the Lorentzian world-sheet, as required by metastring theory [45].

[11] Note that the metastring is bosonic, and that supersymmetry (and the superstring) can be viewed as emergent from this more fundamental formulation (see [49]). In some sense, fermions can be viewed as defects in modular spacetime. The bosonic degrees of freedom associated with quanta of interactions can be viewed as deformations of modular spacetime cells.

for $\lambda\mathbb{K}, \lambda\mathbb{K}' \in \Lambda$. Second, the metric η and the 2-form ω must be compatible, in the sense that $\eta^{-1}\omega := K$ is a product structure, that satisfies the condition $K^2 = 1$.

These two conditions are a consequence of mutual locality on the worldsheet (i.e. worldsheet causality). It is clear that if (Λ, η, ω) is a para-hermitian lattice, then $(\Lambda, 2\eta)$ is a Narain lattice, so the kinematical structure that we highlight is a refinement of the usual one. The extra information is contained in the 2-form ω. This form does not enter expressions for the spectrum or the partition function and this why it is usually ignored. It does enter however crucially in the definition of the vertex operator algebra and parameterizes what is usually referred to as a cocycle. The role of ω is to promote the zero mode double space $\mathcal{P} \simeq \mathbb{R}^{2d}$ dual to $\mathbb{R}[\Lambda]$ to the status of phase space: $\mathcal{P}$ should be viewed as a symplectic manifold. At the quantum level, both geometrical structures η and ω enter in the commutation relations of string operators. ω controls the non-commutativity of the zero-modes while η controls the non-commutativity of the string oscillator modes. This can be seen if one introduces a double notation for the string coordinate $\mathbb{X}(\sigma)$ that includes the string map X and its dual $\tilde{X}$. The string commutation relations, were derived in [7, 8]

$$[\mathbb{X}^A(\sigma), \mathbb{X}^B(\sigma')] = 2i\lambda^2 \left[\pi\omega^{AB} - \eta^{AB}\theta(\sigma - \sigma')\right], \qquad (26)$$

where $\theta(\sigma)$ is the staircase distribution, i.e., a solution of $\theta'(\sigma) = 2\pi\delta(\sigma)$; it is odd and quasi-periodic with period 2π.

Following standard practice, all indices are raised and lowered using η and η^{-1}. The momentum density operator is given by

$$\mathbb{P}_A(\sigma) = \frac{1}{2\pi\alpha'}\eta_{AB}\partial_\sigma\mathbb{X}^B(\sigma)$$

and the previous commutation relation implies that it is conjugate to $\mathbb{X}^A(\sigma)$. The two-form ω appears when one integrates this canonical commutation relation to include the zero-modes, the integration constant being uniquely determined by worldsheet causality. Denoting by $(\hat{\mathbb{X}}, \hat{\mathbb{P}})$ the zero mode components of the string operators $\mathbb{X}(\sigma)$ and $\mathbb{P}(\sigma)$ we simply have that

$$[\hat{\mathbb{P}}_A, \hat{\mathbb{P}}_B] = 0, [\hat{\mathbb{X}}^A, \hat{\mathbb{P}}_B] = i\hbar\delta^A{}_B, [\hat{\mathbb{X}}^A, \hat{\mathbb{X}}^B] = 2\pi i\lambda^2\omega^{AB}. \qquad (27)$$

This is a deformation of the doubled Heisenberg algebra involving the string length λ as a deformation parameter.

So far we have assumed that the background is trivial, with the fields (η, ω) *constant* and given by $\eta(\mathbb{K}, \mathbb{K}') = k \cdot \tilde{k}' + \tilde{k} \cdot k'$, and $\omega(\mathbb{K}, \mathbb{K}') = k \cdot \tilde{k}' - \tilde{k} \cdot k'$. As shown in [7], we can turn on non-trivial backgrounds encoded into ω by changing the $\mathrm{O}(d, d)$ frame $\mathbb{X} \to O\mathbb{X}$. This change of frame preserves η but transforms ω. Any constant ω can be obtained this way. Since ω has an interpretation as the symplectic form on the space of $\mathbb{X}$'s, modifying ω affects the commutation relations[12] $[\hat{\mathbb{X}}^A, \hat{\mathbb{X}}^B] = 2\pi i\lambda^2\Pi^{AB}$,, with $\Pi^{AB}\omega_{BC} = \delta^A{}_C$, where we have introduced the Poisson tensor $\Pi = \omega^{-1}$.

For instance, under a constant B-field transformation $\mathbb{X} = (x^a, \tilde{x}_a) \mapsto (x^a, \tilde{x}_a + B_{ab}x^b)$, the trivial symplectic form $\omega(\mathbb{K}, \mathbb{K}') = k \cdot \tilde{k}' - \tilde{k} \cdot k'$ is mapped onto $\omega(\mathbb{K}, \mathbb{K}') = k_a\tilde{k}'^a - k'_a\tilde{k}^a - 2B_{ab}\tilde{k}^a\tilde{k}'^b$, and the commutators read

$$[\hat{x}^a, \hat{x}^b] = 0, \qquad [\hat{x}^a, \hat{\tilde{x}}_b] = 2\pi i\lambda^2\delta^a{}_b, \qquad [\hat{\tilde{x}}_a, \hat{\tilde{x}}_b] = -4\pi i\lambda^2 B_{ab}. \tag{28}$$

We see that the effect of the B-field is to render the dual coordinates non-commutative (and that the B-field originates from the symplectic structure ω). More generally, we can parameterize an arbitrary $O(d,d)$ transformation as $g = e^{\hat{B}}\hat{A}e^{\hat{\beta}}$, where $\hat{A} \in GL(d)$ and $e^{\hat{B}} = \left(\begin{smallmatrix} 1 & 0 \\ B & 1 \end{smallmatrix}\right)$ and $e^{\hat{\beta}} = \left(\begin{smallmatrix} 1 & \beta \\ 0 & 1 \end{smallmatrix}\right)$ are nilpotent. $e^{\hat{B}}$ is the B-field transformation discussed above, and is associated with the usual B-field deformation in string theory. We note that the transformation of $(x^a, \tilde{x}_a)$ given above does not modify x^a, and thus fields that depend only on x^a are unmodified. The β-transformation on the other hand corresponds to the map $(x^a, \tilde{x}_a) \mapsto (x^a + \beta^{ab}\tilde{x}_b, \tilde{x}_a)$. Equivalently, it has the effect of mapping the symplectic structure to $\omega(\mathbb{K}, \mathbb{K}') = k_a\tilde{k}'^a - k'_a\tilde{k}^a + 2\beta^{ab}k_a k'_b$, and yields commutation relations

$$[\hat{x}^a, \hat{x}^b] = 4\pi i\lambda^2\beta^{ab}, \qquad [\hat{x}^a, \hat{\tilde{x}}_b] = 2\pi i\lambda^2\delta^a{}_b, \qquad [\hat{\tilde{x}}_a, \hat{\tilde{x}}_b] = 0. \tag{29}$$

Dramatically, the coordinates that are usually thought of as the spacetime coordinates have become themselves non-commutative. Since this is the result of an $O(d,d)$ transformation, we know that it can be thought of in similar terms as the B-field; these are related by T-duality. We are familiar with the B-field background because we have, in the non-compact case, a fixed notion of locality in the target space theory. However, in the non-geometric β-field background, we do not have such a notion of locality but we can access it through T-duality[13].

We note that this intrinsic non-commutativity of string theory can be also explicitly illustrated via a simple closed string product, equivalent to the splitting-joining interaction of the pants diagram, that respects this non-commutativity and is covariant with respect to T-duality [8]. This offers new insights on the relationship between closed and open strings,

[12] The algebraic structure that we are working with here has an analogy in electromagnetism in the presence of monopoles. In that analogy, the string length becomes the magnetic length, and the form ω becomes the magnetic field. Another analogy occurs in quantum Hall liquids, the algebra being the magnetic algebra of the lowest Landau level.

[13] Note that the dilaton can be understood as coming from the volume of phase space (see [52]). In general, for varying B-backgrounds we encounter non-associativity as well [53], and the proper closure of such non-commutative and non-associative structure is ensured by the equations of motion for the data of Born geometry, that is, the generalized Einstein equations.

and the non-perturbative formulation of closed string theory in terms of open strings and even more fundamental (and non-commutative) partonic degrees of freedom. Given the mechanism of tachyon condensation in the open string sector [54] and this fundamental relation between open and closed strings, we expect that a similar solution of the "tachyon problem" should exist in the closed string sector as well.

4.2. Non-perturbative formulation of quantum gravity

Let us recapitulate: How does the metastring approach [1, 2, 3, 4, 5, 6, 7, 8, 9, 10] compare to the usual view of string theory? From the classic textbook treatment of string theory [50] we know that there exists a fundamental relation between world-sheet conformal field theory (CFT) and target spacetime geometry, and, in particular, that the beta function for the background spacetime metric is the Ricci tensor to leading order, and so, world-sheet conformal invariance implies the vacuum Einstein equations. But string theory has other background field and fluxes and moduli. Can we write the general CFT as a generalized Ricci flow? This is precisely the achievement of Double Field Theory (DFT) [38].

In the metastring formulation the target space is not the index space (the labels of the background effective field). Effective fields are associated with the level matching constraints or mutual locality (world sheet causality). In DFT that requirement appears as the section condition. Instead of this section condition, in the metastring we have a more general background spacetime (modular spacetime) for the string, because of the intrinsic non-commutativity, stemming from the general quasi-periodicity of the worldsheet X fields. Thus even though the currents dX are periodic, the fields X are not, and this leads to edge modes in the evaluation of the string symplectic form, which ultimately leads to intrinsic non-commutativity of the metastring. Similarly, mutual locality of string vertex operators implies that, in general, they furnish a representation of a Weyl-Heisenberg algebra. Note that the metastring treats the Hamiltonian and diffeomorphism constraints together, on the same footing, and thus instead of solving the differomorphism constraint (via level matching and the strong constraint) we arrive at a completely new background interpretation: the modular spacetime.

The new ingredient of the metastring, as compared to DFT is the symplectic structure that controls intrinsic non-commutativity. This in turn, with the $O(d, d)$ structure (also associated with level matching and the string diffeomorphism constraint) and the conformal (double metric) structure (associated with the string Hamiltonian constraint) defines the Born geometry of the metastring. Note that now we have a new spacetime that defines the habitat of string theory, associated with the maximally commuting subalgebras of the Heisenberg algebra of the intrinsic string non-commutativity. This quantum Lagrangian is modular spacetime of the metastring. This in turn directly relates to the quantum spacetime formulation (modular polarization) of quantum theory in general. The 1d modular line is the 2d torus that is compact and not simply connected. Note the Lorentz covariance of modular spacetime, because the intersection of the symplectic, $O(d,d)$ and the double metric structures leads to the Lorentz group.

In the context of general curved Born geometry the symplectic structure is not closed, and we have non-associativity [53]. Finally, supersymmetry (SUSY) is in principle emergent (as in the constructions of the superstring from the bosonic string [49]). The uniqueness of the connection in generalized geometry can be fixed by the phase space structure, instead of SUSY.

In thinking about this new framework for quantum gravity, the teleparallel formulation [55] of gravity appears as the natural language, because of the inherent "flatness" of T-duality (which might be important in the cosmological context for the geometry of classical spacetime at large scales). In this formulation, gravity is described as a Yang-Mills theory, and it is put on the same footing as the matter sector. (It is an interesting question to understand whether teleparallel equations follow from the requirement of the associativity of the symplectic form? That Einstein equations, as well as other equations for the massless modes of the string, follow from the closure of a symplectic-like form is the hallmark of the Bowick-Rajeev geometric quantization approach to string theory [56].)

The metastring offers a new view on the fundamental question of a non-perturbative formulation of quantum gravity [1, 2, 3, 4, 5, 6, 7, 8, 9, 10]. Note that the world-sheet can be made modular in our formulation, with the doubling of τ and σ, so that $\mathbb{X}(\tau,\sigma)$ can be in general viewed as an infinite dimensional matrix (the matrix indices coming from the Fourier components of the doubles of τ and σ). Then the corresponding metastring action should look like

$$\int Tr[\partial_\tau \mathbb{X}^A \partial_\sigma \mathbb{X}^B(\omega_{AB}+\eta_{AB}) - \partial_\sigma \mathbb{X}^A H_{AB} \partial_\sigma \mathbb{X}^B] d\tau d\sigma, \qquad (30)$$

where the trace is over the matrix indices. Then we could associate the natural partonic degrees of freedom with matrix entries. We arrive at a *non-perturbative quantum gravity* by replacing the sigma derivative with a commutator involving one extra $\mathbb{X}^{26}$ (with $A = 0, 1, 2, ..., 25$)[14]

$$\partial_\sigma \mathbb{X}^A \to [\mathbb{X}^{26}, \mathbb{X}^A]. \qquad (31)$$

This dictionary suggests the following fully interactive and non-perturbative formulation of metastring theory in terms of a (M-theory-like) matrix model form of the above metastring action (with $a, b, c = 0, 1, 2, ..., 25, 26$)

$$\int Tr(\partial_\tau \mathbb{X}^a [\mathbb{X}^b, \mathbb{X}^c] \eta_{abc} - H_{ac}[\mathbb{X}^a, \mathbb{X}^b][\mathbb{X}^c, \mathbb{X}^d] H_{bd}) d\tau, \qquad (32)$$

[14]That the canonical world-sheet of string theory might become non-commutative in a deeper, non-perturbative formulation, was suggested in [57].

where the first term is of a Chern-Simons form and the second of the Yang-Mills form, and η_{abc} contains both ω_{AB} and η_{AB}. *This is then the non-perturbative gravitization of the quantum.*[15]

When discussing a non-commutative phase of string theory it is natural to invoke the IIB matrix model [61], which describes N D-instantons (and is by T-duality related to the Matrix model of M-theory [62]). Given our new viewpoint we can suggest a *new covariant* non-commutative matrix model formulation of string theory, as a theory of quantum gravity, by writing in the large N limit $\partial_\sigma \mathbb{X}^C = [\mathbb{X}, \mathbb{X}^C]$ (and similarly for $\partial_\tau \mathbb{X}^B$) in terms of commutators of two (one for $\partial_\sigma \mathbb{X}^C$ and one for $\partial_\tau \mathbb{X}^C$) extra $N \times N$ matrix valued chiral $\mathbb{X}$'s. *Notice that, in general, we do not need an overall trace, and so the action can be viewed as a matrix, rendering the entire non-perturbative formulation as purely quantum in the sense of the original matrix formulation of quantum mechanics (Born-Jordan and Born-Heisenberg-Jordan [26])* :

$$\S_{\mathrm{ncF}} = \frac{1}{4\pi}[\mathbb{X}^a, \mathbb{X}^b][\mathbb{X}^c, \mathbb{X}^d] f_{abcd}, \tag{33}$$

where instead of 26 bosonic $\mathbb{X}$ matrices one would have 28, with supersymmetry emerging in 10(+2) dimensions from this underlying bosonic formulation. (This would be a non-commutative matrix model formulation of F-theory.) By T-duality, the new covariant M-theory matrix model reads as

$$\S_{\mathrm{ncM}} = \frac{1}{4\pi}\int_\tau \left(\partial_\tau \mathbb{X}^i [\mathbb{X}^j, \mathbb{X}^k] g_{ijk} - [\mathbb{X}^i, \mathbb{X}^j][\mathbb{X}^k, \mathbb{X}^l] h_{ijkl}\right), \tag{34}$$

with 27 bosonic $\mathbb{X}$ matrices, with supersymmetry emerging in 11 dimensions[16]. The relevant information about ω_{AB}, η_{AB} and H_{AB} is now contained in the new dynamical backgrounds f_{abcd} in F-theory, and g_{ijk} and h_{ijkl} in M-theory[17].

[15] In the case of the non-perturbative matrix theory like formulation of the metastring (and quantum gravity) the matrices emerge from the modular world-sheet, and the fundamental commutator from the Poisson bracket with respect to the dual world sheet coordinates (of the modular/quantum world sheet) - that is, quantum gravity "quantizes" itself, and thus quantum mechanics originates in quantum gravity (see also, [58]). This formulation should be distinguished from Penrose's "gravitization of the quantum" and gravity induced "collapse of the wave function" [59]. Also note some similarity of the metastring formulation, in its intrinsic non-commutative form, to the most recent proposal by Penrose regarding "palatial" twistor theory [60].)

[16] In this approach holography [63] (such as AdS/CFT [64], which can be viewed as a "quantum Jarzynski equality on the space of geometrized RG flows" [65]) is emergent in a particular "extensification" of quantum spacetime.

[17] This offers a new formulation of covariant Matrix theory in the M-theory limit[66], which is essentially a partonic formulation - strings emerge from partonic constituents in a certain limit. This new matrix formulation is fundamentally bosonic and thus it is reminiscent of bosonic M-theory [67]. The relevant backgrounds g_{ijk} and h_{ijkl} should be determined by the matrix RG equations. Also, there are lessons here for the new concept of "gravitization of quantum theory" as well as the idea that dynamical Hilbert spaces

This matrix like formulation should be understood as a general non-perturbative formulation of string theory. In this partonic formulation closed strings are collective excitations, in turn constructed from the product of open string fields. The observed classical spacetime emerges as an "extensification" [1, 2, 3, 4, 5, 6, 7, 8, 9, 10], in a particular limit, out of the basic building blocks of quantum spacetime. Their remnants can be found in the low energy bi-local quantum fields, with bi-local (metaparticle) quanta, to which we now turn.

4.3. Metastrings, quantum fields and metaparticles

What is the effective description of closed strings that incorporates the above intrinsic non-commutativity? For a closed sting on a circle of radius R (where the dual radius $\tilde{R}$, is defined as $R\tilde{R} = 2\lambda^2$ and the respective winding integers are n and w) this effective description is captured by the generalized field [7, 8]

$$\Phi(x, \tilde{x}) \equiv \sum_w \Phi_w(x) e^{iw\tilde{x}/\tilde{R}}. \tag{35}$$

This meshes well with the observation [7, 8] that the string product is essentially a representation of the Heisenberg group, which suggests that one should consider the "quantization" map

$$\Phi(x, \tilde{x}) \to \hat{\Phi} = \sum_w \Phi_w(\hat{x}) e^{iw\hat{\tilde{x}}/\tilde{R}}, \tag{36}$$

from generalized fields to non-commutative fields.[18] Under this map the T-duality transformation becomes "localized" and is expressed as the exchange of $\hat{x}$ with $\hat{\tilde{x}}$. The T-dual expression is given by [7, 8]

$$\hat{\Phi} = \sum_n e^{in\hat{x}/R} \Phi_n(\hat{\tilde{x}} - \pi n \tilde{R}) = \sum_n \Phi_n(\hat{\tilde{x}}) e^{in\hat{x}/R}, \tag{37}$$

which has a similar form to (36). We see that the non-commutativity of $\hat{x}$ with $\hat{\tilde{x}}$ allows one to reabsorb all the shifts in terms of a simple reordering that exchanges $\hat{x}$ with $\hat{\tilde{x}}$ and is the expression of T-duality. The "quantized" field is simply expanded in terms of modes as

$$\hat{\Phi} \equiv \sum_{w,n} e^{in\hat{x}/R} \Phi(n, w) e^{iw\hat{\tilde{x}}/\tilde{R}}. \tag{38}$$

or 2-Hilbert spaces (here represented by matrices) are fundamentally needed in quantum gravity [68].

[18] Here, we have chosen a specific operator ordering. Given this ordering, the mapping is well-defined and consistent with the string product.

It is useful at this point to generalize the construction to higher dimensional tori. This can be done in a straightforward manner by introducing the modes $\mathbb{K}^A = (\tilde{k}^a, k_a)$, generalizing $(w/\tilde{R}, n/R)$. The integrality condition for the lattice Λ of admissible modes $\mathbb{K}, \mathbb{K}' \in \Lambda$ reads in this notation as[19] $(\eta \pm \omega)(\lambda\mathbb{K}, \lambda\mathbb{K}') \in \mathbb{Z}$. We now write $\Phi(\mathbb{K}) = \langle \mathbb{K} | \Phi \rangle$ with the ordering chosen as $\langle \mathbb{K} | = \langle 0 | \hat{U}_{-\mathbb{K}}$, where $\hat{U}_{\mathbb{K}} \equiv e^{ik\cdot\hat{x}} e^{i\tilde{k}\cdot\hat{\tilde{x}}}$. This ordering can be seen to be related to the choice of an $\mathrm{O}(d,d)$ frame, where we place the operator associated with x on the left and the operator associated with the dual space $\tilde{x}$ on the right. The key point is that this choice of frame is entirely encoded into the choice of symplectic potential ω and the vertex operator can be covariantly written in terms of $\mathbb{K} = (\tilde{k}, k)$ and $\mathbb{X} = (x, \tilde{x})$ as

$$\hat{U}_{\mathbb{K}} = e^{\frac{i}{2}(\eta+\omega)(\mathbb{K},\hat{\mathbb{X}})} e^{\frac{i}{2}(\eta-\omega)(\mathbb{K},\hat{\mathbb{X}})}. \tag{39}$$

Given this notation we can write the string product covariantly as [7, 8]

$$(\Phi \circ \Psi)(\mathbb{K}) = \sum_{\mathbb{K}'+\mathbb{K}''=\mathbb{K}} \Phi(\mathbb{K}') e^{i\pi(\eta-\omega)(\lambda\mathbb{K}',\lambda\mathbb{K}'')} \Psi(\mathbb{K}''). \tag{40}$$

The non-commutativity of the string product is encoded in terms of a π-flux due to ω. As it turns out the phase factor is exactly the same as the cocycle factor $\epsilon(\mathbb{K}, \mathbb{K}') = e^{i\pi(\eta-\omega)(\lambda\mathbb{K},\lambda\mathbb{K}')}$ that appears in the definition of the vertex operator product [7, 8].

Thus, quantum gravity, in the guise of metastring theory, produces at low energy bi-local quantum fields, with intrinsic non-commutatitivity $\phi(x, \tilde{x})$ where $[x, \tilde{x}] = i\lambda^2$. As we have already discussed in section 3, and as we will see more explicitly in the following section 5, these bi-local fields are doubled, and their proper formulation requires a double scale renormalization group (RG) found in the context of non-commutative field theory [32] (which in certain models lead to a finite non-perturbative renormalization). Such doubled bi-local fields ($\phi(x, \tilde{x})$ and $\tilde{\phi}(x, \tilde{x})$) have metaparticle excitations to be discussed in the next section. One can view such (meta)fields as low energy manifestations of the metastring field. These intrinsically non-commutative quantum fields can also be understood to arise from the representation of the symmetry groups associated with Born geometry. They should be relevant both in the high energy context (see the discussion that follows on dark matter and dark energy[20]) as well as in condensed matter physics (as new quantum order parameters for highly entangled and strongly correlated phases of quantum matter).

[19] In the one dimensional case where $\mathbb{K} = (w/\tilde{R}, n/R)$ this follows directly from $(\eta + \omega)(\lambda\mathbb{K}, \lambda\mathbb{K}') = nw'$ and similarly $(\eta - \omega)(\lambda\mathbb{K}, \lambda\mathbb{K}') = wn'$, given that $n, n', w, w' \in \mathbb{Z}$.

[20] Even the Standard Model of particle physics (coupled to Einstein's gravity) exhibits hidden non-commutative geometry (NCG), as discussed in [22] (for recent reviews and references, consult [69]), and this implies some unique phenomenological consequences [70]. The NCG action should be compared to (34), by replacing the commutators with the NCG Dirac operator.

To summarize, the above manifestly T-duality covariant formulation of closed strings (i.e. the metastring) implies intrinsic non-commutativity of zero-modes. It is thus instructive to formulate a particle-like limit of the metastring that we call the *metaparticle* [9]. Given the form for the symplectic structure of the zero modes derived in section 4. of [7] (equation (67) of that paper, without the contribution coming from string oscillators), the action $S \equiv \int d\tau L$ of the metaparticle is governed by the following Lagrangian (implied by the symplectic structure of the closed string) [7, 8]

$$L = p_\mu \dot{x}^\mu + \tilde{p}^\mu \dot{\tilde{x}}_\mu + \alpha' p_\mu \dot{\tilde{p}}^\mu - \frac{N}{2}\left(p_\mu p^\mu + \tilde{p}_\mu \tilde{p}^\mu - m^2\right) + \tilde{N}\left(p_\mu \tilde{p}^\mu - \mu\right), \tag{41}$$

where N and $\tilde{N}$ are the Lagrange multipliers for the two constraints that follow from the Hamiltonian ($H \equiv \partial_\sigma \mathbb{X}^A H_{AB} \partial_\sigma \mathbb{X}^B = 0$) and diffeomorphism constraints ($D \equiv \partial_\sigma \mathbb{X}^A \eta_{AB} \partial_\sigma \mathbb{X}^B = 0$) of the metastring [3, 6].

Note that the usual particle limit is obtained, at least classically, by taking $\mu \to 0$ and $\tilde{p} \to 0$. The theory of metaparticles can be viewed as the theory of the zero modes of the closed string, which fully takes into account its intrinsic non-commutativity. Given the form of the above Lagrangian, the metaparticle looks like two particles that are entangled through a Berry phase-like $p_\mu \dot{\tilde{p}}^\mu$ factor. The metaparticle is fundamentally non-local, and thus it should not be associated with effective local field theory. In particular, by looking at the metaparticle constraints $p^2 + \tilde{p}^2 = m^2$ and $p\tilde{p} = \mu$, we note that the momenta p and $\tilde{p}$ can be, in principle, widely separated. For example, if m is of the order of the Planck energy, and μ of the order of one TeV (which could be understood as a characteristic particle physics scale), then the momentum p can be of the order of the Planck energy, and the momentum $\tilde{p}$ of the vacuum energy scale. Thus the metaparticle theory is able to naturally relate widely separated scales, which transcends the usual reasoning based on Wilsonian effective field theory (and should be relevant for the naturalness and hierarchy problems).

5. Quantum gravity, metaparticles and dark matter

The theory of metaparticles (the low energy remnants of the metastring, and as such, the low energy remnants of quantum gravity) can thus be defined by the following world-line action [9]

$$S \equiv \int_0^1 d\tau [p\dot{x} + \tilde{p}\dot{\tilde{x}} + \alpha' p\dot{\tilde{p}} - \frac{N}{2}\left(p^2 + \tilde{p}^2 + m^2\right) + \tilde{N}\left(p\tilde{p} - \mu\right)]. \tag{42}$$

Here the signature $(+, -, \ldots, -)$ and the contraction of indices are implicitly assumed. At the classical level, theory of metaparticles is a world-line theory with the usual reparameterization invariance and two additional features [9]. The first new feature of the model is the presence of an additional local symmetry, which from the string point of view corresponds to the completion of worldsheet diffeomorphism invariance. From the particle world-line point of view, this symmetry is associated with an additional local constraint. The second new feature is the presence of a non-trivial symplectic form on the metaparticle phase space, also motivated by string theory [7, 8]. Because of its interpretation as a particle model on Born geometry, associated with the modular representation of quantum theory, the space-time on which the metaparticle propagates is ambiguous, with different choices related by what in string theory we would call T-duality. The attractive feature of this model include world-line causality and unitarity, as well as an explicit mixing of widely separated energy-momentum scales. The metaparticle propagator follows from the world-line path-integral defined by the above action and it has the following form in momentum space [9]

$$G(p, \tilde{p}; p_i, \tilde{p}_i) \sim \delta^{(d)}(p - p_i)\delta^{(d)}(\tilde{p} - \tilde{p}_i)\frac{\delta(p \cdot \tilde{p} - \mu)}{p^2 + \tilde{p}^2 + m^2 - i\varepsilon}. \tag{43}$$

The canonical particle propagator is a highly singular $\tilde{p} \to 0$ (and $\mu \to 0$) limit of this expression. This propagator also predicts the following dispersion relation (in a particular gauge [9]) that can be tested in various experiments and with various probes

$$E_p^2 + \frac{\mu^2}{E_p^2} = \vec{p}^2 + m^2. \tag{44}$$

This formulation is fully compatible with Lorentz covariance, and is a direct consequence of the consistency of quantum theory and a minimal length (and thus Born geometry). In general, for each particle at energy E there exists a dual particle at energy $\frac{\mu}{E}$. (This is complete analogy for the well-known prediction of antiparticles in the union of special relativity and quantum theory.)[21]

We can also discuss the background fields that couple to the metaparticle quanta. Following the well-known procedure of introducing the background fields in the case of particles, by shifting the canonical momentum by a gauge field, we might try to extend the gauging procedure to the metaparticle counterpart. There is a possible ambiguity in this gauging which depends on which configuration variables one decides to work with. If one takes $(x, \tilde{x})$ as configuration variables, one obtains a gauging which could also be motivated by the presence of a "stringy gauge field" in

[21] Regarding the phenomenology of quantum gravity (including the phenomenology of the minimal length) another generic feature presents itself in the context of the quantum version of the gravitational memory effect [71], which should involve "modular supermomentum". For some other tests of quantum gravity, and especially intrinsic non-locality and non-commutativity of quantum spacetime, see [72].

metastring theory [3]

$$S \to \int \Big((p_\mu + A_\mu(x,\tilde{x}))\dot{x}^\mu + (\tilde{p}^\mu + \tilde{A}^\mu(x,\tilde{x}))\dot{\tilde{x}}_\mu \\ + 2\pi\alpha' p_\mu \dot{\tilde{p}}^\mu - e\mathcal{H}(p,\tilde{p}) - \tilde{e}\mathcal{D}(p,\tilde{p}) \Big). \tag{45}$$

Indeed, if we introduce canonical momenta

$$P_\mu = p_\mu + A_\mu(x,\tilde{x}), \qquad \tilde{P}^\mu = \tilde{p}^\mu + \tilde{A}^\mu(x,\tilde{x}), \tag{46}$$

we obtain then

$$S \to \int \Big(P_\mu \dot{x}^\mu + \tilde{P}^\mu \dot{\tilde{x}}_\mu + 2\pi\alpha'(P_\mu - A_\mu(x,\tilde{x}))(\dot{\tilde{P}}^\mu - \frac{d}{dt}\tilde{A}^\mu(x,\tilde{x})) \\ - e\mathcal{H}(P - A(x,\tilde{x}), \tilde{P} - \tilde{A}(x,\tilde{x})) \\ - \tilde{e}\mathcal{D}(P - A(x,\tilde{x}), \tilde{P} - \tilde{A}(x,\tilde{x}))\Big), \tag{47}$$

but, now, because of the α' term we see that $\dot{\tilde{A}}$ contains $\dot{\tilde{x}}$. The background fields $A_\mu(x,\tilde{x})$ and $\tilde{A}^\mu(x,\tilde{x})$ are the natural modular fields in this case. (Note that this procedure explicitly realizes the general comments made in section 3 regarding quantum field theory in the modular polarization.) Thus, the generic prediction here is the existence of a dual field $\tilde{A}$, which is entangled, with the original A field[22].

We expect that the correct field theoretic description of the metaparticle is in terms of the above general non-commutative (modular) field theory $\Phi(x,\tilde{x})$ limit of the metastring [1, 2, 3, 4, 5, 6, 7, 8, 9, 10]. Such effective non-commutative field theory is similar in spirit to [32]. Also, we note that the concept of metaparticles might be argued from the compatibility of the quantum spacetime that underlies the generic representations of quantum theory, as discussed in [5], and thus the metaparticle might be as ubiquitous as the concept of antiparticles which is demanded by the compatibility of relativity and quantum theory. The metaparticles also provide a natural route to the problem of dark matter. To lowest (zeroth) order of the expansion in the non-commutative parameter λ the effective action for Standard

[22]Note that the metaparticle propagator leads to a "Friedel-like" bi-local static potential [73] as well as non-local generalizations of quantum statistics. The traditional statement concerning quantum gravitational effects is that they are tied to the Planck scale. However, quantum gravity can be revealed at macroscopic scales via quantum statistics. In particular, it was argued in [74] that black hole statistics is infinite statistics [75] (which is consistent with non-locality and Lorentz symmetry). Also, in [76] a statistical argument was used to argue for probable values of the cosmological constant. More recently, such statistical arguments were used in [77] to analyze the black hole spin in gravitational wave observations. See [78] for the relevance of infinite statistics for the fine structure of dark energy. As non-local objects, metaparticles can exhibit such quantum statistics.

Model matter Lagrangian (L_m) and their duals (that could be interpreted as dark matter Lagrangian L_{dm}) S_{eff} takes the following form (where we have included the gravitational background as non-dynamical):

$$S_{eff} = -\iint \sqrt{g(x)\tilde{g}(\tilde{x})}[L_m(A(x,\tilde{x})) + \tilde{L}_{dm}(\tilde{A}(x,\tilde{x})) + ...]. \tag{48}$$

Note that after integrating over the "hidden variable parameters $\tilde{x}$ we get an effective theory of visible and dual (dark) matter in the observed spacetime x

$$S_{eff} = -\iint \sqrt{g(x)}[L_m(A(x)) + \tilde{L}_{dm}(\tilde{A}(x)) + ...]. \tag{49}$$

Thus, the metaparticle can be understood as a generic message of string theory/quantum gravity for low energy physics. Like their visible particle cousins, dark matter quanta should be detectable through their particular metaparticle entanglement to visible matter, as indicated by equation (47): $\alpha' A_\mu \frac{d}{dt}\tilde{A}^\mu$ (say, for a photon and its dual). This is a Berry-phase like term that comes from a fully covariant description, and is uniquely different from the usual effective field theory interaction terms between visible and dark matter particles.

Such dark matter quanta are correlated to visible matter and have been discussed in the literature as Modified dark matter [12]. Modified dark matter (MDM) is, at the moment, a phenomenological model of dark matter, inspired by gravitational thermodynamics. For an accelerating Universe with positive cosmological constant Λ, certain phenomenological considerations lead to the emergence of a critical acceleration parameter related to Λ (essentially that "fundamental acceleration" is just the value of Λ expressed as acceleration $\sim cH$, where H is the Hubble parameter, and thus, it is of the order of $10^{-10}m/s^2$). Such a critical acceleration is an effective phenomenological manifestation of MDM, and it is found in correlations between dark matter and baryonic matter in galaxy rotation curves. The resulting MDM mass profiles, which are sensitive to Λ, are consistent with observational data at both the galactic and cluster scales. In particular, the same critical acceleration appears both in the galactic and cluster data fits based on MDM [12]. Furthermore, using some robust qualitative arguments, MDM appears to work well on cosmological scales. If the quanta of modified dark matter are metaparticles, this may explain why, so far, dark matter detection experiments have failed to detect dark matter particles. In particular, the natural model for MDM quanta could be provided by the metaparticle realizations of the Standard Model particles, associated with bi-local extensions of all Standard Model fields. Thus the baryon matter described by the Standard Model fields (the A backgrounds in the above discussion), would have natural cousins (the $\tilde{A}$ backgrounds in the above discussion) in the dark matter sector, which in turn would be sensitive to the dark energy modeled by the cosmological constant Λ. This leads us naturally to the last topic of this talk - the realization of dark energy in the metastring approach to quantum gravity.

6. Quantum gravity, metastrings and dark energy

In this section we give a new interpretation [11] of dark energy from this novel point of view of string theory (and quantum gravity) [1, 2, 3, 4, 5, 6, 7, 8, 9, 10]. Ever since the seminal discovery of dark energy in the late 1990s [79], string theory (viewed as a consistent theory of quantum gravity and matter) has been attempting to deal with this central ingredient of fundamental physics. (For the most recent measurements of the Hubble constant and the associated discrepancies (see [80]). The existence of de Sitter space (dS) as a solution in string theory (and dark energy in the observable universe) is still considered an outstanding open question [81], and the interest in this fundamental issue has been recently reignited in [82].

6.1. Metastring theory and dark energy

We now explain how the generalized geometric formulation of string theory discussed above provides for an effective description of dark energy that is consistent with de Sitter spacetime. This is essentially due to the theory's chirally and non-commutatively doubled realization of the target space and the stringy effective action on the doubled non-commutative spacetime $(x^a, \tilde{x}^a)$

$$S^{nc}_{\text{eff}} = \iint \operatorname{Tr}\sqrt{g(x,\tilde{x})}\,\big[R(x,\tilde{x}) + L_m(x,\tilde{x}) + \ldots\big], \tag{50}$$

where the ellipses denote higher-order curvature terms induced by string theory. (Here we have included the matter Lagrangian L_m as well.) This S^{nc}_{eff} clearly expands into numerous terms with different powers of λ, which upon $\tilde{x}$-integration and from the x-space vantage point produce various effective terms. To lowest (zeroth) order of the expansion in the non-commutative parameter λ of S^{nc}_{eff} takes the form:

$$S_d = -\iint \sqrt{-g(x)}\sqrt{-\tilde{g}(\tilde{x})}[R(x) + \tilde{R}(\tilde{x})], \tag{51}$$

a result which first was obtained almost three decades ago, effectively neglecting ω_{AB} by assuming that $[\hat{x}^a, \hat{\tilde{x}}_b] = 0$ [83]. In this leading limit, the $\tilde{x}$-integration in the first term defines the gravitational constant G_N, and in the second term produces a *positive* cosmological constant constant $\Lambda > 0$. In particular, we are lead to the following low energy effective action valid at long distances of the observed accelerated universe (focusing on the relevant 3+1-dimensional spacetime X, of the $+---$ signature):

$$S_{\text{eff}} = \frac{-1}{8\pi G}\int_X \sqrt{-g}\big(\Lambda + \tfrac{1}{2}R + \mathcal{O}(R^2)\big), \tag{52}$$

with Λ the positive cosmological constant (corresponding to the scale of 10^{-3} eV) and the $\mathcal{O}(R^2)$ denote higher order corrections (which are also required by the sigma model of string theory [84]).

It also follows from this construction that the weakness of gravity is determined by the size of the canonically conjugate dual space, while the smallness of the cosmological constant is given by its curvature. (Higher order terms in λ produce various forms of dark energy [85] and this may even provide for a way of addressing the recent conflicting measurements of the Hubble constant [80].) Given this action, we may proceed reinterpreting [83]: integrate out the dual spacetime coordinates, write the effective action as $\bar{S} \sim \tilde{V} \int_X \sqrt{-g(x)} R(x) + ...$, where $\tilde{V} = \int_{\tilde{X}} \sqrt{-\tilde{g}(\tilde{x})}$, and then relate the dual spacetime volume to the observed spacetime volume as $\tilde{V} \sim V^{-1}$ (T-duality). This produces an "intensive" effective action [83]

$$\bar{S} = \frac{\int_X \sqrt{-g(x)}\big(R(x) + L_m(x)\big)}{\int_X \sqrt{-g(x)}} + \dots \tag{53}$$

By concentrating on the classical description first (we discuss below quantum corrections and the central role of intrinsic non-commutativity in string theory) we get the following Einstein equations [83]

$$R_{ab} - \frac{1}{2} R g_{ab} + T_{ab} + \frac{1}{2} \bar{S}\, g_{ab} = 0, \qquad T_{ab} \stackrel{\text{def}}{=} \frac{\partial L_m}{\partial g^{ab}} - \frac{1}{2} L_m\, g_{ab}. \tag{54}$$

We emphasize that our reinterpretation of [83] does not follow the original presentation and intention. In particular, we directly relate the intensive action (54) to the cosmological constant, $\bar{S} \sim \Lambda$.

Note that this new approach to the question of dark energy (viewed as a cosmological constant) in quantum gravity is realized in certain stringy-cosmic-string-like [86] toy models [87], which can be viewed as illustrative of a generic non-commutative phase of F-theory [88]. In particular, the "see-saw" formula (54) is directly realized in [87] as $M_\Lambda \sim M^2/M_P$, where M_Λ is the dark energy scale, M_P the Planck scale and M, and intermediate scale, coming from the matter sector.

6.2. Dark energy and dark matter from metastring theory

Note that in general, to lowest (zeroth) order of the expansion in the non-commutative parameter λ of S^{nc}_{eff} takes the following form (that also includes the matter sector and its dual) [89]:

$$S_d = -\iint \sqrt{g(x)\tilde{g}(\tilde{x})}[R(x) + \tilde{R}(\tilde{x}) + L_m(A(x,\tilde{x})) + \tilde{L}_{dm}(\tilde{A}(x,\tilde{x}))], \tag{55}$$

Here the A fields denote the usual Standard Model fields, and the $\tilde{A}$ are their duals, as predicted by the general formulation of quantum theory that is sensitive to the minimal length. Note that after integrating over the dual spacetime, and after taking into account T-duality, the equation (54) now reads

$$\bar{S} = \frac{\int_X \sqrt{-g(x)}\big(R(x) + L_m(x) + \tilde{L}_{dm}(x)\big)}{\int_X \sqrt{-g(x)}} + \dots \tag{56}$$

The proposal here is that the dual sector (as already indicated in the previous section) should be interpreted as the dark matter sector, which is correlated to the visible sector via the dark energy sector, as discussed in [12]. We emphasize the unity of the description of the entire dark sector based on the properties of the dual spacetime, as predicted by the generic formulation of string theory (as a quantum theory with a dynamical Born geometry).

6.3. Dark energy and radiative stability

Let us return to the discussion of the dark energy sector. The above results from the commutative limit are not stable under loop corrections. which has been addressed in the recent work of Kaloper and Padilla (called the sequester mechanism) who also extended these results to loops of arbitrary order, in the effective field theory [90]. In that context, the effective field theory expansion has to have another global scale, s, so that the sequestering action is proportional to

$$\int_X \sqrt{-g}\Big[\frac{R}{2G} + s^4 L(s^{-2} g^{ab}) + \frac{\Lambda}{G}\Big] + \sigma\Big(\frac{\Lambda}{s^4 \mu^4}\Big), \tag{57}$$

where L denotes the combined Lagrangians for the matter and dark matter sectors, μ is a mass scale and $\sigma(\frac{\Lambda}{s^4\mu^4})$ is a global interaction that is not integrated over [90]. This can be provided by our set up: Start with bilocal fields $\phi(x, \tilde{x})$ [5], and replace the dual labels $\tilde{x}$ and also λ (in a coarsest approximation) by the global dynamical scale $s \sim \Delta\tilde{x} \sim \lambda^2 \Delta x^{-1}$. Also, normal ordering produces σ. This is an effective realization of the sequester mechanism in a non-commutative phase of string theory. Furthermore, the intrinsic non-commutativity of the zero modes x and $\tilde{x}$ in $[x, \tilde{x}] = i\lambda^2$ corrects the zeroth order results in λ in several ways. In particular, it is natural to ask whether the non-zero ω^{AB} in $[\hat{\mathbb{X}}^A, \hat{\mathbb{X}}^B] = i\omega^{AB}$ stabilizes the cosmological constant directly on the level of the effective non-commutative action. The fully non-commutative analysis is intricate, but for conformally flat metrics, $g_{\mu\nu} = \phi^2 \eta_{\mu\nu}$, the action (50)–(51) produces a non-commutative $\Lambda\phi^4$ theory, which is a natural non-commutative generalization of the effective action for conformal metrics $\int_X (\partial_\mu \phi \partial^\mu \phi + \frac{\Lambda}{3}\phi^4)$, with the non-commutative product depending on λ[23]. Unlike the commutative limit of the theory, the beautiful results of Grosse and Wulkenhaar [92]

[23]See also [91].

demonstrate the non-perturbative solvability of the above non-commutative $\Lambda\phi^4$ theory, explicitly showing the finite renormalization of Λ in terms of the bare coupling. At least in this highly simplified, conformal degree limit, non-commutativity thus can afford a small, radiatively and perhaps even non-perturbatively stable cosmological constant for the non-commutative form of the "doubled" effective action.

However, non-commutative field theories have both UV and IR scales and the effective description is defined by expanding around self-dual fixed points, and it is organized by keeping both the Wilsonian UV cutoff as well as the IR scale. This clearly meshes nicely with the UV and IR aspects of the see-saw formula. Identifying M_Λ and M_P as the IR and UV cut-offs, respectively, the double-scale RG flow identifies a self-dual fixed point [32]. Given that the phase-space formulation [1, 4, 5, 6, 7] is a T-duality covariant description of string theory, this naturally relates $M_P \to M^2/M_P$ under T-duality. The prediction of our effective stringy [93] cosmic-string-like models [87] $M_\Lambda \sim M^2/M_P$ then satisfies these conditions, with $M_P \sim \epsilon = 1/\lambda$ the fundamental energy scale corresponding to the fundamental length λ, which is consistent with observations provided M is a TeV scale. We emphasize that the usual spacetime discussion of string theory is compatible with local effective field theory, which does not account for the radiative stability of vacuum energy. What we argue is that this feature of string theory is an artifact of a spacetime description which is not generic. The generic formulation of string theory is doubled and generalized-geometric, and intrinsically non-commutative, and it leads to an effective field theory that is sequestered, and thus, to leading order, to a radiatively stable vacuum energy. (Including further corrections due to intrinsic non-commutativity.) Only in a singular limit in which one neglects the intrinsic non-commutativity and works only within a spacetime section of the general doubled description does one find the usual effective field theory, with a spacetime interpretation, and the usual questions regarding the existence of dS background in string theory [81] (and the related issues related to holography in the context of asymptotically de Sitter spaces [94] as well as supersymmetry breaking and the existence of a small and positive cosmological constant [95])[24]

7. Conclusion: Quantum gravity and the real world

In this talk we have summarized the recent work on quantum foundations of string theory and quantum gravity [1, 2, 3, 4, 5, 6, 7, 8, 9, 10]. In particular, we have discussed intrinsic non-commutativity in quantum gravity related to a new concept of quantum spacetime, called modular spacetime

[24]Note also that Starobinsky inflation [96] may appear as a natural product of the higher order terms in the λ expansion that, after integrating over $\tilde{x}$ can result in $\int_X \sqrt{-g}(R + aR^2)$, at the next to leading order in λ. Starobinsky inflation beautifully fits the observed data [97], and is non-supersymmetric — which is consistent with the supersymmetry-breaking nature of our discussion.

that also appears as a habitat for metastring theory and that is deeply rooted in the foundations of quantum theory (and, especially, in the concept of modular variables that goes back to the work of Weyl, Schwinger and Aharonov). Note that this concept stems from a quantization of spacetime, and not from quantization of gravitational field/metric. Even the flat space is quantized according to our approach to quantum gravity. This allows for superposition and entanglement of spacetimes. Also, this formulation provides for an explicit construction of spacetime quanta or qubits (the fully "compactified" bosonic string), and a new non-perturbative definition of quantum gravity as "gravitization of the quantum". Such a fully dynamical (or "curved") construction of Born geometry can be approached from the point of view of "teleparallel gravity" in which one utilizes the at (zero-curvature) connection and crucially introduces non-zero torsion [55]. This viewpoint is natural for the rigid structure of Born geometry and it allows for "curving" of T-duality. In some sense, by going from our new formulation of quantum mechanics in terms of modular, or quantum spacetime, with hidden but fixed Born geometry, and its application to quantum field theory, to an explicit formulation of quantum gravity that involves dynamical Born geometry, as is the case in metastring theory, we are retracing (in a purely quantum context) the line of development that led from special relativity (and fixed Minkowski geometry) and its application to classical relativistic field theory, to general theory of relativity with a dynamical spacetime geometry[25].

The underlying physical principle here is relative locality - different observers probe different spacetimes; these different spacetimes are sections of a quantum (modular) spacetime - implying, in general, a dynamical momentum space[26]. According to relative locality, quantum mechanics follows from non-locality that is consistent with causality. In particular, it follows from the existence of fundamental length and fundamental time that are consistent with Lorentz symmetry. The geometry of quantum non-locality is Born geometry. In particular, if the relevant physics lives on a spacetime lattice, the full set of quantum numbers involves the lattice and its dual. This leads to unitary (modular) variables, and modular spacetime (spacetime and its dual being tied via intrinsic non-commutativity that involves the fundamental length/time.) Then it follows that quantum gravity is dynamical Born geometry found in the metastring approach to quantum gravity. The effective description is given in terms of quantum field theory with intrinsic non-commutativity - with metaparticle excitations. The renormalization group has both UV and IR scales - leading to a double

[25]There is an interesting connection here to quantum logic [98], "Logic being to quantum theory what geometry is to gravity" (R. Sorkin). Quantum gravity then appears as "warped logic" (D. Finkelstein), and as "third relativity" (J. A. Wheeler); first relativity being represented as Galilean, special and general relativity and second relativity as quantum theory.

[26]Momentum space geometry is relevant for the recent progress in scattering amplitudes [99].

RG, with direct relevance for the recent discussion of the infrared limit of QFT [71]. This new understanding of quantum field theory with manifest quantum non-locality is relevant both for particle physics and also for condensed matter physics. In particular, some of the implications for the real world are dark energy as curvature of the dual spacetime [11], and dark matter correlated to visible matter and dark energy [12] and represented by metaparticles [10].

Finally, we comment on the problem of vacuum selection. Born reciprocity [16] demands a symmetric dynamical geometric structures in spacetime and energy momentum space. Thus matter and spacetime are put on equal footing (which reminiscent of with some intuitions from F-theory). The question is whether Born reciprocity can be used as a criterion for vacuum selection in quantum gravity, which selects "maximally symmetric solutions" both in spacetime (de Sitter space) and in the matter sector (Standard Model and its dual, describing the dark matter sector). Then the apparent robustness of the "genetic code" (masses and couplings) of particle physics (see, for example, [100]) might be the consequence of an attractor mechanism that makes the observed cosmology and particle physics "universal" (this is similar to what happens in universal biology of the "genetic code" based on horizontal gene transfer [101]). The idea here is that a universal "genetic code" for particle physics (that is, the particle masses and couplings, as well as cosmological parameters) can be obtained via a horizontal information transfer mechanism that apparently leads to a universal genetic code. In the case of universal cosmology, such horizontal information transfer can be provided by an interaction between spacetime quanta via the gravitization of the quantum, leading to a "maximally symmetric" attractor solution. It would be fascinated to explore this idea more concretely in the future.

References

[1] L. Freidel, R. G. Leigh and D. Minic, Phys. Lett. B **730**, 302 (2014).

[2] L. Freidel, R. G. Leigh and D. Minic, Int. J. Mod. Phys. D **23**, no. 12, 1442006 (2014).

[3] L. Freidel, R. G. Leigh and D. Minic, JHEP **1506**, 006 (2015).

[4] L. Freidel, R. G. Leigh and D. Minic, Int. J. Mod. Phys. D **24**, no. 12, 1544028 (2015).

[5] L. Freidel, R. G. Leigh and D. Minic, Phys. Rev. **D 94**, 104052 (2016).

[6] L. Freidel, R. G. Leigh and D. Minic, J. Phys. Conf. Ser. **804**, no. 1, 012032 (2017).

[7] L. Freidel, R. G. Leigh and D. Minic, JHEP **1709**, 060 (2017)¿

[8] L. Freidel, R. G. Leigh and D. Minic, Phys. Rev. D **96**, no. 6, 066003 (2017).

[9] L. Freidel, J. Kowalski-Glikman, R. G. Leigh and D. Minic, Phys. Rev. D **99**, no. 6, 066011 (2019).

[10] L. Freidel, R. G. Leigh and D. Minic, Int. J. Mod. Phys. A **34**, no. 28, 1941004 (2019).

[11] P. Berglund, T. Hubsch and D. Minic, Phys. Lett. B **798**, 134950 (2019) Int. J. Mod. Phys. D **28**, 1902003 (2019). JHEP **1912**, 166 (2019) [JHEP **2019**, 166 (2020)].

[12] C. M. Ho, D. Minic and Y. J. Ng, Phys. Lett. B **693**, 567 (2010). Gen. Rel. Grav. **43**, 2567 (2011) [Int. J. Mod. Phys. D **20**, 2887 (2011)]. Phys. Rev. D **85**, 104033 (2012). D. Edmonds, D. Farrah, C. M. Ho, D. Minic, Y. J. Ng and T. Takeuchi, Astrophys. J. **793**, 41 (2014). Int. J. Mod. Phys. A **32**, no. 18, 1750108 (2017). D. Edmonds, D. Farrah, D. Minic, Y. J. Ng and T. Takeuchi, Int. J. Mod. Phys. D **27**, no. 02, 1830001 (2017). D. Edmonds, D. Minic and T. Takeuchi, in preparation.

[13] K. G. Wilson and J. B. Kogut, Phys. Rept. **12**, 75 (1974); K. G. Wilson, Rev. Mod. Phys. **47**, 773 (1975).

[14] J. Zak, Phys. Rev. **168**, 686 (1968). Also, J. Zak, Phys. Rev. Lett. 19 (1967) 1385 - 1387.

[15] L. Freidel, F. J. Rudolph and D. Svoboda, JHEP **1711**, 175 (2017); also, arXiv:1806.05992 [hep-th].

[16] M. Born, Nature 136 (1935) 952; Rev. Mod. Phys. 21 (1949) 463; Proc. Roy. Soc. London A165 (1938) 291.

[17] G. Amelino-Camelia, L. Freidel, J. Kowalski-Glikman and L. Smolin, Phys. Rev. D **84**, 084010 (2011); G. Amelino-Camelia, L. Freidel, J. Kowalski-Glikman and L. Smolin, Gen. Rel. Grav. **43**, 2547 (2011) [Int. J. Mod. Phys. D **20**, 2867 (2011)].

[18] Y. Aharonov, A. Petersen and H. Pendleton, Int. J. Theor. Phys. **2**, 213 (1969).

[19] J. Schwinger, *Quantum Kinematics and Dynamics*, Frontiers in Physics, Perseus, 2000; G. 't Hooft, "The Cellular Automaton Interpretation of Quantum Mechanics. A View on the Quantum Nature of our Universe, Compulsory or Impossible?", arXiv:1405.1548 [quant-ph]

[20] M. Gell-Mann and J. B. Hartle, Phys. Rev. D **47**, 3345 (1993).

[21] H. Weyl, *The theory of groups and quantum mechanics* (New York: Dover) (1931).

[22] A. Connes, *Noncommutative Geometry*, Academic Press, San Diego, 1994; A. Connes and M. Marcolli, *Noncommutative Geometry, Quantum Fields and Motives*, AMS, 2007.

[23] I. M. Gelfand, and M. A. Naimark, *Mat.Sbornik* **12** 197 (1943).

[24] Y. Aharonov, and D. Rohrlich, *Quantum Paradoxes: Quantum Theory for the Perplexed* (New York: Wiley) (2005).

[25] L. N. Chang, Z. Lewis, D. Minic and T. Takeuchi, Mod. Phys. Lett. B **27**, 1350064 (2013); J. Phys. A **46**, 065304 (2013); J. Phys. A **46**, 485306 (2013); Int. J. Mod. Phys. A **29**, 1430006 (2014). T. Takeuchi, L. N. Chang, Z. Lewis and D. Minic, AIP Conf. Proc. **1508**, no. 1, 502 (2012) L. N. Chang, D. Minic and T. Takeuchi, J. Phys. Conf. Ser. **1275**, no. 1, 012036 (2019).

[26] B. L. van der Waerden, *Sources of Quantum Mechanics* (New York: Dover) (1968).

[27] S. L. Adler, *"Quaternionic quantum mechanics and quantum fields,"* New York, USA: Oxford Univ. Pr. (1995) 586 p.

[28] B. O. Koopman, Proceedings of the National Academy of Sciences. 17 (5), (1931) 315; J. von Neumann, Annals of Mathematics. 33 (3), (1932) 587.

[29] G. Mackey, *Acta Math.* **99** 265 (1958); G. Mackey, *Annals Math.* **55** 101 (1952).

[30] M. Stone, *Proc.Nat.Acad.Sci.* **16** 172 (1930): J. von Neuman, *Math.Ann.* **104** 570 (1931): G. Mackey, *Proc.Nat.Acad.Sci.* **35** 537 (1949).

[31] Y. Aharonov and D. Bohm, Phys. Rev. **115**, 485 (1959).

[32] M. R. Douglas and N. A. Nekrasov, Rev. Mod. Phys. **73**, 977 (2001); R. J. Szabo, Phys. Rept. **378**, 207 (2003); H. Grosse and R. Wulkenhaar, Commun. Math. Phys. **256**, 305 (2005).

[33] A. M. Polyakov, Phys. Lett. B **59**, 82 (1975) [Phys. Lett. **59B**, 82 (1975)]; Nucl. Phys. B **120**, 429 (1977); *Gauge Fields and Strings*, Harwood, 1987; I. I. Kogan and A. Kovner, Phys. Rev. D **51**, 1948 (1995).

[34] R. P. Feynman, Nucl. Phys. B **188**, 479 (1981); D. Karabali, C. j. Kim and V. P. Nair, Phys. Lett. B **434**, 103 (1998); R. G. Leigh, D. Minic and A. Yelnikov, Phys. Rev. Lett. **96**, 222001 (2006); Phys. Rev. D **76**, 065018 (2007); L. Freidel, R. G. Leigh and D. Minic, Phys. Lett. B **641**, 105 (2006); L. Freidel, R. G. Leigh, D. Minic and A. Yelnikov, AMS/IP Stud. Adv. Math. **44**, 109 (2008).

[35] A. A. Tseytlin, *Nucl. Phys.* B **350** 395 (1991); *Phys. Lett.* B **242** 163 (1990); R. Floreanini, and R. Jackiw, *Phys. Rev. Lett.* **59** 1873 (1987).

[36] I. R. Klebanov and L. Susskind, Nucl. Phys. B **309**, 175 (1988).

[37] J. Hughes, J. Liu and J. Polchinski, Nucl. Phys. B **316**, 15 (1989).

[38] W. Siegel, *Phys. Rev.* D **47** 5453 (1993); Phys. Rev. D **48**, 2826 (1993) C. Hull, and B. Zwiebach, *JHEP* **0909** 099 (2009); O. Hohm, C. Hull and B. Zwiebach, JHEP **1007**, 016 (2010); O. Hohm, D. Lust and B. Zwiebach, Fortsch. Phys. **61**, 926 (2013).

[39] D. Minic, J. Polchinski and Z. Yang, Nucl. Phys. B **369**, 324 (1992); J. Polchinski, "What is string theory?", hep-th/9411028.

[40] G. W. Gibbons, *J. Geom. Phys.* **8** 147 (1992); A. Ashtekar and T. Schilling, *Geometrical formulation of quantum mechanics*, gr-qc/9706069; D. Minic, and H. C. Tze, *Phys. Rev.* D **68** 061501 (2003); *Phys. Lett.* B **581** 111 (2004); V. Jejjala, M. Kavic, and D. Minic, *Int. J. Mod. Phys.* A **22** 3317 (2007).

[41] B. Fedosov, *Deformation quantization and index theory* (Berlin: Akademie-Verlag) (1996).

[42] N. Woodhouse, *Geometric Quantization* (Oxford: Clarendon Press) (1991).

[43] R. E. Borcherds, *Proc. Nat. Acad. Sci.* **83** 3068 (1986); *Inventiones Mathematicae* **109** 405 (1992).

[44] G. W. Moore, "Finite in all directions", [hep-th/9305139]

[45] E. Witten, JHEP **1504**, 055 (2015).

[46] S. B. Giddings, and S. A. Wolpert, *Commun. Math. Phys.* **109** 177 (1987); I. M. Krichever, and S. P. Novikov, *Funct.Anal.Appl.* **21** 294 (1987).

[47] S. Nakamura, *Tokyo J. Math.* **23** 87 (2000).

[48] L. Freidel, D. Garner, and S. Ramgoolam, *Phys. Rev.* D **91** 126001 (2015).

[49] F. Englert, H. Nicolai and A. Schellekens, Nucl. Phys. B **274**, 315 (1986).

[50] J. Polchinski, *String theory. Vol. 1: An introduction to the bosonic string.* Cambridge University Press, 1998. M. B. Green, J. H. Schwarz and E. Witten, *Superstring theory: volume 1*, Cambridge University Press, 1987.

[51] K. S. Narain, Phys. Lett. **169B**, 41 (1986).

[52] E. Boffo and P. Schupp, JHEP **2001**, 007 (2020)

[53] R. Blumenhagen and E. Plauschinn, J. Phys. A **44**, 015401 (2011); R. Blumenhagen, A. Deser, D. Lust, E. Plauschinn and F. Rennecke, J. Phys. A **44**, 385401 (2011); D. Lust, PoS CORFU **2011**, 086 (2011); D. Andriot, M. Larfors, D. Lust and P. Patalong, JHEP **1306**, 021 (2013) M. Gunaydin and D. Minic, Fortsch. Phys. **61**, 873 (2013); D. Mylonas, P. Schupp and R. J. Szabo, PoS ICMP **2013**, 007 (2013); P. Aschieri, M. Dimitrijevic Ciric and R. J. Szabo, JHEP **1802**, 036 (2018); R. J. Szabo, PoS CORFU **2018**, 100 (2019).

[54] A. Sen, JHEP **9808**, 012 (1998).

[55] A. Einstein, Preussische Akademie der Wissenschaftern, Phys-math. Klasse, Sitzungsberichte, 217221 (1928).

[56] M. J. Bowick and S. G. Rajeev, Phys. Rev. Lett. **58**, 535 (1987) Erratum: [Phys. Rev. Lett. **58**, 1158 (1987)]; Nucl. Phys. B **293**, 348 (1987).

[57] J. J. Atick and E. Witten, Nucl. Phys. B **310**, 291 (1988).

[58] S. L. Adler, *Quantum theory as an emergent phenomenon: The statistical mechanics of matrix models as the precursor of quantum field theory,* hep-th/0206120.

[59] R. Penrose, Found. Phys. **44**, 557 (2014).

[60] R. Penrose, Phil. Trans. Roy. Soc. Lond. A **373**, 20140237 (2015).

[61] N. Ishibashi, H. Kawai, Y. Kitazawa, and A. Tsuchiya, *Nucl. Phys.* **B498** (1997) 467–491.

[62] T. Banks, W. Fischler, S. H. Shenker, and L. Susskind, *Phys. Rev.* **D55** (1997) 5112–5128.

[63] G. 't Hooft, Conf. Proc. C **930308**, 284 (1993); L. Susskind, J. Math. Phys. **36**, 6377 (1995).

[64] J. M. Maldacena, Int. J. Theor. Phys. **38**, 1113 (1999) [Adv. Theor. Math. Phys. **2**, 231 (1998)]; S. S. Gubser, I. R. Klebanov and A. M. Polyakov, Phys. Lett. B **428**, 105 (1998); E. Witten, Adv. Theor. Math. Phys. **2**, 253 (1998).

[65] D. Minic and M. Pleimling, Phys. Lett. B **700**, 277 (2011); N. Gray, D. Minic and M. Pleimling, Int. J. Mod. Phys. A **28**, 1330009 (2013).

[66] D. Minic, hep-th/9909022. hep-th/0009131. H. Awata, M. Li, D. Minic and T. Yoneya, JHEP **0102**, 013 (2001). D. Minic and H. C. Tze, Phys. Lett. B **536**, 305 (2002); Phys. Lett. B **581**, 111 (2004).

[67] G. T. Horowitz and L. Susskind, J. Math. Phys. **42**, 3152 (2001).

[68] S. Bunk, L. Muller and R. J. Szabo, Lett. Math. Phys. **109**, no. 8, 1827 (2019); V. Balasubramanian, J. de Boer and D. Minic, gr-qc/0211003; L. Smolin, J. Math. Phys. **36**, 6417 (1995); L. Crane, J. Math. Phys. **36**, 6180 (1995). D. Minic, Phys. Lett. B **442**, 102 (1998); T. Banks and W. Fischler, "M theory observables for cosmological space-times," hep-th/0102077.

[69] A. H. Chamseddine and W. D. Van Suijlekom, "A survey of spectral models of gravity coupled to matter," arXiv:1904.12392 [hep-th]. A. Devastato, M. Kurkov and F. Lizzi, Int. J. Mod. Phys. A **34**, no. 19, 1930010 (2019).

[70] U. Aydemir, D. Minic and T. Takeuchi, Phys. Lett. B **724**, 301 (2013); U. Aydemir, D. Minic, C. Sun and T. Takeuchi, Phys. Rev. D **91**, 045020 (2015); Int. J. Mod. Phys. A **31**, no. 01, 1550223 (2016); Mod. Phys. Lett. A **31**, no. 18, 1650101 (2016); JHEP **1809**, 117 (2018).

[71] A. Strominger, "Lectures on the Infrared Structure of Gravity and Gauge Theory," arXiv:1703.05448 [hep-th], Princeton University Press, 2018.

[72] M. Kavic, J. H. Simonetti, S. E. Cutchin, S. W. Ellingson and C. D. Patterson, JCAP **0811**, 017 (2008); M. Kavic, D. Minic and J. Simonetti, Int. J. Mod. Phys. D **17**, 2495 (2009); Int. J. Mod. Phys. D **27**, no. 14, 1847007 (2018); J. H. Simonetti, M. Kavic, D. Minic, U. Surani and V. Vijayan, Astrophys. J. **737**, L28 (2011); J. Estes, M. Kavic, M. Lippert and J. H. Simonetti, Int. J. Mod. Phys. D **26**, no. 12, 1743004 (2017); S. L. Liebling, M. Lippert and M. Kavic, JHEP **1803**, 176 (2018); D. C. Dai, D. Minic and D. Stojkovic, Phys. Rev. D **98**, no. 12, 124026 (2018); D. C. Dai, D. Minic, D. Stojkovic and Changbo Fu, "Testing ER=EPR", [arXiv:2002.08178 [hep-th]]; D. C. Dai, M. J. Kavic, D. Minic, J. Simonetti and D. Stojkovic, in preparation.

[73] L. Freidel, J. Kowalski-Glikman, R. G. Leigh and D. Minic, in preparation.

[74] A. Strominger, Phys. Rev. Lett. **71**, 3397 (1993).

[75] O. W. Greenberg, Phys. Rev. Lett. **64**, 705 (1990).

[76] P. Horava and D. Minic, Phys. Rev. Lett. **85**, 1610 (2000).

[77] E. Bianchi, A. Gupta, H. M. Haggard and B. S. Sathyaprakash, "Quantum gravity and black hole spin in gravitational wave observations: a test of the Bekenstein-Hawking entropy," arXiv:1812.05127 [gr-qc].

[78] V. Jejjala, M. Kavic and D. Minic, Adv. High Energy Phys. **2007**, 21586 (2007); V. Jejjala, M. Kavic, D. Minic and T Takeuchi, in preparation.

[79] A. G. Riess *et al.*, *Astron. J.* **116** (1998) 1009–1038. S. Perlmutter *et al.*, *Astrophys. J.* **517** (1999) 565–586.

[80] A. G. Riess *et al.*, *Astrophys. J.* **826** no. 1, (2016) 56; *Astrophys. J.* **861** no. 2, (2018) 126.
N. Aghanim *et al.*, "Planck 2018 results. VI. Cosmological parameters," arXiv:1807.06209 [astro-ph.CO].

[81] M. Grana, Phys. Rept. **423** (2006) 91. M. R. Douglas and S. Kachru, Rev. Mod. Phys. **79** (2007) 733. S. R. Green et al. Class. Quant. Grav. **29** (2012) 075006. D. Kutasov et al. Phys. Rev. Lett. **115**, no. 7 (2015) 071305. U. H. Danielsson and T. Van Riet, *Int. J. Mod. Phys.* **D27** no. 12, (2018) 1830007.

[82] G. Obied, H. Ooguri, L. Spodyneiko, and C. Vafa, "De Sitter Space and the Swampland," arXiv:1806.08362 [hep-th]. P. Agrawal *et al.*, *Phys. Lett.* **B784** (2018) 271–276.D. Andriot, Fortsch. Phys. **67**, no. 7, 1900026 (2019) E. Palti, Fortsch. Phys. **67**, no. 6, 1900037 (2019).

[83] A. A. Tseytlin, *Phys. Rev. Lett.* **66** (1991) 545–548.

[84] D. H. Friedan, *Ann. Phys.* **163** (1985) 318–419; *Phys. Rev. Lett.* **45** (1980) 1057.

[85] A. Joyce, B. Jain, J. Khoury, and M. Trodden, *Phys. Rept.* **568** (2015) 1–98.

[86] B. R. Greene, A. D. Shapere, C. Vafa, and S.-T. Yau, *Nucl. Phys.* **B337** (1990) 1.

[87] P. Berglund, T. Hübsch, and D. Minic, *JHEP* **09** (2000) 015; *JHEP* **02** (2001) 010; *JHEP* **01** (2001) 041; *Phys. Lett.* **B512** (2001) 155–160.*Phys. Lett.* **B534** (2002) 147–154.*Phys. Rev.* **D67** (2003) 041901.

[88] C. Vafa, *Nucl. Phys.* **B469** no. 3, (1996) 403–415.

[89] P. Berglund, T. Hübsch, and D. Minic, in preparation.

[90] N. Kaloper and A. Padilla, Phys. Rev. Lett. **112**, no. 9, 091304 (2014). *Phys. Rev.* **D90** no. 8, (2014) 084023; [ibid, no. 10 (2014) 109901]; A. Padilla, "Lectures on the Cosmological Constant Problem," arXiv:1502.05296 [hep-th].

[91] A. M. Polyakov, "A Few projects in string theory," hep-th/9304146.

[92] H. Grosse and R. Wulkenhaar, *Commun. Math. Phys.* **329** (2014) 1069–1130; *RIMS Kokyuroku* **1904** (2013) 67–105.

[93] J. Polchinski and A. Strominger, *Phys. Rev. Lett.* **67** (1991) 1681–1684.

[94] A. Strominger, JHEP **0110**, 034 (2001); E. Witten, hep-th/0106109; V. Balasubramanian, J. de Boer and D. Minic, Class. Quant. Grav. **19**, 5655 (2002) [Annals Phys. **303**, 59 (2003)]; Phys. Rev. D **65**, 123508 (2002) V. Balasubramanian, P. Horava and D. Minic, JHEP **0105**, 043 (2001).

[95] E. Witten, *Int. J. Mod. Phys.* **A10** (1995) 1247–1248.V. Jejjala, R. G. Leigh, and D. Minić, *Phys. Lett.* **B556** (2003) 71–79.K. Becker, M. Becker, and A. Strominger, *Phys. Rev.* **D51** (1995) R6603–R6607.

[96] A. A. Starobinsky, *Phys. Lett.* **B91** (1980) 99–102.

[97] Planck Collaboration, Y. Akrami *et al.*, "Planck 2018 results. X. Constraints on inflation," arXiv:1807.06211 [astro-ph.CO].

[98] D. R. Finkelstein, *Quantum Relativity*, Springer, 1996. R. D. Sorkin, "Logic is to the quantum as geometry is to gravity," arXiv:1004.1226 [quant-ph], in G.F.R. Ellis, J. Murugan and A. Weltman (eds), *Foundations of Space and Time* (Cambridge University Press), 2010.

[99] N. Arkani-Hamed, J. L. Bourjaily, F. Cachazo, A. B. Goncharov, A. Postnikov and J. Trnka, "Grassmannian Geometry of Scattering Amplitudes," arXiv:1212.5605 [hep-th], Cambridge University Press 2012.

[100] H. Chan, S. Tsou and S. T. Tsou, Int. J. Mod. Phys. A **30**, no. 30, 1530060 (2015).

[101] K. Vetsigian, C. Woese,†‡§ and N. Goldenfeld, PNAS, 103(28): 10696 (2006); J. A. Argyriadis, Y. H. He, V. Jejjala and D. Minic, arXiv:1909.10405 [q-bio.OT].

Entropy production in systems with spontaneously broken time-reversal*

Mihail Mintchev†
Istituto Nazionale di Fisica Nucleare and Dipartimento di Fisica,
Largo Pontecorvo 3, 56127 Pisa, Italy

Paul Sorba‡
Laboratoire d'Annecy-le-Vieux de Physique Théorique, CNRS,
Université de Savoie, BP 110, 74941 Annecy-le-Vieux, France

Abstract

We study the entropy production in non-equilibrium quantum systems without dissipation, which is generated exclusively by the spontaneous breaking of time-reversal invariance. Systems which preserve the total energy and particle number and are in contact with two heat reservoirs are analysed. Focussing on point-like interactions, we derive the probability distribution induced by the entropy production operator. We show that all its moments are positive in the zero frequency limit. The analysis covers both Fermi and Bose statistics.

1. Introduction

We investigate below non-equilibrium quantum systems with infinite degrees of freedom, where the time reversal symmetry T is *spontaneously* broken. In particular, we explore the impact of the spontaneous T-breaking on the entropy production, which represents the key indicator for the departure from equilibrium. The presentation summarises our previous results [1, 2] and pursues further our study of the entropy production in systems with the structure shown in Fig. 1. One has two heat reservoirs R_i with (inverse) temperatures $\beta_i \geq 0$ and chemical potentials μ_i, which communicate via the gates G_i with the interaction domain $\mathbb{D}$. Particles are emitted and absorbed by the reservoirs R_i trough the gates G_i and propagate along the leads L_i to the interaction domain $\mathbb{D}$. The interaction in $\mathbb{D}$ drives the

* One of us (P.S.) would like to thank the organisers Prof. B. Dragovich and Dr. M. Vojinovic for their kind invitation to present our results at MPHYS10, Belgrade, Sept. 2019. The scientific quality of the meeting as well as the perfect organisation have been particularly appreciated.
† e-mail address: mintchev@df.unipi.it
‡ e-mail address: paul.sorba@lapth.cnrs.fr

Figure 1: System connected to two heat reservoirs.

system away from equilibrium. The capacity of R_i is assumed large enough so that the particle emission and absorption processes do not change their parameters.

Systems of this type in one space dimension are widely investigated. They are successfully applied for studying the transport properties of quantum wire junctions [3]-[8], quantum Hall edges [9] and recently are also engineered in laboratory by ultracold Bose gases [10]-[12]. The remarkable precision reached in such experiments allows to explore fundamental aspects of non-equilibrium many-body quantum physics.

In the next section we establish some general universal features of non-equilibrium systems, which communicate with two heat reservoirs. Afterwards, in section 3 we provide a microscopic picture for the entropy production in such systems and propose a quantum field theory framework for the derivation of the probability distribution generated by the entropy production operator. In section 4 we illustrate this approach at work exploring two examples - the fermionic and bosonic Schrödinger junctions with point-like interaction. The last section is devoted to the conclusions.

2. Spontaneous breaking of time-reversal invariance

Quantum systems in contact with two heat reservoirs exhibit in general a complex behaviour. In order to simplify the picture, we assume in what follows the conservation of the particle number and the total energy

$$N = \int_{G_1}^{G_2} \mathrm{d}x \, j_t(t,x) \, , \quad H = \int_{G_1}^{G_2} \mathrm{d}x \, \vartheta_t(t,x) \, , \tag{1}$$

j_t and ϑ_t being the particle and energy densities. In other words there is no particle and energy dissipation. Let j_x and ϑ_x be the associated local conserved currents which obey the continuity equations

$$\partial_t j_t(t,x) - \partial_x j_x(t,x) = 0 \, , \tag{2}$$
$$\partial_t \vartheta_t(t,x) - \partial_x \vartheta_x(t,x) = 0 \, . \tag{3}$$

Combining (1) with (2,3) and taking into account the orientation of the leads L_i one finds

$$\dot{N} = 0 \quad \Rightarrow \quad j_x(t,G_1) + j_x(t,G_2) = 0 \, , \tag{4}$$
$$\dot{H} = 0 \quad \Rightarrow \quad \vartheta_x(t,G_1) + \vartheta_x(t,G_2) = 0 \, , \tag{5}$$

the dot indicating the time derivative. Therefore the particle and energy inflows are equal to the outflows as expected.

Let us consider now the operation of time-reversal. It is well known that this operation is implemented by an anti-unitary operator T, acting in the state space $\mathcal{H}$ of the system. We recall [13] that

$$T\, j_x(t, x)\, T^{-1} = -j_x(-t, x) \tag{6}$$

and assume that the dynamics preserves time-reversal invariance, namely

$$T\, H\, T^{-1} = H\,. \tag{7}$$

Now, let $\Omega \in \mathcal{H}$ be any time-translation invariant state of the system with non-vanishing particle flow between the two reservoirs

$$\langle j_x(t, x)\rangle_\Omega \equiv (\Omega\,,\, j_x(t, x)\Omega) \neq 0\,, \tag{8}$$

where $(\cdot\,,\,\cdot)$ is the scalar product in $\mathcal{H}$. Taking the expectation value of (6) one has

$$\langle T\, j_x(t, x)\, T^{-1}\rangle_\Omega = -\langle j_x(-t, x)\rangle_\Omega\,, \tag{9}$$

which, combined with the fact that $\langle j_x(t, x)\rangle_\Omega$ is actually t-independent due to the time-translation invariance of Ω, implies that

$$T\,\Omega \neq \Omega\,. \tag{10}$$

We conclude that time-reversal is broken in the state $\Omega \in \mathcal{H}$ in spite of the fact that the dynamics is time-reversal invariant (7). This result can be compactly formulated as follows.

Proposition: Any state, which is invariant under time translations and generates a non-vanishing expectation value for the particle current, breaks time-reversal symmetry.

This is a genuine quantum field theory phenomenon of spontaneous breaking of a discrete symmetry, whose order parameter is $\langle j_x(t, x)\rangle_\Omega$.

The state Ω has another remarkable feature. Let $q_x(t, G_i)$ be the heat current flowing through the gate G_i. Since the value of the chemical potential in G_i is μ_i, one has [14]

$$q_x(t, G_i) = \vartheta_x(t, G_i) - \mu_i j_x(t, G_i)\,. \tag{11}$$

Notice that we adopt here only the value of the heat current in the gates G_i. The point is that the heat current in the interaction domain $\mathbb{D}$ is not known, because the temperature and the chemical potential are not defined in this region. In order to introduce the concept of local parameters $\beta(x)$ and $\mu(x)$ for $x \in \mathbb{D}$ one needs further model dependent assumptions, which are not relevant and not needed for our construction.

From (11) it follows that for $\mu_1 \neq \mu_2$ the heat flow through G_1 differs from that through G_2. In fact,

$$\dot{Q} \equiv -q_x(t, G_2) - q_x(t, G_1) = (\mu_2 - \mu_1) j_x(t, G_1) \neq 0\,. \tag{12}$$

Now we recall that the total energy H of the system has two different components - *heat* and *chemical* energies. Since H is conserved, (12) implies that these components are not separately conserved and can be converted one into the other. This process, discovered in [15], depends on the state Ω and more precisely on the expectation value

$$\langle \dot{Q} \rangle_\Omega = (\Omega\,, \dot{Q}\Omega)\,. \tag{13}$$

Chemical energy is converted to heat energy if $\langle \dot{Q} \rangle_\Omega > 0$. The opposite process takes place for $\langle \dot{Q} \rangle_\Omega < 0$ and energy transmutation is absent only if $\langle \dot{Q} \rangle_\Omega = 0$. It is worth stressing that there is *no dissipation* during the energy conversion.

Summarising, the physical consequences of particle and energy conservation in any state Ω, satisfying (8), are:

(i) spontaneous breaking of time-reversal;

(ii) conversion of heat to chemical energy or vice versa.

It is worth stressing that these deeply related features follow exclusively from symmetry considerations and do *not* depend on the interaction taking place in $\mathbb{D}$. In this sense they are *universal* and hold for all systems of the type shown in Fig. 1.

3. Entropy production

The properties (i)-(ii), established in the previous section, have relevant physical implications. Among others, they provide a simple and remarkable mechanism for non-vanishing entropy production $\dot{S}$ in absence of dissipation. Our main goal in what follows will be to study this aspect in detail. According to the second law of classical thermodynamics $\dot{S} \geq 0$. In the quantum case $\dot{S}$ is an operator, which has the following form [14]

$$\dot{S} = -\beta_1\, q_x(t, G_1) - \beta_2\, q_x(t, G_2) \tag{14}$$

in terms of the heat currents flowing through the gates G_i. In order to investigate the properties of this operator, one needs a microscopic approach [1, 2] to the particle transport in the system. It is based on the observation that (due to particle number conservation), there are three fundamental elementary processes which take place:

(a) emission and reabsorption of any number of particles from the same reservoir;

(b) emission of n particles from the "hot" reservoir R_2 and their absorption by the "cold" one R_1;

(c) emission of n particles from the "cold" reservoir R_1 and their absorption by the "hot" one R_2.

Let us denote by p_0, p_n and p_{-n} with $n = 1, 2, ...$ the probabilities of the events (a), (b) and (c) respectively. A substantial difference with respect to the classical case is that in the quantum world the processes of the type (c) have a priori non-vanishing probabilities. At this stage it is useful to introduce the sequences

$$P = \{p_n \, : \, n = 0, \pm 1, \pm 2, ...\}\,, \qquad \Sigma = \{\sigma_n \, : \, n = 0, \pm 1, \pm 2, ...\}\,, \tag{15}$$

where σ_n is the entropy production associated with the process with probability p_n. It is natural to expect that $\sigma_0 = 0$, $\{\sigma_n > 0,\, n = 1, 2, ...\}$ and $\{\sigma_n < 0,\, n = -1, -2, ...\}$.

Summarising, both processes with positive and negative entropy production occur at the microscopic level. It is clear that the sequences (15) fully codify the entropy production in the system. So, the problem is to determine P and Σ. We will show below that both P and Σ can be extracted from the probability distribution $\varrho[\dot{S}]$ generated by the correlation functions

$$\mathcal{M}_n[\dot{S}] = \langle \dot{S}(t_1) \cdots \dot{S}(t_n) \rangle_\Omega \,. \tag{16}$$

The based observation now is that $\mathcal{M}_n[\dot{S}]$ are the moments of $\varrho[\dot{S}]$, namely

$$\mathcal{M}_n[\dot{S}] = \int_{\mathcal{D}} \mathrm{d}\sigma \, \sigma^n \varrho[\dot{S}](\sigma)\,, \tag{17}$$

where $\mathcal{D}$ is the range of entropy production. The strategy at this point consists of three steps:

- derivation of the correlation functions $\mathcal{M}_n[\dot{S}]$ given by (16);
- reconstruction of the distribution $\varrho[\dot{S}]$ from its moments $\mathcal{M}_n[\dot{S}]$ (also known as moment problem [16]);
- derivation of the sequences P and Σ from $\varrho[\dot{S}]$.

Although conceptually very clear, the first two steps are practically quite involved. They include the derivation of the infinite sequence of correlation functions (16), followed by the solution of the moment problem. Nevertheless we give in the next section two examples where this program can be taken to the end. We describe one fermionic and one bosonic model, where the interaction domain $\mathbb{D}$ is reduced to a point $x = 0$. We show that with such point-like interaction one can determine $\varrho[\dot{S}]$ and consequently P and Σ in exact and explicit form and discuss the physical implications of the solution.

4. Fermionic/bosonic Schrödinger junction

We analyse in this section two examples with point-like interaction described by a unitary scattering matrix $\mathbb{S}$ as shown in Fig. 2. The dynamics along the leads L_i with coordinates $\{(x,i) : x < 0\,, i = 1,2\}$, where $|x|$ is the distance from the interaction and i labels the lead, is defined by the Schrödinger equation

$$\left(\mathrm{i}\partial_t + \frac{1}{2m}\partial_x^2\right)\psi(t,x,i) = 0\,, \tag{18}$$

where ψ is a quantum field satisfying the equal-time canonical (anti)com-

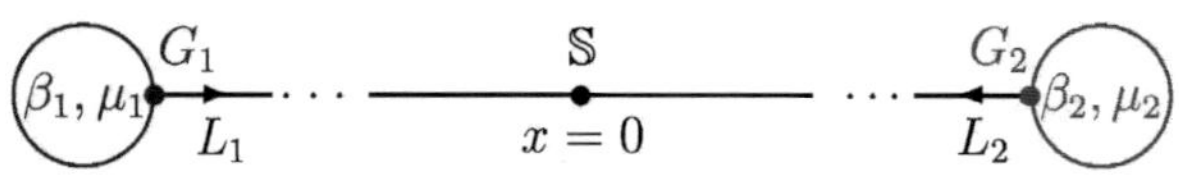

Figure 2: System with point-like interaction $\mathbb{S}$ at $x = 0$.

mutation relations

$$[\psi(t,x,i)\,,\,\psi^*(t,y,j)]_\pm = \delta_{ij}\,\delta(x-y) \tag{19}$$

and * stands for Hermitian conjugation. The interaction is fixed by the boundary condition

$$\lim_{x\to 0^-}\sum_{j=1}^{2}[\lambda(\mathbb{I}-\mathbb{U})_{ij} + \mathrm{i}(\mathbb{I}+\mathbb{U})_{ij}\partial_x]\,\psi(t,x,j) = 0\,, \tag{20}$$

where λ is a free parameter with dimension of mass and $\mathbb{U}$ is an arbitrary unitary matrix $\mathbb{U} \in U(2)$. This is [17] the *most general* boundary condition ensuring the *self-adjointness* of the Hamiltonian. In momentum space the associated scattering matrix is given by [17, 18]

$$\mathbb{S}(k) = -\frac{[\lambda(\mathbb{I}-\mathbb{U}) - k(\mathbb{I}+\mathbb{U})]}{[\lambda(\mathbb{I}-\mathbb{U}) + k(\mathbb{I}+\mathbb{U})]}\,, \tag{21}$$

where the matrices in the numerator and denominator commute. One can easily verify that this scattering matrix is unitary $\mathbb{S}(k)\mathbb{S}^*(k) = \mathbb{I}$ and satisfies Hermitian analyticity $\mathbb{S}^*(k) = \mathbb{S}(-k)$. The dynamics fixed by (18-20) is invariant under time-reversal if and only if $\mathbb{U}$ and consequently $\mathbb{S}(k)$ are *symmetric*.

We will show below that the above system, called in what follows Schrödinger junction, nicely illustrates the program from the previous section and works simultaneously for both Fermi (+) and Bose (−) statistics.

For simplicity we assume in what follows that $\mathbb{S}(k)$ has no bound states, referring for the general case to [19]. Then, the general solution of (18-20) is given by

$$\psi(t,x,i) = \sum_{j=1}^{2} \int_0^\infty \frac{dk}{2\pi} \mathrm{e}^{-\mathrm{i}\omega(k)t} \left[\mathrm{e}^{-\mathrm{i}kx}\,\delta_{ij} + \mathrm{e}^{\mathrm{i}kx}\,\mathbb{S}^*_{ij}(k)\right] a_j(k)\,, \tag{22}$$

where $\omega(k) = \frac{k^2}{2m}$ is the dispersion relation and the oscillators

$$\{a_i(k),\, a_i^*(k),\, :, k > 0,\, i = 1, 2\} \tag{23}$$

generate a standard canonical (anti)commutation relation algebra $\mathcal{A}_\pm$

$$[a_i(k)\,,\, a_j^*(p)]_\pm = \delta_{ij}\, 2\pi\delta(k-p)\,. \tag{24}$$

Notice that (23) annihilate and create only in-coming particles because $k > 0$. The contribution of the out-going excitations is generated by the second term in the integrand of (22), which involves the scattering matrix $\mathbb{S}(k)$.

The basic physical observables, we are interested in, are

$$j_x(t,x,i) = \frac{\mathrm{i}}{2m}\left[\psi^*(\partial_x\psi) - (\partial_x\psi^*)\psi\right](t,x,i)\,, \tag{25}$$

$$\begin{aligned}\vartheta_x(t,x,i) = \frac{1}{4m}[&(\partial_t\psi^*)\,(\partial_x\psi) + (\partial_x\psi^*)\,(\partial_t\psi) \\ &- (\partial_t\partial_x\psi^*)\,\psi - \psi^*\,(\partial_t\partial_x\psi)](t,x,i)\,,\end{aligned} \tag{26}$$

$$\dot{S}(t,x) = -\sum_{i=1}^{2} \beta_i\, q(t,x,i)\,, \qquad q_x(t,x,i) = \vartheta_x(t,x,i) - \mu_i j_x(t,x,i)\,. \tag{27}$$

In order to compute the correlation functions of the entropy production operator $\dot{S}(t,x)$ one must fix a representation of the oscillator algebras $\mathcal{A}_\pm$. We choose the Landauer-Büttiker (LB) representation [20, 21] $\{\mathcal{H}_{\rm LB},\, \Omega_{\rm LB}\}$, which represents a non-equilibrium generalisation of the Gibbs representation [22] to the case of systems which exchange particles and energy with more then one heat reservoir. Adopting the field-theoretical construction [23] of $\{\mathcal{H}_{\rm LB},\, \Omega_{\rm LB}\}$, one can derive in explicit form the expectation values

$$\langle a^*_{l_1}(k_1) a_{m_1}(p_1) \cdots a^*_{l_n}(k_n) a_{m_n}(p_n)\rangle^\pm_{\rm LB}\,, \qquad k_i, p_i > 0\,, \tag{28}$$

in the LB *steady* state $\Omega_{\rm LB}$. It is convenient for this purpose to introduce the matrix

$$\mathbb{M}^\pm_{ij} = \begin{cases} 2\pi\delta(k_i - p_j)\delta_{l_i m_j} d^\pm_{l_i}[\omega(k_i)]\,, & i \leq j\,, \\ \mp 2\pi\delta(k_i - p_j)\delta_{l_i m_j}\left(1 \mp d^\pm_{l_i}[\omega(k_i)]\right)\,, & i > j\,, \end{cases} \tag{29}$$

where

$$d_l^{\pm}(\omega) = \frac{1}{e^{\beta_l(\omega-\mu_l)} \pm 1}\, , \qquad \text{(for bosons } \mu_l < 0) \tag{30}$$

are the Fermi/Bose distribution of the reservoir R_l. Then one has [23]

$$\langle a_{l_1}^*(k_1)a_{m_1}(p_1)\cdots a_{l_n}^*(k_n)a_{m_n}(p_n)\rangle_{\rm LB}^{\pm} = \begin{cases} \mathbf{det}\,[\mathbb{M}^+]\, , & k_i, p_i > 0\, , \\ \mathbf{perm}\,[\mathbb{M}^-]\, , & k_i, p_i > 0\, , \end{cases} \tag{31}$$

where **det** and **perm** indicate the determinant and the permanent of the corresponding matrices. It is perhaps useful to recall that

$$\mathbf{perm}\,[\mathbb{M}] = \sum_{\sigma_i \in \mathcal{P}_n} \prod_{i=1}^n \mathbb{M}_{i\sigma_i}\, , \qquad \mathcal{P}_n\text{−set of all permutations of } n \text{ elements}\, . \tag{32}$$

Equations (29-31) are the basic ingredients for deriving the correlation functions in the LB representation. Our first step in this direction will be to compute some one-point functions. In particular, we will check that time-reversal is spontaneously broken, namely that $T\Omega_{\rm LB} \neq \Omega_{\rm LB}$. Following the argument in section 1, it is enough to control the mean value of the particle current (25). One finds

$$\langle j(t,x,i)\rangle_{\rm LB}^{\pm} = \int_0^\infty \frac{{\rm d}\omega}{2\pi} \sum_{l=1}^2 \left[\delta_{il} - |\mathbb{S}_{il}(\sqrt{2m\omega})|^2\right] d_l^{\pm}(\omega) \neq 0\, , \tag{33}$$

which confirms the spontaneous T-breaking.

For the mean value of the entropy production one has

$$\langle \dot{S}(t,x)\rangle_{\rm LB}^{\pm} = \int_0^\infty \frac{{\rm d}\omega}{2\pi} |\mathbb{S}_{12}(\sqrt{2m\omega})|^2 [\gamma_2(\omega) - \gamma_1(\omega)][d_1^{\pm}(\omega) - d_2^{\pm}(\omega)] \geq 0\, , \tag{34}$$

where

$$\gamma_i(\omega) \equiv \beta_i(\omega - \mu_i)\, . \tag{35}$$

The positivity follows directly from the fact that the two square brackets in the integrand have always the same sign. The bound (34) is a special case of the general result of Nenciu [24] for systems in the LB state and in contact with arbitrary number of heat reservoirs.

We turn now to the n-point correlation functions

$$\langle \dot{S}(t_1,x_1)\cdots \dot{S}(t_n,x_n)\rangle_{\rm LB}^{\pm}\, , \tag{36}$$

which, due to the total energy conservation, depend only on the time differences $\{\hat{t}_i \equiv t_i - t_{i+1} \, : \, i = 1,...,n-1\}$. This fact allows to introduce for

$n \geq 2$ the frequency ν via the Fourier transform

$$w_n^{\pm}[\dot{S}](x_1, ..., x_n; \nu) =$$
$$\int_{-\infty}^{\infty} \mathrm{d}\hat{t}_1 \cdots \int_{-\infty}^{\infty} \mathrm{d}\hat{t}_{n-1} \mathrm{e}^{-\mathrm{i}\nu(\hat{t}_1+\cdots\hat{t}_{n-1})} \langle \dot{S}(t_1, x_1) \cdots \dot{S}(t_n, x_n) \rangle^{\pm}_{\mathrm{LB}} . \tag{37}$$

Following the classical studies [25]-[29] in quantum noise and full counting statistics, we perform the zero-frequency limit in which the quantum fluctuations are integrated over the whole time axes. It turns out [1, 2] that in this limit the position dependence in (37) drops out and one finds

$$\lim_{\nu \to 0^+} w_n^{\pm}[\dot{S}](x_1, ..., x_n; \nu) = \int_0^{\infty} \frac{\mathrm{d}\omega}{2\pi} \mathcal{M}_n^{\pm}[\dot{S}](\omega) , \tag{38}$$

where $\mathcal{M}_n^{\pm}[\dot{S}]$ are precisely the moments (17) of the probability distribution $\varrho_n^{\pm}[\dot{S}]$ we are looking for. Using (29-31), one gets

$$\mathcal{M}_n^{\pm}[\dot{S}] = \begin{cases} \gamma_{21}^n(\omega)\, \mathbf{det}[\mathbb{D}^+(\omega; l_1, ..., l_n)] , \\ \gamma_{21}^n(\omega)\, \mathbf{perm}[\mathbb{D}^-(\omega; l_1, ..., l_n)] , \end{cases} \tag{39}$$

where

$$\gamma_{ij}(\omega) \equiv \gamma_i(\omega) - \gamma_j(\omega) = (\beta_i - \beta_j)\omega - (\beta_i \mu_i - \beta_j \mu_j) , \tag{40}$$

is a basic dimensionless parameter defining the entropy generated by transporting a particle with energy ω from the reservoir R_i to R_j. Moreover the $\mathbb{D}^{\pm}$-matrices are given by

$$\mathbb{D}_{ij}^{\pm}(\omega; l_1, ..., l_n) = \begin{cases} \mathbb{J}_{l_j l_i}(\omega) d_{l_j}^{\pm}(\omega) , & i \leq j , \\ \mp \mathbb{J}_{l_j l_i}(\omega) \left[1 \mp d_{l_j}^{\pm}(\omega)\right] , & i > j , \end{cases} \tag{41}$$

with

$$\mathbb{J}_{11}(\omega) = -\mathbb{J}_{22}(\omega) = |\mathbb{S}_{12}(\sqrt{2m\omega})|^2 \equiv \tau(\omega) , \tag{42}$$
$$\mathbb{J}_{12}(\omega) = \overline{\mathbb{J}}_{21}(\omega) = -\mathbb{S}_{11}(\sqrt{2m\omega})\, \overline{\mathbb{S}}_{12}(\sqrt{2m\omega}) , \tag{43}$$

where $\tau(\omega)$ is the *transmission probability.* For $\tau(\omega) = 0$ the leads are isolated and the system is in equilibrium.

At this stage, using (39-43) one can derive the inequality

$$\mathcal{M}_n^{\pm}[\dot{S}] \geq 0 , \qquad n =, 1, 2, ... \tag{44}$$

which is one of our main results. The rigorous proof of the bound (44) can be found in [1] for fermions and in [2] for bosons. Referring to these papers for the details, we will give below an alternative intuitive explanation, which has the advantage of providing a simple physical interpretation in terms of the sequence of probabilities P in (15). In order to derive P we introduce the moment generating function [16]

$$\chi^{\pm}[\dot{S}](\lambda) = \sum_{n=0}^{\infty} \frac{(\mathrm{i}\lambda)^n}{n!} \mathcal{M}_n^{\pm}[\dot{S}] , \tag{45}$$

where the ω-dependence is implicit. After some algebra one finds

$$\chi^{+}[\dot{S}](\lambda) = 1 + \mathrm{i}c_1^{+}\sqrt{\tau}\sin(\lambda\gamma_{21}\sqrt{\tau}) + c_2^{+}\left[\cos(\lambda\gamma_{21}\sqrt{\tau}) - 1\right] , \tag{46}$$

$$\chi^{-}[\dot{S}](\lambda) = \frac{1}{1 - \mathrm{i}c_1^{-}\sqrt{\tau}\sin(\lambda\gamma_{21}\sqrt{\tau}) - c_2^{-}\left[\cos(\lambda\gamma_{21}\sqrt{\tau}) - 1\right]} , \tag{47}$$

where $c_i^{\pm}$ are expresses via the reservoir distributions $d_i^{\pm}$ as follows

$$c_1^{\pm} \equiv d_1^{\pm} - d_2^{\pm} , \qquad c_2^{\pm} \equiv d_1^{\pm} + d_2^{\pm} \mp 2d_1^{\pm}d_2^{\pm} . \tag{48}$$

The final step towards the entropy production distributions $\varrho^{\pm}[\dot{S}]$ is the Fourier transform

$$\varrho^{\pm}[\dot{S}](\sigma) = \int_0^{\infty} \frac{\mathrm{d}\lambda}{2\pi} \mathrm{e}^{-\mathrm{i}\lambda\sigma} \chi^{\pm}[\dot{S}](\lambda) . \tag{49}$$

Since according to (46,47) $\chi^{\pm}[\dot{S}]$ are periodic functions in λ with period $2\pi/\sqrt{\tau}$, the Fourier transform is a superposition of δ-functions - the so called "Dirac comb". The fermionic comb has only three "teeth"

$$\varrho^{+}[\dot{S}](\sigma) = \sum_{k=-1}^{1} p_k^{+}\, \delta(\sigma - k\gamma_{21}\sqrt{\tau}) , \tag{50}$$

because only single particle processes are allowed by Pauli's principle since the energy ω is fixed and there is no degeneracy in spin and momentum in the Schrödinger junction. In (50)

$$p_{\pm 1}^{+} = \frac{1}{2}\left(c_2^{+} \mp c_1^{+}\sqrt{\tau}\right) , \qquad p_0^{+} = 1 - c_2^{+} , \tag{51}$$

and

$$\sigma_{\pm 1} = \pm\gamma_{21}\sqrt{\tau} , \qquad \sigma_0 = 0 , \tag{52}$$

are precisely the elements of the sequences P^{+} and Σ^{+} we are looking for. In fact one can easily verify that

$$p_k^{+} \geq 0 , \qquad \sum_{k=-1}^{1} p_k^{+} = 1 . \tag{53}$$

The bosonic case is technically more involved since the comb has infinite "teeth"

$$\varrho^-[\dot{S}](\sigma) = \sum_{k=-\infty}^{\infty} p_k^- \,\delta(\sigma - k\gamma_{21}\sqrt{\tau})\,, \tag{54}$$

because multi-particle processes are allowed. The computation, performed in [2], gives

$$p_{\pm n}^- = \frac{(c_2^- \pm c_1^- \sqrt{\tau})^n}{2^n(1+c_2^-)^{n+1}}\, {}_2F_1\left[\frac{1+n}{2}, \frac{2+n}{2}, n+1, \frac{(c_2^-)^2 - \tau(c_1^-)^2}{(1+c_2^-)^2}\right], \tag{55}$$

$$\sigma_{\pm n} = \pm n\gamma_{21}\sqrt{\tau}\,, \qquad n = 0, \pm 1, \pm 2, \ldots \tag{56}$$

where ${}_2F_1$ is the Gauss hypergeometric function. Equations (55,56) define the sequences P^- and Σ^- in the bosonic case. In analogy with (53) one has [2]

$$p_k^- \geq 0\,, \qquad \sum_{k=-\infty}^{\infty} p_k^- = 1\,. \tag{57}$$

Summarising, (53,57) imply that both $\varrho^{\pm}[\dot{S}]$ are well defined probability distributions. Equations (51,55) provide in explicit form the probabilities $p_k^{\pm}$ for the basic emission and absorption processes in terms of the Dirac/Bose distributions $d_i^{\pm}$ and the interaction τ. All quantum fluctuations in the zero frequency limit are taken into account. The relative simplicity of p_k^+ with respect to p_k^- is a consequence of the Pauli principle. In order to compare $p_k^{\pm}$ for different k and to get a more precise idea about the distributions $\varrho^{\pm}[\dot{S}]$ it is convenient to introduce the smeared versions $\varrho_\alpha^{\pm}[\dot{S}]$ by the substitution

$$\delta(\sigma) \longmapsto \delta_\alpha(\sigma) \equiv \frac{\alpha}{\sqrt{\pi}}\, \mathrm{e}^{-\alpha^2\sigma^2}\,, \qquad \alpha > 0\,, \tag{58}$$

in (50,54). By construction $\varrho_\alpha^{\pm}[\dot{S}] \to \varrho^{\pm}[\dot{S}]$ for $\alpha \to \infty$ in the sense of generalised functions. The advantage of $\varrho_\alpha^{\pm}[\dot{S}]$ is that they are not singular and can be easily plotted for some value of α and the parameters β_i, μ_i and τ. Typical plots are shown in Fig. 3, where the left and right panel display the fermionic and bosonic distribution respectively.

The relative height of the peaks in Fig. 3 allows to compare the different probabilities of the fundamental emission-absorption processes described in points (a)-(c) in section 3. We see that the predominant process is the emission and reabsorption by the same reservoir with vanishing entropy production $\sigma = 0$. For $\sigma \neq 0$ the peaks are symmetric with respect to the origin and the right ones ($\sigma > 0$) always dominate the left ones ($\sigma < 0$). This observation provides the physical explanation for the positivity bound (44) on the moments $\mathcal{M}_n^{\pm}[\dot{S}]$.

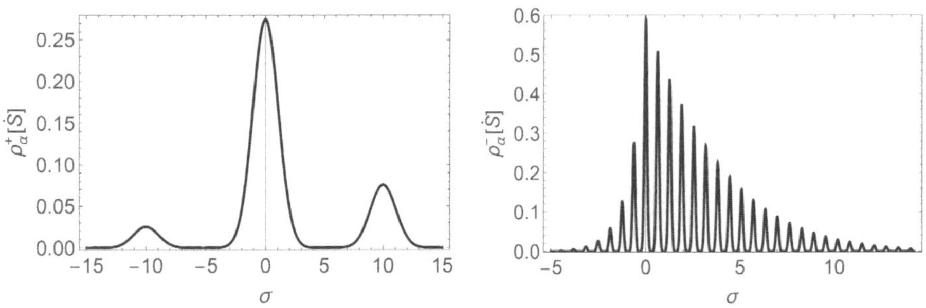

Figure 3: Fermionic and bosonic entropy production distributions.

The knowledge of the sequences $P^{\pm}$ and $\Sigma^{\pm}$ in explicit form (51, 52, 55, 56) has relevant physical applications. Using this microscopic information one can establish the fluctuation relations [30]-[32] governing the entropy production in the LB state. Moreover, one can introduce a concept of efficiency of the quantum transport, which goes beyond the meal value description and takes into account all quantum fluctuations. For more details about these applications we refer the reader to [1, 2].

In conclusion, for non-equilibrium quantum systems in the LB state microscopic processes with negative entropy production occur with some non-vanishing probability. For each such process however, there exists a more probable one with the opposite positive entropy production, which dominates. This feature can be interpreted in our context as a *quantum* version of the second law of thermodynamics.

5. Conclusions

The above study develops a field theoretic approach to the entropy production in non-equilibrium quantum systems in a state Ω, which breaks spontaneously time-reversal invariance. The parameter which controls this phenomenon is the expectation value of the particle current in Ω. The mechanism allows for non-trivial entropy production even in absence of dissipation. The basic idea of the framework is to derive and investigate the probability distribution $\varrho[\dot{S}]$, generated by the n-point correlation functions of the entropy production operator $\dot{S}$ in the state Ω. One can extract from $\varrho[\dot{S}]$ the sequence P of probabilities, associated with the fundamental processes of emission and absorption of particles from the heat reservoirs, driving the system away from equilibrium. In this way one obtains a microscopic picture of the quantum transport and entropy production, which takes into account all quantum fluctuations. These general ideas have been illustrated in the paper on the example of two exactly solvable models - the fermionic and bosonic Schrödinger junctions with point-like interaction. We show that in these two cases one can derive in exact and closed form the probability distribution $\varrho[\dot{S}]$ and prove that all its moments $\mathcal{M}_n[\dot{S}]$ are non-negative in the zero frequency limit, which provides a bridge with the second law of non-equilibrium classical thermodynamics.

Our investigation demonstrates that the entropy production operator plays a fundamental role in non-equilibrium quantum physics. For this reason it will be important to test the above ideas in other models and within alternative frameworks. Among others, non-equilibrium conformal field theory [33]-[35], generalised quantum hydrodynamics [36]-[39] and the theory of periodically driven quantum systems [40] could be adopted. Studies in this direction will certainly contribute for better understanding the fascinating properties of the quantum world away from equilibrium.

References

[1] M. Mintchev, L. Santoni and P. Sorba, *Phys. Rev. E* **96** (2017) 052124.

[2] M. Mintchev, L. Santoni and P. Sorba, *Annalen der Physik* **530** (2018) 1800170.

[3] C. L. Kane and M. P. A. Fisher, *Phys. Rev. B* **46** (1992) 15233.

[4] C. Nayak, M. P. A. Fisher, A. W. W. Ludwig and H. H. Lin, *Phys. Rev. B* **59** (1999) 15694.

[5] S. Chen, B. Trauzettel, and R. Egger, *Phys. Rev. Lett.* **89** (2002) 226404.

[6] C. Chamon, M. Oshikawa, and I. Affleck, *Phys. Rev. Lett.* **91** (2003) 206403.

[7] B. Bellazzini, M. Mintchev and P. Sorba, *J. Phys. A: Math. and Theor.* **40** (2007) 2485.

[8] B. Bellazzini, P. Calabrese and M. Mintchev, *Phys. Rev. B* **79** (2009) 085122.

[9] A. Stern, *Annals of Physics* **323** (2008) 204.

[10] I. Bloch, J. Dalibard and W. Zwerger, *Rev. Mod. Phys.* **80** (2008) 885.

[11] M. A. Cazalilla, R. Citro, T. Giamarchi, E. Orignac and M. Rigol, *Rev. Mod. Phys.* **83** (2011) 1405.

[12] F. Chevy and C. Salomon, *J. Phys. B: At. Mol. Opt. Phys.* **49** (2016) 192001.

[13] S. Weinberg, *The Quantum Theory of Fields: Vol. I Foundations*, Cambridge University Press, Cambridge 2000.

[14] H. B. Callen, *Thermodynamics and an Introduction to Thermostatistics*, John Wiley and Sons, 1960.

[15] M. Mintchev, L. Santoni and P. Sorba, *J. Phys. A: Math. and Theor.* **48** (2015) 055003.

[16] J. A. Shohat and J. D. Tamarkin, *The Problem of Moments*, American Mathematical Society, Providence, Rhode Island 1970.

[17] V. Kostrykin and R. Schrader, *Fortschr. Phys.* **48** (2000) 703.

[18] M. Harmer, *J. Phys. A: Math. and Gen.* **33** (2000) 9015.

[19] M. Mintchev, L. Santoni and P. Sorba, *Annalen der Physik* **529** (2017) 1600274.

[20] R. Landauer, IBM J. Res. Dev. **1**, 233 (1957); Philos. Mag. **21**, 863 (1970).

[21] M. Büttiker, Phys. Rev. Lett. **57**,1761 (1986); IBM J. Res. Dev. **32**, 317 (1988).

[22] O. Bratteli and D. W. Robinson, *Operator Algebras and Quantum Statistical Mechanics 2: Equilibrium States. Models in Quantum Statistical Mechanics*, Springer-Verlag, Berlin 1997.

[23] M. Mintchev, *J. Phys. A: Math. and Theor.* **44** (2011) 415201.

[24] G. Nenciu, *J. Math. Phys.* **48** (2007) 033302.

[25] Th. Martin and R. Landauer, *Phys. Rev. B* **45** (1992) 1742.

[26] M. Büttiker, *Phys. Rev. B* **46** (1992) 12485.

[27] M. J. M. de Jong and C. W. J. Beenakker, *Shot noise in mesoscopic systems*, in *Mesoscopic Electron Transport*, edited by L. L. Sohn, L. P. Kouwenhoven and G. Schoen, NATO ASI Series 345, Kluwer Academic Publishers, Dordrecht 1997, 225.

[28] Ya. M. Blanter and M. Büttiker, *Phys. Rep.* **336** (2000) 1.

[29] M. Mintchev, L. Santoni and P. Sorba, *J. Phys. A: Math. and Theor.* **48** (2015) 285002.

[30] G. E. Crooks, *Phys. Rev. E* **60** (1999) 2721.

[31] C. Jarzynski, *J. Statist. Phys.* **98** (2000) 77.

[32] D. J. Evans and D. J. Searles, *Adv. Phys.* **51** (2002) 1529.

[33] D. Bernard and B. Doyon, *Ann. Henri Poincaré* **16** (2015) 113.

[34] S. Hollands and R. Longo, *Commun. Math. Phys.* **357** (2018) 43.

[35] M. Mintchev and P. Sorba, *J. Phys. A: Math. and Theor.* **46** (2013) 095006.

[36] M. Fagotti, *Phys. Rev. B* **96** (2017) 220302.

[37] B. Bertini, M. Fagotti, L. Piroli, and P. Calabrese, *J. Phys. A: Math. and Theor.* **51** (2018) 39LT01.

[38] V. Alba, B. Bertini and M. Fagotti, *SciPost Phys.* **6** (2019) 059.

[39] B. Doyon, *Lecture notes on Generalised Hydrodynamics*, arXiv:1912.08496.

[40] K. Brandner, *Coherent Transport in Periodically Driven Mesoscopic Conductors: From Scattering Matrices to Quantum Thermodynamics*, arXiv:2002.11063.

From 3D torus with H-flux to torus with R-flux and back*

Bojan Nikolić†
Institute of Physics Belgrade, University of Belgrade
Pregrevica 118, Belgrade, Serbia

Danijel Obrić‡
Institute of Physics Belgrade, University of Belgrade
Pregrevica 118, Belgrade, Serbia

Abstract

In this article we study 3D closed bosonic string propagating in the constant metric and Kalb-Ramond field with one non-zero component, $B_{xy} = Hz$, where field strength H is infinitesimal. We will T-dualize along line $x \to y \to z$, which means that we T-dualize first along x coordinate, then along y and, finally, along z coordinate. After first two T-dualizations we obtain Q flux theory which is just locally well defined, while after all three T-dualizations we obtain nonlocal R flux theory. The Q flux theory is commutative one and the R flux theory is noncommutative and nonassociative one. After that we reverse the T-dualization line and T-dualize along $z \to y \to x$. All three theories are nonlocal, but after the first T-dualization we obtain commutative and associative theory, while after we T-dualize along y, we get noncommutative and associative theory. T-dualizing along x, we come to the theory which is both noncommutative and nonassociative. The form of the final T-dual action does not depend on the order of T-dualization while noncommutativity and nonassociativity relations could be obtained from those in the $x \to y \to z$ case by replacing $H \to -H$.

1. Introduction

Heisenberg suggested coordinate noncommutativity in order to solve the problem of infinities before developing of renormalization procedure. In the paper [1] discrete Lorentz invariant space-time is constructed, which means that coordinates are noncommutative.

Noncommutativity came into the focus of interest with the appearance of the paper [2], where it is shown that open string endpoints in the presence

---

* This work has been supported by the Serbian Ministry of Science and Technological Development, under contract No. 171031.

† e-mail address: bnikolic@ipb.ac.rs
‡ e-mail address: dobric@ipb.ac.rs

of the constant metric and Kalb-Ramond field became noncommutative. After this article many others [3] appeared addressing the same subject but using different approaches.

The closed bosonic string in the presence of constant metric and Kalb-Ramond filed remains commutative, because there are no boundary conditions constraining string dynamics. Noncommutativity could be achieved [4] but using T-duality procedure and coordinate dependent Kalb-Ramond field.

T-duality as a fundamental feature of string theory [5, 6, 7, 8, 9, 10, 11] is realized within Buscher T-dualization procedure [6] which can considered as definition of T-dualization. It is applicable along directions on which background fields do not depend. In order to work with coordinate dependent backgrounds generalized T-dualization procedure is developed [12, 13, 14].

Here we will study closed bosonic string in the presence of the constant metric and linear dependent Kalb-Ramond field with just one nonzero component, $B_{xy} = Hz$, the background already analyzed in [15]. In all calculations we keep constant and linear terms in infinitesimal field strength H. We will use transformation laws, relations which connect initial and T-dual variables, in canonical form, expressed in terms of the coordinates and momenta. Our task is to T-dualize along T-dualization chain $x \to y \to z$ and in opposite direction and examine the influence of the T-dualization sequence on the form of the final theory (theory obtained after three T-dualizations) as well as on the noncommutativity and nonassociativity parameters.

T-dualizations along x and y produces the Q-flux background [15], which is still locally well defined, but the theory is commutative. Applying the generalized T-duality procedure [12, 13, 14], z T-dualization gives R flux theory which is nonlocal one because it depends on the non-locally defined variable ΔV. Nonzero Poisson brackets of the T-dual coordinates shows that there is a connection of non-locality and closed string noncommutativity.

The form of noncommutativity is proportional to the infinitesimal field strength H and difference of the initial coordinates. When arguments of the coordinates are different, $\sigma \neq \bar{\sigma}$, there exists noncommutativity. The consequence of the coordinate dependent noncommutativity relations is broken Jacobi identity - nonassociativity occurs. Nonassociativity parameter is proportional to the field strength H.

In the second part of the article we will T-dualize first along z and then along isometry directions y and finally along x. After first T-dualization we get commutative and associative theory as in xyz case. The second T-dualization produces noncommutative and associative theory. In the xyz case, theory second in the T-dualization chain is both commutative and associative. The action of the final theory is the same as in xyz case which is nonassociative and noncommutative. The noncommutativity and nonassociativity parameters have one additional "$-$" sign comparing with the corresponding ones in [16].

2. Action and T-dualization procedure

In this section we will present the construction of the model and give some important details of the T-dualization procedure.

2.1. Model

The closed bosonic string action is of the form [5]

$$S = \kappa \int_{\Sigma} d^2\xi \sqrt{-g} \left\{ \left[\frac{1}{2} g^{\alpha\beta} G_{\mu\nu}(x) + \frac{\varepsilon^{\alpha\beta}}{\sqrt{-g}} B_{\mu\nu}(x) \right] \partial_\alpha x^\mu \partial_\beta x^\nu + \Phi(x) R^{(2)} \right\}, \tag{1}$$

where world-sheet surface Σ is parameterized by $\xi^\alpha = (\tau\,,\sigma)$ $[(\alpha = 0\,,1)$, $\sigma \in (0\,,\pi)]$, while x^μ $(\mu = 0, 1, 2, \dots, D-1)$ are space-time coordinates. Intrinsic world sheet metric is denoted by $g_{\alpha\beta}$, and the corresponding scalar curvature with $R^{(2)}$. Here $G_{\mu\nu}$ is, in the general case, coordinate dependent metric, $B_{\mu\nu}$ is coordinate dependent Kalb-Ramond field, while Φ is dilaton field.

If we intend to have conformal symmetry on the quantum level, background fields are not arbitrarily chosen i.e. they must obey the space-time field equations [17]

$$\beta^G_{\mu\nu} \equiv R_{\mu\nu} - \frac{1}{4} B_{\mu\rho\sigma} B_\nu{}^{\rho\sigma} + 2 D_\mu a_\nu = 0\,, \tag{2}$$

$$\beta^B_{\mu\nu} \equiv D_\rho B^\rho{}_{\mu\nu} - 2 a_\rho B^\rho{}_{\mu\nu} = 0\,, \tag{3}$$

$$\beta^\Phi \equiv 2\pi\kappa \frac{D-26}{6} - R - \frac{1}{24} B_{\mu\rho\sigma} B^{\mu\rho\sigma} - D_\mu a^\mu + 4a^2 = c\,, \tag{4}$$

where c is an arbitrary constant. It holds

$$D^\nu \beta^G_{\nu\mu} + \partial_\mu \beta^\Phi = 0\,, \tag{5}$$

which means that third beta function, β^Φ, can be zero or nonzero constant. The notation is in the standard form: $R_{\mu\nu}$ and D_μ are Ricci tensor and covariant derivative with respect to the space-time metric $G_{\mu\nu}$, while field strength for Kalb-Ramond field $B_{\mu\nu}$ and dilaton gradient are defined as

$$B_{\mu\nu\rho} = \partial_\mu B_{\nu\rho} + \partial_\nu B_{\rho\mu} + \partial_\rho B_{\mu\nu}\,, \quad a_\mu = \partial_\mu \Phi\,. \tag{6}$$

Choosing Kalb-Ramond field to be linearly coordinate dependent and dilaton field to be constant, we obtain (2)

$$R_{\mu\nu} - \frac{1}{4} B_{\mu\rho\sigma} B_\nu{}^{\rho\sigma} = 0\,. \tag{7}$$

Assuming that Kalb-Ramond field strength is infinitesimal we can take $G_{\mu\nu}$ to be constant but in approximation linear in $B_{\mu\nu\rho}$. As a consequence the third equation (4) becomes

$$2\pi\kappa\frac{D-26}{6}=c\,. \tag{8}$$

The arbitrary constant c can be fixed, $c=-\frac{23\pi}{3}$, which gives $D=3$.

The background fields are of the form

$$G_{\mu\nu}=\begin{pmatrix} R_1^2 & 0 & 0\\ 0 & R_2^2 & 0\\ 0 & 0 & R_3^2\end{pmatrix}, \quad B_{\mu\nu}=\begin{pmatrix} 0 & Hz & 0\\ -Hz & 0 & 0\\ 0 & 0 & 0\end{pmatrix}, \tag{9}$$

where $R_\mu(\mu=1,2,3)$ are radii of the compact dimensions. If we rescale the coordinates

$$x^\mu \longmapsto x'^\mu = R_\mu x^\mu\,, \tag{10}$$

the form of the metric simplifies

$$G_{\mu\nu}=\begin{pmatrix} 1 & 0 & 0\\ 0 & 1 & 0\\ 0 & 0 & 1\end{pmatrix}. \tag{11}$$

After all the action gets the form

$$\begin{aligned} S &= \kappa\int_\Sigma d^2\xi\partial_+x^\mu\Pi_{+\mu\nu}\partial_-x^\nu \qquad (12)\\ &= \kappa\int_\Sigma d^2\xi\left[\frac{1}{2}\left(\partial_+x\partial_-x+\partial_+y\partial_-y+\partial_+z\partial_-z\right)\right.\\ &+ \left.\partial_+xHz\partial_-y-\partial_+yHz\partial_-x\right]\,, \end{aligned}$$

where $\partial_\pm=\partial_\tau\pm\partial_\sigma$, and

$$\Pi_{\pm\mu\nu}=B_{\mu\nu}\pm\frac{1}{2}G_{\mu\nu}\,,\quad x^\mu=\begin{pmatrix} x\\ y\\ z\end{pmatrix}. \tag{13}$$

2.2. T-dualization procedure

In the standard Buscher procedure, the starting point is assumption that the target space has isometries. Our background is coordinate dependent so it is useful to examine if the isometry exists. Let us start with the coordinate shift

$$\delta x^\mu=\lambda^\mu=const, \tag{14}$$

and assume that all the coordinates are compact. As $B_{\mu\nu}$ is linear in coordinate, we have

$$\begin{aligned} \delta S &= \frac{\kappa}{3} B_{\mu\nu\rho}\lambda^\rho \int d^2\xi \partial_+ x^\mu \partial_- x^\nu \\ &= \frac{\kappa}{3} B_{\mu\nu\rho}\lambda^\rho \epsilon^{\alpha\beta} \int d^2\xi \partial_\alpha x^\mu \partial_\beta x^\nu . \end{aligned} \tag{15}$$

This is proportional to the total divergence

$$\delta S = \frac{\kappa}{3} B_{\mu\nu\rho}\lambda^\rho \epsilon^{\alpha\beta} \int d^2\xi \partial_\alpha (x^\mu \partial_\beta x^\nu) = 0, \tag{16}$$

which vanishes in the case of the closed string and the topologically trivial mapping of the world-sheet into the space-time. So, the isometry exists even in the case we have chosen.

To localize the global symmetry, we introduce the gauge fields v^μ_α and substitute the ordinary derivatives with the covariant ones

$$\partial_\alpha x^\mu \to D_\alpha x^\mu = \partial_\alpha x^\mu + v^\mu_\alpha . \tag{17}$$

The covariant derivatives are gauge invariant under the following transformation law for the gauge fields

$$\delta v^\mu_\alpha = -\partial_\alpha \lambda^\mu, \quad (\lambda^\mu = \lambda^\mu(\tau,\sigma)). \tag{18}$$

This replacement is not sufficient to make the action locally invariant because the background field $B_{\mu\nu}$ is coordinate dependent. The coordinate x^μ, should be replaced with the invariant coordinate

$$\begin{aligned} \Delta x^\mu_{inv} &\equiv \int_P d\xi^\alpha \, D_\alpha x^\mu = \int_P (d\xi^+ D_+ x^\mu + d\xi^- D_- x^\nu) \\ &= x^\mu - x^\mu(\xi_0) + \Delta V^\mu , \end{aligned} \tag{19}$$

where

$$\Delta V^\mu \equiv \int_P d\xi^\alpha v^\mu_\alpha = \int_P (d\xi^+ v^\mu_+ + d\xi^- v^\mu_-). \tag{20}$$

In order to make gauge fields v^μ_α nonphysical degrees of freedom, the corresponding field strength

$$F^\mu_{\alpha\beta} \equiv \partial_\alpha v^\mu_\beta - \partial_\beta v^\mu_\alpha , \tag{21}$$

must vanish. Technically, this can be achieved by introducing the Lagrange multiplier y_μ, and the appropriate additional term in the Lagrangian

$$S_{add} = \frac{\kappa}{2} \int d^2 (v^\mu_+ \partial_- y_\mu - v^\mu_- \partial_+ y_\mu), \tag{22}$$

where the last term is equal $\frac{1}{2} y_\mu F^\mu_{+-}$ up to the total divergence. At the end of procedure we fix gauge freedom in the way that $x^\mu(\xi) = x^\mu(\xi_0)$.

3. T-dualization along chain $x \to y \to z$

In this section we will make T-dualization along chain $x \to y \to z$, step by step. Our goal is to find transformation laws, and using them we will calculate noncommutativity and nonassociativity relations.

3.1. Twisted torus geometry from torus with H-flux

Because background fields do not depend on coordinate x, T-dualization along direction x is performed within standard Buscher procedure (without introduction of invariant coordinate). Here we will repeat the standard and generalized Buscher procedure explained above, while in other cases we will just give the final results.

Since x direction is an isometry one, action has a global shift symmetry, $x \longrightarrow x+a$. Localizing this symmetry we replace ordinary derivatives with the covariant ones

$$\partial_\pm x \longrightarrow D_\pm x = \partial_\pm x + v_\pm \,, \tag{23}$$

where $v_\pm$ are gauge field. In order to have T-dual action with the same number of degrees of freedom as initial one, we have to add following term to the action

$$S_{add} = \frac{\kappa}{2}\int_\Sigma d^2\xi y_1(\partial_+ v_- - \partial_- v_+)\,, \tag{24}$$

where y_1 is a Lagrange multiplier. Symmetry enables us to fix gauge, $x = const.$, which produces

$$\begin{aligned} S_{fix} &= \kappa\int d^2\xi\left[\frac{1}{2}(v_+v_- + \partial_+ y\partial_- y + \partial_+ z\partial_- z)\right.\\ &+ v_+ Hz\partial_- y - \partial_+ y Hz v_-\\ &+ \left.\frac{1}{2}y_1(\partial_+ v_- - \partial_- v_+)\right]. \end{aligned} \tag{25}$$

On the equations of motion for y_1 field strength for the gauge field $v_\pm$ is equal to zero

$$F_{+-} = \partial_+ v_- - \partial_- v_+ = 0\,. \tag{26}$$

Vanishing of the field strength gives us the solution

$$v_\pm = \partial_\pm x\,. \tag{27}$$

Applying this solution from gauge fixed action (25) we restore initial action (12). Varying the gauge fixed action with respect to the gauge fields we get

$$v_- = -\partial_- y_1 - 2Hz\partial_- y\,, \tag{28}$$

$$v_+ = \partial_+ y_1 + 2Hz\partial_+ y\,. \tag{29}$$

Using (28) and (29) from gauge fixed action (25) we get the T-dual action

$$ {}_xS = \kappa \int_\Sigma d^2\xi \partial_+ ({}_xX)^\mu {}_x\Pi_{+\mu\nu} \partial_- ({}_xX)^\nu \,, \tag{30} $$

where

$$ {}_xX^\mu = \begin{pmatrix} y_1 \\ y \\ z \end{pmatrix} , \tag{31} $$

and T-dual background fields

$$ {}_x\Pi_{+\mu\nu} = {}_xB_{\mu\nu} + \frac{1}{2} {}_xG_{\mu\nu} \,, \quad {}_xB_{\mu\nu} = 0 \,, \quad {}_xG_{\mu\nu} = \begin{pmatrix} 1 & 2Hz & 0 \\ 2Hz & 1 & 0 \\ 0 & 0 & 1 \end{pmatrix} . \tag{32} $$

Geometry described by background fields (32) defines so called *twisted torus geometry*. String theory after one T-dualization is geometrical (flux H takes the role of connection).

Combining two sets of equations of motion, (27), (28) and (29), we get the transformation laws

$$ \partial_\pm x \cong \pm \partial_\pm y_1 \pm 2Hz \partial_\pm y \,, \tag{33} $$

where $\cong$ denotes T-duality relation. From the initial action (12) we can easily find canonical momentum π_x

$$ \pi_x = \frac{\delta S}{\delta \dot{x}} = \kappa (\dot{x} - 2Hzy') \,, \tag{34} $$

where $\dot{A} \equiv \partial_\tau A$ and $A' \equiv \partial_\sigma A$. Transformation law (33) produces the relation

$$ \dot{x} \cong y_1' + 2Hzy' \,, \tag{35} $$

which, combined with the expression for π_x, enables us to find transformation law in canonical form

$$ \pi_x \cong \kappa y_1' \,. \tag{36} $$

The initial theory described by action (12) is geometrical one and their coordinates and canonical momenta satisfy standard Poisson algebra

$$ \{x^\mu(\sigma), \pi_\nu(\bar{\sigma})\} = \delta^\mu{}_\nu \delta(\sigma - \bar{\sigma}) \,, \quad \{x^\mu, x^\nu\} = \{\pi_\mu, \pi_\nu\} = 0 \,. \tag{37} $$

Using this algebra it is simply to show that obtained theory (30) is commutative one

$$ \{{}_xX^\mu, {}_xX^\nu\} = 0 \,. \tag{38} $$

3.2. The second step - Q-flux theory

In this subsection the starting action is the action obtained after T-dualization along x (30). Because y is an isometry direction T-dualization along y direction will be performed according to standard Buscher procedure.

Let us construct the gauge fixed action starting with (30)

$$\begin{aligned} S_{fix} &= \kappa \int_\Sigma d^2\xi \left[\frac{1}{2}(\partial_+ y_1 \partial_- y_1 + v_+ v_- + \partial_+ z \partial_- z) \right. \\ &+ \left. \partial_+ y_1 H z v_- + v_+ H z \partial_- y_1 + \frac{1}{2} y_2 (\partial_+ v_- - \partial_- v_+) \right] . \end{aligned} \quad (39)$$

The equation of motion for Lagrange multiplier y_2 produces

$$\partial_+ v_- - \partial_- v_+ = 0 \longrightarrow v_\pm = \partial_\pm y . \quad (40)$$

Application of these equations of motions transfers action (39) to (30). Varying the gauge fixed action with respect to the gauge fields are

$$v_\pm = \pm \partial_\pm y_2 - 2Hz\partial_\pm y_1 . \quad (41)$$

Putting these expressions into gauge fixed action, we get T-dual action

$${}_{xy}S = \kappa \int d^2\xi \partial_+ ({}_{xy}X)^\mu \, {}_{xy}\Pi_{+\mu\nu} \partial_- ({}_{xy}X)^\nu , \quad (42)$$

where the background fields are

$${}_{xy}B_{\mu\nu} = \begin{pmatrix} 0 & -Hz & 0 \\ Hz & 0 & 0 \\ 0 & 0 & 0 \end{pmatrix} = -B_{\mu\nu} , \quad {}_{xy}G_{\mu\nu} = \begin{pmatrix} 1 & 0 & 0 \\ 0 & 1 & 0 \\ 0 & 0 & 1 \end{pmatrix} , \quad (43)$$

and

$$({}_{xy}X)^\mu = \begin{pmatrix} y_1 \\ y_2 \\ z \end{pmatrix} , \quad {}_{xy}\Pi_{+\mu\nu} = {}_{xy}B_{\mu\nu} + \frac{1}{2} {}_{xy}G_{\mu\nu} = \begin{pmatrix} \frac{1}{2} & -Hz & 0 \\ Hz & \frac{1}{2} & 0 \\ 0 & 0 & \frac{1}{2} \end{pmatrix} . \quad (44)$$

We see that background fields look like those of torus with H flux (12). Their global properties are non-trivial and that is a reason why the term nongeometry is introduced.

Using T-dual transformation laws in Lagrangian form

$$\partial_\pm y \cong \pm \partial_\pm y_2 - 2Hz\partial_\pm y_1 , \quad (45)$$

in combination with the expression for canonical momentum of the initial theory

$$\pi_y = \frac{\delta S}{\delta \dot{y}} = \kappa(\dot{y} + 2Hzx') . \quad (46)$$

we get T-dual transformation law in canonical form

$$\pi_y \cong \kappa y_2' . \tag{47}$$

From standard Poisson algebra (37) it follows that theory obtained by two T-dualizations along isometry directions is still commutative

$$\{_{xy}X^\mu , {}_{xy}X^\nu\} = 0 . \tag{48}$$

3.3. Full T-dualized theory

Kalb-Ramond field is dependent on z but this direction is still isometry one (see subsection 2.2). For T-dualization along z we use generalized T-dualization procedure [12, 13, 14].

Starting action is the one obtained in the previous step (42). The first step in the T-dualization procedure is localizing shift symmetry of the action (42) along z direction. This means that we have to introduce covariant derivative

$$\partial_\pm z \longrightarrow D_\pm z = \partial_\pm z + v_\pm . \tag{49}$$

Then we introduce the invariant coordinate as line integral

$$z^{inv} = \int_P d\xi^\alpha D_\alpha z = \int_P d\xi^+ D_+ z + \int_P d\xi^- D_- z = z(\xi) - z(\xi_0) + \Delta V , \tag{50}$$

where

$$\Delta V = \int_P d\xi^\alpha v_\alpha = \int_P (d\xi^+ v_+ + d\xi^- v_-) . \tag{51}$$

In order to make $v_\pm$ to be nonphysical degrees of freedom we add to the action term with Lagrange multiplier

$$S_{add} = \frac{\kappa}{2} \int_\Sigma d^2\xi \; y_3 (\partial_+ v_- - \partial_+ v_-) . \tag{52}$$

The final form of the action is

$$\begin{aligned} \bar{S} &= \kappa \int_\Sigma d^2\xi \Big[-H z^{inv} (\partial_+ y_1 \partial_- y_2 - \partial_+ y_2 \partial_- y_1) \qquad (53) \\ &+ \frac{1}{2} (\partial_+ y_1 \partial_- y_1 + \partial_+ y_2 \partial_- y_2 + D_+ z D_- z) + \frac{1}{2}\, y_3 (\partial_+ v_- - \partial_- v_+) \Big] . \end{aligned}$$

The final step in the procedure is gauge fixing, $z(\xi) = z(\xi_0)$, and then the gauge fixed action is of the form

$$\begin{aligned} S_{fix} &= \kappa \int_\Sigma d^2\xi \, [-H \Delta V (\partial_+ y_1 \partial_- y_2 - \partial_+ y_2 \partial_- y_1) \qquad (54) \\ &+ \frac{1}{2} (\partial_+ y_1 \partial_- y_1 + \partial_+ y_2 \partial_- y_2 + v_+ v_-) + \frac{1}{2} y_3 (\partial_+ v_- - \partial_- v_+) \Big] . \end{aligned}$$

We restore initial theory (42) from the gauge fixed action using equation of motion for Lagrange multiplier y_3

$$\partial_+ v_- - \partial_- v_+ = 0 \Longrightarrow v_\pm = \partial_\pm z\,, \quad \Delta V = \Delta z\,. \tag{55}$$

Varying the gauge fixed action (54) with respect to the gauge fields we get

$$v_\pm = \pm\partial_\pm y_3 - 2\beta^\mp\,, \tag{56}$$

where $\beta^\pm$ are functions defined as

$$\beta^\pm = \pm\frac{1}{2}H(y_1\partial_\mp y_2 - y_2\partial_\mp y_1)\,. \tag{57}$$

They are a result of the variation of the term containing ΔV. Using relations (56) from the gauge fixed action, we obtain the T-dual action

$$_{xyz}S = \kappa\int_\Sigma d^2\xi\,\partial_{+\,xyz}X^\mu\,{}_{xyz}\Pi_{+\mu\nu}\partial_{-\,xyz}X^\nu\,, \tag{58}$$

where background fields are

$$_{xyz}B_{\mu\nu} = \begin{pmatrix} 0 & -H\Delta\tilde{y}_3 & 0 \\ H\Delta\tilde{y}_3 & 0 & 0 \\ 0 & 0 & 0 \end{pmatrix}, \quad {}_{xyz}G_{\mu\nu} = \begin{pmatrix} 1 & 0 & 0 \\ 0 & 1 & 0 \\ 0 & 0 & 1 \end{pmatrix}, \tag{59}$$

and

$$_{xyz}X^\mu = \begin{pmatrix} y_1 \\ y_2 \\ y_3 \end{pmatrix}. \tag{60}$$

The double coordinate $\tilde{y}_3$ is defined as

$$\partial_\pm y_3 \equiv \pm\partial_\pm\tilde{y}_3\,, \tag{61}$$

while, because it stands together with H, we calculate ΔV in the zeroth order

$$\Delta V = \int d\xi^+\partial_+ y_3 - \int d\xi^-\partial_- y_3\,. \tag{62}$$

Because ΔV is defined as line integral, the nonlocality occures in the T-dual theory. The result of the three T-dualization is a theory with R flux.

Lagrangian form of the T-dual transformation law is

$$\partial_\pm z \cong \pm\partial_\pm y_3 - 2\beta^\mp\,. \tag{63}$$

while its canonical form is

$$y_3' \cong \frac{1}{\kappa}\pi_z - H(xy' - yx')\,. \tag{64}$$

4. Noncommutativity and nonassociativity

In this section we will calculate noncommutativity and nonassociativity relations using canonical forms of the transformation laws.

4.1. Noncommutativity relations

Rewriting the relations (36), (47) and (64) as

$$y_1' \cong \frac{1}{\kappa}\pi_x\,, \quad y_2' \cong \frac{1}{\kappa}\pi_y\,, \quad y_3' \cong \frac{1}{\kappa}\pi_z - H(xy' - yx')\,. \tag{65}$$

we conclude that nontrivial Poisson brackets will be $\{y_1(\sigma), y_3(\bar\sigma)\}$ and $\{y_2(\sigma), y_3(\bar\sigma)\}$. Using the result presented in the Appendix A and the relations

$$\{y_1'(\sigma), y_3'(\bar\sigma)\} \cong \frac{2}{\kappa}Hy'(\sigma)\delta(\sigma-\bar\sigma) + \frac{1}{\kappa}Hy(\sigma)\delta'(\sigma-\bar\sigma)\,, \tag{66}$$

$$\{y_2'(\sigma), y_3'(\bar\sigma)\} \cong -\frac{2}{\kappa}Hx'(\sigma)\delta(\sigma-\bar\sigma) - \frac{1}{\kappa}Hx(\sigma)\delta'(\sigma-\bar\sigma)\,, \tag{67}$$

we get the Poisson brackets of the T-dual coordinates

$$\{y_1(\sigma), y_3(\bar\sigma)\} \cong -\frac{H}{\kappa}\left[2y(\sigma) - y(\bar\sigma)\right]\theta(\sigma-\bar\sigma)\,, \tag{68}$$

$$\{y_2(\sigma), y_3(\bar\sigma)\} \cong \frac{H}{\kappa}\left[2x(\sigma) - x(\bar\sigma)\right]\theta(\sigma-\bar\sigma)\,. \tag{69}$$

If $\sigma = \bar\sigma$ then these two Poisson brackets are zero. But if we choose that $\sigma - \bar\sigma = 2\pi$ then $\theta(2\pi) = 1$ and it follows

$$\{y_1(\sigma+2\pi), y_3(\sigma)\} \cong -\frac{H}{\kappa}\left[4\pi N_y + y(\sigma)\right]\,, \tag{70}$$

$$\{y_2(\sigma+2\pi), y_3(\sigma)\} \cong \frac{H}{\kappa}\left[4\pi N_x + x(\sigma)\right]\,, \tag{71}$$

where N_x and N_y are winding numbers defined as

$$x(\sigma+2\pi) - x(\sigma) = 2\pi N_x\,, \quad y(\sigma+2\pi) - y(\sigma) = 2\pi N_y\,. \tag{72}$$

4.2. Nonassociativity

Let us start calculating Poisson brackets $\{y_1(\sigma), x(\bar\sigma)\}$ and $\{y_2(\sigma), y(\bar\sigma)\}$. Similar to the calculations presented in Appendix A, we start with

$$\{\Delta y_1(\sigma,\sigma_0), x(\bar\sigma)\} = \{\int_{\sigma_0}^{\sigma} d\eta y_1'(\eta), x(\bar\sigma)\}\,. \tag{73}$$

Using the T-dual transformation in canonical form (36), we obtain

$$\{\Delta y_1(\sigma,\sigma_0), x(\bar\sigma)\} \cong \frac{1}{\kappa}\{\int_{\sigma_0}^{\sigma} d\eta \pi_x(\eta), x(\bar\sigma)\}\,, \tag{74}$$

which produces further

$$\{\Delta y_1(\sigma,\sigma_0), x(\bar\sigma)\} \cong -\frac{1}{\kappa}\left[\theta(\sigma-\bar\sigma)-\theta(\sigma_0-\bar\sigma)\right]$$
$$\Longrightarrow \quad \{y_1(\sigma), x(\bar\sigma)\} \cong -\frac{1}{\kappa}\theta(\sigma-\bar\sigma)\,. \tag{75}$$

The relation $\{y_2(\sigma), y(\bar\sigma)\}$ can be obtained in the same way

$$\{y_2(\sigma), y(\bar\sigma)\} \cong -\frac{1}{\kappa}\theta(\sigma-\bar\sigma)\,. \tag{76}$$

Now it is straightforward to calculate Jacobi identity using (68), (69) and above Poisson brackets

$$\begin{aligned} & \{y_1(\sigma_1), y_2(\sigma_2), y_3(\sigma_3)\} \equiv \\ & \{y_1(\sigma_1), \{y_2(\sigma_2), y_3(\sigma_3)\}\} + \{y_2(\sigma_2), \{y_3(\sigma_3), y_1(\sigma_1)\}\} \\ + \; & \{y_3(\sigma_3), \{y_1(\sigma_1), y_2(\sigma_2)\}\} \cong \\ & -\frac{2H}{\kappa^2}\left[\theta(\sigma_1-\sigma_2)\theta(\sigma_2-\sigma_3)+\theta(\sigma_2-\sigma_1)\theta(\sigma_1-\sigma_3)\right. \\ + \; & \left.\theta(\sigma_1-\sigma_3)\theta(\sigma_3-\sigma_2)\right]\,. \end{aligned} \tag{77}$$

Jacobi identity is nonzero which means that theory with R-flux is in general nonassociative. For $\sigma_2=\sigma_3=\sigma$ and $\sigma_1=\sigma+2\pi$ we get

$$\{y_1(\sigma+2\pi), y_2(\sigma), y_3(\sigma)\} \cong \frac{2H}{\kappa^2}\,. \tag{78}$$

5. Reversed T-dualization chain $z \to y \to x$

Our intention is to T-dualize (12) in opposite direction to find the form of noncommutativity and nonassociativity relations. Because we T-dualize first along z, all three theories are R-flux ones.

The T-dualization procedure is presented and already applied three times. So, omitting all technical steps, the T-dual action is of the form

$${}_zS = \kappa\int_\Sigma d^2\xi \partial_+ {}_zX^\mu {}_z\Pi_{+\mu\nu}\partial_- {}_zX^\nu\,, \tag{79}$$

where

$${}_zX^\mu = \begin{pmatrix} x \\ y \\ y_3 \end{pmatrix}, \quad {}_z\Pi_{+\mu\nu} = {}_zB_{\mu\nu} + \frac{1}{2}{}_zG_{\mu\nu}\,, \tag{80}$$

$$
{}_zB_{\mu\nu} = \begin{pmatrix} 0 & H\Delta V & 0 \\ -H\Delta V & 0 & 0 \\ 0 & 0 & 0 \end{pmatrix}, \quad {}_zG_{\mu\nu} = \begin{pmatrix} 1 & 0 & 0 \\ 0 & 1 & 0 \\ 0 & 0 & 1 \end{pmatrix}. \tag{81}
$$

T-dual transformation laws are

$$
\partial_\pm z \cong \pm\partial_\pm y_3 \mp H(x\partial_\pm y - y\partial_\pm x). \tag{82}
$$

Momentum of the initial theory (12) canonically conjugated to the coordinate z is of the form

$$
\pi_z = \kappa\dot{z}, \tag{83}
$$

so, the T-dual transformation law in canonical form is

$$
y_3' \cong \frac{1}{\kappa}\pi_z + H(xy' - yx'). \tag{84}
$$

From the expressions (84) and (37), we get that coordinates ${}_zX^\mu$ are commutative. Consequently, Jacobiator is equal to zero, which means that theory is associative.

After T-dualization along y direction the T-dual action is

$$
{}_{zy}S = \kappa\int_\Sigma d^2\xi\, \partial_+\, {}_{zy}X^\mu\, {}_{zy}\Pi_{+\mu\nu}\, \partial_-\, {}_{zy}X^\nu, \tag{85}
$$

where

$$
{}_{zy}X^\mu = \begin{pmatrix} x \\ y_2 \\ y_3 \end{pmatrix}, \quad {}_{zy}\Pi_{+\mu\nu} = {}_{zy}B_{\mu\nu} + \frac{1}{2}\, {}_{zy}G_{\mu\nu}, \tag{86}
$$

$$
{}_{zy}B_{\mu\nu} = 0, \quad {}_{zy}G_{\mu\nu} = \begin{pmatrix} 1 & -2H\Delta V & 0 \\ -2H\Delta V & 1 & 0 \\ 0 & 0 & 1 \end{pmatrix}. \tag{87}
$$

The canonical form of the T-dual transformation law is

$$
y_2' \cong \frac{1}{\kappa}\pi_y. \tag{88}
$$

The only non-zero Poisson bracket is

$$
\{y_2'(\sigma), y_3'(\bar\sigma)\} \cong \frac{H}{\kappa}\left[2x'(\sigma)\delta(\sigma-\bar\sigma) + x(\sigma)\delta'(\sigma-\bar\sigma)\right]. \tag{89}
$$

Using the instructions from Appendix A, the solution is of the form (107)

$$
\{y_2(\sigma), y_3(\bar\sigma)\} \cong -\frac{H}{\kappa}\left[2x(\sigma) - x(\bar\sigma)\right]\theta(\sigma-\bar\sigma). \tag{90}
$$

For $\sigma - \bar{\sigma} = 2\pi$, we have

$$\{y_2(\sigma + 2\pi), y_3(\sigma)\} \cong -\frac{H}{\kappa}\left[x(\sigma) + 4\pi N_x\right], \tag{91}$$

where N_x is winding number for x coordinate defined as

$$x(\sigma + 2\pi) - x(\sigma) = 2\pi N_x . \tag{92}$$

It is straightforward to calculate the Jacobiator

$$\begin{aligned} & \{x(\sigma_1), \{y_2(\sigma_2), y_3(\sigma_3)\}\} + \{y_2(\sigma_2), \{y_3(\sigma_3), x(\sigma_1)\}\} \\ + \;& \{y_3(\sigma_3), \{x(\sigma_1), y_2(\sigma_2)\}\} \cong 0 . \end{aligned} \tag{93}$$

Because it is zero, we conclude that this R-flux theory is noncommutative and associative one.

The full T-dualozed action is

$$_{zyx}S = \kappa \int_\Sigma d^2\xi \partial_+ \,_{zyx}X^\mu \,_{zyx}\Pi_{+\mu\nu} \,_{zyx}\, X^\nu , \tag{94}$$

where

$$_{zyx}X^\mu = \begin{pmatrix} y_1 \\ y_2 \\ y_3 \end{pmatrix}, \quad _{zyx}\Pi_{+\mu\nu} = \,_{zyx}B_{\mu\nu} + \frac{1}{2}\,_{zyx}G_{\mu\nu} \tag{95}$$

$$_{zyx}B_{\mu\nu} = \begin{pmatrix} 0 & -H\Delta V & 0 \\ H\Delta V & 0 & 0 \\ 0 & 0 & 0 \end{pmatrix}, \quad _{zyx}G_{\mu\nu} = \begin{pmatrix} 1 & 0 & 0 \\ 0 & 1 & 0 \\ 0 & 0 & 1 \end{pmatrix}. \tag{96}$$

The canonical form of the T-dual transformation law is the same as in the case $x \to y \to z$

$$y_1' \cong \frac{1}{\kappa}\pi_x . \tag{97}$$

The set of T-dual transformation laws, (84), (88) and (97), is the same as in the xyz case up to $H \to -H$. Consequently, we have

$$\{y_1(\sigma), y_3(\bar{\sigma})\} \cong \frac{H}{\kappa}\left[2y(\sigma) - y(\bar{\sigma})\right]\theta(\sigma - \bar{\sigma}), \tag{98}$$

$$\{y_2(\sigma), y_3(\bar{\sigma})\} \cong -\frac{H}{\kappa}\left[2x(\sigma) - x(\bar{\sigma})\right]\theta(\sigma - \bar{\sigma}), \tag{99}$$

and nonassociativity

$$\begin{aligned} & \{y_1(\sigma_1), y_2(\sigma_2), y_3(\sigma_3)\} \equiv \\ & \{y_1(\sigma_1), \{y_2(\sigma_2), y_3(\sigma_3)\}\} + \{y_2(\sigma_2), \{y_3(\sigma_3), y_1(\sigma_1)\}\} \\ + \;& \{y_3(\sigma_3), \{y_1(\sigma_1), y_2(\sigma_2)\}\} \cong \\ & \frac{2H}{\kappa^2}\left[\theta(\sigma_1 - \sigma_2)\theta(\sigma_2 - \sigma_3) + \theta(\sigma_2 - \sigma_1)\theta(\sigma_1 - \sigma_3)\right. \\ + \;& \left.\theta(\sigma_1 - \sigma_3)\theta(\sigma_3 - \sigma_2)\right] . \end{aligned} \tag{100}$$

6. Conclusion

In this article we considered closed bosonic string propagating in the constant metric and Kalb-Ramond field with just one nonzero component, $B_{xy} = -B_{yx} = Hz$. We worked in the approximation linear in H. We T-dualized along chain xyz and in opposite direction. Our goal was to examine the influence of the sequence of T-dualizations on the full T-dualized theory, geometrical features and noncommutativity and nonassociativity of the coordinates.

T-dualizing along chain xyz we concluded that the first theory is geometrical, commutative and associative, the second one is nongeometrical, commutative and associative, while the full T-dualized theory is nongeometrical R flux theory which is noncommuattive and nonassociative. When we reversed the direction of T-dualizations we obtained that all three theories are non geometrical R flux ones. But the first one is both commutative and associative, the second one is noncommutative and associative, while the third one is both nonassociative and noncommutative. The corresponding parameters can be obtained from those in xyz case by replacing $H \to -H$. The form of the full T-dualized theory does not depend on the sequence of T-dualizations.

A Important Poisson bracket

Let us start with the Poisson bracket of the σ derivatives of two arbitrary functions in the form

$$\{A'(\sigma), B'(\bar{\sigma})\} = U'(\sigma)\delta(\sigma-\bar{\sigma}) + V(\sigma)\delta'(\sigma-\bar{\sigma}) \,, \tag{101}$$

where $\delta'(\sigma-\bar{\sigma}) \equiv \partial_\sigma \delta(\sigma-\bar{\sigma})$. Our task is to find

$$\{A(\sigma), B(\bar{\sigma})\} \,,$$

from

$$\{\Delta A(\sigma,\sigma_0), \Delta B(\bar{\sigma},\bar{\sigma}_0)\} \,,$$

where

$$\Delta A(\sigma,\sigma_0) = \int_{\sigma_0}^{\sigma} dx A'(x) = A(\sigma) - A(\sigma_0) \,,$$
$$\Delta B(\bar{\sigma},\bar{\sigma}_0) = \int_{\bar{\sigma}_0}^{\bar{\sigma}} dx B'(x) = B(\bar{\sigma}) - B(\bar{\sigma}_0) \,. \tag{102}$$

It is obvious that

$$\{\Delta A(\sigma,\sigma_0), \Delta B(\bar{\sigma},\bar{\sigma}_0)\} = \int_{\sigma_0}^{\sigma} dx \int_{\bar{\sigma}_0}^{\bar{\sigma}} dy \; \left[U'(x)\delta(x-y) + V(x)\delta'(x-y)\right] \,, \tag{103}$$

and after integration over y we get

$$\begin{aligned}&\{\Delta A(\sigma,\sigma_0),\Delta B(\bar\sigma,\bar\sigma_0)\} \\ = &\int_{\sigma_0}^{\sigma} dx\{U'(x)\,[\theta(x-\bar\sigma_0)-\theta(x-\bar\sigma)]+V(x)\,[\delta(x-\bar\sigma_0)-\delta(x-\bar\sigma)]\},\end{aligned} \tag{104}$$

where $\theta(x)$ is defined as

$$\theta(x)=\int_0^x d\eta\delta(\eta)=\frac{1}{2\pi}\left[x+2\sum_{n\geq 1}\frac{1}{n}\sin(nx)\right]=\begin{cases}0 & \text{if } x=0\\ 1/2 & \text{if } 0<x<2\pi\,.\\ 1 & \text{if } x=2\pi\end{cases} \tag{105}$$

Integrating over x, finally we get

$$\begin{aligned}&\{\Delta A(\sigma,\sigma_0),\Delta B(\bar\sigma,\bar\sigma_0)\}= \\ &U(\sigma)[\theta(\sigma-\bar\sigma_0)-\theta(\sigma-\bar\sigma)]-U(\sigma_0)[\theta(\sigma_0-\bar\sigma_0)-\theta(\sigma_0-\bar\sigma)] \\ - &U(\bar\sigma_0)[\theta(\sigma-\bar\sigma_0)-\theta(\sigma_0-\bar\sigma_0)]+U(\bar\sigma)[\theta(\sigma-\bar\sigma)-\theta(\sigma_0-\bar\sigma)] \\ + &V(\bar\sigma_0)[\theta(\sigma-\bar\sigma_0)-\theta(\sigma_0-\bar\sigma_0]-V(\bar\sigma)[\theta(\sigma-\bar\sigma)-\theta(\sigma_0-\bar\sigma)]\end{aligned} \tag{106}$$

From the last expression, we extract the desired Poisson bracket

$$\{A(\sigma),B(\bar\sigma)\}=-[U(\sigma)-U(\bar\sigma)+V(\bar\sigma)]\theta(\sigma-\bar\sigma)\,. \tag{107}$$

References

[1] H. S. Snyder, *Phys.Rev.* **71** (1947) 38.

[2] N. Seiberg and E. Witten, *JHEP* **09** (1999) 032.

[3] A. Connes, M. R. Douglas and A. Schwarz, *JHEP* **02** (1998) 003; M. R. Douglas and C. Hull, *JHEP* **02** (1998) 008; V. Schomerus, *JHEP* **06** (1999) 030; F. Ardalan, H.Arfaei and M. M. Sheikh-Jabbari, *JHEP* **02** (1999) 016; C. S. Chu and P. M. Ho, *Nucl. Phys.* **B550** (1999) 151; F. Ardalan, H. Arfaei and M. M. Sheikh-Jabbari, *Nucl. Phys.* **B576** (2000) 578; C. S. Chu and P. M. Ho, *Nucl. Phys.* **B568** (2000) 447; T.Lee, *Phys. Rev.* **D62** (2000) 024022; B. Nikolić and B. Sazdović, *Phys.Rev.* **D74** (2006) 045024; *Phys.Rev.* **D75** (2007) 085011.

[4] D. Lust *JHEP* **12** (2010) 084.

[5] K. Becker, M. Becker and J. Schwarz *String Theory and M-Theory: A Modern Introduction*; B. Zwiebach, *A First Course in String Theory*, Cambridge University Press, 2004; J. Polchinski, *String theory - Volume II*, Cambridge University Press, 1998.

[6] T. Buscher, *Phys. Lett.* **B 194** (1987) 59; **201** (1988) 466.

[7] M. Roček and E. Verlinde, *Nucl.Phys.* **B 373** (1992) 630.

[8] A. Giveon, M. Porrati and E. Rabinovici, *Phys. Rep.* **244** (1994) 77.

[9] E. Alvarez, L. Alvarez-Gaume, J. Barbon and Y. Lozano, *Nucl. Phys.* **B 415** (1994) 71.

[10] B. Nikolić and B. Sazdović, *Nucl. Phys.* **B 836** (2010) 100.

[11] D. S. Berman, D. C. Thompson, *Phys.Rept.* **566** (2014) 1-60.

[12] Lj. Davidović and B. Sazdović, *EPJ* **C 74** (2014) 2683.

[13] Lj. Davidović, B. Nikolić and B. Sazdović, *EPJ* **C 75** (2015) 576.

[14] Lj. Davidović, B. Sazdović, *JHEP* **11** (2015) 119.

[15] D. Andriot, M. Larfors, D. Luest, P. Patalong, *JHEP* **06** (2013) 021.

[16] B. Nikolić and D. Obrić, *Fortschr. Phys.* **66** (2018) 040009.

[17] C. G. Callan, D. Friedan, E. J. Martinec and M. J. Perry, *Nucl. Phys.* **B262** (1985) 593-609.

Construction and examples of higher gauge theories*

Tijana Radenković[†]
Institute of Physics, University of Belgrade,
Pregrevica 118, 11080 Belgrade, Serbia

Marko Vojinović[‡]
Institute of Physics, University of Belgrade,
Pregrevica 118, 11080 Belgrade, Serbia

Abstract

We provide several examples of higher gauge theories, constructed as generalizations of a BF model to $2BF$ and $3BF$ models with constraints. Using the framework of higher category theory, we introduce appropriate 2-groups and 3-groups, and construct the actions for the corresponding constrained $2BF$ and $3BF$ theories. In this way, we can construct actions which describe the correct dynamics of Yang-Mills, Klein-Gordon, Dirac, Weyl, and Majorana fields coupled to Einstein-Cartan gravity. Each action is naturally split into a topological sector and a sector with simplicity constraints. The properties of the higher gauge group structure opens up a possibility of a nontrivial unification of all fields.

1. Introduction

The quantization of the gravitational field is one of the fundamental open problems in modern physics. There are various approaches to this problem, some of which have developed into vast research frameworks. One of such frameworks is the Loop Quantum Gravity approach, which aims to establish a nonperturbative quantization of gravity, both canonically and covariantly [1, 2, 3]. The covariant approach is slightly more general, and

* This work was supported by the project ON171031 of the Ministry of Education, Science and Technological Development (MPNTR) of the Republic of Serbia, and partially by the bilateral scientific cooperation between Austria and Serbia through the project "Causality in Quantum Mechanics and Quantum Gravity - 2018-2019", no. 451-03-02141/2017-09/02, supported by the Federal Ministry of Science, Research and Economy (BMWFW) of the Republic of Austria, and the Ministry of Education, Science and Technological Development (MPNTR) of the Republic of Serbia.

† e-mail address: rtijana@ipb.ac.rs
‡ e-mail address: vmarko@ipb.ac.rs

focuses on providing a possible rigorous definition of the path integral for the gravitational field,

$$Z = \int \mathcal{D}g \, e^{iS[g]} \,. \tag{1}$$

This is done by considering a triangulation of a spacetime manifold, and defining the path integral as a discrete state sum of the gravitational field configurations living on the simplices in the triangulation. This quantization technique is known as the *spinfoam* quantization method, and is performed via the following three steps:

(1) one writes the classical action $S[g]$ as a constrained BF action;

(2) one uses the Lie group structure, underlying the topological sector of the action, to define a triangulation-independent state sum Z;

(3) one imposes the simplicity constraints on the state sum, promoting it into a triangulation-dependent state sum, which serves as a definition for the path integral (1).

So far, this quantization prescription has been implemented for various choices of the gravitational action, of the Lie group, and of the spacetime dimension. For example, in 3 dimensions, historically the first spinfoam model is known as the Ponzano-Regge model [4]. In 4 dimensions there are multiple models, depending on the choice of the Lie group and the way one imposes the simplicity constraints [5, 6, 7, 8, 9]. While these models do give a definition for the gravitational path integral, none of them are able to consistently include matter fields. Including the matter fields has so far had limited success [10], mainly due to the absence of the tetrad fields from the topological sector of the theory.

In order to resolve this issue, a new approach has been developed, using the framework of *higher gauge theory* (see [11] for a review). In particular, one uses the idea of a *categorical ladder* to generalize the BF action (based on a Lie group) into a $2BF$ action (based on the so-called 2-group structure). A suitable choice of the *Poincaré 2-group* introduces the needed tetrad fields into the topological sector of the action [12]. While this result opened up a possibility to couple matter fields to gravity, the matter fields could not be naturally expressed using the underlying algebraic structure of a 2-group, rendering the spinfoam quantization method inapplicable. Namely, the matter sector could indeed be added to the classical action, but could not be expressed itself as a constrained $2BF$ theory, which means that the steps 1–3 above could not be performed for the matter sector of the action, but only for gravity.

This final issue has recently been resolved in [13], by passing from the 2-group structure to the 3-group structure, generalizing the action one step further in the categorical ladder. This generalization naturally gives rise to the so-called $3BF$ action, which turns out to be suitable for a unified description of both gravity and matter fields. The steps of the categorical ladder and their corresponding structures are summarized as follows:

categorical structure	algebraic structure	linear structure	topological action	degrees of freedom
Lie group	Lie group	Lie algebra	BF theory	gauge fields
Lie 2-group	Lie crossed module	differential Lie crossed module	$2BF$ theory	tetrad fields
Lie 3-group	Lie 2-crossed module	differential Lie 2-crossed module	$3BF$ theory	scalar and fermion fields

The purpose of this paper is to give a systematic overview of the constructions of classical BF, $2BF$ and $3BF$ actions, both pure and constrained, in order to demonstrate the categorical ladder procedure and the construction of higher gauge theories. In other words, we focus on the step 1 of the spinfoam quantization programme.

The layout of the paper is as follows. Section 2 deals with models based on a BF theory. First we discuss the pure, topological BF theory, and then pass on to the the physically more interesting Yang-Mills theory in Minkowski spacetime and the Plebanski formulation of general relativity. In Section 3 we study the first step in the categorical ladder, namely models based on the $2BF$ theory. After introducing the pure $2BF$ theory, we study the relevant formulation of general relativity [12], and then the coupled Einstein-Yang-Mills theory. Then, in Section 4 we perform the second step in the categorical ladder, passing on to models based on the $3BF$ theory. After the introduction of the pure $3BF$ model, we construct constrained $3BF$ actions for the cases of Klein-Gordon, Dirac, Weyl and Majorana fields, all coupled to the Einstein-Cartan gravity in the standard way. As we shall see, the scalar and fermion fields will be *naturally associated to a new gauge group*, generalizing the purpose of a gauge group in the Yang-Mills theory, which opens up a possibility of an algebraic classification of matter fields. Finally, Section 5 contains a discussion and conclusions.

The notation and conventions are as follows. The local Lorentz indices are denoted by the Latin letters $a, b, c, \dots$, take values $0, 1, 2, 3$, and are raised and lowered using the Minkowski metric η_{ab} with signature $(-,+,+,+)$. Spacetime indices are denoted by the Greek letters $\mu, \nu, \dots$, and are raised and lowered by the spacetime metric $g_{\mu\nu} = \eta_{ab} e^a{}_\mu e^b{}_\nu$, where $e^a{}_\mu$ are the tetrad fields. The inverse tetrad is denoted as $e^\mu{}_a$. All other indices that appear in the paper are dependent on the context, and their usage is explicitly defined in the text where they appear. We work in the natural system of units where $c = \hbar = 1$, and $G = l_p^2$, where l_p is the Planck length.

2. BF theory

We begin with a short review of BF theories. See [14, 15, 16] for additional information.

2.1. Pure BF theory

Given a Lie group G, and denoting its corresponding Lie algebra as $\mathfrak{g}$, one introduces the pure BF action as follows (we limit ourselves to the physically relevant case of 4-dimensional spacetime manifolds $\mathcal{M}_4$):

$$S_{BF} = \int_{\mathcal{M}_4} \langle B \wedge \mathcal{F} \rangle_{\mathfrak{g}} \,. \tag{2}$$

Here, $\mathcal{F} \equiv \mathrm{d}\alpha + \alpha \wedge \alpha$ is the curvature 2-form for the algebra-valued connection 1-form $\alpha \in \mathcal{A}^1(\mathcal{M}_4 \,, \mathfrak{g})$, and $B \in \mathcal{A}^2(\mathcal{M}_4 \,, \mathfrak{g})$ is a Lagrange multiplier 2-form, while $\langle \_\,, \_ \rangle_{\mathfrak{g}}$ denotes a G-invariant bilinear symmetric nondegenerate form.

One can see from (2) that the action is diffeomorphism invariant, and it is also gauge invariant with respect to G, provided that B transforms as a scalar with respect to G.

Varying the action (2) with respect to B^β and α^β, where the index β is the group G index (which counts the generators of $\mathfrak{g}$), one obtains the following equations of motion,

$$\mathcal{F}^\beta = 0 \,, \qquad \nabla B^\beta \equiv \mathrm{d}B^\beta + f_{\gamma\delta}{}^\beta \alpha^\gamma \wedge B^\delta = 0 \,, \tag{3}$$

where $f_{\gamma\delta}{}^\beta$ are the structure constants of the Lie group G. From the first equation of motion, one immediately sees that α is a flat connection, meaning that $\alpha = 0$ up to gauge transformations. Given this, the second equation of motion implies that B is constant. Therefore, there are no local propagating degrees of freedom, and the theory is called *topological*.

2.2. Yang-Mills theory

In physics one is usually interested in theories which are not topological, i.e., which have local propagating degrees of freedom. As a rule of thumb, one recognizes that the theory does have local propagating degrees of freedom if one of the equations of motion is a second-order partial differential equation, usually featuring a D'Alambertian operator $\Box$ in some form. In order to transform the pure BF action into such a theory, one adds an additional term to the action, commonly called the *simplicity constraint*. The resulting action is called a *constrained BF theory*. A nice example is the Yang-Mills theory for the $SU(N)$ group in Minkowski spacetime, which can be rewritten as a constrained BF theory in the following way:

$$\begin{aligned} S \;=\; & \int B_I \wedge F^I + \lambda^I \wedge \left(B_I - \frac{12}{g} M_{abI} \delta^a \wedge \delta^b \right) \\ & + \zeta^{abI} \left(M_{abI} \varepsilon_{cdef} \delta^c \wedge \delta^d \wedge \delta^e \wedge \delta^f - g_{IJ} F^J \wedge \delta_a \wedge \delta_b \right) . \end{aligned} \tag{4}$$

Here $F \equiv \mathrm{d}A + A \wedge A$ is again the curvature 2-form for the connection $A \in \mathcal{A}^1(\mathcal{M}_4 \,, \mathfrak{su}(N))$, and $B \in \mathcal{A}^2(\mathcal{M}_4 \,, \mathfrak{su}(N))$ is the Lagrange multiplier

2-form. The Killing form $g_{IJ} \equiv \langle \tau_I , \tau_J \rangle_{\mathfrak{su}(N)} \propto f_{IK}{}^L f_{JL}{}^K$ is used to raise and lower the indices $I, J, \dots$ which count the generators of $SU(N)$, while $f_{IJ}{}^K$ are the structure constants for the $\mathfrak{su}(N)$ algebra. In addition to the topological $B \wedge F$ term, there are also two simplicity constraint terms present, featuring two Lagrange multipliers, a 2-form λ^I and a 0-form ζ^{abI}. The 0-form M_{abI} is also a Lagrange multiplier, while g is the coupling constant for the Yang-Mills theory.

Finally, δ^a is a nondynamical 1-form, such that there exists a global coordinate frame in which its components are equal to the Kronecker symbol $\delta^a{}_\mu$ (hence the notation δ^a). The 1-form δ^a plays the role of a background field, and defines the global spacetime metric, via the equation

$$\eta_{\mu\nu} = \eta_{ab} \delta^a{}_\mu \delta^b{}_\nu \,, \tag{5}$$

where $\eta_{ab} \equiv \mathrm{diag}(-1, +1, +1, +1)$ is the Minkowski metric. Since the coordinate system is global, the spacetime manifold $\mathcal{M}_4$ is understood to be flat. The indices $a, b, \dots$ are local Lorentz indices, taking values $0, \dots, 3$. Note that the field δ^a has all the properties of the tetrad 1-form e^a in the flat Minkowski spacetime. Also note that the action (4) is manifestly diffeomorphism invariant and gauge invariant with respect to $SU(N)$, but not background independent, due to the presence of δ^a.

Varying the action (4) with respect to the variables ζ^{abI}, M_{abI}, A^I, B_I, and λ^I, respectively (but not with respect to the background field δ^a), we obtain the equations of motion:

$$M_{abI} \varepsilon_{cdef} \delta^c \wedge \delta^d \wedge \delta^e \wedge \delta^f - F_I \wedge \delta_a \wedge \delta_b = 0 \,, \tag{6}$$

$$-\frac{12}{g} \lambda^I \wedge \delta^a \wedge \delta^b + \zeta^{abI} \varepsilon_{cdef} \delta^c \wedge \delta^d \wedge \delta^e \wedge \delta^f = 0 \,, \tag{7}$$

$$-\mathrm{d}B_I + f_{JI}{}^K B_K \wedge A^J + \mathrm{d}(\zeta^{ab}{}_I \delta_a \wedge \delta_b) - f_{JI}{}^K \zeta^{ab}{}_K \delta_a \wedge \delta_b \wedge A^J = 0 \,, \tag{8}$$

$$F_I + \lambda_I = 0 \,, \tag{9}$$

$$B_I - \frac{12}{g} M_{abI} \delta^a \wedge \delta^b = 0 \,, \tag{10}$$

From the equations (6), (7), (9) and (10) one obtains the multipliers as algebraic functions of the field strength $F^I{}_{\mu\nu}$ for the dynamical field A^I:

$$\begin{aligned} M_{abI} &= \frac{1}{48} \varepsilon_{abcd} F_I{}^{cd} \,, & \zeta^{abI} &= \frac{1}{4g} \varepsilon^{abcd} F^I{}_{cd} \,, \\ \lambda_{Iab} &= F_{Iab} \,, & B_{Iab} &= \frac{1}{2g} \varepsilon_{abcd} F_I{}^{cd} \,. \end{aligned} \tag{11}$$

Here we used the notation $F_{Iab} = F_{I\mu\nu}\delta_a{}^\mu \delta_b{}^\nu$, and similarly for other variables, where we exploited the fact that $\delta^a{}_\mu$ is invertible. Using these equations and the differential equation (8) one obtains the equation of motion for gauge field $A^I{}_\mu$,

$$\nabla_\rho F^{I\rho\mu} \equiv \partial_\rho F^{I\rho\mu} + f_{JK}{}^I A^J{}_\rho F^{K\rho\mu} = 0\,. \tag{12}$$

This is precisely the classical equation of motion for the free Yang-Mills theory. Note that this is a second-order partial differential equation for the field $A^I{}_\mu$, and moreover contains the $\Box$ operator in the first term.

In addition to the Yang-Mills theory, one can easily extend the action (4) in order to describe the massive vector field and obtain the Proca equation of motion. This is done by adding a mass term

$$-\frac{1}{4!} m^2 A_{I\mu} A^I{}_\nu \eta^{\mu\nu} \varepsilon_{abcd} \delta^a \wedge \delta^b \wedge \delta^c \wedge \delta^d \tag{13}$$

to the action (4). Of course, this term explicitly breaks the $SU(N)$ gauge symmetry of the action.

2.3. Plebanski general relativity

The second example of the constrained BF theory is the Plebanski action for general relativity [16, 14]. Using the Lorentz group $SO(3,1)$ as a gauge group, one constructs a constrained BF action as

$$S = \int_{\mathcal{M}_4} B_{ab} \wedge R^{ab} + \phi_{abcd} B^{ab} \wedge B^{cd}\,. \tag{14}$$

Here R^{ab} is the curvature 2-form for the spin connection ω^{ab}, B_{ab} is the usual Lagrange multiplier 2-form, while ϕ_{abcd} is the additional Lagrange multiplier 0-form multiplying the term $B^{ab} \wedge B^{cd}$ to form a simplicity constraint. It can be shown that the variation of this action with respect to B_{ab}, ω^{ab} and ϕ_{abcd} gives rise to the equations of motion of vacuum general relativity. However, in this model the tetrad fields appear only as a solution of the simplicity constraint equation of motion $B^{ab} \wedge B^{cd} = 0$. Therefore, being intrinsically on-shell objects, the tetrad fields are not present in the action itself and cannot be quantized. This renders the Plebanski model unsuitable for coupling of matter fields to gravity [10, 12, 20]. Nevertheless, regarded as a model for pure gravity, the Plebanski model has been successfully quantized in the context of spinfoam models [8, 9, 1, 2].

3. $2BF$ theory

In this section we perform the first step of the *categorical ladder*, generalizing the algebraic notion of a group to the notion of a 2-group. This leads to the generalization of the BF theory to the $2BF$ theory, also sometimes called $BFCG$ theory [11, 17, 18, 19].

3.1. Pure $2BF$ theory

In order to circumvent the issue of tetrad fields not being present in the Plebanski action, in the context of higher category theory [11] a recent promising approach has been developed [12, 21, 22, 23, 20, 24]. As an essential ingredient, let us first give a short review of the 2-group formalism.

Within the framework of category theory, the group as an algebraic structure can be understood as a category with only one object and invertible morphisms [11]. Additionally, the notion of a category can be generalized to the so-called *higher categories*, which have not only objects and morphisms, but also 2-morphisms (morphisms between morphisms), and so on. This process of generalization is called the *categorical ladder*. Using this process, one can introduce the notion of a *2-group* as a 2-category consisting of only one object, where all the morphisms and all 2-morphisms are invertible. It has been shown that every strict 2-group is equivalent to a *crossed module* $(H \xrightarrow{\partial} G\,,\triangleright)$, see [13] for detailed definitions. Here G and H are groups, ∂ is a homomorphism from H to G, while $\triangleright : G\times H \to H$ is an action of G on H.

Similarly to the case of an ordinary Lie group G which has a naturally associated notion of a connection α, giving rise to a BF theory, the 2-group structure has a naturally associated notion of a 2-connection $(\alpha\,,\beta)$, described by the usual $\mathfrak{g}$-valued 1-form $\alpha \in \mathcal{A}^1(\mathcal{M}_4\,,\mathfrak{g})$ and an $\mathfrak{h}$-valued 2-form $\beta \in \mathcal{A}^2(\mathcal{M}_4\,,\mathfrak{h})$, where $\mathfrak{h}$ is a Lie algebra of the Lie group H. The 2-connection gives rise to the so-called *fake 2-curvature* $(\mathcal{F},\mathcal{G})$, given as

$$\mathcal{F} = \mathrm{d}\alpha + \alpha\wedge\alpha - \partial\beta\,, \qquad \mathcal{G} = d\beta + \alpha\wedge^{\triangleright}\beta\,. \tag{15}$$

Here $\alpha\wedge^{\triangleright}\beta$ means that α and β are multiplied as forms using $\wedge$, and simultaneously multiplied as algebra elements using $\triangleright$, see [13]. The curvature pair $(\mathcal{F},\mathcal{G})$ is called "fake" because of the presence of the additional term $\partial\beta$ in the definition of $\mathcal{F}$ [11].

Using the structure of a 2-group, or equivalently the crossed module, one can generalize the BF action to the so-called $2BF$ action, defined as follows [17, 18]:

$$S_{2BF} = \int_{\mathcal{M}_4} \langle B\wedge\mathcal{F}\rangle_{\mathfrak{g}} + \langle C\wedge\mathcal{G}\rangle_{\mathfrak{h}}\,. \tag{16}$$

Here the 2-form $B \in \mathcal{A}^2(\mathcal{M}_4\,,\mathfrak{g})$ and the 1-form $C \in \mathcal{A}^1(\mathcal{M}_4\,,\mathfrak{h})$ are Lagrange multipliers. Also, $\langle\_\,,\_\rangle_{\mathfrak{g}}$ and $\langle\_\,,\_\rangle_{\mathfrak{h}}$ denote the G-invariant bilinear symmetric nondegenerate forms for the algebras $\mathfrak{g}$ and $\mathfrak{h}$, respectively. As a consequence of the axiomatic structure of a crossed module (see [13]), the bilinear form $\langle\_\,,\_\rangle_{\mathfrak{h}}$ is H-invariant as well. See [17, 18] for review and references.

Similarly to the BF action, the $2BF$ action is also topological, which can be seen from equations of motion. Varying with respect to B^α and C^a one obtains

$$\mathcal{F}^\alpha = 0\,, \qquad \mathcal{G}^a = 0\,, \tag{17}$$

where indices a count the generators of the group H. Varying with respect to α^α and β^a one obtains the equations for the multipliers,

$$\mathrm{d}B_\alpha + f_{\alpha\beta}{}^\gamma B_\gamma \wedge \alpha^\beta - \rhd_{\alpha a}{}^b C_b \wedge \beta^a = 0\,, \tag{18}$$

$$\mathrm{d}C_a - \partial_a{}^\alpha B_\alpha + \rhd_{\alpha a}{}^b C_b \wedge \alpha^\alpha = 0\,. \tag{19}$$

We can again see that the equations of motion are only first-order and have only very simple solutions (note that this is not a sufficient argument for the absence of local propagating degrees of freedom — a counterexample is the Dirac equation, being a first-order partial differential equation which *does* have propagating degrees of freedom). One can additionally use the Hamiltonian analysis to rigorously demonstrate that there are no local propagating degrees of freedom [22, 23]. Thus the $2BF$ theory is also topological.

3.2. General relativity

An important example of a crossed module structure is a vector space V equipped with an isometry group O. Namely, V can be regarded as an Abelian Lie group with addition as a group operation, so that a representation of O on V is an action $\rhd$ of O on the group V, giving rise to the crossed module $(V \stackrel{\partial}{\to} O\,, \rhd)$, where the homomorphism ∂ is chosen to be trivial (it maps every element of V into a unit of O).

We can employ this construction to introduce the *Poincaré 2-group*. One constructs a crossed module by choosing

$$G = SO(3,1)\,, \qquad H = \mathbb{R}^4\,. \tag{20}$$

The map ∂ is trivial, while $\rhd$ is a natural action of $SO(3,1)$ on $\mathbb{R}^4$, defined by the equation

$$M_{ab} \rhd P_c = \eta_{[bc}P_{a]}\,, \tag{21}$$

where M_{ab} and P_a are the generators of groups $SO(3,1)$ and $\mathbb{R}^4$, respectively. The action $\rhd$ of $SO(3,1)$ on itself is given via conjugation. At the level of the algebra, conjugation reduces to the action via the adjoint representation, so that

$$M_{ab} \rhd M_{cd} = [\, M_{ab}\,,\, M_{cd}\,] \equiv \eta_{ad}M_{bc} - \eta_{ac}M_{bd} + \eta_{bc}M_{ad} - \eta_{bd}M_{ac}\,. \tag{22}$$

The 2-connection (α, β) is given by the algebra-valued differential forms

$$\alpha = \omega^{ab}M_{ab}\,, \qquad \beta = \beta^a P_a\,, \tag{23}$$

where ω^{ab} is called the spin connection. The corresponding 2-curvature in this case is given by

$$\begin{aligned} \mathcal{F} &= (\mathrm{d}\omega^{ab} + \omega^a{}_c \wedge \omega^{cb})M_{ab} &&\equiv R^{ab}M_{ab}\,, \\ \mathcal{G} &= (\mathrm{d}\beta^a + \omega^a{}_b \wedge \beta^b)P_a &&\equiv \nabla\beta^a P_a \equiv G^a P_a\,, \end{aligned} \tag{24}$$

Note that, since ∂ is trivial, the fake curvature is the same as ordinary curvature. Introducing the bilinear forms

$$\langle M_{ab}\,, M_{cd}\rangle_{\mathfrak{g}} = \eta_{a[c}\eta_{bd]}\,, \qquad \langle P_a\,, P_b\rangle_{\mathfrak{h}} = \eta_{ab}\,, \tag{25}$$

one can show that 1-forms C^a transform in the same way as the tetrad 1-forms e^a under the Lorentz transformations and diffeomorphisms, so the fields C^a can be identified with the tetrads. Then one can rewrite the pure $2BF$ action (16) for the Poincaré 2-group as

$$S_{2BF} = \int_{\mathcal{M}_4} B^{ab} \wedge R_{ab} + e_a \wedge \nabla\beta^a\,. \tag{26}$$

Note that the above step of recognizing that $C^a \equiv e^a$ was crucial, since we now see that the tetrad fields are explicitly present in the $2BF$ action for the Poincaré 2-group.

In order to promote (26) to an action for general relativity, we add a convenient simplicity constraint term:

$$S = \int_{\mathcal{M}_4} B^{ab} \wedge R_{ab} + e_a \wedge \nabla\beta^a - \lambda_{ab} \wedge \left(B^{ab} - \frac{1}{16\pi l_p^2}\varepsilon^{abcd}e_c \wedge e_d\right). \tag{27}$$

Here λ_{ab} is a Lagrange multiplier 2-form associated to the simplicity constraint term, and l_p is the Planck length. Note that the term "simplicity constraint" derives its name from the fact that the constraint imposes the property of *simplicity* on B^{ab} — a 2-form is said to be *simple* if it can be written as an exterior product of two 1-forms.

Varying the action (27) with respect to B_{ab}, e_a, ω_{ab}, β_a and λ_{ab}, we obtain the following equations of motion:

$$R_{ab} - \lambda_{ab} = 0\,, \tag{28}$$

$$\nabla\beta_a + \frac{1}{8\pi l_p^2}\varepsilon_{abcd}\lambda^{bc} \wedge e^d = 0\,, \tag{29}$$

$$\nabla B_{ab} - e_{[a} \wedge \beta_{b]} = 0\,, \tag{30}$$

$$\nabla e_a = 0\,, \tag{31}$$

$$B^{ab} - \frac{1}{16\pi l_p^2}\varepsilon^{abcd}e_c \wedge e_d = 0\,. \tag{32}$$

Given this system of equations, all fields can be algebraically determined in terms of the tetrads $e^a{}_\mu$, as follows. From the equations (31) and (32) we obtain that $\nabla B^{ab} = 0$, from which it follows, using the equation (30), that

$e_{[a} \wedge \beta_{b]} = 0$. Assuming that the tetrads are nondegenerate, $e \equiv \det(e^a{}_\mu) \neq 0$, it can be shown that this is equivalent to $\beta^a = 0$ [12]. Therefore, from the equations (28), (30), (31) and (32) we obtain

$$\lambda^{ab}{}_{\mu\nu} = R^{ab}{}_{\mu\nu}\,, \quad \beta^a{}_{\mu\nu} = 0\,, \quad B_{ab\mu\nu} = \frac{1}{8\pi l_p^2} \varepsilon_{abcd} e^c{}_\mu e^d{}_\nu\,, \quad \omega^{ab}{}_\mu = \triangle^{ab}{}_\mu\,. \tag{33}$$

Here the Ricci rotation coefficients are defined as

$$\triangle^{ab}{}_\mu \equiv \frac{1}{2}(c^{abc} - c^{cab} + c^{bca}) e_{c\mu}\,, \tag{34}$$

where

$$c^{abc} = e^\mu{}_b e^\nu{}_c (\partial_\mu e^a{}_\nu - \partial_\nu e^a{}_\mu)\,. \tag{35}$$

The last equation establishes that the spin connection 1-form ω^{ab} is expressed as a function of the tetrads, which then implies the same for the curvature 2-form R^{ab}. Finally, the remaining equation (29) then reduces to

$$\varepsilon_{abcd} R^{bc} \wedge e^d = 0\,, \tag{36}$$

which is nothing but the vacuum Einstein field equation,

$$R_{\mu\nu} - \frac{1}{2} g_{\mu\nu} R = 0\,.$$

Therefore, the action (27) is classically equivalent to general relativity.

3.3. Einstein-Yang-Mills theory

As we have already mentioned above, the main advantage of the action (27) over the Plebanski model lies in the fact that the tetrad fields are explicitly present in the topological sector of the action. This allows one to couple matter fields in a straightforward way [12]. However, one can do even more [13], and couple the $SU(N)$ Yang-Mills fields to gravity within a unified framework of 2-group formalism.

Namely, we can modify the Poincaré 2-group structure to include the $SU(N)$ gauge group, as follows. We choose the two Lie groups as

$$G = SO(3,1) \times SU(N)\,, \qquad H = \mathbb{R}^4\,, \tag{37}$$

and we define the action $\triangleright$ of the group G in the following fashion. As in the case of the Poincaré 2-group, it acts on itself via conjugation. Next, it acts on H such that the $SO(3,1)$ subgroup acts on $\mathbb{R}^4$ via the vector representation (21), while the action of the $SU(N)$ subgroup is trivial,

$$\tau_I \triangleright P_a = 0\,, \tag{38}$$

where τ_I are the $SU(N)$ generators. The map ∂ also remains trivial, as before. The form of the 2-connection (α, β) now reflects the structure of the group G,

$$\alpha = \omega^{ab} M_{ab} + A^I \tau_I \,, \qquad \beta = \beta^a P_a \,, \tag{39}$$

where A^I is the gauge connection 1-form. Next, the curvature for α then becomes

$$\mathcal{F} = R^{ab} M_{ab} + F^I \tau_I \,, \qquad F^I \equiv \mathrm{d}A^I + f_{JK}{}^I A^J \wedge A^K \,. \tag{40}$$

The curvature for β remains the same as before, because of (38). Finally, the product structure of the group G implies that its Killing form $\langle \_\,, \_\rangle_{\mathfrak{g}}$ reduces to the Killing forms for the $SO(3,1)$ and $SU(N)$, along with the identity $\langle M_{ab}\,, \tau_I\rangle_{\mathfrak{g}} = 0$.

Given a crossed module defined in this way, its corresponding pure $2BF$ action (16) becomes

$$S_{2BF} = \int_{\mathcal{M}_4} B^{ab} \wedge R_{ab} + B^I \wedge F_I + e_a \wedge \nabla \beta^a \,, \tag{41}$$

where $B^I \in \mathcal{A}^2(\mathcal{M}_4\,, \mathfrak{su}(N))$ is the new Lagrange multiplier. The action (41) is topological, and again we add appropriate simplicity constraint terms, in order to transform it into action with nontrivial dynamics. The constraint giving rise to gravity is the same as in (27), while the constraint for the gauge fields is given as in the action (4) with the substitution $\delta^a \to e^a$. Putting everything together, we obtain:

$$\begin{aligned} S = \int_{\mathcal{M}_4} & B^{ab} \wedge R_{ab} + B^I \wedge F_I + e_a \wedge \nabla \beta^a \\ & - \lambda_{ab} \wedge \left(B^{ab} - \frac{1}{16\pi l_p^2} \varepsilon^{abcd} e_c \wedge e_d \right) + \lambda^I \wedge \left(B_I - \frac{12}{g} M_{abI} e^a \wedge e^b \right) \\ & + \zeta^{abI} \Big(M_{abI} \varepsilon_{cdef} e^c \wedge e^d \wedge e^e \wedge e^f - g_{IJ} F^J \wedge e_a \wedge e_b \Big) \,. \end{aligned} \tag{42}$$

It is crucial to note that the Yang-Mills simplicity constraints in (42) are obtained from the Yang-Mills action (4) by substituting the nondynamical background field δ^a from (4) with a dynamical field e^a. The relationship between these fields has already been hinted at in the equation (5), which describes the connection between δ^a and the flat spacetime metric $\eta_{\mu\nu}$. Once promoted to e^a, this field becomes dynamical due to the presence of gravitational terms, while the equation (5) becomes the usual relation between the tetrad and the metric,

$$g_{\mu\nu} = \eta_{ab} e^a{}_\mu e^b{}_\nu \,, \tag{43}$$

further confirming the identification $C^a = e^a$. Moreover, the total action (42) now becomes background independent, as expected in general relativity. All this is a consequence of the fact that the tetrad field is explicitly present in the topological sector of the action (27), and represents a clear improvement over the Plebanski model.

Taking the variations of the action (42) with respect to the variables B_{ab}, ω_{ab}, β_a, λ_{ab}, ζ^{abI}, M_{abI}, B_I, λ^I, A^I, and e^a, we obtain equations of motion. Similarly as before, all variables can be algebraically expressed as functions of A^I and e^a and their derivatives:

$$\begin{gathered}\lambda_{ab\mu\nu} = R_{ab\mu\nu}\,, \qquad \beta_{a\mu\nu} = 0\,, \qquad \omega_{ab\mu} = \triangle_{ab\mu}\,, \qquad \lambda_{abI} = F_{abI}\,, \\ B_{\mu\nu I} = -\frac{e}{2g}\varepsilon_{\mu\nu\rho\sigma}F^{\rho\sigma}{}_I\,, \qquad B_{ab\mu\nu} = \frac{1}{8\pi l_p^2}\varepsilon_{abcd}e^c{}_\mu e^d{}_\nu\,, \\ M_{abI} = -\frac{1}{4eg}\varepsilon^{\mu\nu\rho\sigma}F_{\mu\nu}{}^I e^a{}_\rho e^b{}_\sigma\,, \qquad \zeta^{abI} = \frac{1}{4eg}\varepsilon^{\mu\nu\rho\sigma}F_{\mu\nu}{}^I e^a{}_\rho e^b{}_\sigma\,.\end{gathered} \tag{44}$$

In addition, we obtain two differential equations — An equation for A^I,

$$\nabla_\rho F^{I\rho\mu} \equiv \partial_\rho F^{I\rho\mu} + \Gamma^\rho{}_{\lambda\rho}F^{I\lambda\mu} + f_{JK}{}^I A^J{}_\rho F^{K\rho\mu} = 0\,, \tag{45}$$

where $\Gamma^\lambda{}_{\mu\nu}$ is the standard Levi-Civita connection, and an equation for e^a,

$$R^{\mu\nu} - \frac{1}{2}g^{\mu\nu}R = 8\pi l_p^2\, T^{\mu\nu}\,, \tag{46}$$

where

$$T^{\mu\nu} \equiv -\frac{1}{4g}\left(F_{\rho\sigma}{}^I F^{\rho\sigma}{}_I g^{\mu\nu} + 4F^{\mu\rho}{}_I F_\rho{}^{\nu I}\right)\,. \tag{47}$$

In this way, we see that both gravity and gauge fields can be successfully represented within a unified framework of higher gauge theory, based on a 2-group structure. A generalization from $SU(N)$ Yang-Mills case to more complicated cases such as $SU(3) \times SU(2) \times U(1)$ is completely straightforward.

4. $3BF$ theory

While the structure of a 2-group can successfully describe both gravitational and gauge fields, unfortunately it cannot accommodate other matter fields, such as scalars or fermions. In order to remedy this drawback, we make one further step in the categorical ladder, passing from the notion of a 2-group to the notion of a 3-group. As it turns out, the 3-group structure is excellent for the description of all fields that are present in the Standard Model, coupled to gravity. Moreover, a 3-group contains one more gauge group, which is novel and corresponds to the choice of the scalar and fermion fields present in the theory. This is an unexpected and beautiful result, not present in ordinary gauge theory.

As before, we will begin by introducing the notion of a 3-group, and constructing the corresponding $3BF$ action. Afterwards, we will modify this action by adding appropriate simplicity constraints, giving rise to theories with expected nontrivial dynamics. Along the way, we shall see that scalar and fermion fields are being treated pretty much on an equal footing with gravity and gauge fields.

4.1. Pure $3BF$ theory

Similarly to the concepts of a group and a 2-group, one can introduce the notion of a 3-group in the framework of higher category theory, as a 3-category with only one object where all the morphisms, 2-morphisms and 3-morphisms are invertible. Also, in the same way as a 2-group is equivalent to a crossed module, it was proved that a strict 3-group is equivalent to a 2-*crossed module* [25].

A Lie 2-crossed module, denoted as $(L \overset{\delta}{\to} H \overset{\partial}{\to} G\,, \triangleright\,, \{\_\,,\_\})$, is an algebraic structure specified by three Lie groups G, H and L, together with the homomorphisms δ and ∂, an action $\triangleright$ of the group G on all three groups, and a G-equivariant map

$$\{\_\,,\_\} : H \times H \to L\,.$$

called the Peiffer lifting. The maps ∂, δ, $\triangleright$ and the Peiffer lifting satisfy certain axioms, so that the resulting structure is equivalent to a 3-group [13].

Like in the cases of BF and $2BF$ actions, we can introduce a gauge invariant topological $3BF$ action over the manifold $\mathcal{M}_4$ for a given 2-crossed module $(L \overset{\delta}{\to} H \overset{\partial}{\to} G\,, \triangleright\,, \{\_\,,\_\})$. Denoting $\mathfrak{g}$, $\mathfrak{h}$ and $\mathfrak{l}$ as Lie algebras corresponding to the groups G, H and L, respectively, one can introduce a 3-connection (α, β, γ) given by the algebra-valued differential forms $\alpha \in \mathcal{A}^1(\mathcal{M}_4\,, \mathfrak{g})$, $\beta \in \mathcal{A}^2(\mathcal{M}_4\,, \mathfrak{h})$ and $\gamma \in \mathcal{A}^3(\mathcal{M}_4\,, \mathfrak{l})$. The corresponding fake 3-curvature $(\mathcal{F}\,, \mathcal{G}\,, \mathcal{H})$ is then defined as

$$\begin{gathered}\mathcal{F} = \mathrm{d}\alpha + \alpha \wedge \alpha - \partial\beta\,, \qquad \mathcal{G} = \mathrm{d}\beta + \alpha \wedge^{\triangleright} \beta - \delta\gamma\,, \\ \mathcal{H} = \mathrm{d}\gamma + \alpha \wedge^{\triangleright} \gamma + \{\beta \wedge \beta\}\,,\end{gathered} \tag{48}$$

see [25, 26] for details. Note that γ is a 3-form, while its corresponding field strength $\mathcal{H}$ is a 4-form, necessitating that the spacetime manifold be at least 4-dimensional. Then, a $3BF$ action is defined as

$$S_{3BF} = \int_{\mathcal{M}_4} \langle B \wedge \mathcal{F} \rangle_{\mathfrak{g}} + \langle C \wedge \mathcal{G} \rangle_{\mathfrak{h}} + \langle D \wedge \mathcal{H} \rangle_{\mathfrak{l}}\,, \tag{49}$$

where $B \in \mathcal{A}^2(\mathcal{M}_4, \mathfrak{g})$, $C \in \mathcal{A}^1(\mathcal{M}_4, \mathfrak{h})$ and $D \in \mathcal{A}^0(\mathcal{M}_4, \mathfrak{l})$ are Lagrange multipliers. Note that in precisely 4 spacetime dimensions the Lagrange multiplier D corresponding to $\mathcal{H}$ is a 0-form, i.e. a scalar function. The functionals $\langle \_ , \_ \rangle_{\mathfrak{g}}$, $\langle \_ , \_ \rangle_{\mathfrak{h}}$ and $\langle \_ , \_ \rangle_{\mathfrak{l}}$ are G-invariant bilinear symmetric nondegenerate forms on $\mathfrak{g}$, $\mathfrak{h}$ and $\mathfrak{l}$, respectively. Under certain conditions, the forms $\langle \_ , \_ \rangle_{\mathfrak{h}}$ and $\langle \_ , \_ \rangle_{\mathfrak{l}}$ are also H-invariant and L-invariant.

One can see that varying the action with respect to the variables B^α, C^a and D^A (where indices A count the generators of the group L), one obtains the equations of motion

$$\mathcal{F}^\alpha = 0 \,, \qquad \mathcal{G}^a = 0 \,, \qquad \mathcal{H}^A = 0 \,, \tag{50}$$

while varying with respect to α^α, β^a, γ^A one obtains

$$\mathrm{d}B_\alpha + f_{\alpha\beta}{}^\gamma B_\gamma \wedge \alpha^\beta - \rhd_{\alpha a}{}^b C_b \wedge \beta^a + \rhd_{\alpha B}{}^A D_A \wedge \gamma^B = 0 \,, \tag{51}$$

$$\mathrm{d}C_a - \partial_a{}^\alpha B_\alpha + \rhd_{\alpha a}{}^b C_b \wedge \alpha^\alpha + 2X_{\{ab\}}{}^A D_A \wedge \beta^b = 0 \,, \tag{52}$$

$$\mathrm{d}D_A - \rhd_{\alpha A}{}^B D_B \wedge \alpha^\alpha + \delta_A{}^a C_a = 0 \,. \tag{53}$$

4.2. Klein-Gordon theory

Now we proceed to demonstrate that one can use the 3-group structure and the corresponding $3BF$ theory to describe the Klein-Gordon field coupled to general relativity. We begin by specifying a 2-crossed module, which is used to construct the topological $3BF$ theory, and then we impose appropriate simplicity constraints to obtain the desired equations of motion.

We specify a 2-crossed module $(L \overset{\delta}{\to} H \overset{\partial}{\to} G \,, \rhd \,, \{\_ , \_\})$, as follows. The groups are given as

$$G = SO(3,1) \,, \qquad H = \mathbb{R}^4 \,, \qquad L = \mathbb{R} \,. \tag{54}$$

The group G acts on itself via conjugation, on H via the vector representation, and on L via the trivial representation. This specifies the definition of the action $\rhd$. The map ∂ is chosen to be trivial, as before. The map δ is also trivial, that is, every element of L is mapped to the identity element of H. Finally, the Peiffer lifting is trivial as well, mapping every ordered pair of elements in H to an identity element in L. This specifies one concrete 2-crossed module which, as we shall see below, corresponds to gravity and one real scalar field.

Given this choice of a 2-crossed module, the 3-connection $(\alpha \,, \beta \,, \gamma)$ takes the form

$$\alpha = \omega^{ab} M_{ab} \,, \qquad \beta = \beta^a P_a \,, \qquad \gamma = \gamma \mathbb{I} \,, \tag{55}$$

where $\mathbb{I}$ is the sole generator of the Lie group $\mathbb{R}$. Since the homomorphisms ∂ and δ are trivial, as well as the Peiffer lifting, the fake 3-curvature (48) reduces to the ordinary 3-curvature,

$$\mathcal{F} = R^{ab} M_{ab} \,, \qquad \mathcal{G} = \nabla \beta^a P_a \,, \qquad \mathcal{H} = \mathrm{d}\gamma \,, \tag{56}$$

where we used the fact that G acts trivially on L, that is, $M_{ab} \triangleright \mathbb{I} = 0$. This means that the 3-form γ transforms as a scalar with respect to Lorentz symmetry. Consequently, its Lagrange multiplier D also transforms as a scalar, since it also belongs to the algebra $\mathfrak{l}$. Since D is also a 0-form, it transforms as a scalar with respect to diffeomorphisms as well. In other words, D completely behaves as a real scalar field, so we relabel it into more traditional notation, $D \equiv \phi$, and write the pure $3BF$ action (49) as:

$$S_{3BF} = \int_{\mathcal{M}_4} B^{ab} \wedge R_{ab} + e_a \wedge \nabla \beta^a + \phi \, \mathrm{d}\gamma \,, \tag{57}$$

where the bilinear form for L is $\langle \mathbb{I} \,, \mathbb{I} \rangle_{\mathfrak{l}} = 1$.

The existence of a scalar field in the $3BF$ action is a crucial property of a 3-group in a 4-dimensional spacetime, just like identifying the Lagrange multiplier C^a with a tetrad field e^a was a crucial property of the $2BF$ action and the Poincaré 2-group. We can also see that the choice of the third gauge group, L, dictates the number and the structure of the matter fields present in the action. In this case, $L = \mathbb{R}$ implies that we have only one real scalar field, corresponding to a single generator $\mathbb{I}$ of $\mathbb{R}$. The trivial nature of the action $\triangleright$ of $SO(3,1)$ on $\mathbb{R}$ implies that ϕ transforms as a scalar field. Finally, the scalar field appears in the topological sector of the action, making the quantization procedure feasible.

As in the case of BF and $2BF$ theories, we need to add appropriate simplicity constraints to the action (57). In order to obtain the Klein-Gordon field ϕ of mass m coupled to gravity in the standard way, the action takes the form:

$$\begin{aligned} S = \int_{\mathcal{M}_4} & B^{ab} \wedge R_{ab} + e_a \wedge \nabla \beta^a + \phi \, \mathrm{d}\gamma \\ & - \lambda_{ab} \wedge \left(B^{ab} - \frac{1}{16\pi l_p^2} \varepsilon^{abcd} e_c \wedge e_d \right) \\ & + \lambda \wedge \left(\gamma - \frac{1}{2} H_{abc} e^a \wedge e^b \wedge e^c \right) \\ & + \Lambda^{ab} \wedge \left(H_{abc} \varepsilon^{cdef} e_d \wedge e_e \wedge e_f - \mathrm{d}\phi \wedge e_a \wedge e_b \right) \\ & - \frac{1}{2 \cdot 4!} m^2 \phi^2 \varepsilon_{abcd} e^a \wedge e^b \wedge e^c \wedge e^d \,. \end{aligned} \tag{58}$$

The first row is the topological sector (57), the second row is the familiar simplicity constraint for gravity from the action (27), the third and fourth rows contain the new simplicity constraints featuring the Lagrange multiplier 1-forms λ and Λ^{ab} and the 0-form H_{abc}, while the fifth row is the mass term for the scalar field.

The variation of (58) with respect to the variables B_{ab}, ω_{ab}, β_a, λ_{ab}, Λ_{ab}, γ, λ, H_{abc}, ϕ and e^a gives us the equations of motion. As before, all variables can be algebraically expressed in terms of the tetrads e^a and the scalar field ϕ:

$$\begin{aligned} \lambda_{ab\mu\nu} &= R_{ab\mu\nu}\,, \qquad \omega^{ab}{}_{\mu} = \triangle^{ab}{}_{\mu}\,, \qquad \gamma_{\mu\nu\rho} = -\frac{e}{2}\varepsilon_{\mu\nu\rho\sigma}\partial^{\sigma}\phi\,, \\ \beta^a{}_{\mu\nu} &= 0\,, \qquad \Lambda^{ab}{}_{\mu} = \frac{1}{12e}g_{\mu\lambda}\varepsilon^{\lambda\nu\rho\sigma}\partial_{\nu}\phi e^a{}_{\rho}e^b{}_{\sigma}\,, \qquad \lambda_{\mu} = \partial_{\mu}\phi\,, \\ H^{abc} &= \frac{1}{6e}\varepsilon^{\mu\nu\rho\sigma}\partial_{\mu}\phi e^a{}_{\nu}e^b{}_{\rho}e^c{}_{\sigma}\,, \qquad B_{ab\mu\nu} = \frac{1}{8\pi l_p^2}\varepsilon_{abcd}e^c{}_{\mu}e^d{}_{\nu}\,. \end{aligned} \tag{59}$$

The equations of motion for e^a and ϕ, however, are differential equations. The equation for the scalar field becomes the covariant Klein-Gordon equation,

$$\left(\nabla_{\mu}\nabla^{\mu} - m^2\right)\phi = 0\,, \tag{60}$$

while the equation for the tetrads is

$$R^{\mu\nu} - \frac{1}{2}g^{\mu\nu}R = 8\pi l_p^2\, T^{\mu\nu}\,, \tag{61}$$

where

$$T^{\mu\nu} \equiv \partial^{\mu}\phi\partial^{\nu}\phi - \frac{1}{2}g^{\mu\nu}\left(\partial_{\rho}\phi\partial^{\rho}\phi + m^2\phi^2\right) \tag{62}$$

is the stress-energy tensor for a single real scalar field.

4.3. Einstein-Cartan-Dirac theory

In order to describe the Dirac field coupled to Einstein-Cartan gravity, we follow the same procedure as for the case of the scalar field, but now we choose the 2-crossed module $(L \overset{\delta}{\to} H \overset{\partial}{\to} G\,, \triangleright\,, \{\_\,,\_\})$ in a different way, as follows. The groups are:

$$G = SO(3,1)\,, \qquad H = \mathbb{R}^4\,, \qquad L = \mathbb{R}^8(\mathbb{G})\,, \tag{63}$$

where $\mathbb{G}$ is the algebra of complex Grassmann numbers. The maps ∂, δ and the Peiffer lifting are trivial, as before. The action of the group G on itself is given via conjugation, on H via vector representation, and on L via spinor representation, in the following way. Denoting the 8 generators of the Lie group $\mathbb{R}^8(\mathbb{G})$ as P_α and P^α, where the index α takes the values $1,\dots,4$, the action $\triangleright$ of G on L is thus given explicitly as

$$M_{ab} \triangleright P_{\alpha} = \frac{1}{2}(\sigma_{ab})^{\beta}{}_{\alpha}P_{\beta}\,, \qquad M_{ab} \triangleright P^{\alpha} = -\frac{1}{2}(\sigma_{ab})^{\alpha}{}_{\beta}P^{\beta}\,, \tag{64}$$

where $\sigma_{ab} = \frac{1}{4}[\gamma_a, \gamma_b]$, and γ_a are the usual Dirac matrices, satisfying the anticommutation rule $\{\gamma_a\,, \gamma_b\} = -2\eta_{ab}$.

As in the case of the scalar field, the choice of the group L dictates the matter content of the theory, while the action $\triangleright$ of G on L specifies its transformation properties.

Let us now proceed to construct the $3BF$ action. The 3-connection $(\alpha\,,\beta\,,\gamma)$ takes the form

$$\alpha = \omega^{ab} M_{ab}\,, \qquad \beta = \beta^a P_a\,, \qquad \gamma = \gamma^\alpha P_\alpha + \bar{\gamma}_\alpha P^\alpha\,, \tag{65}$$

while the 3-curvature $(\mathcal{F}\,,\mathcal{G}\,,\mathcal{H})$ is given as

$$\begin{aligned} &\mathcal{F} = R^{ab} M_{ab}\,, \qquad \mathcal{G} = \nabla \beta^a P_a\,, \\ \mathcal{H} &= \left(\mathrm{d}\gamma^\alpha + \frac{1}{2}\omega^{ab}(\sigma_{ab})^\alpha{}_\beta \gamma^\beta\right) P_\alpha + \left(\mathrm{d}\bar{\gamma}_\alpha - \frac{1}{2}\omega^{ab}\bar{\gamma}_\beta (\sigma_{ab})^\beta{}_\alpha\right) P^\alpha \\ &\equiv (\overrightarrow{\nabla}\gamma)^\alpha P_\alpha + (\bar{\gamma}\overleftarrow{\nabla})_\alpha P^\alpha\,, \end{aligned} \tag{66}$$

where we have used (64). The bilinear form $\langle\_\,,\_\rangle_{\mathfrak{l}}$ is defined via its action on the generators:

$$\begin{aligned} \langle P_\alpha\,, P_\beta\rangle_{\mathfrak{l}} &= 0\,, \qquad \langle P^\alpha\,, P^\beta\rangle_{\mathfrak{l}} = 0\,, \\ \langle P_\alpha\,, P^\beta\rangle_{\mathfrak{l}} &= -\delta_\alpha^\beta\,, \qquad \langle P^\alpha\,, P_\beta\rangle_{\mathfrak{l}} = \delta^\alpha_\beta\,. \end{aligned} \tag{67}$$

Note that the bilinear form defined in this way is antisymmetric, rather than symmetric, when it acts on the generators. The reason for this is the following. For general $A, B \in \mathfrak{l}$, we want the bilinear form to be symmetric. Expanding A and B into components, we can write

$$\langle A\,, B\rangle_{\mathfrak{l}} = A^I B^J g_{IJ}\,, \qquad \langle B\,, A\rangle_{\mathfrak{l}} = B^J A^I g_{JI}\,. \tag{68}$$

Since we require the bilinear form to be symmetric, the two expressions must be equal. However, since the coefficients in $\mathfrak{l}$ are Grassmann numbers, we have $A^I B^J = -B^J A^I$, so it follows that $g_{IJ} = -g_{JI}$. Hence the antisymmetry of (67) — it compensates for the anticommutativity property of the Grassman coefficients, making the bilinear form symmetric for general algebra elements $A, B \in \mathfrak{l}$.

Now we employ the action $\triangleright$ of G on L to determine the transformation properties of the Lagrange multiplier D in (49). Indeed, the choice of the group L dictates that D contains 8 independent complex Grassmannian matter fields as its components. Moreover, due to the fact that D is a 0-form and that it transforms according to the spinorial representation of $SO(3,1)$, we can identify its components with the Dirac bispinor fields, and write

$$D = \psi^\alpha P_\alpha + \bar{\psi}_\alpha P^\alpha\,. \tag{69}$$

This is again an illustration of the fact that information about the structure of the matter sector in the theory is specified by the choice of the group L

in the 2-crossed module, and its transformation properties with respect to the Lorentz group are fixed by the action $\triangleright$.

Given all of the above, we write the corresponding pure $3BF$ action as:

$$S_{3BF} = \int_{\mathcal{M}_4} B^{ab} \wedge R_{ab} + e_a \wedge \nabla \beta^a + (\bar{\gamma}\overleftarrow{\nabla})_\alpha \psi^\alpha + \bar{\psi}_\alpha (\overrightarrow{\nabla}\gamma)^\alpha \,. \tag{70}$$

In order to obtain the action that gives us the dynamics of Einstein-Cartan theory of gravity coupled to a Dirac field, we add the following simplicity constraints:

$$\begin{aligned} S = \int_{\mathcal{M}_4} & B^{ab} \wedge R_{ab} + e_a \wedge \nabla \beta^a + (\bar{\gamma}\overleftarrow{\nabla})_\alpha \psi^\alpha + \bar{\psi}_\alpha (\overrightarrow{\nabla}\gamma)^\alpha \\ & - \lambda_{ab} \wedge \left(B^{ab} - \frac{1}{16\pi l_p^2} \varepsilon^{abcd} e_c \wedge e_d \right) \\ & - \lambda^\alpha \wedge \left(\bar{\gamma}_\alpha - \frac{i}{6} \varepsilon_{abcd} e^a \wedge e^b \wedge e^c (\bar{\psi}\gamma^d)_\alpha \right) \\ & + \bar{\lambda}_\alpha \wedge \left(\gamma^\alpha + \frac{i}{6} \varepsilon_{abcd} e^a \wedge e^b \wedge e^c (\gamma^d \psi)^\alpha \right) \\ & - \frac{1}{12} m \, \bar{\psi}\psi \, \varepsilon_{abcd} e^a \wedge e^b \wedge e^c \wedge e^d + 2\pi i l_p^2 \, \bar{\psi}\gamma_5 \gamma^a \psi \, \varepsilon_{abcd} e^b \wedge e^c \wedge \beta^d . \end{aligned} \tag{71}$$

Similarly to the previous case of the scalar field, we recognize the topological sector in the first row, the gravitational simplicity constraint in the second row, while the third and fourth rows contain the new simplicity constraints for the Dirac field, featuring the Lagrange multiplier 1-forms λ^α and $\bar{\lambda}_\alpha$. The fifth row contains the mass term for the Dirac field, and a term which ensures the correct coupling between the torsion and the spin of the Dirac field. In particular, we want to obtain

$$T_a \equiv \nabla e_a = 2\pi l_p^2 s_a \,, \tag{72}$$

as one of the equations of motion, where

$$s_a = i \varepsilon_{abcd} e^b \wedge e^c \bar{\psi} \gamma_5 \gamma^d \psi \tag{73}$$

is the Dirac spin 2-form. Of course, other alternative coupling choices are possible, but we choose this one since this is the traditional coupling most often discussed in textbooks.

The variation of the action (71) with respect to B_{ab}, λ^{ab}, $\bar{\gamma}_\alpha$, γ^α, λ^α, $\bar{\lambda}_\alpha$, $\bar{\psi}_\alpha$, ψ^α, e^a, β^a and ω^{ab}, again gives us equations of motion, which can be algebraically solved for all fields as functions of e^a, ψ and $\bar{\psi}$:

$$B_{ab\mu\nu} = \frac{1}{8\pi l_p^2}\varepsilon_{abcd}e^c{}_\mu e^d{}_\nu\,, \qquad \lambda^\alpha{}_\mu = (\overrightarrow{\nabla}_\mu\psi)^\alpha\,, \qquad \bar{\lambda}_{\alpha\mu} = (\bar{\psi}\overleftarrow{\nabla}_\mu)_\alpha\,,$$
$$\bar{\gamma}_{\alpha\mu\nu\rho} = i\varepsilon_{abcd}e^a{}_\mu e^b{}_\nu e^c{}_\rho(\bar{\psi}\gamma^d)_\alpha\,, \qquad \gamma^\alpha{}_{\mu\nu\rho} = -i\varepsilon_{abcd}e^a{}_\mu e^b{}_\nu e^c{}_\rho(\gamma^d\psi)^\alpha\,,$$
$$\beta^a{}_{\mu\nu} = 0\,, \qquad \lambda_{ab\mu\nu} = R_{ab\mu\nu}\,, \qquad \omega^{ab}{}_\mu = \triangle^{ab}{}_\mu + K^{ab}{}_\mu\,. \tag{74}$$

Here $K^{ab}{}_\mu$ is the contorsion tensor, constructed in the standard way from the torsion tensor. In addition, we also obtain

$$T_a \equiv \nabla e_a = 2\pi l_p^2 s_a\,, \tag{75}$$

which is precisely the desired equation (72) for the torsion. Finally, the differential equations of motion for ψ and $\bar{\psi}$ are the standard covariant Dirac equation,

$$(i\gamma^a e^\mu{}_a\overrightarrow{\nabla}_\mu - m)\psi = 0\,, \tag{76}$$

and its conjugate,

$$\bar{\psi}(i\overleftarrow{\nabla}_\mu e^\mu{}_a\gamma^a + m) = 0\,, \tag{77}$$

where $e^\mu{}_a$ is the inverse tetrad. The differential equation of motion for e^a is

$$R^{\mu\nu} - \frac{1}{2}g^{\mu\nu}R = 8\pi l_p^2\, T^{\mu\nu}\,, \tag{78}$$

where

$$T^{\mu\nu} \equiv \frac{i}{2}\bar{\psi}\gamma^a\overleftrightarrow{\nabla}^\nu e^\mu{}_a\psi - \frac{1}{2}g^{\mu\nu}\bar{\psi}\Big(i\gamma^a\overleftrightarrow{\nabla}_\rho e^\rho{}_a - 2m\Big)\psi\,, \tag{79}$$

Here, we used the notation $\overleftrightarrow{\nabla} = \overrightarrow{\nabla} - \overleftarrow{\nabla}$. As expected, the equations of motion (75), (76), (77) and (78) are precisely the equations of motion of the Einstein-Cartan-Dirac theory.

4.4. Weyl and Majorana fields coupled to Einstein-Cartan gravity

As is well known, the Dirac fermions are not an irreducible representation of the Lorentz group, and one can rewrite them as left-chiral and right-chiral irreducible Weyl fermion fields. Hence, it is useful to construct the 2-crossed module and a constrained $3BF$ action for left and right Weyl spinors. For simplicity, we will discuss only the left-chiral spinor field (the right-chiral can be studied analogously). Additionally, we can also describe Majorana fermions using the same formalism, the only difference being the presence of an additional mass term in the Majorana action.

We soecify a 2-crossed module $(L \stackrel{\delta}{\to} H \stackrel{\partial}{\to} G\,, \rhd\,, \{\_\,,\_\})$, in a way similar to the Dirac case, as follows. The groups are:

$$G = SO(3,1)\,, \qquad H = \mathbb{R}^4\,, \qquad L = \mathbb{R}^4(\mathbb{G})\,. \tag{80}$$

The maps ∂, δ and the Peiffer lifting are trivial. The action $\rhd$ of the group G on G, H and L is given in the same way as for the Dirac case, whereas the spinorial representation reduces to

$$M_{ab} \rhd P^\alpha = \frac{1}{2}(\sigma_{ab})^\alpha{}_\beta P^\beta\,, \qquad M_{ab} \rhd P_{\dot\alpha} = \frac{1}{2}(\bar\sigma_{ab})^{\dot\beta}{}_{\dot\alpha} P_{\dot\beta}\,, \tag{81}$$

where $\sigma^{ab} = -\bar\sigma^{ab} = \frac{1}{4}(\sigma^a\bar\sigma^b - \sigma^b\bar\sigma^a)$, for $\sigma^a = (1, \vec\sigma)$ and $\bar\sigma^a = (1, -\vec\sigma)$, in which $\vec\sigma$ denotes the set of three Pauli matrices. The four generators of the group L are denoted as P^α and $P_{\dot\alpha}$, where the Weyl indices $\alpha, \dot\alpha$ take values $1, 2$.

The 3-connection $(\alpha\,, \beta\,, \gamma)$ takes the form

$$\alpha = \omega^{ab} M_{ab}\,, \qquad \beta = \beta^a P_a\,, \qquad \gamma = \gamma_\alpha P^\alpha + \bar\gamma^{\dot\alpha} P_{\dot\alpha}\,, \tag{82}$$

while the 3-curvature $(\mathcal{F}\,, \mathcal{G}\,, \mathcal{H})$ is

$$\begin{aligned} \mathcal{F} &= R^{ab} M_{ab}\,, \qquad \mathcal{G} = \nabla\beta^a P_a\,, \\ \mathcal{H} &= \left(\mathrm{d}\gamma_\alpha + \frac{1}{2}\omega^{ab}(\sigma^{ab})^\beta{}_\alpha \gamma_\beta\right) P^\alpha + \left(\mathrm{d}\bar\gamma^{\dot\alpha} + \frac{1}{2}\omega_{ab}(\bar\sigma^{ab})^{\dot\alpha}{}_{\dot\beta}\bar\gamma^{\dot\beta}\right) P_{\dot\alpha} \\ &\equiv (\vec\nabla\gamma)_\alpha P^\alpha + (\bar\gamma\overleftarrow{\nabla})^{\dot\alpha} P_{\dot\alpha}\,. \end{aligned} \tag{83}$$

The Lagrange multiplier D now contains as coefficients the spinor fields ψ_α and $\bar\psi^{\dot\alpha}$,

$$D = \psi_\alpha P^\alpha + \bar\psi^{\dot\alpha} P_{\dot\alpha}\,, \tag{84}$$

and the bilinear form $\langle \_\,, \_\rangle_{\mathfrak{l}}$ for the group L is

$$\begin{aligned} \langle P^\alpha\,, P^\beta\rangle_{\mathfrak{l}} &= \varepsilon^{\alpha\beta}\,, \qquad \langle P_{\dot\alpha}\,, P_{\dot\beta}\rangle_{\mathfrak{l}} = \varepsilon_{\dot\alpha\dot\beta}\,, \\ \langle P^\alpha\,, P_{\dot\beta}\rangle_{\mathfrak{l}} &= 0\,, \qquad \langle P_{\dot\alpha}\,, P^\beta\rangle_{\mathfrak{l}} = 0\,, \end{aligned} \tag{85}$$

where $\varepsilon^{\alpha\beta}$ and $\varepsilon_{\dot\alpha\dot\beta}$ are the usual two-dimensional antisymmetric Levi-Civita symbols.

The pure $3BF$ action (49) now becomes

$$S_{3BF} = \int_{\mathcal{M}_4} B^{ab} \wedge R_{ab} + e_a \wedge \nabla\beta^a + \psi^\alpha \wedge (\vec\nabla\gamma)_\alpha + \bar\psi_{\dot\alpha} \wedge (\bar\gamma\overleftarrow{\nabla})^{\dot\alpha}\,. \tag{86}$$

In order to obtain the suitable equations of motion for the Weyl spinors, we again introduce appropriate simplicity constraints, to obtain:

$$
\begin{aligned}
S = \int_{\mathcal{M}_4} & B^{ab} \wedge R_{ab} + e_a \wedge \nabla \beta^a + \psi^\alpha \wedge (\overrightarrow{\nabla}\gamma)_\alpha + \bar{\psi}_{\dot{\alpha}} \wedge (\bar{\gamma}\overleftarrow{\nabla})^{\dot{\alpha}} \\
& - \lambda_{ab} \wedge (B^{ab} - \frac{1}{16\pi l_p^2}\varepsilon^{abcd} e_c \wedge e_d) \\
& - \lambda^\alpha \wedge (\gamma_\alpha + \frac{i}{6}\varepsilon_{abcd} e^a \wedge e^b \wedge e^c \sigma^d{}_{\alpha\dot{\beta}} \bar{\psi}^{\dot{\beta}}) \\
& - \bar{\lambda}_{\dot{\alpha}} \wedge (\bar{\gamma}^{\dot{\alpha}} + \frac{i}{6}\varepsilon_{abcd} e^a \wedge e^b \wedge e^c \bar{\sigma}^{d\dot{\alpha}\beta} \psi_\beta) \\
& - 4\pi l_p^2 \varepsilon_{abcd} e^a \wedge e^b \wedge \beta^c (\bar{\psi}_{\dot{\alpha}} \bar{\sigma}^{d\dot{\alpha}\beta} \psi_\beta) \,.
\end{aligned} \tag{87}
$$

The new simplicity constraints, in the third and fourth rows, feature the Lagrange multiplier 1-forms λ_α and $\bar{\lambda}^{\dot{\alpha}}$. Also, in analogy to the coupling between the spin and the torsion in Einstein-Cartan-Dirac theory, the term in the fifth row is chosen to ensure that the coupling between the Weyl spin tensor

$$
s_a \equiv i\varepsilon_{abcd} e^b \wedge e^c \; \psi^\alpha \sigma^d{}_{\alpha\dot{\beta}} \bar{\psi}^{\dot{\beta}} \tag{88}
$$

and torsion is given as:

$$
T_a = 4\pi l_p^2 s_a \,. \tag{89}
$$

The action for the Majorana field is precisely the same, but for an additional mass term in the action:

$$
-\frac{1}{12} m \varepsilon_{abcd} e^a \wedge e^b \wedge e^c \wedge e^d (\psi^\alpha \psi_\alpha + \bar{\psi}_{\dot{\alpha}} \bar{\psi}^{\dot{\alpha}}) \,. \tag{90}
$$

The variation of the action (87) with respect to the variables B_{ab}, λ^{ab}, γ_α, $\bar{\gamma}^{\dot{\alpha}}$, λ_α, $\bar{\lambda}^{\dot{\alpha}}$, ψ_α, $\bar{\psi}^{\dot{\alpha}}$, e^a, β^a and ω^{ab} gives us the equations of motion, which can be algebraically solved for all variables as functions of ψ_α, $\bar{\psi}^{\dot{\alpha}}$ and e^a:

$$
\begin{gathered}
\beta^a{}_{\mu\nu} = 0 \,, \quad \lambda^{ab}{}_{\mu\nu} = R^{ab}{}_{\mu\nu} \,, \quad \lambda_{\alpha\mu} = \nabla_\mu \psi_\alpha \,, \quad \bar{\lambda}^{\dot{\alpha}}{}_\mu = \nabla_\mu \bar{\psi}^{\dot{\alpha}} \,, \\
B_{ab\mu\nu} = \frac{1}{8\pi l_p^2} \varepsilon_{abcd} e^c{}_\mu e^d{}_\nu \,, \quad \omega_{ab\mu} = \triangle_{ab\mu} + K_{ab\mu} \,, \\
\gamma_{\alpha\mu\nu\rho} = i\varepsilon_{abcd} e^a{}_\mu e^b{}_\nu e^c{}_\rho \sigma^d{}_{\alpha\dot{\beta}} \bar{\psi}^{\dot{\beta}} \,, \quad \bar{\gamma}^{\dot{\alpha}}{}_{\mu\nu\rho} = i\varepsilon_{abcd} e^a{}_\mu e^b{}_\nu e^c{}_\rho \bar{\sigma}^{d\dot{\alpha}\beta} \psi_\beta \,.
\end{gathered} \tag{91}
$$

In addition, one also obtains (89). Finally, the differential equations of motion for the spinor and tetrad fields are

$$
\bar{\sigma}^{a\dot{\alpha}\beta} e^\mu{}_a \nabla_\mu \psi_\beta = 0 \,, \qquad \sigma^a{}_{\alpha\dot{\beta}} e^\mu{}_a \nabla_\mu \bar{\psi}^{\dot{\beta}} = 0 \,, \tag{92}
$$

and

$$R^{\mu\nu} - \frac{1}{2}g^{\mu\nu}R = 8\pi l_p^2 \, T^{\mu\nu} \,, \tag{93}$$

where

$$\begin{aligned} T^{\mu\nu} \;\equiv\;& \frac{i}{2}\bar{\psi}\bar{\sigma}^b e^\nu{}_b \nabla^\mu \psi + \frac{i}{2}\psi\sigma^b e^\nu{}_b \nabla^\mu \bar{\psi} \\ & -\frac{1}{2}g^{\mu\nu}\Big(i\bar{\psi}\bar{\sigma}^a e^\lambda{}_a \nabla_\lambda \psi + i\psi\sigma^a e^\lambda{}_a \nabla_\lambda \bar{\psi} \Big) \,. \end{aligned} \tag{94}$$

Here we have suppressed the spinor indices, for simplicity. In the case of the Majorana field, the equations of motion (91) remain the same. The equations of motion for ψ_α and $\bar{\psi}^{\dot{\alpha}}$ obtain the additional mass term,

$$i\sigma^a{}_{\alpha\dot{\beta}} e^\mu{}_a \nabla_\mu \bar{\psi}^{\dot{\beta}} - m\psi_\alpha = 0 \,, \qquad i\bar{\sigma}^{a\dot{\alpha}\beta} e^\mu{}_a \nabla_\mu \psi_\beta - m\bar{\psi}^{\dot{\alpha}} = 0 \,, \tag{95}$$

while the stress-energy tensor becomes

$$\begin{aligned} T^{\mu\nu} \;\equiv\;& \frac{i}{2}\bar{\psi}\bar{\sigma}^b e^\nu{}_b \nabla^\mu \psi + \frac{i}{2}\psi\sigma^b e^\nu{}_b \nabla^\mu \bar{\psi} \\ & -g^{\mu\nu}\frac{1}{2}\left[i\bar{\psi}\bar{\sigma}^a e^\lambda{}_a \nabla_\lambda \psi + i\psi\sigma^a e^\lambda{}_a \nabla_\lambda \bar{\psi} - \frac{1}{2}m\left(\psi\psi + \bar{\psi}\bar{\psi}\right) \right] \,. \end{aligned} \tag{96}$$

5. Conclusions

Let us summarize the results of the paper. In Section 2 we have introduced the BF theory and discussed models based on constrained BF action, in particular the Yang-Mills theory in Minkowski spacetime and the Plebanski formulation of general relativity. Section 3 was devoted to the first step in the categorical ladder and the $2BF$ theory. After introducing the notions of a 2-group, a crossed module, and the corresponding $2BF$ theory, we have studied the $2BF$ formulation of general relativity and the Einstein-Yang-Mills theory. Then, in Section 4 we have performed one more step in the categorical ladder, and introduced the notions of a 3-group, 2-crossed module, and the $3BF$ theory. This structure was employed to construct the constrained $3BF$ actions for the cases of Klein-Gordon, Dirac, Weyl and Majorana fields, each coupled to the Einstein-Cartan gravity in the standard way. In those descriptions, it turned out that the scalar and fermion fields are associated to a *new gauge group*, similar to the gauge fields being associated to a gauge group in the Yang-Mills theory. This opens up a possibility of a classification of matter fields based on an algebraic structure of a 3-group.

All the obtained results serve to complete the first step of the spinfoam quantization programme, as outlined in the Introduction. This paves the way to the study of steps 2 and 3 of the programme. Namely, the full action for gravity, gauge fields and matter is written completely in the langulage of differential forms, which can be easily adapted to a triangulated spacetime manifold, in the sense of Regge calculus. This can be seen in the following table:

d	triangulation	dual triangulation	form	fields	field strengths
0	vertex	4-polytope	0-form	ϕ, $\psi_{\tilde{\alpha}}$, $\bar{\psi}^{\tilde{\alpha}}$	
1	edge	3-polyhedron	1-form	ω^{ab}, A^I, e^a	
2	triangle	face	2-form	β^a, B^{ab}	R^{ab}, F^I, T^a
3	tetrahedron	edge	3-form	γ, $\gamma_{\tilde{\alpha}}$, $\bar{\gamma}^{\tilde{\alpha}}$	$\mathcal{G}^a$
4	4-simplex	vertex	4-form		$\mathcal{H}$, $\mathcal{H}_{\tilde{\alpha}}$, $\bar{\mathcal{H}}^{\tilde{\alpha}}$

This data can be utilized to construct a Regge-discretized topological $3BF$ action, and from that a state sum Z, giving rise to a rigorous definition of the path integral

$$Z = \int \mathcal{D}g \int \mathcal{D}\phi \; e^{iS[g,\phi]} \,, \tag{97}$$

which is a generalization of (1) in the sense that it adds matter fields (including the gauge boson sector) to gravity at the quantum level. Being a topological theory, and given the underlying structure of the 3-group, a pure $3BF$ action ought to ensure the topological invariance of the state sum Z, i.e., Z should be triangulation independent. This step, however, requires the generalizations of the Peter-Weyl and Plancharel theorems to 2-groups and 3-groups, which are unfortunately still missing (though there are some attempts to circumvent them at least in the 2-group case [27, 28]). Namely, the purpose of the Peter-Weyl and Plancharel theorems is to provide a decomposition of a function on a group into a sum over the corresponding irreducible representations, which then specifies the spectrum of labels for the simplices in the triangulation, and fixes the domain of values for the fields living on those simplices. In the absence of the two theorems, one can still try to *guess* the irreducible representations of the 2- and 3-groups, as was done for example in the *spincube model* of quantum gravity [12], or to try to construct the state sum using other techniques, as was done in [27, 28]).

Of course, when building a realistic theory, we are not interested in a topological theory, but instead in one which contains local propagating degrees of freedom. Thus the state sum Z need not be a topological invariant. This is obtained via the step 3 of the spinfoam quantization programme, by imposing the simplicity constraints on Z. The classical actions discussed in this paper manifestly distinguish the topological sector from the simplicity constraints, which have been explicitly determined. Imposing them should thus be a straightforward procedure for a given Z. Completing this programme would ultimately lead us to a tentative state sum describing both gravity and matter at a quantum level, which is a topic for future research.

In addition to the construction of a full quantum theory of gravity, there are also many additional possible studies of the classical constrained $3BF$ action. For example, a Hamiltonian analysis of the theory could be interesting for the canonical quantization programme, and some work has begun in this area [29]. Also, it is worth looking into the idea of imposing the simplicity constraints using a spontaneous symmetry breaking mechanism. Finally, one can also study in more depth the mathematical structure and properties of the simplicity constraints. The list is not conclusive, and there may be many other interesting topics to study.

References

[1] C. Rovelli, *Quantum Gravity*, Cambridge University Press, Cambridge (2004).

[2] C. Rovelli and F. Vidotto, *Covariant Loop Quantum Gravity*, Cambridge University Press, Cambridge (2014).

[3] T. Thiemann, *Modern Canonical Quantum General Relativity*, Cambridge University Press, Cambridge (2007).

[4] G. Ponzano and T. Regge, *Semiclassical limit of Racah coefficients*, *Spectroscopic and Group Theoretical Methods in Physics*, edited by F. Block, North Holland, Amsterdam (1968).

[5] J. W. Barrett and L. Crane, *Relativistic spin networks and quantum gravity*, *J. Math. Phys.* **39**, 3296 (1998), `arXiv:gr-qc/9709028`.

[6] J. W. Barrett and L. Crane, *A Lorentzian Signature Model for Quantum General Relativity*, *Class. Quant. Grav.* **17**, 3101 (2000), `arXiv:gr-qc/9904025`.

[7] H. Ooguri, *Topological Lattice Models in Four Dimensions*, *Mod. Phys. Lett. A* **7**, 279 (1992), `arXiv:hep-th/9205090`.

[8] J. Engle, E. R. Livine, R. Pereira and C. Rovelli, *LQG vertex with finite Immirzi parameter*, *Nucl. Phys.* **B799**, 136 (2008), `arXiv:0711.0146`.

[9] L. Freidel and K. Krasnov, *A New Spin Foam Model for 4d Gravity*, *Class. Quant. Grav.* **25**, 125018 (2008), `arXiv:0708.1595`.

[10] E. Bianchi, M, Han, E. Magliaro, C. Perini, C. Rovelli and W. Wieland, *Spinfoam fermions*, *Class. Quantum Grav.* **30**, 235023 (2013), `arXiv:1012.4719`.

[11] J. C. Baez and J. Huerta, *An Invitation to Higher Gauge Theory*, *Gen. Relativ. Gravit.* **43**, 2335 (2011), `arXiv:1003.4485`.

[12] A. Miković and M. Vojinović, *Poincare 2-group and quantum gravity*, *Class. Quant. Grav.* **29**, 165003 (2012), `arXiv:1110.4694`.

[13] T. Radenković and M. Vojinović, *Higher Gauge Theories Based on 3-groups*, *JHEP* **10**, 222 (2019), `arXiv:1904.07566`.

[14] M. Celada, D. González and M. Montesinos, *BF gravity*, *Class. Quant. Grav.* **33**, 213001 (2016), `arXiv:1610.02020`.

[15] C. Rovelli, *Zakopane lectures on loop gravity,* `arXiv:1102.3660`.

[16] J. F. Plebanski, *On the separation of Einsteinian substructures*, *J. Math. Phys.* **18**, 2511 (1977).

[17] F. Girelli, H. Pfeiffer and E. M. Popescu, *Topological Higher Gauge Theory - from BF to BFCG theory*, *Jour. Math. Phys.* **49**, 032503 (2008), `arXiv:0708.3051`.

[18] J. F. Martins and A. Miković, *Lie crossed modules and gauge-invariant actions for 2-BF theories*, Adv. Theor. Math. Phys. **15**, 1059 (2011), arXiv:1006.0903.

[19] L. Crane and M. D. Sheppeard, *2-categorical Poincare Representations and State Sum Applications*, arXiv:math/0306440.

[20] M. Vojinović, *Causal Dynamical Triangulations in the Spincube Model of Quantum Gravity*, Phys. Rev. D **94**, 024058 (2016), arXiv:1506.06839.

[21] A. Miković, *Spin-cube Models of Quantum Gravity*, Rev. Math. Phys. **25**, 1343008 (2013), arXiv:1302.5564.

[22] A. Miković and M. A. Oliveira, *Canonical formulation of Poincare BFCG theory and its quantization*, Gen. Relativ. Gravit. **47**, 58 (2015), arXiv:1409.3751.

[23] A. Miković, M. A. Oliveira and M. Vojinović, *Hamiltonian analysis of the BFCG theory for a generic Lie 2-group*, arXiv:1610.09621.

[24] A. Miković, M. A. Oliveira and M. Vojinović, *Hamiltonian analysis of the BFCG formulation of General Relativity*, Class. Quant. Grav. **36**, 015005 (2019), arXiv:1807.06354.

[25] J. F. Martins and R. Picken, *The fundamental Gray 3-groupoid of a smooth manifold and local 3-dimensional holonomy based on a 2-crossed module*, Differ. Geom. Appl. Journal **29**, 179 (2011), arXiv:0907.2566.

[26] W. Wang, *On 3-gauge transformations, 3-curvature and Gray-categories*, Jour. Math. Phys. **55**, 043506 (2014), arXiv:1311.3796.

[27] A. Baratin and L. Freidel, *A 2-categorical state sum model*, J. Math. Phys. **56**, 011705 (2015), arXiv:1409.3526.

[28] S. K. Asante, B. Dittrich, F. Girelli, A. Riello and P. Tsimiklis, *Quantum geometry from higher gauge theory*, arXiv:1908.05970.

[29] T. Radenković and M. Vojinović, *Hamiltonian Analysis for the Scalar Electrodynamics as 3BF Theory*, Symmetry **12**, 620 (2020), arXiv:2004.06901.

The spin 1 XXZ Gaudin model with boundary*

Igor Salom†
Institute of Physics, University of Belgrade
Pregrevica 118, 11080 Belgrade, Serbia

Nenad Manojlović‡
Departamento de Matemática, Faculdade de Ciências e Tecnologia
Universidade do Algarve, Campus de Gambelas,
8005-139 Faro, Portugal
and
Grupo de Física Matemática da Universidade de Lisboa
Departamento de Matemática, Faculdade de Ciências
Campo Grande, Edifício C6, 1749-016 Lisboa, Portugal

Nuno Cirilo António§
Centro de Análise Funcional e Aplicações
Instituto Superior Técnico, Universidade de Lisboa
Av. Rovisco Pais, 1049-001 Lisboa, Portugal

Abstract

We obtain the non-unitary classical r-matrix of the spin 1 trigonometric Gaudin model with boundary terms. Starting from the R-matrix and corresponding K-matrix of the spin $\frac{1}{2}$ XXZ Heisenberg chain the so-called fusion procedure yields the R and K matrices of the spin 1 XXZ Heisenberg chain. We demonstrate that the corresponding classical r and K matrices satisfy, respectively, the classical Yang-Baxter equation and the classical reflection equation. Consequently, we show that the relevant non-unitary classical r-matrix satisfies the generalized classical Yang-Baxter equation and therefore defines the non-periodic spin 1 trigonometric Gaudin model.

1. Introduction

An approach to study periodic Gaudin models [1, 2, 3] is based on the unitary classical r-matrices [4, 5, 6, 7] and the corresponding Sklyanin linear

* N.M. gratefully acknowledges useful discussions with Eric Ragoucy. I.S. was supported in part by the Serbian Ministry of Education, Science and Technological Development under grant number ON 171031.

† e-mail address: isalom@ipb.ac.rs
‡ e-mail address: nmanoj@ualg.pt
§ e-mail address: nantonio@math.ist.utl.pt

bracket [5, 6, 7, 8, 9, 10, 11]. The algebra generated by the entries of the relevant Lax operators can be used to completely solve the system through the algebraic Bethe ansatz, or the separation of variable method [5, 6, 7, 8, 9, 10, 11].

Similar approach to the non-periodic Gaudin models, based on the non-unitary classical r-matrices and corresponding linear bracket, enabled full implementation of the algebraic Bethe ansatz in the spin $\frac{1}{2}$ case [12, 13, 14, 15, 16, 17, 18, 19]. In this paper we initiate the study of the spin 1 non-periodic Gaudin models by deriving the relevant non-unitary classical r-matrix. This classical r-matrix is a solution to the generalized classical Yang-Baxter equation. An important advantage of this approach is that the generalized classical Yang-Baxter equation is equivalent to both the classical Yang-Baxter equation and classical reflection equation [15]. To obtain this non-unitary r-matrix we use the fusion procedure starting from the quantum R-matrix and the reflection K-matrix of the spin $\frac{1}{2}$ XXZ Heisenberg chain. The classical r-matrix we obtain defines completely the non-periodic spin 1 Gaudin model.

This paper is organized as follows. In the section 2. we describe briefly the fusion procedure as applied to the R and K matrices of the spin $\frac{1}{2}$ XXZ Heisenberg chain. The results of the second section are then presented and analysed in the section 3.. The quasi-classical limit which leads to the spin 1 non-unitary r-matrix is presented in the section 4..

2. Spin $\frac{1}{2}$ XXZ Heisenberg chain

In our study of the spin $\frac{1}{2}$ XXZ Heisenberg spin chain [16] the starting point is always the R-matrix [20, 21, 22, 23]

$$R(\lambda,\eta) = \begin{pmatrix} \sinh(\lambda+\eta) & 0 & 0 & 0 \\ 0 & \sinh(\lambda) & \sinh(\eta) & 0 \\ 0 & \sinh(\eta) & \sinh(\lambda) & 0 \\ 0 & 0 & 0 & \sinh(\lambda+\eta) \end{pmatrix}. \tag{1}$$

This R-matrix satisfies the Yang-Baxter equation [24, 22, 23, 20, 21] in the space $\mathbb{C}^2 \otimes \mathbb{C}^2 \otimes \mathbb{C}^2$

$$R_{12}(\lambda-\mu)R_{13}(\lambda)R_{23}(\mu) = R_{23}(\mu)R_{13}(\lambda)R_{12}(\lambda-\mu), \tag{2}$$

and it also has other relevant properties such as

$U(1)$ symmetry	$\left[\sigma_1^3+\sigma_2^3, R_{12}(\lambda)\right] = 0;$
unitarity	$R_{12}(\lambda)R_{21}(-\lambda) = \sinh(\eta-\lambda)\sinh(\eta+\lambda)\mathbb{1};$
parity invariance	$R_{21}(\lambda) = R_{12}(\lambda);$
temporal invariance	$R_{12}^t(\lambda) = R_{12}(\lambda);$
crossing symmetry	$R(\lambda) = \mathcal{J}_1 R^{t_2}(-\lambda-\eta)\mathcal{J}_1,$

where t_2 denotes the transpose in the second space and the two-by-two matrix $\mathcal{J}$ is proportional to the Pauli matrix σ^2, i.e. $\mathcal{J} = \imath\sigma^2$.

A way to introduce non-periodic boundary conditions which are compatible with the integrability of the bulk model, was developed in [25]. Boundary conditions on the left and right sites of the chain are encoded in the left and right reflection matrices K^- and K^+. The compatibility condition between the bulk and the boundary of the system takes the form of the so-called reflection equation. It is written in the following form for the left reflection matrix acting on the space $\mathbb{C}^2$ at the first site $K^-(\lambda) \in \mathrm{End}(\mathbb{C}^2)$

$$R_{12}(\lambda-\mu)K_1^-(\lambda)R_{21}(\lambda+\mu)K_2^-(\mu) = K_2^-(\mu)R_{12}(\lambda+\mu)K_1^-(\lambda)R_{21}(\lambda-\mu). \tag{3}$$

Due to the properties of the R-matrix (1) the dual reflection equation can be presented in the following form

$$\begin{aligned} R_{12}(\mu-\lambda)K_1^+(\lambda)R_{21}(-\lambda-\mu-2\eta)K_2^+(\mu) = \\ K_2^+(\mu)R_{12}(-\lambda-\mu-2\eta)K_1^+(\lambda)R_{21}(\mu-\lambda). \end{aligned} \tag{4}$$

One can then verify that the mapping

$$K^+(\lambda) = K^-(-\lambda-\eta) \tag{5}$$

is a bijection between solutions of the reflection equation and the dual reflection equation. After substitution of (5) into the dual reflection equation (4) one gets the reflection equation (3) with shifted arguments.

The general, spectral parameter dependent, solutions of the reflection equation (3) and the dual reflection equation (4) can be written as follows [26, 27, 28]

$$K^-(\lambda) = \begin{pmatrix} \kappa^- \sinh(\xi^- + \lambda) & \psi^- \sinh(2\lambda) \\ \phi^- \sinh(2\lambda) & \kappa^- \sinh(\xi^- - \lambda) \end{pmatrix}, \tag{6}$$

$$K^+(\lambda) = \begin{pmatrix} \kappa^+ \sinh(\xi^+ - \lambda - \eta) & -\psi^+ \sinh(2(\lambda+\eta)) \\ -\phi^+ \sinh(2(\lambda+\eta)) & \kappa^+ \sinh(\xi^+ + \lambda + \eta) \end{pmatrix}. \tag{7}$$

Due to the fact that the reflection matrices $K^{\mp}(\lambda)$ are defined up to multiplicative constants the values of parameters $\kappa^{\mp}$ are not essential, as long as they are different from zero. Although the R-matrix (1) has the $U(1)$ symmetry the reflection matrices $K^{\mp}(\lambda)$ (6) and (7) cannot be brought to the upper triangular form by the symmetry transformations like in the case of the XXX Heisenberg spin chain [14].

We will not discuss the Lax operators and the corresponding monodromies, relevant for the study of the spin $\frac{1}{2}$ XXZ Heisenberg spin chain [16]. Our aim is to obtain the R and K matrices corresponding the spin 1 XXZ Heisenberg chain. As our first step we use the fusion procedure [29, 21] in the space $\mathbb{C}^2 \otimes \mathbb{C}^2 \otimes \mathbb{C}^2 \otimes \mathbb{C}^2$ with the aim of deriving the spin 1

XXZ R-matrix. To this end we observe that the R-matrix (1) at $\lambda = -\eta$ is proportional to the projector onto the antisymmetric subspace of the space $\mathbb{C}^2 \otimes \mathbb{C}^2$,

$$\mathrm{P}^- = \frac{-1}{2\sinh(\eta)} \begin{pmatrix} 0 & 0 & 0 & 0 \\ 0 & \frac{1}{2} & -\frac{1}{2} & 0 \\ 0 & -\frac{1}{2} & \frac{1}{2} & 0 \\ 0 & 0 & 0 & 0 \end{pmatrix}. \tag{8}$$

For our purpose, the relevant projector is the complementary one, which projects onto the symmetric subspace of the space $\mathbb{C}^2 \otimes \mathbb{C}^2$,

$$\mathrm{P}^+ = \mathbb{1} - \mathrm{P}^-. \tag{9}$$

We obtain the spin 1 representation in the tensor product of the last two spaces,

$$R_{1(34)}(\lambda) = P^+_{34} R_{14}(\lambda - \eta) R_{13}(\lambda) P^+_{34}. \tag{10}$$

It straightforward the check that the corresponding Yang-Baxter equation is satisfied

$$R_{12}(\lambda - \mu) R_{1(34)}(\lambda) R_{2(34)}(\mu) = R_{2(34)}(\mu) R_{1(34)}(\lambda) R_{12}(\lambda - \mu). \tag{11}$$

Analogously, we have

$$R_{(12)3}(\lambda) = P^+_{12} R_{13}(\lambda) R_{23}(\lambda + \eta) P^+_{12}. \tag{12}$$

Finally, we obtain the relevant R-matrix

$$R_{(12)(34)}(\lambda) = P^+_{34} R_{(12)4}(\lambda - \eta) R_{(12)3}(\lambda) P^+_{34}. \tag{13}$$

In this case, the fusion procedure for the K-matrix [30] yields

$$K^-_{(12)}(\lambda) = P^+_{12} K^-_1(\lambda) R_{12}(2\lambda + \eta) K^-_2(\lambda + \eta) P^+_{12}. \tag{14}$$

A direct calculation shows that the corresponding reflection equation is satisfied

$$\begin{aligned} R_{(12)(34)}(\lambda - \mu) K^-_{(12)}(\lambda) R_{(34)(12)}(\lambda + \mu) K^-_{(34)}(\mu) = \\ = K^-_{(34)}(\mu) R_{(12)(34)}(\lambda + \mu) K^-_{(12)}(\lambda) R_{(34)(12)}(\lambda - \mu). \end{aligned} \tag{15}$$

Therefore we have achieved our main objective in obtaining the relevant R and K matrices. In the following section we will look at their explicit form and some important properties.

3. Spin 1 XXZ Heisenberg chain

In the appropriate bases the R-matrix (13) of the spin one XXZ Heisenberg chain can be represented as the following 9×9 matrix

$$R(\lambda,\eta) = \left(\begin{array}{ccc|ccc|ccc} a_1 & & & & & & & & \\ & a_2 & & b_1 & & & & & \\ & & a_3 & & b_2 & & b_3 & & \\ \hline & b_1 & & a_2 & & & & & \\ & & b_2 & & a_4 & & b_2 & & \\ & & & & & a_2 & & b_1 & \\ \hline & & b_3 & & b_2 & & a_3 & & \\ & & & & & b_1 & & a_2 & \\ & & & & & & & & a_1 \end{array}\right), \tag{16}$$

where the entries are given by

$$\begin{aligned} a_1 &= \sinh(\lambda+\eta)\sinh(\lambda+2\eta), \quad b_1 = \sinh(\lambda+\eta)\sinh(2\eta),\\ a_2 &= \sinh(\lambda)\sinh(\lambda+\eta), \quad b_2 = \sinh^2(\lambda)\sinh(2\eta)\cosh(\eta),\\ a_3 &= \sinh(\lambda)\sinh(\lambda-\eta), \quad b_3 = \sinh(\eta)\sinh(2\eta)\sinh(\lambda)\sinh(\lambda+\eta),\\ a_4 &= \sinh(\lambda)\sinh(\lambda+\eta)+\sinh(\eta)\sinh(2\eta). \end{aligned}$$

This R-matrix satisfies the Yang-Baxter equation in the space $\mathbb{C}^3 \otimes \mathbb{C}^3 \otimes \mathbb{C}^3$

$$R_{12}(\lambda-\mu)R_{13}(\lambda)R_{23}(\mu) = R_{23}(\mu)R_{13}(\lambda)R_{12}(\lambda-\mu), \tag{17}$$

and has the $U(1)$ symmetry

$$[h_1+h_2, R_{12}(\lambda,\eta)] = 0, \tag{18}$$

where $h = \mathrm{diag}(1,0,-1)$. The similarity transformation

$$\mathrm{Ad}\exp(\alpha\lambda(h_1-h_2))R_{12}(\lambda,\eta), \tag{19}$$

with $\alpha = \frac{1}{2}$, yields the $O(3)$ invariant form of this R-matrix [31, 32, 33]. The R-matrix 16 has some important properties such as regularity, unitarity, PT-symmetry and crossing symmetry. The regularity condition at $\lambda = 0$ reads

$$R(0,\eta) = \sinh(\eta)\sinh(2\eta)\mathcal{P}, \tag{20}$$

where $\mathcal{P}$ is the permutation matrix of $\mathbb{C}^3 \otimes \mathbb{C}^3$. The unitarity relation is

$$R_{12}(\lambda)R_{12}(-\lambda) = \rho(\lambda)\mathbb{1}, \tag{21}$$

here ρ is the following function

$$\rho(\lambda) = \sinh(\lambda+\eta)\sinh(\lambda+2\eta)\sinh(\lambda-\eta)\sinh(\lambda-2\eta). \tag{22}$$

The so-called PT-symmetry states

$$R_{12}^{t}(\lambda) = R_{12}(\lambda). \tag{23}$$

Finally, the R-matrix (16) has the following crossing symmetry property:

$$R(\lambda) = (\mathcal{J} \otimes \mathbb{1})\, R^{t_2}(-\lambda - \eta)\, \left(\mathcal{J}^{-1} \otimes \mathbb{1}\right), \tag{24}$$

where t_2 denotes the transpose in the second space and the matrix $\mathcal{J}$ is given by

$$\mathcal{J} = \begin{pmatrix} 0 & 0 & -1 \\ 0 & 1 & 0 \\ -1 & 0 & 0 \end{pmatrix}. \tag{25}$$

In the basis in which the R-matrix (13) takes the form (16) the K-matrix (14) is given by

$$K^{-}(\lambda, \eta) = \sinh(2\lambda + \eta) \begin{pmatrix} \mathrm{k}_{11} & \mathrm{k}_{12} & \mathrm{k}_{13} \\ \mathrm{k}_{21} & \mathrm{k}_{22} & \mathrm{k}_{23} \\ \mathrm{k}_{31} & \mathrm{k}_{32} & \mathrm{k}_{33} \end{pmatrix}, \tag{26}$$

where

$$\begin{aligned}
\mathrm{k}_{11} &= {\kappa^{-}}^{2} \sinh(\xi^{-} + \lambda + \tfrac{\eta}{2}) \sinh(\xi^{-} + \lambda - \tfrac{\eta}{2}) + \psi^{-}\phi^{-} \sinh(2\lambda - \eta) \sinh(\eta), \\
\mathrm{k}_{22} &= {\kappa^{-}}^{2} \sinh(\xi^{-} - \lambda + \tfrac{\eta}{2}) \sinh(\xi^{-} + \lambda - \tfrac{\eta}{2}) + \psi^{-}\phi^{-} \sinh(2\lambda - \eta) \sinh(2\lambda + \eta), \\
\mathrm{k}_{33} &= {\kappa^{-}}^{2} \sinh(\xi^{-} - \lambda + \tfrac{\eta}{2}) \sinh(\xi^{-} - \lambda - \tfrac{\eta}{2}) + \psi^{-}\phi^{-} \sinh(2\lambda - \eta) \sinh(\eta), \\
\mathrm{k}_{12} &= \kappa^{-}\psi^{-} \sqrt{2}\sqrt{\cosh(\eta)} \sinh(\xi^{-} + \lambda - \tfrac{\eta}{2}) \sinh(2\lambda), \\
\mathrm{k}_{23} &= \kappa^{-}\psi^{-} \sqrt{2}\sqrt{\cosh(\eta)} \sinh(\xi^{-} - \lambda + \tfrac{\eta}{2}) \sinh(2\lambda), \\
\mathrm{k}_{13} &= {\psi^{-}}^{2} \sinh(2\lambda - \eta) \sinh(2\lambda), \\
\mathrm{k}_{21} &= \kappa^{-}\phi^{-} \sqrt{2}\sqrt{\cosh(\eta)} \sinh(\xi^{-} + \lambda - \tfrac{\eta}{2}) \sinh(2\lambda), \\
\mathrm{k}_{32} &= \kappa^{-}\phi^{-} \sqrt{2}\sqrt{\cosh(\eta)} \sinh(\xi^{-} - \lambda + \tfrac{\eta}{2}) \sinh(2\lambda), \\
\mathrm{k}_{31} &= {\phi^{-}}^{2} \sinh(2\lambda - \eta) \sinh(2\lambda).
\end{aligned} \tag{27}$$

The R-matrix (16) and the above K-matrix satisfy the reflection equation

$$R_{12}(\lambda - \mu) K_1^{-}(\lambda) R_{21}(\lambda + \mu) K_2^{-}(\mu) = K_2^{-}(\mu) R_{12}(\lambda + \mu) K_1^{-}(\lambda) R_{21}(\lambda - \mu). \tag{28}$$

The study of the spin 1 XXZ Heisenberg chain would require the dual reflection equation and the relevant K^{+} matrix. Although, now this would be straightforward we will not proceed in this direction since our main aim is to derive the corresponding Gaudin model through the so-called quasi-classical limit [14, 16].

4. Spin 1 XXZ Gaudin model

The classical r-matrices are essential tools in the study of the Gaudin models [5, 4, 6, 7, 8, 9, 10]. To this end we observe the quasi-classical property of the R-matrix (16)

$$\frac{1}{\sinh^2(\lambda)} R(\lambda, \eta) = \mathbb{1} + \eta \mathrm{r}(\lambda) + \mathcal{O}(\eta^2), \tag{29}$$

where the classical r-matrix is given by [4, 6]

$$\mathrm{r}(\lambda) = \frac{2}{\sinh(\lambda)} \left(\begin{array}{ccc|ccc|ccc} \cosh(\lambda) & & & & & & & & \\ & & & 1 & & & & & \\ & & -\cosh(\lambda) & & 1 & & & & \\ \hline & 1 & & & & & & & \\ & & 1 & & & & 1 & & \\ & & & & & & & 1 & \\ \hline & & & & 1 & & -\cosh(\lambda) & & \\ & & & & & 1 & & & \\ & & & & & & & & \cosh(\lambda) \end{array} \right) + \coth(\lambda)\mathbb{1}. \tag{30}$$

As it is very well known[4, 6], the above r-matrix has the unitarity property

$$\mathrm{r}_{21}(-\lambda) = -\mathrm{r}_{12}(\lambda), \tag{31}$$

and it satisfies the classical Yang-Baxter equation

$$[\mathrm{r}_{13}(\lambda), \mathrm{r}_{23}(\mu)] + [\mathrm{r}_{12}(\lambda - \mu), \mathrm{r}_{13}(\lambda) + \mathrm{r}_{23}(\mu)] = 0. \tag{32}$$

Our objective is the non-periodic Gaudin model and therefore we also need the classical K-matrix [14, 15, 16] which is obtained from the K-matrix (26) by setting $\eta = 0$,

$$\mathrm{K}(\lambda) \equiv K^-(\lambda, 0). \tag{33}$$

A direct consequence of the equation (28) is the classical reflection equation [34, 35, 15]:

$$\begin{aligned} &\mathrm{r}_{12}(\lambda - \mu)\mathrm{K}_1(\lambda)\mathrm{K}_2(\mu) + \mathrm{K}_1(\lambda)\mathrm{r}_{21}(\lambda + \mu)\mathrm{K}_2(\mu) = \\ &= \mathrm{K}_2(\mu)\mathrm{r}_{12}(\lambda + \mu)\mathrm{K}_1(\lambda) + \mathrm{K}_2(\mu)\mathrm{K}_1(\lambda)\mathrm{r}_{21}(\lambda - \mu). \end{aligned} \tag{34}$$

It can be shown [15, 12] that by defining the non-unitary classical r-matrix

$$\mathrm{r}_{12}^K(\lambda, \mu) = \mathrm{r}_{12}(\lambda - \mu) - \mathrm{K}_2(\mu)\mathrm{r}_{12}(\lambda + \mu)\mathrm{K}_2^{-1}(\mu), \tag{35}$$

the classical Yang-Baxter equation (32) and (34) combine into one equation, the so-called generalized classical Yang-Baxter equation

$$\left[\mathrm{r}_{32}^{\mathrm{K}}(\nu, \mu), \mathrm{r}_{13}^{\mathrm{K}}(\lambda, \nu)\right] + \left[\mathrm{r}_{12}^{\mathrm{K}}(\lambda, \mu), \mathrm{r}_{13}^{\mathrm{K}}(\lambda, \nu) + \mathrm{r}_{23}^{\mathrm{K}}(\mu, \nu)\right] = 0. \tag{36}$$

As the classical r-matrix defines the Sklyanin linear bracket, an essential tool in the study of the periodic Gaudin model [5, 6, 7, 8, 9, 10], the non-unitary r-matrix (34), through the relevant linear bracket, defines the so-called generalized Gaudin algebra essential in the study of the non-periodic model [12, 15, 17, 18, 19] .

References

[1] M. Gaudin, *Diagonalisation d'une classe d'hamiltoniens de spin*, J. Physique 37 (1976) 1087–1098.

[2] M. Gaudin, *La fonction d'onde de Bethe*, chapter 13 Masson, Paris, 1983.

[3] M. Gaudin, *The Bethe Wavefunction*, Cambridge University Press, 2014.

[4] A. A. Belavin and V. G. Drinfeld. *Solutions of the classical Yang-Baxter equation for simple Lie algebras* (in Russian), Funktsional. Anal. i Prilozhen. 16 (1982), no. 3, 1–29; translation in Funct. Anal. Appl. 16 (1982) no. 3, 159-180.

[5] E. K. Sklyanin, *Separation of variables in the Gaudin model*, Zap. Nauchn. Sem. Leningrad. Otdel. Mat. Inst. Steklov. (LOMI) 164 (1987) 151–169; translation in J. Soviet Math. 47 (1989) 2473–2488.

[6] M. A. Semenov-Tian-Shansky, *Quantum and classical integrable systems*, in Integrability of Nonlinear Systems, Lecture Notes in Physics Volume 495 (1997) 314-377.

[7] B. Jurčo, *Classical Yang-Baxter equations and quantum integrable systems (Gaudin models)*, in Quantum groups (Clausthal, 1989), Lecture Notes in Phys. Volume 370 (1990) 219–227.

[8] P. P. Kulish and N. Manojlović, *Bethe vectors of the $osp(1|2)$ Gaudin model*, Letters in Mathematical Physics **Vol. 55** (2001) 77-95.

[9] P. P. Kulish and N. Manojlović, *Creation operators and Bethe vectors of the $osp(1|2)$ Gaudin model*, Journal of Mathematical Physics **Vol. 42** No. 10 (2001) 4757-4778.

[10] P. P. Kulish and N. Manojlović, *Trigonometric $osp(1|2)$ Gaudin model*, Journal of Mathematical Physics **Vol. 44** No. 2 (2003) 676-700.

[11] N. Manojlović, Z. Nagy and I. Salom, *Derivation of the trigonometric Gaudin Hamiltonians*, Proceedings of the 8th Mathematical Physics meeting: Summer School and Conference on Modern Mathematical Physics, 24 - 31 August 2014, Belgrade, Serbia, SFIN **XXVIII** Series A: Conferences No. A1, ISBN: 978-86-82441-43-4, (2015) 127–135.

[12] T. Skrypnyk, *Non-skew-symmetric classical r-matrix, algebraic Bethe ansatz, and Bardeen-Cooper-Schrieffer-type integrable systems*, J. Math. Phys. 50 (2009) 033540, 28 pages.

[13] N. Cirilo António, N. Manojlović and Z. Nagy, *Trigonometric $s\ell(2)$ Gaudin model with boundary terms*, Reviews in Mathematical Physics **Vol. 25** No. 10 (2013) 1343004 (14 pages); `arXiv:1303.2481`.

[14] N. Cirilo António, N. Manojlović and I. Salom, *Algebraic Bethe ansatz for the XXX chain with triangular boundaries and Gaudin model*, Nuclear Physics **B 889** (2014) 87-108; `arXiv:1405.7398`.

[15] N. Cirilo António, N. Manojlović, E. Ragoucy and I. Salom, *Algebraic Bethe ansatz for the $s\ell(2)$ Gaudin model with boundary*, Nuclear Physics **B 893** (2015) 305-331; `arXiv:1412.1396`.

[16] N. Manojlović and I. Salom, *Algebraic Bethe ansatz for the XXZ Heisenberg spin chain with triangular boundaries and the corresponding Gaudin model*, Nuclear Physics **B 923** (2017) 73-106; arXiv:1705.02235.

[17] I. Salom, N. Manojlović and N. Cirilo António, *Generalized $s\ell(2)$ Gaudin algebra and corresponding Knizhnik-Zamolodchikov equation*, Nuclear Physics **B 939** (2019) 358-371.

[18] N. Manojlović and I. Salom, *Algebraic Bethe ansatz for the trigonometric $s\ell(2)$ Gaudin model with triangular boundary*, Symmetry 12 (2020) 352; arXiv:1709.06419.

[19] I. Salom and N. Manojlović, *Creation operators of the non-periodic $sl(2)$ Gaudin model*, Proceedings of the 8th Mathematical Physics meeting: Summer School and Conference on Modern Mathematical Physics, 24 - 31 August 2014, Belgrade, Serbia, SFIN **XXVIII** Series A: Conferences No. A1, ISBN: 978-86-82441-43-4, (2015) 149–155.

[20] L. A. Takhtajan and L. D. Faddeev, *The quantum method for the inverse problem and the XYZ Heisenberg model*, (in Russian) Uspekhi Mat. Nauk **34** No. 5 (1979) 13–63; translation in Russian Math. Surveys **34** No.5 (1979) 11–68.

[21] P. P. Kulish and E. K. Sklyanin, *Quantum spectral transform method. Recent developments*, Lecture Notes in Physics **151** (1982), 61–119.

[22] R. J. Baxter, *Partition function of the Eight-Vertex lattice model*, Annals of Physics 70 (1972) 193–228.

[23] R. J. Baxter, *Exactly solved models in statistical mechanics*, Academic Press, London (1982).

[24] C. N. Yang, *Some exact results for the many-body problem in one dimension with repulsive delta-function interaction*, Physical Review Letters 19 (1967) 1312-1315.

[25] E. K. Sklyanin, *Boundary conditions for integrable quantum systems*, J. Phys. A: Math. Gen. **21** (1988) 2375–2389.

[26] H. J. de Vega and A. González Ruiz, *Boundary K-matrices for the XYZ, XXZ, XXX spin chains*, J. Phys. A: Math. Gen. **27** (1994), 6129–6137.

[27] S. Ghoshal and A. B. Zamolodchikov, *Boundary S-matrix and boundary state in two-dimensional integrable quantum field theory*, International Journal of Modern Physics A **09**, 3841 (1994) 3841–3885.

[28] S. Ghoshal and A. B. Zamolodchikov, *Errata: Boundary S-matrix and boundary state in two-dimensional integrable quantum field theory*, International Journal of Modern Physics A **09**, 4353 (1994) 4353.

[29] P. P. Kulish, N. Yu. Reshetikhin and E. K. Sklyanin, *Yang-Baxter equation and representation theory I*, Letters in Mathematical Physics **5** (1981) 393–403.

[30] L. Mezincescu and R. Nepomechie, *Fusion procedure for open chains*, Journal of Physics A: Mathematical and Theoretical 25 (1992) 2533-2543.

[31] Alexander B. Zamolodchikov and Alexey B. Zamolodchikov, *Relativistic factorized S-matrix in two dimensions having $O(N)$ isotopic symmetry* Nuclear Phys. B 133 (1978) 525–535.

[32] Alexander B. Zamolodchikov and Alexey B. Zamolodchikov, *Factorized S-matrices in two dimensions as the exact solutions of certain relativistic quantum field theory models* Annals of Physics Vol. 120 (1979) 253–291.

[33] P. P. Kulish and E. K. Sklyanin, *Solutions of the Yang-Baxter equation*, (Russian) Zap. Nauchn. Sem. Leningrad. Otdel. Mat. Inst. Steklov. (LOMI) 95 (1980), 129–160; translation in J. Soviet Math. Vol. 19 (1982), 1596–1620.

[34] E. K. Sklyanin, *Boundary conditions for integrable equations*, (Russian) Funktsional. Anal. i Prilozhen. 21 (1987) 86–87; translation in Functional Analysis and Its Applications Volume 21, Issue 2 (1987) 164–166.

[35] E. K. Sklyanin, *Boundary conditions for integrable systems*, in the Proceedings of the VIIIth international congress on mathematical physics (Marseille, 1986), World Sci. Publishing, Singapore, (1987) 402–408.

Velocity memory effect for gravitational waves with torsion*

Dejan Simić[†]
Institute of Physics, University of Belgrade,
Pregrevica 118, 11080 Belgrade, Serbia

Abstract

We study the geodesic motion of massive test particle in the presense of the torsional plane-fronted (pp) wave in the three-dimensional (3D) gravity. The idea of this investigation is to test the appearance of the memory effect for the torsional waves. Our analysis discovers that the *velocity memory effect* happens for all waves that go to zero as, retarded time, u goes to infinity at sufficient fast rate.

1. Introduction

Does some observable change occur when a gravitational wave passes through a system of test particles in Minkowski spacetime? The answer is affirmative and is know as the gravitational memory [1, 2]. This is the effect that happens when a gravitational wave passes through a system of test particles, in asymptotically flat spacetime, which are initially at rest. If, after the passage of a gravitational wave, permanent displacement of particles occur we call this displacement memory effect [1, 2, 3, 4] and if particles have non-zero relative velocity we call it velocity memory effect [5, 6, 7]. This is important effect because it represents a possible experimental set up for the detection of gravitational waves and investigation of their properties.

In this article we will analyze the geodesic motion of massive test particles in the background of the pp wave with torsion. The reason why we undertook this investigation is to see is there a memory effect for gravitational waves with torsion in the Poincaré gauge theory [8, 9, 10]. To investigate this we will use the solutions with propagating torsion [11]. Important thing to note is that gravitational pp waves in 3D are solutions which do not exist without torsion [11], meaning that in the absence of torsion metric becomes trivial. This offers us an interesting opportunity to study the effects of torsion at the level of geodesic motion.

---

* This work was partially supported by the Serbian Science Foundation under Grant No. 171031.

† e-mail address: dsimic@ipb.ac.rs

The paper is organized as follows. First, we review pp waves without torsion in three dimensions and show that this solutions do not exist in 3D general relativity. After that, we analyze the pp waves with torsion in 3D Poincaré gauge theory of gravity. Next, we derive the geodesic equations for the metric of the pp wave with torsion. Unfortunately, the geodesic equations cannot be solved analytically except in a very special case, which is not interesting from the aspect of memory effect because it is not asymptotically flat. Due to this technical problem we had to solve geodesic equations numerically and results are given as plots of velocity in a function of retarded time u.

Conventions we are using are the following. The Latin indices $(i, j, ...)$ refer to the local Lorentz coordinates and run over $(0, 1, 2)$. The spacetime indices are denoted with letters of Greek alphabet. The contraction of vector with a form we label with $\rfloor$. The e^i is triad 1-form and the dual basis h_i is defined by the following equation $h_i \rfloor e^j = \delta^j_i$. For the Hodge dual we use the standard symbol *, and the Hodge dual of triad is $^\star e^i = \frac{1}{2}\varepsilon^{ijk} e_j e_k$. The exterior product of forms is implicit in all formulas.

2. The pp waves without torsion in three-dimensions

To better understand the nature of the pp waves with torsion we start with the Riemannian pp waves. For more details see Ref. [11].

The metric of the pp waves in Brinkmann coordinates is

$$ds^2 = 2du(Sdu + dv) - dy^2 \,, \tag{1a}$$

we, also, introduce an auxiliary function H

$$S = \frac{1}{2} H(u, y) \,. \tag{1b}$$

From the metric it is easy to derive the form of the triad e^i so that $ds^2 = \eta_{ij} e^i \otimes e^j$ holds, where η_{ij} is the half-null flat metric

$$\eta_{ij} = \begin{pmatrix} 0 & 1 & 0 \\ 1 & 0 & 0 \\ 0 & 0 & -1 \end{pmatrix} .$$

The triad is given by

$$e^0 := du \,, \qquad e^1 := Sdu + dv \,, \qquad e^2 := dy \,, \tag{2a}$$

The dual frame h_i define by the requirement $h_i \rfloor e^j = \delta^j_i$, where $\rfloor$ is a label for contraction, reads

$$h_0 = \partial_u - S\partial_v \,, \qquad h_1 = \partial_v \,, \qquad h_2 = \partial_y \,. \tag{2b}$$

Starting from the general formula for the Riemann connection

$$\omega^{ij} := -\frac{1}{2}\Big[h^i \rfloor de^j - h^j \rfloor de^i - (h^i \rfloor h^j \rfloor de^k)e_k\Big]\,,$$

we derive that the only non-zero component is

$$\omega^{12} = -\partial_y S\, e^0\,. \tag{3a}$$

From the above connection the Riemannian curvature is easily obtained

$$R^{ij} = 2e^0 k^{[i}Q^{j]}\,, \tag{4a}$$

where $k^i = (0,1,0)$ is a null vector and

$$Q^i = \partial_y^2 S e^2 \delta_2^i\,. \tag{4b}$$

The Ricci 1-form $Ric^i := h_m \rfloor Ric^{mi}$ is given by

$$Ric^i = e^0 k^i Q\,, \qquad Q = h_i \rfloor Q^i = \frac{1}{2}\partial_y^2 H, \tag{5a}$$

while the scalar curvature is zero

$$R = 0\,. \tag{5b}$$

Up to this point discussion was valid for any theory of gravity, now we want to look at what happens in general relativity in 3D. Action of general relativity $I = -a_0 \int d^3xR$ leads to the vacuum field equations

$$2a_0 G^n{}_i = 0\,, \tag{6}$$

where $G^n{}_i$ is the Einstein tensor. Einstein equations, after substitution of the metric, give

$$\partial_y^2 H = 0\,, \tag{7}$$

which only has a trivial solution

$$H = C(u) + yD(u)\,,$$

due to the fact that the curvature identically vanishes and this solution is diffeomorphic to Minkowski spacetime. This is to be expected because general relativity in three-dimensions does not have propagating local degrees of freedom and consequently should not have a wave solution.

3. The pp waves with torsion

The theory we will consider is a quadratic Poincaré theory of gravity which generalizes general relativity with 6 additional quadratic terms and as many free parameters $a_0, \ldots, a_2$ and $b_0, \ldots, b_2$. We will not write the action interested reader can find more details in Ref. [11].

The pp wave with torsion is a generalization of Riemannian pp wave which is obtained under assumption that the triad field (2) remains unchanged, while the connection takes the form

$$\omega^{ij} = \tilde{\omega}^{ij} + \frac{1}{2}\varepsilon^{ij}{}_m k^m k_n e^n G\,, \tag{8a}$$

$$G := S' + K\,. \tag{8b}$$

The function $K = K(u, y)$ is added to account for the effect of torsion, which is seen from the following expression for torsion

$$T^i := \nabla e^i = \frac{1}{2} K k^i k_m {}^\star e^m\,. \tag{9}$$

The curvature 2-form, Ricci 1- form and Ricci scalar are given by

$$\begin{aligned} R^{ij} &= \varepsilon^{ijm} k_m k^{n\star} e_n\, G'\,, \\ Ric^i &= \frac{1}{2} k^i k_m e^m G'\,, \\ R &= 0\,. \end{aligned} \tag{10}$$

The geometric configuration defined by the triad field (2) and the connection (8) represents a generalized gravitational plane-fronted wave of GR_Λ, or the *torsion wave* for short. The vector field $k = \partial_v$ is the Killing vector for both the metric and the torsion; moreover, it is a null and covariantly constant vector field. This allows us to consider the solution (12) as a generalized pp-wave.

The field equations [11] are given by

$$\begin{aligned} &a_0 G' - a_1 K' = 0\,, \qquad && \Lambda = 0\,, \\ &K'' + m^2 K = 0\,, && m^2 = \frac{a_0(a_1 - a_0)}{b_4 a_1}\,, \end{aligned} \tag{11}$$

with $G = S' + K$ and $S = H/2$. The solution of this equations is

$$\begin{aligned} K &= A(u)\cos my + B(u)\sin my\,, \\ \frac{1}{2}H &= \frac{a_1 - a_0}{a_0 m}\left(A(u)\sin my - B(u)\cos my\right) + h_1(u) + h_2(u)y\,. \end{aligned} \tag{12}$$

As we already said the h_1 and h_2 do not contribute to the radiation part of the curvature and can be discarded as trivial solution. Consequently, when the torsion is not present the metric becomes trivial. This is to be expected since general relativity in three-dimensions is a theory without propagating local degrees of freedom. Because the metric is crucially related to the torsion we can extract information about the torsion already on the level of the metric and geodesic motion.

4. Motion of massive test particle

In this section we investigate the geodesic motion of massive test particle in the presence of the massive pp wave with torsion described in the previous section. The geodesic motion of the test particle is obtained by solving a geodesic equation in which appear Christoffel (Riemannian) connection. So first we have to find the Christoffel connection for the metric of the masive pp wave with torsion.

4.1. Christoffel connection

The Christoffel connection is easily derived from the metric using the well known formula

$$\tilde{\Gamma}^{\mu}_{\nu\rho} = \frac{1}{2} g^{\mu\alpha} \left(\partial_\nu g_{\alpha\rho} + \partial_\rho g_{\alpha\nu} - \partial_\alpha g_{\nu\rho} \right) , \tag{13}$$

and its non-zero components are given by

$$\begin{aligned} \tilde{\Gamma}^{v}{}_{uu} &= \frac{1}{2}\partial_u H \,, \\ \tilde{\Gamma}^{v}{}_{uy} &= \frac{1}{2}H' \,, \qquad \tilde{\Gamma}^{v}{}_{yu} = \frac{1}{2}\partial_y H \,, \\ \tilde{\Gamma}^{y}{}_{uu} &= \frac{1}{2}\partial_y H \,. \end{aligned} \tag{14}$$

Let us note that existence of a non-trivial metric of the pp wave is due to the presence of torsion. This allows us to see effects of torsion on the level of metric and, consequently, in geodesic motion of test particles [12].

4.2. Geodesic equations

The geodesic equation for u is

$$\frac{d^2 u}{d\lambda^2} = 0 \,. \tag{15}$$

Consequently, we take $u \equiv \lambda$.

The equation for y reads

$$\ddot{y} + \frac{1}{2}\partial_y H = 0 \,, \tag{16a}$$

after substitution of explicit form of the function H it becomes

$$\ddot{y} + \frac{a_1 - a_0}{a_0}(A(u)\cos my + B(u)\sin my) = 0\,. \tag{16b}$$

The equation for v is

$$\ddot{v} + \frac{1}{2}\partial_u H + \partial_y H \dot{y} = 0\,, \tag{17}$$

or explicitly

$$\ddot{v} + \frac{a_1 - a_0}{m a_0}(A'(u)\sin my - B'(u)\cos my)$$
$$+\frac{a_1 - a_0}{a_0}(A(u)\cos my + B(u)\sin my)\dot{y} = 0\,, \tag{18}$$

4.3. Velocity memory effect

The velocity memory effect is present in the case when functions $A(u)$ and $B(u)$ vanish for large u. For numerical calculations it is better to introduce functions $\bar{A}(u) = \frac{a_1 - a_0}{a_0}A$ and $\bar{B}(u) = \frac{a_1 - a_0}{a_0}B(u)$ which we will use later in the text instead of $A(u)$ and $B(u)$. The velocity changes as one changes initial conditions, so this is a true observable effect. We do not show plots for different initial conditions because we wanted the presentation to be as short as possible.

4.4. Shockwave case

In the shock wave case when functions $\bar{A}(u) = 0$ and $\bar{B}(u)$ vanishes exponentially $\bar{B}(u) = e^{-(u-10)^2}$ numerical solutions of the geodesic equations gives the plots [12]for the particle velocities $\dot{y}$ and $\dot{v}$ shown in the Figure 1.

4.5. Slow fall off

In the case when $\bar{A}(u) = 0$ and $\bar{B}(u) = 1/u$ numerical solutions lead to the following plots [12] for the particle velocities $\dot{y}$ and $\dot{v}$ shown in the Figure 2.

5. Conclusion

We investigated a motion of massive test particles in asymptotically flat pp wave spacetime with torsion. The meaning of this is that test particle is initially well described by a particle in Minkowski spacetime and at some point a pp wave passes by and at time-like infinity a particle is again well described by its motion in Minkowski spacetime. Consequently, properties of particles motion at initial time and at infinity can be consistently compared.

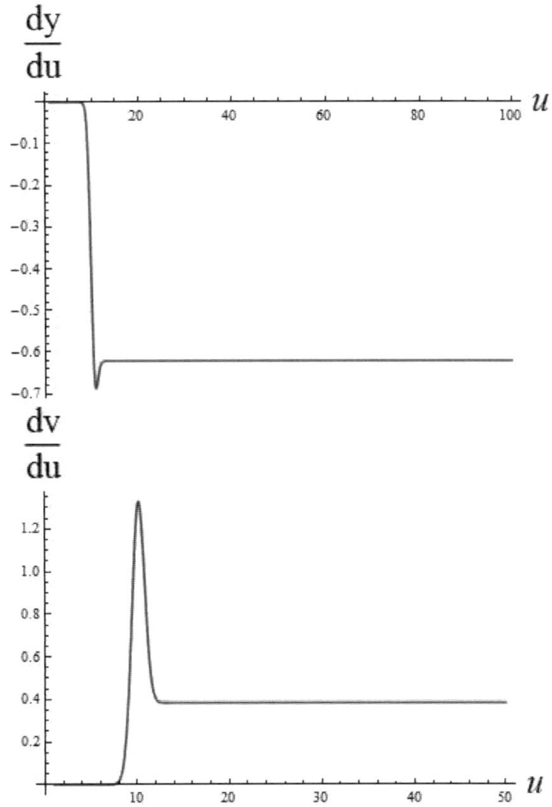

Figure 1: The plot for the particle velocity $\dot{y}$ and $\dot{v}$ in units $m = 1$, for $\bar{B} = -e^{-(u-10)^2}$

The conclusion is that velocity memory effect happens both for the exponentially fast fall-off of the gravitational wave as well as for the arbitrary polynomial fall-off. For the related work on memory effect for massive gravitons see Ref. [13]. This is the first time the memory effect for gravitational waves with torsion is analyzed. To authors knowledge, this is also the first example of the memory effect in three-dimensional gravity.

In the last few years there was a lot of effort on connecting asymptotic symmetries, soft theorems and memory effect [14]. This approach based on BMS symmetry offers a new perspective on the black hole microstates and information loss [15]. It is an open problem to connect the memory effect described in this article with asymptotic symmetry of the theory.

It is very interesting to generalize the analysis of this paper to the pp waves in four dimensions [16]. Preliminary results [17] show that most of the conclusions of this paper transfer to the four-dimensional case. This is important effect because it offers a possible experimental set up for the

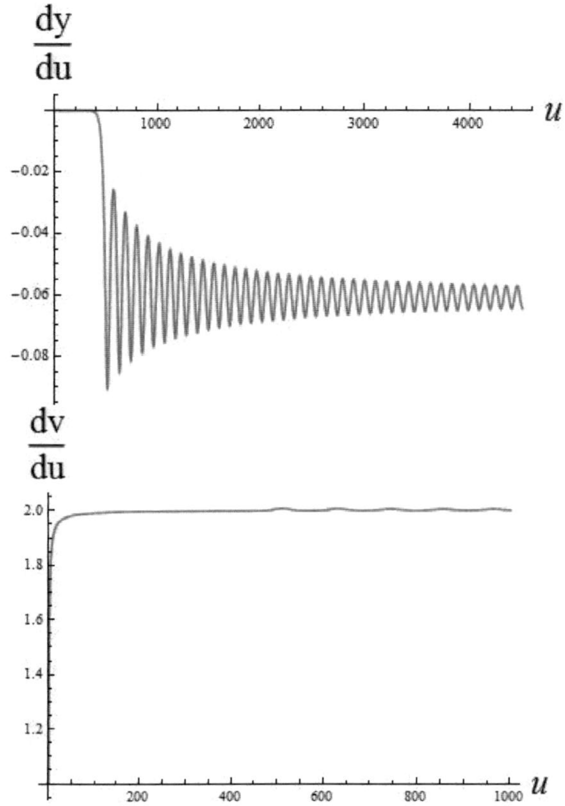

Figure 2: The plot for the particle velocity $\dot{y}$ and $\dot{v}$ in units $m = 1$, for $\bar{B} = -1/u$

detection of torsion.

References

[1] Ya. B. Zeldovich and A. G. Polnarev, Radiation of gravitational waves by a cluster of superdense stars, Astron. Zh. **51**, 30 (1974) [Sov. Astron. **18** 17 (1974)].

[2] V. B. Braginsky and L P Grishchuk, Kinematic resonance and the memory effect in free mass gravitational antennas, Zh. Eksp. Teor. Fiz. **89** 744 (1985) [Sov. Phys. JETP **62**, 427 (1985)].

[3] D. Christodoulou, Nonlinear nature of gravitation and gravitational wave experiments, Phys. Rev. Lett. **67** (1991) 1486.

[4] K. S. Thorne, Gravitational-wave bursts with memory: The Christodoulou effect, Phys. Rev. D **45** 520 (1992).

[5] H. Bondi and F. A. E. Pirani, Gravitational waves in general relativity III. Exact plane waves, Proc. Roy. Soc. Lond. A **251** (1959) 519-533.

[6] L. P. Grishchuk and A. G. Polnarev, Gravitational wave pulses with velocity coded memory, Sov. Phys. JETP **69** (1989) 653 [Zh. Eksp. Teor. Fiz. **96** (1989) 1153]

[7] P.-M Zhang, C. Duval, G.W. Gibbons and P.A. Horvathy, The Memory Effect for Plane Gravitational Waves, Phys. Lett. B **772** (2017) 743-746;
P. M Zhang, C. Duval, G.W. Gibbons and P.A. Horvathy, Soft gravitons and the memory effect for plane gravitational waves, Phys. Rev. D **96** (2017) no.6, 064013;
P. M. Zhang, C. Duval, G.W. Gibbons and P. A. Horvathy, Velocity Memory Effect for Polarized Gravitational Waves, JCAP **1805** (2018) no.05, 030.

[8] M. Blagojević, *Gravitation and Gauge Symmetries* (IoP Publishing, Bristol, 2002);
T. Ortín, *Gravity and Strings* (Cambridge University Press, Cambridge, 2004).

[9] Yu. N. Obukhov, Poincaré gauge gravity: Selected topics, Int. J. Geom. Meth. Mod. Phys. **3**, 95–138 (2006).

[10] M. Blagojević and F. W. Hehl (eds.), *Gauge Theories of Gravitation, A Reader with Commentaries* (Imperial College Press, London, 2013).

[11] M. Blagojević and B. Cvetković, Gravitational waves with torsion in 3D, Phys. Rev. D **90** (2014) 044006.

[12] B. Cvetković and D. Simić, Velocity memory effect without soft particles, Phys. Rev. D**101** (2020).

[13] E. Kilicarslan and B. Tekin, Graviton mass and memory, Eur. Phys. J. C **79** (2019) no 2, 114.

[14] Temple He, V. Lysov, P. Mitra and A. Strominger, BMS supertranslations and Weinbergs soft graviton theorem, JHEP **1505** (2015) 151;
A. Strominger and A. Zhiboedov, Gravitational Memory, BMS Supertranslations and Soft Theorems, JHEP **1601** (2016) 086.

[15] S. W. Hawking, M. J. Perry and A. Strominger, Soft Hair on Black Holes, Phys. Rev. Lett. **116** (2016) no.23, 231301.

[16] M. Blagojević and B. Cvetković, Generalized pp waves in Poincaré gauge theory, Phys. Rev. D **95** (2017) no.10, 104018;
M. Blagojević, B. Cvetković and Y. N. Obukhov, Generalized plane waves in Poincaré gauge theory of gravity, Phys. Rev. D **96** (2017) no.6, 064031.

[17] B. Cvetković and D. Simić, in preparation.

Chiral asymmetry in the weak interaction via Clifford algebras

Ovidiu Cristinel Stoica*
Dept. of Theoretical Physics, NIPNE—HH, Bucharest, Romania.

Abstract

Despite its tremendous success, the Standard Model of particle physics does not explain why the weak interaction breaks chiral symmetry. Various unified theories got us closer to an answer, but too often the explanation consists of labeling the $\mathrm{SU}(2)_w$ singlet representations as right-handed, and the doublet ones as left handed. This by itself does not ensure a chiral preference, because chirality itself, arising in the Dirac spinors, is not a property of the internal gauge group representations. Something deeper than mere labeling is required. In this paper, some of the progress using exterior and Clifford algebras is reviewed, and a possible explanation for chiral asymmetry is presented. It is shown how such a solution is present, rather implicitly, in a model proposed in a previous article.

1. Introduction

An important question, unanswered in the Standard Model of particle physics, is *Why is the electroweak interaction chiral?*

A general procedure to answer this question is to try to identify the rules behind the symmetry groups and their representations present in the Standard Model (SM), and then identify the mathematical structure behind these rules. This heuristic led to various models aiming to explain the properties of the SM from more fundamental structures.

In particular, a model that solves the problem of chirality of the electroweak interaction should solve at least three main problems.

The first problem is, of course,

Problem 1. *The model must include the doublet and singlet representations of* $\mathrm{SU}(2)_\mathrm{W}$*, as known from the Standard Model.*

The main problem to be solved to answer the question of the chiral preference of the electroweak interaction, is the following:

* e-mail address: cristi.stoica@theory.nipne.ro

Problem 2. *The model must lead inevitably to the conclusion that leptons and quarks transform under* $\mathrm{SU}(2)_{\mathrm{W}}$ *as doublets, in the left-handed case, and as singlets, in the right-handed case. Their anti-particles should transform under* $\mathrm{SU}(2)_{\mathrm{W}}$ *as singlets, in the left-handed case, and as doublets, in the right-handed case.*

I used the notation $\mathrm{SU}(2)_{\mathrm{W}}$ rather than the common one $\mathrm{SU}(2)_{\mathrm{L}}$, because the latter may "smuggle" into the model the idea that the weak interaction is automatically chiral.

While weak interaction seems to be easier to understand in the case of leptons, also quarks take part of it, so there is a third problem that should be solved along with these ones:

Problem 3. *The model must inevitably include the proper* $\mathrm{SU}(2)_{\mathrm{W}}$ *as doublet or singlet representations for the up and down types of quarks, as well as for the charged and neutral leptons.*

Models addressing Problem 1 include *Grand Unified Theories* (GUT) based on larger simple Lie groups like SU(5) [19] and Spin(10) (sometimes referred to as SO(10)) [18, 15], models based on Clifford algebras [1, 3, 4, 6, 2, 11, 8, 10, 30, 9, 24, 31, 20], models based on octonions [22, 12, 16], sedenions [21], and the exceptional real Jordan algebra of dimension 27 [14, 25], without exhausting the list. The models are different, but there are common structures appearing in different forms – basically making use of the representations of $\mathrm{SU}(2)_{\mathrm{W}}$ on the exterior algebra of a two-dimensional complex space, often this being expressed in terms of Clifford algebras. Most proposed models simply combine the $\mathrm{SU}(2)_{\mathrm{W}}$ doublet representation with the left-handed component of the Dirac spinor, and the $\mathrm{SU}(2)_{\mathrm{W}}$ singlet representations with its right-handed component, and are seen sometimes as explaining chirality. A review of the common structure in terms of Clifford algebras, and of the general method for addressing Problem 1 can be found in [17], for the leptonic case. However, getting the singlet and doublet $\mathrm{SU}(2)_{\mathrm{W}}$ representations and labeling them as "left-handed" and "right-handed", and merely coupling these representations of the internal $\mathrm{SU}(2)_{\mathrm{W}}$ with the correct chiral components of the Dirac spinor, is not enough to explain chiral asymmetry. Chirality is an observable of the Dirac spinors, not of the representations of $\mathrm{SU}(2)_{\mathrm{W}}$, so a true explanation of the chiral preference of the electroweak interaction should also solve Problem 2.

Some of the models mentioned above solve Problem 3 automatically, in particular the SU(5) and Spin(10) Grand Unified Theories[1], the model in [3, 4], and [16].

An important step in answering Problem 2, together with Problems 1 and 3, is made in [4], in which it is shown how chiral spinors corresponding

[1]Although they require an additional symmetry breaking, additional interaction, and predict the proton decay, which was ruled out for reasonable parameters.

to both spacetime and internal degrees of freedom can be made to connect in the proper way the $\mathrm{SU}(2)_{\mathrm{W}}$ representations with chirality. A model addressing Problems 1 and 3, and partially Problem 2, can be found in [26, 28, 29, 27], although this model uses a representation of the Dirac spinors in which time is a scalar, and the chiral preference is obtained by imposing that right-handed neutrinos do not take part in any Standard Model interactions. A closely related model addressing Problems 1, 2, and 3, and which at the same time includes Dirac spinors in the standard form, which is naturally Lorentz invariant, is proposed in [24]. There, the action of the Dirac algebra and of the spin group $\mathrm{Spin}_{1,3}$ break the symmetry, favoring a chiral component, and at the same time leads to a Weinberg angle given by $\sin^2\theta_W = 0.25$. A central part of the present article consists of explaining the structures behind the model presented merely implicitly and in an undetailed form in [24], and showing how this model address Problems 1, 2, and 3.

Another problem, which will not be discussed here, is the existence of three generations, and some models deal with it, for example [4, 16], and [21, 20], which extends [24] to three generations.

This article includes a discussion of the Clifford-algebra-based structures which allow to address the Problems 1, 2, and 3, how various models do this partially, and a more explicit and detailed description of how this was implemented in [24] in an implicit way. Section §2. contains the relations between the representation of the Dirac algebra, and of the groups $\mathrm{SU}(2)_{\mathrm{W}}$ and $\mathrm{SU}(3)_{\mathrm{C}}$, as they appear in the Standard Model, and the Clifford algebra representations. Section §3. explains how they connect to address the Problems 1, 2, and 3. Section §4. describes how these problems are implicitly addressed by the model proposed in [24]. The article concludes with Section §5..

2. Clifford algebra patterns in the Standard Model

This section explains some patterns in the Standard Model, which will be used to address Problems 1, 2, and 3. We will see that the same underlying algebraic structure appears in three different forms, for each of the three problems. In the case of Problems 1 and 3, the structure arises from the fundamental and the trivial representations of $\mathrm{SU}(2)_{\mathrm{W}}$ and $\mathrm{SU}(3)_{\mathrm{C}}$, while in the case of Problem 2 from the representation of the Dirac algebra on one of its ideals. In a first form, the common structure is an exterior algebra, but more refinement leads to the idea that it is in fact a Clifford algebra. The fact that the same type of algebraic structure appears in all three instances, and the fact that these structures appear in one form or another in most proposed models, is not accidental.

2.1. Exterior algebra patterns in the Standard Model

Both the weak symmetry group and the color symmetry group are instances of the special unitary group $\mathrm{SU}(n)$, for $n = 2$, respectively $n = 3$. The standard representation of $\mathrm{SU}(n)$ is the complex vector space $\mathbb{C}_n \cong \mathbb{C}^n$ endowed with the canonical Hermitian metric. The fundamental and the trivial representations of $\mathrm{SU}(n)$ are $\bigwedge^k \mathbb{C}_n$, and they are classified by charges of $\mathrm{U}(1) \subset \mathrm{U}(n)$. By examining the patterns of weak isospin and color of the leptons and quarks, we see that all of the fundamental and trivial representations of $\mathrm{SU}(2)_{\mathrm{w}}$ and $\mathrm{SU}(3)_{\mathrm{c}}$ appear in the Standard Model.

The representations of the weak symmetry group $\mathrm{SU}(2)_{\mathrm{w}}$ present in the Standard Model are displayed in Table 1. They are all present, maybe with the exception of $\bigwedge^0 \mathbb{C}_2$, which corresponds to the still hypothetical right-handed neutrino.

Representation	Particles	Hypercharge
$\bigwedge^0 \mathbb{C}_2$	$(\nu_e)_R$ (?)	0
$\bigwedge^1 \mathbb{C}_2$	$(\nu_e, e^-)_L$	-1
$\bigwedge^2 \mathbb{C}_2$	$(e^-)_R$	-2

Table 1: The representations of the weak symmetry group $\mathrm{SU}(2)_{\mathrm{w}}$.

All of the fundamental and the trivial representations of the color symmetry group $\mathrm{SU}(3)_{\mathrm{c}}$ are present in the Standard Model, as seen in Table 2.

Representation	Particles	Electric charge
$\bigwedge^0 \mathbb{C}_3$	$\overline{\nu}_e$	0
$\bigwedge^1 \mathbb{C}_3$	d^r, d^y, d^b	$-\frac{1}{3}$
$\bigwedge^2 \mathbb{C}_3$	$\overline{u}^{\overline{r}}, \overline{u}^{\overline{y}}, \overline{u}^{\overline{b}}$	$-\frac{2}{3}$
$\bigwedge^3 \mathbb{C}_3$	e^-	-1

Table 2: The representations of the color symmetry group $\mathrm{SU}(3)_{\mathrm{c}}$.

An interesting difference between the representations listed in Table 1 and Table 2 is that in the former, only particles appear, while the second one contains particles alternated with antiparticles. This remark is important to address Problem 3. In both cases, the conjugate representations are also valid, and they apply to the antiparticles of the particles from the tables.

2.2. Dirac spinors and chirality

Let γ^μ be the Clifford basis of the Dirac algebra $\mathbb{C}\ell_4 = C\ell_{1,3} \otimes \mathbb{C}$.

Following [7], we define

$$\begin{cases} e_1 = \frac{1}{2}(\gamma_0 + \gamma_3), e_2 = \frac{1}{2}(-i\gamma_2 + \gamma_1) \\ f_1 = \frac{1}{2}(\gamma_0 - \gamma_3), f_2 = \frac{1}{2}(-i\gamma_2 - \gamma_1). \end{cases} \tag{1}$$

Then, $f_1 f_2$ is nilpotent, and defines a minimal left ideal $\mathbb{C}\ell_4 f_1 f_2$.

In the basis

$$(e_1 f_1 f_2, e_2 f_1 f_2, 1 f_1 f_2, e_1 e_2 f_1 f_2) \tag{2}$$

of the minimal left ideal $\mathbb{C}\ell_4 f_1 f_2$, the matrix form of γ^μ is the Weyl representation.

Let $\mathcal{W}$ be the vector space spanned by (e_1, e_2). Then, the spinors from $\bigwedge^- \mathcal{W} f_1 f_2$ are Weyl spinors of left chirality, and those from $\bigwedge^+ \mathcal{W} f_1 f_2$ are Weyl spinors of right chirality.

Note that the basis (2) is in fact the same as the exterior algebra basis

$$(e_1 f_1 f_2, e_2 f_1 f_2, 1 f_1 f_2, e_1 \wedge e_2 f_1 f_2). \tag{3}$$

This happens because the inner product between Dirac spinors vanishes on the space $\mathcal{W}$ spanned by (e_1, e_2). In fact, this is the reason why we can take as Dirac spinors the exterior algebra $\bigwedge^\bullet \mathcal{W}$, with the basis

$$(e_1, e_2, 1, e_1 \wedge e_2). \tag{4}$$

This fact will turn out to be relevant in the following.

2.3. Representations of complex Clifford algebras, even dimension

We already noticed in subsection §2.2. that there is an equivalence between the representation of the Dirac algebra $\mathbb{C}\ell_4 = C\ell_{1,3} \otimes \mathbb{C}$ by Clifford multiplication on one of its minimal left ideals, and a representation on the exterior algebra $\bigwedge^\bullet \mathcal{W}$. This is in fact true for any complex Clifford algebra $\mathbb{C}\ell_{2r}$ [5, 7].

To see this, let us consider the orthonormal basis of a complex n-dimensional vector space V endowed with a complex inner product,

$$(e_1, \dots, e_r, e_{r+1}, \dots, e_{2r}), \tag{5}$$

where $e_j^2 = 1$.

Then, we can build the *Witt basis* basis

$$\begin{cases} a_j := \frac{1}{2}(e_j + i e_{r+j}) \\ a_j^\dagger := \frac{1}{2}(e_j - i e_{r+j}). \end{cases} \tag{6}$$

The *Witt decomposition* of the vector space V is $V := W \oplus W^\dagger$, where W is spanned by (a_j), and W is spanned by $(a_j^\dagger)$.

Let $a \in \bigwedge^r W$, $a := a_1 \wedge \ldots \wedge a_r = a_1 \ldots a_r$. Then, a is *nilpotent* ($a^2 = 0$), so $\bigwedge^\bullet W^\dagger a$ is a left ideal. In fact, $\bigwedge^\bullet W^\dagger a$ is a minimal left ideal, and its elements are called *algebraic spinors*. It provides an irreducible representation of the Clifford algebra $\mathbb{C}\ell_{2r}$, as we will see.

On the vector space $\bigwedge^\bullet W^\dagger a$ (hence also on the exterior algebra $\bigwedge^\bullet W^\dagger$), $a^\dagger$ and a act like *creation* and *annihilation* operators:

$$\begin{cases} \{a_j, a_k\} = \{a_j^\dagger, a_k^\dagger\} = 0 \\ \{a_j, a_k^\dagger\} = \delta_{jk}. \end{cases} \tag{7}$$

Let $\phi \in \bigwedge^\bullet W^\dagger$. Then,

$$\begin{cases} a_j^\dagger \phi a = a_j^\dagger \wedge \phi a \\ a_j \phi a = i_{a_j} \phi a, \end{cases} \tag{8}$$

where i_{a_j} is the *interior product.*

Since $\dim W^\dagger = 2^r$, this is an irreducible representation of $\mathbb{C}\ell_{2r}$.

Because the inner product vanishes on W, the Clifford product on the subalgebra generated by W coincides with the exterior product. The same holds for $W^\dagger$.

The algebra $\mathbb{C}\ell_{2r}$ is spanned by elements of the form

$$a_{j_1}^\dagger \dots a_{j_p}^\dagger a a^\dagger a_{k_1} \dots a_{k_q}, \tag{9}$$

$p, q \in \{0, \dots, r\}$, $1 \leq j_1 < \dots < j_p \leq r$, $1 \leq k_1 < \dots < k_q \leq r$ [24]. Hence, the Clifford algebra $\mathbb{C}\ell_{2r}$ is isomorphic to the matrix algebra $\mathbb{C}(2^r)$, according to the classification theory [5, 7, 23].

Since the elements of the form $a_{j_1}^\dagger \dots a_{j_p}^\dagger$ span $\bigwedge^\bullet W^\dagger$, $\mathbb{C}\ell_{2r}$ is the direct sum of the minimal left ideals of the form

$$\bigwedge{}^\bullet W^\dagger a a^\dagger a_{k_1} \dots a_{k_q}. \tag{10}$$

On these ideals, $\mathbb{C}\ell_{2r}$ is represented just like on $\bigwedge^\bullet W^\dagger a$ (8).

2.4. From the exterior algebra $\bigwedge^\bullet \mathbb{C}_r$ to Clifford algebra $\mathbb{C}\ell_{2r}$

Conversely, we can start with the complex r-dimensional vector space $\mathbb{C}_r$ endowed with a Hermitian inner product $\mathfrak{h}$. Then, $\mathfrak{h}$ gives a canonical isomorphism $\overline{\mathbb{C}_r} \cong \mathbb{C}_r^*$.

On $\bigwedge^\bullet(\overline{\mathbb{C}_r} \oplus \mathbb{C}_r) \cong \bigwedge^\bullet \overline{\mathbb{C}_r} \otimes_{\mathbb{C}} \bigwedge^\bullet \mathbb{C}_r$, define an associative product by

$$\begin{cases} uv := u \wedge v, & u^\dagger v = u^\dagger \wedge v + \frac{1}{2} u^\dagger(v), \\ u^\dagger v^\dagger := u^\dagger \wedge v^\dagger, & u v^\dagger = u \wedge v^\dagger + \frac{1}{2} v^\dagger(u), \end{cases} \tag{11}$$

for any $u, v \in \bigwedge^\bullet \mathbb{C}_r$.

Then, if (a_j) is an orthonormal basis of $\mathbb{C}_r$,

$$\begin{cases} \{a_j, a_k\} = \{a_j^\dagger, a_k^\dagger\} = 0 \\ \{a_j, a_k^\dagger\} = \delta_{jk}. \end{cases} \tag{12}$$

One obtains a Clifford algebra $\mathbb{C}\ell(\overline{\mathbb{C}_r} \oplus \mathbb{C}_r) \cong \mathbb{C}\ell_{2r}$.

The role of the Hermitian inner product $\mathfrak{h}$ consists in allowing the definition of the adjoint of an element $u \in \bigwedge^{\bullet}\mathbb{C}_r$.

Note that we don't need a Hermitian inner product $\mathfrak{h}$ if we start from $\bigwedge^{\bullet}(\mathbb{C}_r^* \oplus \mathbb{C}_r)$ rather than from $\bigwedge^{\bullet}(\overline{\mathbb{C}_r} \oplus \mathbb{C}_r)$, because we can define the Clifford product, instead of (11), as

$$(u+\xi)(v+\eta) := (u+\xi)\wedge(v+\eta) + \frac{1}{2}\left(\xi(v)+\eta(u)\right), \qquad (13)$$

for any $u, v \in \bigwedge^{\bullet}\mathbb{C}_r$ and $\xi, \eta \in \bigwedge^{\bullet}\mathbb{C}_r^*$.

3. The common structure of the weak, color, and spin group representations

Let us make some remarks surrounding the discussion from Section §2., which will lead to a solution of Problems 1, 2, and 3.

Remark 1. The Dirac spinors are represented by $\bigwedge^{\bullet}\mathbb{C}_2$. Left chiral spinors are represented by $\bigwedge^{-}\mathbb{C}_2$, and right chiral spinors by $\bigwedge^{+}\mathbb{C}_2$.

Remark 2. As seen in Table 1, the weak force acts on the odd part $\bigwedge^{1}\mathbb{C}_2 = \bigwedge^{-}\mathbb{C}_2$ of $\bigwedge^{\bullet}\mathbb{C}_2$ (which is not the same instance of $\bigwedge^{\bullet}\mathbb{C}_2$ as the one in Remark 1). A review of the occurrences of this pattern for the weak interaction in a few models, addressing Problem 1, but not Problems 2 and 3, was presented in [17]

Remark 3. Remarks 1 and 2 suggest that the exterior algebras used to represent Dirac spinors and the weak interaction are subalgebras of a larger space. On this space, the representation of $SU(2)_{\mathrm{w}}$ should leave unchanged the elements representing right chiral particles. Since the representation of $SU(2)_{\mathrm{w}}$ on $\bigwedge^{\bullet}\mathbb{C}_2$ gives the correct doublet and singlets, to represent Dirac spinors we need to double the number of degrees of freedom. The representation of the Lorentz group, which is included in the Dirac algebra, should commute with the representation of $SU(2)_{\mathrm{w}}$.

Remark 4. The representations of $SU(3)_{\mathrm{c}}$ and $U(1)_{\mathrm{em}}$ on the exterior algebra $\bigwedge^{\bullet}\mathbb{C}_3$ and its conjugate are alternating particles with antiparticles (*cf.* Table 2).

Remark 5. From Remarks 2 and 4, the representations of the Lie groups $SU(2)_{\mathrm{w}}$ and of $SU(3)_{\mathrm{c}}$ should be combined into a larger space. The representations of $SU(2)_{\mathrm{w}}$ and $SU(3)_{\mathrm{c}}$ on this larger space should commute. The natural combination is the exterior algebra $\bigwedge^{\bullet}\mathbb{C}_5$, which was used in [19]. The groups $SU(2)_{\mathrm{w}}$ and $SU(3)_{\mathrm{c}}$ were extended to the larger group $SU(5)$, which is included in $\mathrm{Spin}(10)$. The group $\mathrm{Spin}(10)$ was used in [18, 15] with the same $\bigwedge^{\bullet}\mathbb{C}_5$ seen as a representation of $\mathrm{Spin}(10)$.

Remark 6. From Remarks 3 and 4, the representations of the Dirac algebra, of $SU(2)_{\mathrm{w}}$, and of $SU(3)_{\mathrm{c}}$, should be combined into a larger space. The representations of the Lorentz group, $\mathrm{SU}(2)_{\mathrm{w}}$, and $\mathrm{SU}(3)_{\mathrm{c}}$ on this larger space should commute.

Remark 7. The space in Remark 3 can be taken to be $\mathbb{C}_2 \times \bigwedge^\bullet \mathbb{C}_2$, but also $\bigwedge^\bullet \mathbb{C}_1 \times \bigwedge^\bullet \mathbb{C}_2$, which has the same dimension. The space in Remark 6 can be taken to be $\bigwedge^\bullet \mathbb{C}_{1+2+3}$, in order to include all representations (which is larger than $\bigwedge^\bullet \mathbb{C}_5$ of the GUT models SU(5) and Spin(10)). Note that, since the Dirac spinors include the antiparticles, it is enough to take for the colors either $\bigwedge^\bullet \mathbb{C}_3$, or its conjugate, and the same holds for $\bigwedge^\bullet \mathbb{C}_2$. The resulting dimension is the correct one for a generation of leptons and quarks, including their spin degrees of freedom.

Remark 8. A way to obtain the extended representation from Remark 7 is to extend $\mathbb{C}\ell_4$ to $\mathbb{C}\ell_6$ (see Appendix B from [24]).

Remark 9. One way to include the exterior algebra $\bigwedge^\bullet \mathbb{C}_6$ from Remark 7 is to extend the Dirac algebra from $\mathbb{C}\ell_4$ to $\mathbb{C}\ell_{12}$, and to use the representation of the spinors of $\mathbb{C}\ell_{2r}$, where $r = 6$, as in subsection §2.3.. This was proposed in [4]. The representation on the space of spinors of $\mathbb{C}\ell_{12}$ was also used later in [16], where a justification was given in terms of the Dixon algebra $\mathbb{C} \otimes \mathbb{H} \otimes \mathbb{O}$ (also see [12]).

Remark 10. There is a more compact way than the Clifford algebra $\mathbb{C}\ell_{12}$ from Remark 9, which is still as natural. The Clifford algebra $\mathbb{C}\ell_{12}$ is represented as complex linear transformations of a $2^{\frac{12}{2}}$ complex vector space, the space of $\mathbb{C}\ell_{12}$ spinors. But, according to the algebraic definition of spinors, spinors are elements of a minimal left ideal of the Clifford algebra. Since a Clifford algebra $\mathbb{C}\ell_{2r}$ can be decomposed as a direct sum of 2^r minimal left ideals, *cf.* (10), it follows that the total number of degrees of freedom representable as algebraic spinors is $2^r \times 2^r$. Hence, the number of degrees of freedom contained in the spinor of $\mathbb{C}\ell_{12}$ is equal to the number of degrees of freedom contained in the sum of algebraic spinors (10) of the Clifford algebra $\mathbb{C}\ell_6$.

A model which represents the degrees of freedom of a generation of leptons and quarks was proposed in [6]. It includes the electroweak interactions based on $C\ell_{1,6} \cong C\ell_{1,3} \otimes C\ell_{0,3}$, and the ideals are interpreted as representing leptons and quarks, but it seems to lack explicit $\mathrm{SU}(3)_{\mathrm{c}}$ symmetry. In [26, 28, 29, 27], the full symmetries of the Standard Model were shown to arise in a model based on $C\ell_7 \cong C\ell_3 \otimes C\ell_4$, from the condition to preserve the current and that the right-handed neutrino is non-interacting (which means that chiral asymmetry is imposed as a condition, but only for the neutrino, and it extends automatically to the electron and the quarks). The theory includes time as a scalar, and instead of four-vectors it employs *paravectors* – sums of three-vectors and scalars. While this allows to define Lorentz transformations, it does it by considering time as a scalar. In addition to the three space dimensions and the scalar time, which can be expressed using $C\ell_3$, there are four extra dimensions up to $C\ell_7$. The predicted Weinberg angle is given by $\sin^2 \theta_W = 0.375$.

A model which includes all leptons and quarks, together with their Dirac spin degrees of freedom, in the minimal left ideal decomposition of $\mathbb{C}\ell_6$, was proposed in [24]. It shares some common structures, mentioned throughout this article, with other models, but it is different, and it was obtained independently and from different considerations. On the one hand, extending the Dirac algebra to include the weak symmetry $\mathrm{SU}(2)_{\mathrm{w}}$ was shown in [24] to result in $\mathbb{C}\ell_6$. The two actions require $\bigwedge^{\bullet}\mathbb{C}_3$, which can be obtained as a minimal left ideal of $\mathbb{C}\ell_6$. On the other hand, starting from the remark that leptons and quarks, and their antiparticles, behave under $\mathrm{SU}(3)_{\mathrm{c}}$ like the exterior algebra $\bigwedge^{\bullet}\mathbb{C}_3$ and its conjugate (see Remark 4 and Table 2), the tensor product of the two algebras $\bigwedge^{\bullet}\overline{\mathbb{C}}_3 \otimes \bigwedge^{\bullet}\mathbb{C}_3$ turns out to be $\mathbb{C}\ell_6$. This is because the algebra $\bigwedge^{\bullet}\mathbb{C}_3$ used to represent the color $\mathrm{SU}(3)_{\mathrm{c}}$ can be obtained as a minimal right ideal of the same Clifford algebra $\mathbb{C}\ell_6$. The Dirac algebra and the weak $\mathrm{SU}(2)_{\mathrm{w}}$ act at the left, and the color $\mathrm{SU}(3)_{\mathrm{c}}$ at the right, permuting the minimal left ideals. Since they act at the left and at the right, their actions commute. All minimal left ideals of $\mathbb{C}\ell_6$ are then used to represent an entire generation of leptons and quarks. Then, the two ways to reach $\mathbb{C}\ell_6$ correspond to actions on the left and on the right on the ideal decomposition of the algebra. The chiral preference of the weak interaction emerges due to the fact that the vector space $\overline{\mathbb{C}}_3 \oplus \mathbb{C}_3$ has a natural orientation due to the Hermitian inner product $\mathfrak{h}$, combined with a geometric symmetry breaking introduced by the action of the Dirac algebra and of the spin group $\mathrm{Spin}_{1,3}$. This resolves the necessity of introducing additional Higgs mechanisms of spontaneous symmetry breaking, and gives the correct number of degrees of freedom, and the correct symmetries, for a generation.

At the level of matrix representations of the algebras, the real Clifford algebras $C\ell_{1,6}$ and $C\ell_7$, and the complex Clifford algebra $\mathbb{C}\ell_6$, are all represented by the same matrix algebra $\mathbb{C}(8)$, which explains the relation between the three models [6], [26, 28, 29, 27], and [24]. Note that, while the Clifford algebra $\mathbb{C}\ell_6$ also occurs in octonion-based models like [22, 12] and [16], and its ideals are used to index leptons and quarks, and the $\mathrm{SU}(3)_{\mathrm{c}}$ and $U(1)$ symmetries result from $\mathbb{C}\ell_6$, the minimal left ideals of this algebra are not used there to represent the full degrees of freedom of a generation of leptons and quarks. This makes sense indeed, since in these models octonions are fundamental and dictate the symmetries and the representations, and $\mathbb{C}\ell_6$ occurs as operators acting on complex octonions. This is why the model from [16] has to include the full algebra $\mathbb{C} \otimes \mathbb{H} \otimes \mathbb{O}$, and the spinor representation is in terms of $\mathbb{C}\ell_{12}$, and not on the ideal decomposition of $\mathbb{C}\ell_6$, despite its important role in the model for classifying the leptons and quarks.

In the following, I will show in more detail how the considerations made so far about the relation between chirality and the weak interaction are realized in the model [24].

4. The Standard Model algebra

The construction from equation (11) was used in [24], where we started from a complex three-dimensional vector space χ of the colors and obtained a representation of a generation of the Standard Model leptons and quarks, with the correct Dirac, Lorentz, electromagnetic, weak, and strong symmetries, with all required degrees of freedom, and the correct chiralities.

The representation of the color $SU(3)_c$ used in [24] corresponds to the representation in subsection §2.4. for $r = 3$. This decomposes the Clifford algebra $\mathbb{C}\ell_6$ into minimal left ideals as in equation (10), of the form

$$\bigwedge{}^{\bullet}\chi^{\dagger}\mathfrak{q}\mathfrak{q}^{\dagger}\mathfrak{q}_{k_1}\dots\mathfrak{q}_{k_j}, \tag{14}$$

where $(\mathfrak{q}_1, \mathfrak{q}_2, \mathfrak{q}_3)$ is a basis of the vector space χ, $j \in \{0, 1, 2, 3\}$, $k_j \in \{1, 2, 3\}$, and

$$\begin{cases} \mathfrak{q} := & \mathfrak{q}_1\mathfrak{q}_2\mathfrak{q}_3, \\ \mathfrak{q}^{\dagger} := & \mathfrak{q}^{\dagger}_3\mathfrak{q}^{\dagger}_2\mathfrak{q}^{\dagger}_1, \end{cases} \tag{15}$$

which are nilpotent elements. The elements $\mathfrak{p} := \mathfrak{q}\mathfrak{q}^{\dagger}$ and $\mathfrak{p}' := \mathfrak{q}^{\dagger}\mathfrak{q}$ are idempotent. Both the nilpotent and the idempotent elements defined here are primitive, and can be used to define minimal left or right ideals.

If, in (14), $j = 0$ or $j = 3$, the minimal left ideals of the algebra represent pairs of charged and neutral leptons, and if $j = 1$ or $j = 2$ they represent up or down quarks, respectively their antiparticles. They transform like the correct singlet, respectively triplet representations of $SU(3)_c$ under the subgroup of the spin group of $\overline{\chi} \oplus \chi$ which preserves the Hermitian inner product on χ and $\chi^{\dagger}$. This result was based on the results from [13]. The same was obtained in [16], building on the work of [22, 1, 3], and it is based on preserving the Witt decomposition, which in our case is $\overline{\chi} \oplus \chi$. Both representations of $SU(3)_c$ are equivalent and include, in addition to the color symmetry, the $U(1)_{em}$ electromagnetic symmetry.

Since the right action of $SU(3)_c$ is different than the left actions of the Dirac algebra and of $SU(2)_w$, it is justified to choose, for the minimal left ideal $\bigwedge^{\bullet}\chi^{\dagger}q$, a different basis than the one based on q_j used to index the minimal left ideals in (14). The purpose is to make explicit the left actions of the Dirac algebra and of $SU(2)_w$.

For this, let us define first

$$\begin{cases} \mathfrak{e}_j = \mathfrak{q}_j + \mathfrak{q}^{\dagger}_j \\ \tilde{\mathfrak{e}}_j = i\left(\mathfrak{q}^{\dagger}_j - \mathfrak{q}_j\right), \end{cases} \tag{16}$$

where $j \in \{1, 2, 3\}$, and also define

$$\begin{cases} \mathfrak{e} := & \mathfrak{e}_1\mathfrak{e}_2\mathfrak{e}_3, \\ \tilde{\mathfrak{e}} := & \tilde{\mathfrak{e}}_1\tilde{\mathfrak{e}}_2\tilde{\mathfrak{e}}_3. \end{cases} \tag{17}$$

We now define

$$\begin{cases} \mathfrak{u}_{\mathrm{u}} &= -i\mathfrak{e}_3\tilde{\mathfrak{e}}_1, & \mathfrak{u}'_{\mathrm{u}} &= \tilde{\mathfrak{e}}\mathfrak{e}_2, \\ \mathfrak{u}_{\mathrm{d}} &= i\tilde{\mathfrak{e}}, & \mathfrak{u}'_{\mathrm{d}} &= i\mathfrak{e}, \\ \mathfrak{u}_{\circ} &= i\tilde{\mathfrak{e}}_2\mathfrak{e}_3, & \mathfrak{u}'_{\circ} &= \tilde{\mathfrak{e}}\mathfrak{e}_1, \end{cases} \tag{18}$$

where $j \in \{\mathrm{u}, \mathrm{d}, \circ\}$, and

$$\begin{cases} \omega_j = \frac{1}{2}\left(\mathfrak{u}_j + i\mathfrak{u}'_j\right) \\ \omega^{\dagger}{}_j = \frac{1}{2}\left(\mathfrak{u}_j - i\mathfrak{u}'_j\right), \end{cases} \tag{19}$$

The representations of $\mathrm{SU}(2)_{\mathrm{w}}$ from Table 1 are then spanned by the various products involving $\omega^{\dagger}{}_{\mathrm{u}}$ and $\omega^{\dagger}{}_{\mathrm{d}}$, which are the singlets spanned by 1 and $\omega^{\dagger}{}_{\mathrm{u}} \wedge \omega^{\dagger}{}_{\mathrm{d}} = \omega^{\dagger}{}_{\mathrm{u}}\omega^{\dagger}{}_{\mathrm{d}}$, and the doublet spanned by $\omega^{\dagger}{}_{\mathrm{u}}$ and $\omega^{\dagger}{}_{\mathrm{d}}$. But since to each particle also correspond two degrees of freedom coming from the dimension of the Weyl spinor, one needs a way to double the degrees of freedom, according to Remark 3, and this is provided by 1 and $\omega^{\dagger}{}_{\circ}$. From Remark 7 we learn that the inclusion of both leptons and quarks requires the exterior algebra $\bigwedge^{\bullet}\mathbb{C}_3$, which occurs as $\bigwedge^{\bullet}\chi$, and multiplies the minimal left ideal at the right.

We consider the minimal left ideal determined by the idempotent element $\mathfrak{p}$ as split into two four-dimensional space of Dirac spinors,

$$\begin{cases} \mathbb{W}_0 := \mathrm{span}_{\mathbb{C}}\left(\mathfrak{p}, \omega^{\dagger}{}_{\circ}\mathfrak{p}, \omega^{\dagger}{}_{\mathrm{u}}\mathfrak{p}, \omega^{\dagger}{}_{\mathrm{u}}\omega^{\dagger}{}_{\circ}\mathfrak{p}\right) \\ \mathbb{W}_1 := \mathrm{span}_{\mathbb{C}}\left(\omega^{\dagger}{}_{\mathrm{d}}\mathfrak{p}, \omega^{\dagger}{}_{\circ}\omega^{\dagger}{}_{\mathrm{d}}\mathfrak{p}, \omega^{\dagger}{}_{\mathrm{u}}\omega^{\dagger}{}_{\mathrm{d}}\mathfrak{p}, \omega^{\dagger}{}_{\mathrm{u}}\omega^{\dagger}{}_{\circ}\omega^{\dagger}{}_{\mathrm{d}}\mathfrak{p}\right). \end{cases} \tag{20}$$

We also define the following subspace of Weyl spinors,

$$\mathbb{W}_w := \mathrm{span}_{\mathbb{C}}\left(\mathfrak{p}, \omega^{\dagger}{}_{\circ}\mathfrak{p}\right) \tag{21}$$

Then, a generation of leptons and quarks is classified as in Table 3, as shown in [24].

By inspecting Table 3, we can see that the degrees of the bases from the algebras $\bigwedge^{\bullet}\mathbb{C}_2$ and $\bigwedge^{\bullet}\mathbb{C}_3$ used to represent the symmetries $\mathrm{SU}(2)_{\mathrm{w}}$ and $\mathrm{SU}(3)_{\mathrm{c}}$ implement the Remarks from Section §3., by this addressing the Problems 1, 2, and 3. Full details showing how they transform under the $\mathrm{SU}(3)_{\mathrm{c}}$, $\mathrm{SU}(2)_{\mathrm{w}}$, and $\mathrm{U}(1)_{\mathrm{em}}$ groups are given in [24], which addresses implicitly the Problems 1, 2, and 3. Here we explained the underlying mechanisms leading to the fact that the weak force is chiral as we know it for both leptons and quarks, as well as their antiparticles, which was rather implicit in [24].

Particle		ν	$\overline{\mathrm{d}}$	u	e^+
Spinor space		$\mathbb{W}_0$	$\mathbb{W}_0\mathfrak{q}_j$	$\mathbb{W}_0\mathfrak{q}_{jk}$	$\mathbb{W}_0\mathfrak{q}_{123}$
Classifier		$\mathfrak{p}$	$\mathfrak{p}\mathfrak{q}_j$	$\mathfrak{p}\mathfrak{q}_{jk}$	$\mathfrak{p}\mathfrak{q}_{123}$
Electric charge		0	$+\frac{1}{3}$	$+\frac{2}{3}$	$+1$
Chiral space	**L**	$\omega^\dagger_{\mathrm{u}}\mathbb{W}_w$	$\mathbb{W}_w\mathfrak{q}_j$	$\omega^\dagger_{\mathrm{u}}\mathbb{W}_w\mathfrak{q}_{jk}$	$\mathbb{W}_w\mathfrak{q}_{123}$
	R	$\mathbb{W}_w$	$\omega^\dagger_{\mathrm{u}}\mathbb{W}_w\mathfrak{q}_j$	$\mathbb{W}_w\mathfrak{q}_{jk}$	$\omega^\dagger_{\mathrm{u}}\mathbb{W}_w\mathfrak{q}_{123}$
Weak isospin	**L**	$+\frac{1}{2}$	0	$+\frac{1}{2}$	0
	R	0	$+\frac{1}{2}$	0	$+\frac{1}{2}$
Hypercharge	**L**	-1	$+\frac{2}{3}$	$+\frac{1}{3}$	$+2$
	R	0	$-\frac{1}{3}$	$+\frac{4}{3}$	$+1$
Particle		e^-	$\overline{\mathrm{u}}$	d	$\overline{\nu}$
Spinor space		$\mathbb{W}_1$	$\mathbb{W}_1\mathfrak{q}_j$	$\mathbb{W}_1\mathfrak{q}_{jk}$	$\mathbb{W}_1\mathfrak{q}_{123}$
Classifier		$\mathfrak{q}^\dagger\mathfrak{p}$	$\mathfrak{q}^\dagger\mathfrak{p}\mathfrak{q}_j$	$\mathfrak{q}^\dagger\mathfrak{p}\mathfrak{q}_{jk}$	$\mathfrak{q}^\dagger\mathfrak{p}\mathfrak{q}_{123}$
Electric charge		-1	$-\frac{2}{3}$	$-\frac{1}{3}$	0
Chiral space	**L**	$\omega^\dagger_{\mathrm{d}}\mathbb{W}_w$	$\omega^\dagger_{\mathrm{u}}\omega^\dagger_{\mathrm{d}}\mathbb{W}_w\mathfrak{q}_j$	$\omega^\dagger_{\mathrm{d}}\mathbb{W}_w\mathfrak{q}_{jk}$	$\omega^\dagger_{\mathrm{u}}\omega^\dagger_{\mathrm{d}}\mathbb{W}_w\mathfrak{q}_{123}$
	R	$\omega^\dagger_{\mathrm{u}}\omega^\dagger_{\mathrm{d}}\mathbb{W}_w$	$\omega^\dagger_{\mathrm{d}}\mathbb{W}_w\mathfrak{q}_j$	$\omega^\dagger_{\mathrm{u}}\omega^\dagger_{\mathrm{d}}\mathbb{W}_w\mathfrak{q}_{jk}$	$\omega^\dagger_{\mathrm{d}}\mathbb{W}_w\mathfrak{q}_{123}$
Weak isospin	**L**	$-\frac{1}{2}$	0	$-\frac{1}{2}$	0
	R	0	$-\frac{1}{2}$	0	$-\frac{1}{2}$
Hypercharge	**L**	-1	$-\frac{4}{3}$	$+\frac{1}{3}$	0
	R	-2	$-\frac{1}{3}$	$-\frac{2}{3}$	1

Table 3: Properties of leptons and quarks in the model based on the Clifford algebra $\mathbb{C}\ell_6$ from [24].

5. Conclusions

We have seen the underlying mechanisms connecting chirality with the weak force and the notion of particle-antiparticle. We have seen how these relations are accounted for by representations on exterior algebras and Clifford algebras. The chiral preference of the weak interaction does not simply come from calling the singlet representations right-handed and the doublet ones left-handed, or from merely adding the subscript L to SU(2), or even simply declaring that only the left chiral part of the Dirac spinor has to participate in the weak interaction. This points out to the idea that there should be an intimate relation between the Dirac spinors and the internal degrees of freedom, geometric in nature. We have seen that to have an answer, one has to address Problems 1, 2, and 3, and how these problems were addressed to various degrees in several models. In all these examples, the exterior algebra or the Clifford algebra representations turned out to be essential. An account of how the weak force turns out to be chiral, which is implicit and automatically present in the model proposed in [24] was given more explicitly here.

References

[1] A Barducci, F Buccella, R Casalbuoni, L Lusanna, and E Sorace, *Quantized grassmann variables and unified theories*, Phys. Lett. B **67** (1977), no. 3, 344–346.

[2] J Besprosvany, *Gauge and space-time symmetry unification*, Int. J. Theor. Phys. **39** (2000), no. 12, 2797–2836.

[3] R Casalbuoni and R Gatto, *Unified description of quarks and leptons*, Phys. Lett. B **88** (1979), no. 3-4, 306–310.

[4] R Casalbuoni and R Gatto, *Unified theories for quarks and leptons based on Clifford algebras*, Phys. Lett. B **90** (1980), no. 1-2, 81–86.

[5] C. Chevalley, *The algebraic theory of spinors and Clifford algebras (Collected works)*, vol. 2, Springer, 1997.

[6] JSR Chisholm and RS Farwell, *Properties of Clifford algebras for fundamental particles*, Clifford (Geometric) Algebras: With Applications to Physics, Mathematics, and Engineering (W E Baylis, ed.), Birkhäuser Boston, Boston, MA, 1996, pp. 365–388.

[7] A. Crumeyrolle, *Orthogonal and symplectic Clifford algebras. Spinor structures*, Kluwer Academic Publishers (Dordrecht and Boston), 1990.

[8] C Daviau, *Retour à l'onde de Louis de Broglie*, Ann. Fond. Louis Broglie, vol. 40, 2015, p. 113.

[9] C Daviau, *Gauge group of the Standard Model in* $C\ell_{1,5}$, Adv. Appl. Clifford Algebras **27** (2017), no. 1, 279–290.

[10] C Daviau and J Bertrand, *Electro-weak gauge, Weinberg-Salam angle*, Journal of Modern Physics **6** (2015), no. 14, 2080.

[11] C Daviau and J Bertrand, *The Standard Model of quantum physics in Clifford algebra*, World Scientific, 2015.

[12] G M Dixon, *Division algebras: Octonions, quaternions, complex numbers and the algebraic design of physics*, vol. 290, Springer Science & Business Media, 2013.

[13] C Doran, D Hestenes, F Sommen, and N Van Acker, *Lie groups as spin groups*, J. Math. Phys. **34** (1993), no. 8, 3642–3669.

[14] M Dubois-Violette, *Exceptional quantum geometry and particle physics*, Nucl. Phys. B **912** (2016), 426–449.

[15] H Fritzsch and P Minkowski, *Unified interactions of leptons and hadrons*, Ann. Phys. **93** (1975), no. 1-2, 193–266.

[16] C Furey, *Standard Model physics from an algebra?*, Preprint arXiv:1611.09182 (2016).

[17] C Furey, *A demonstration that electroweak theory can violate parity automatically (leptonic case)*, Int. J. Mod. Phys. A. **33** (2018), no. 04, 1830005.

[18] H Georgi, *State of the art – gauge theories*, AIP (Am. Inst. Phys.) Conf. Proc., no. 23, pp. 575-582, Harvard Univ., Cambridge, MA, 1975.

[19] H Georgi and SL Glashow, *Unity of all elementary-particle forces*, Phys. Rev. Lett. **32** (1974), no. 8, 438.

[20] AB Gillard and NG Gresnigt, *The $\mathbb{C}\ell(8)$ algebra of three fermion generations with spin and full internal symmetries*, arXiv preprint arXiv:1906.05102 (2019).

[21] AB Gillard and NG Gresnigt, *Three fermion generations with two unbroken gauge symmetries from the complex sedenions*, Eur. Phys. J. C **79** (2019), no. 5, 446.

[22] M Günaydin and F Gürsey, *Quark statistics and octonions*, Phys. Rev. D **9** (1974), no. 12, 3387.

[23] H. B. Lawson and M. L. Michelsohn, *Spin geometry*, Princeton University Press, 1989.

[24] O.C. Stoica, *Leptons, quarks, and gauge from the complex Clifford algebra $\mathbb{C}\ell_6$ (the Standard Model Algebra)*, Adv. Appl. Clifford Algebras **28** (2018), no. 3, 52, arXiv:1702.04336.

[25] I Todorov and M Dubois-Violette, *Deducing the symmetry of the Standard Model from the automorphism and structure groups of the exceptional Jordan algebra*, (2017).

[26] G Trayling, *A geometric approach to the Standard Model*, Preprint arXiv:hep-th/9912231 (1999).

[27] G Trayling, *An algebraic basis for the standard-model gauge group. Disertation*, (2000).

[28] G Trayling and WE Baylis, *A geometric basis for the standard-model gauge group*, J. Phys. A: Math. Theor. **34** (2001), no. 15, 3309.

[29] G Trayling and WE. Baylis, *The Cl_7 approach to the Standard Model*, Clifford Algebras: Applications to Mathematics, Physics, and Engineering (Rafał Abłamowicz, ed.), Birkhäuser Boston, Boston, MA, 2004, pp. 547–558.

[30] P Żenczykowski, *Elementary particles and emergent phase space*, World Scientific, 2013.

[31] P Żenczykowski, *From Clifford algebra of nonrelativistic phase space to quarks and leptons of the Standard Model*, Adv. Appl. Clifford Algebras **27** (2017), no. 1, 333–344.

Holographic inflation with tachyon field as an attractor solution*

Marko Stojanović†
Faculty of Medicine, University of Niš, Serbia
Milan Milošević‡
Faculty of Science and Mathematics, University of Niš, Serbia
Goran Đorđević§
Faculty of Science and Mathematics, University of Niš, Serbia
Dragoljub D. Dimitrijević¶
Faculty of Science and Mathematics, University of Niš, Serbia

Abstract

We consider the attractor behaviour of the tachyon field dynamics in the holographic inflation cosmology. Our focus is on the model with exponential potential. The solutions of the dynamical equation in the phase-space are examined and we show they correspond to a dynamical attractor.

1. Introduction

Cosmological inflation is the leading theory for resolving some of the problems in the standard Big Bang model. Inflation predicts a short epoch during which the universe has suddenly expanded in size. The generated quantum fluctuations provided the seeds of cosmic microwave background anisotropy and the large scale structure [1].

One of the main features of the inflationary models, as well as one of the reasons to introduce inflation theory, is the assumption that the evolution of the universe, which can be described by a scalar field, is independent of the initial conditions. This property is regarded to the notion of cosmological

*This work has been supported by the ICTP - SEENET-MTP project NT-03 Cosmology-Classical and Quantum Challenges. The authors acknowledge support provided by the Serbian Ministry for Education, Science and Technological Development, as well as the COST actions CA18108 and CA15177.
† e-mail address: marko.stojanovic@pmf.edu.rs
‡ e-mail address: mmilan@seenet-mtp.info
§ e-mail address: gorandj@junis.ni.ac.rs
¶ e-mail address: ddrag@pmf.ni.ac.rs

attractors: dynamical conditions under which the evolving scalar fields approach a certain kind of behavior without fine tuning [2].

We analyze the properties of the attractor solutions in a model of inflation with tachyonic field in the framework of holographic cosmology. The use of the tachyon field in the inflation models is inspired by string theory [3]. We consider dynamics of the holographic model on a D3-brane located at the holographic boundary of an asymptotic ADS_5 bulk. Dynamics can be describe by effective four-dimensional Einstein equations on the holographic boundary [4].

2. Tachyon dynamics in the holographic braneworld

Under assumption that the holographic braneworld is spatially flat, the line element is of the form

$$ds^2 = g_{\mu\nu}dx^\mu dx^\nu = dt^2 - a^2(t)(dr^2 + r^2 d\Omega^2). \tag{1}$$

In this case, following the procedure given in the ref. [5], the holographic Friedmann equations can be derived from effective four-dimensional Einstein equations

$$h^2 - \frac{1}{4}h^4 = \frac{\kappa^2}{3}\ell^4\rho, \tag{2}$$

$$\dot{h}\left(1 - \frac{1}{2}h^2\right) = -\frac{\kappa^2}{2}\ell^3(p+\rho), \tag{3}$$

where ℓ is the AdS curvature radius and $\kappa^2 = 8\pi G_{\rm N}/\ell^2$ is the fundamental dimensionless coupling. The overdot denotes derivative with respect to a dimensionless time variable $\tau = t/\ell$, while a dimensionless expansion rate is $h = \dot{a}/a$ [5]. The time evolution of the energy density ρ and the pressure p of the tachyon field θ is governed by the Lagrangian of the Dirac-Born-Infeld type [6]

$$\mathcal{L} = -\ell^{-4}V(\theta/\ell)\sqrt{1 - g^{\mu\nu}\theta_{,\mu}\theta_{,\nu}}\,, \tag{4}$$

i.e.

$$\rho = \frac{\ell^{-4}V}{\sqrt{1-\dot{\theta}^2}}, \quad p = -\ell^{-4}V\sqrt{1-\dot{\theta}^2}. \tag{5}$$

Here, V denotes a tachyon potential. The field θ is already redefined, so that its dimension is l. The energy conservation equation

$$\dot{\rho} + 3\frac{h}{l}(\rho + p) = 0, \tag{6}$$

gives a second order differential equation for $\theta(\tau)$

$$\frac{\ddot{\theta}}{1-\dot{\theta}^2} + 3\frac{h}{l}\dot{\theta} + \frac{V_{,\theta}}{V} = 0, \tag{7}$$

where $_{,\theta}$ denotes derivative with respect to theta. The equation (2) gives two solutions, but in the low density limit ($\kappa^2\ell^4\rho \ll 1$) only solution [5, 6]

$$h^2 = 2\left(1 - \sqrt{1 - \frac{\kappa^2}{3}\ell^4\rho}\right), \tag{8}$$

is reducible to the Friedmann equation in the standard cosmology. It follows that the physical range of the expansion rate is $0 \le h^2 \le 2$.

In the slow-role regime, when the field changes slowly under time due to the conditions [5]

$$\dot{\theta}^2 \ll 1, \quad |\ddot{\theta}| \ll 3\frac{h}{\ell}\dot{\theta}, \tag{9}$$

and while $\dot{\theta}$ can be approximated by

$$\dot{\theta} \simeq -\frac{\ell V_{,\theta}}{3hV}, \tag{10}$$

the equation (8) takes the form

$$h^2 \simeq 2\left(1 - \sqrt{1 - \frac{\kappa^2}{3}V}\right). \tag{11}$$

There are several ways to define the slow-roll parameters to ensure that the field changes slowly in time. We use the definition for the horizon-flow parameters ε_j given in the ref. [7]

$$\varepsilon_0 = \frac{h_*}{h}, \qquad \varepsilon_{i+1} = \frac{d\ln|\varepsilon_i|}{dN}, \quad i \ge 0, \tag{12}$$

where h_* is the Hubble rate at some chosen time and N is the number of e-folds. During inflation we have $\varepsilon_1 < 1$, $\varepsilon_2 < 1$ and inflation ends once either of the two parameters exceeds unity.

3. Attractor behavior - general consideration

To analyze the dynamics of the tachyon field it is useful to remind on the properties of an attractor behavior.

The definition of an attractor in the cosmological context is argued in ref. [2]. In cosmology, the concept of an attractor solution is introduced slightly different and less rigorous than in the mathematical literature [8]. Most often the term attractor is related to the specific properties of the depicted solutions in the phase space. Although the properties of the attractors depend on the choice of coordinates, the phase space we are going to work with will be represented as $(\theta, \dot{\theta})$ plane, as it is usually done.

Regarding to the ref. [9] the attractor solution can be found demanding

$$\frac{d\dot{\theta}}{d\theta} \simeq 0. \tag{13}$$

The equation (7) can be rewritten in the form

$$\frac{d\dot{\theta}}{d\theta} \simeq -\left(3\frac{h}{l}\dot{\theta} + \frac{V_{,\theta}}{V}\right)\frac{1-\dot{\theta}^2}{\dot{\theta}}, \tag{14}$$

where $\frac{d\dot{\theta}}{d\theta}$ is equal to zero if the expression in the parenthesis is equal to zero. This requirement leads to the approximate expression (10) and has been fulfilled only during the slow-roll regime. This is consistent with the well known fact that the slow-roll approximation provides an attractor solution ref. [10].

4. Attractor solutions

In a similar manner, as it was proposed in the ref. [11], we demonstrate the attractor behaviour of our model. We consider the expansion rate as a function of the inflaton field $h = h(\theta)$. Using equations (2) and (3), the time derivative of the tachyon field, in terms of the dimensionless expansion rate and its first derivative with respect to tachyon field, is given by

$$\dot{\theta} = -\frac{2}{3}\ell\frac{1-\frac{1}{2}h^2}{h^2-\frac{1}{4}h^4}h_{,\theta}\,, \tag{15}$$

This allows us to rewrite the equation (2) as the first order ordinary differential equations

$$9(h^2-\frac{1}{4}h^4)^2 - 4\ell^2(1-\frac{1}{2}h^2)^2 h_{,\theta}^2 = \kappa^4 V^2. \tag{16}$$

The equation (16) is of the Hamilton-Jacobi type equation. Suppose that $h_0(\theta)$ is its solution in the case of a general potential, which is able to support inflation. In order to examine the attractor behavior of θ, we consider the expansion rate h. Taking the perturbation $\delta h(\theta)$ of the solution $h_0(\theta)$ for the expansion rate

$$h(\theta) = h_0(\theta) + \delta h(\theta), \tag{17}$$

and substituting (17) in the equation (16), keeping only the terms up to the first order of the perturbation, we find

$$\frac{9}{4}h^3\frac{4-h^2}{2-h^2}\frac{d\theta}{h_{,\theta}} = \frac{d\delta h}{\delta h}. \tag{18}$$

Integrating the above expression, using the equation (15) to expres $h_{,\theta}$, and the equation (10) for $\dot{\theta}$, one gets

$$\delta h(\theta) = \delta h(\theta_i) e^{-3N}, \tag{19}$$

where the number of e-folds N is defined as in the reference [5]

$$N \equiv \frac{1}{\ell} \int_{t_i}^{t_f} h dt \simeq -3 \int_{\theta_i}^{\theta_f} \frac{h^2 V}{\ell^2 V_{,\theta}} d\theta. \tag{20}$$

The subscripts i and f are regarded to the beginning and to the end of inflation, respectively. To solve the horizon and the flatness problem, we need about 50-60 e-folds. The perturbation δh decreases as we approach the end of inflation, and the model possesses the attractor behavior. Therefore, the all solutions quickly approach each other and they are independent of initial conditions, as we demonstrate in the following section.

5. Attractor behavior with exponential potential

We consider the following exponential potential

$$V = e^{-\omega|\theta|/\ell}, \tag{21}$$

where ω is a dimensionless parameter. This potential is one of the most frequently used for the tachyon systems [12]. The value of the parameter ω is determined by the initial value of expansion rate h_i and the number of e-folds N. From the equation (20), and using the condition $\varepsilon_{2f} = 1$ for the end of inflation, one gets an approximate expression [5]

$$\omega^2 \simeq \frac{12\left[\frac{h_i^2}{2} + \ln\left(1 - \frac{h_i^2}{4}\right)\right]}{N+1}. \tag{22}$$

The initial value of the field can be estimated from (11), giving [5]

$$\theta_i \simeq -\frac{1}{\omega} \ln\left[\frac{3}{\kappa^2}\left(1 - (1 - \frac{h_i^2}{2})^2\right)\right]. \tag{23}$$

To examine behavior of the inflationary solution in the phase-space we need to find the solution of the equation (14), which for the exponential potential takes the form

$$\frac{d\dot{\theta}}{d\theta} \simeq -\frac{1}{l}\left(3\frac{h}{l}\dot{\theta} - \omega\right)\frac{1 - \dot{\theta}^2}{\dot{\theta}}. \tag{24}$$

The equation (24) is solved numerically for a variety of arbitrary chosen initial values. Using equations (10) and (11) one can find the relation between θ and $\dot{\theta}$ which is valid in the slow-roll regime

$$\dot{\theta} \simeq \frac{\omega}{\sqrt{18(1-\sqrt{1-\frac{\kappa^2}{3}e^{-\omega|\theta|/\ell}})}}. \tag{25}$$

As it is shown in the section 3., slow roll approximation is necessary in order to provide a solution that is an attractor solution. Although the equation (14) is independent of κ, we need to choose a numerical value for κ in order to set the initial value for $\dot{\theta}$, and we choose $\kappa = 1$.

In our model inflation ends when $\varepsilon_{2f} = 1$, and the inflaton field reaches the value [5]

$$\theta_f = \frac{\ell}{\omega} \ln \frac{\kappa^2}{\omega^2}. \tag{26}$$

The obtained diagram, in $(\theta, \dot{\theta})$ plane, is presented in the (Fig. 1).

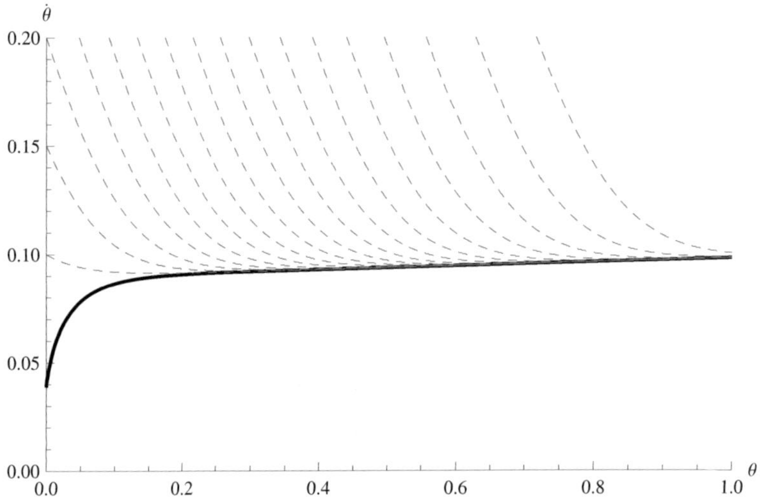

Figure 1: $\dot{\theta}$ versus θ diagram, calculeted numerically for several initial values of $\dot{\theta}$ and fixed parameter $\omega = 0.164$. The solid line represents the solution obtained in respect to the initial values constrained by the equation (25).

For many different choices of the initial conditions the phase space trajectories converge at late time and there exists a curve that attracts most of the trajectories. According to the criteria given in the reference [2] the solution in the model behaves as an attractor.

6. Conclusions

In this paper, we consider the inflation model with a tachyon field in holographic cosmology and it is shown that the solutions for the tachyon field θ have properties of the an attractor. It is shown that the use of the Hamilton-Jacobi equation (16) is very effective in this consideration. Also, in the case of the exponential potential a numerical solution of the dynamical equation is found for a wide interval of the initial values and plotted in a the phase space $(\theta, \dot{\theta})$. The numerical results fully confirm the existence of the attractor behaviour.

It is an interesting task to examine and demonstrate the attractor behaviour for other tachyon-type potentials, and to check if any possible violation of the attractor behaviour may appear.

References

[1] A. H. Guth, *The Inflationary Universe: A Possible Solution to the Horizon and Flatness Problems*, Phys. Rev. D **23** (1981) 347.

[2] G. N. Remmen and S. M. Carroll, *Attractor Solutions in Scalar-Field Cosmology*, Phys. Rev. D **88** (2013) 083518.

[3] A. Feinstein, *Power law inflation from the rolling tachyon*, Phys. Rev. D **66** (2002) 063511.

[4] N. Bilic, *Randall-Sundrum versus holographic cosmology*, Phys. Rev. D **93** (2016) no.6, 066010.

[5] N. Bilic, D. D. Dimitrijevic, G. S. Djordjevic, M. Milosevic and M. Stojanovic, *Tachyon inflation in the holographic braneworld*, JCAP **1908** (2019) 034.

[6] N. Bilic, *Holographic cosmology and tachyon inflation*, Int. J. Mod. Phys. A **33** (2018) no.34, 1845004.

[7] D. A. Steer and F. Vernizzi, *Tachyon inflation: Tests and comparison with single scalar field inflation*, Phys. Rev. D **70** (2004) 043527.

[8] J. Milnor, *On the concept of attractor*, Commun.Math. Phys. **99**, (1985) 177-195.

[9] V. Mukhanov, *Physical Foundations of Cosmology*, Cambridge University Press, 2012.

[10] A. R. Liddle, P. Parsons and J. D. Barrow, *Formalizing the slow roll approximation in inflation*, Phys. Rev. D **50** (1994) 7222.

[11] Z. K. Guo, Y. S. Piao, R. G. Cai and Y. Z. Zhang, *Inflationary attractor from tachyonic matter*, Phys. Rev. D **68** (2003) 043508.

[12] A. Sen, *Field theory of tachyon matter*, Mod. Phys. Lett. A **17** (2002) 1797.

Highly entangled quantum spin chains*

Fumihiko Sugino†
Center for Theoretical Physics of the Universe,
Institute for Basic Science, Expo-ro 55,
Yuseong-gu, Daejeon 34126, Rebublic of Korea

Abstract

We mainly discuss a highly entangled quantum spin chain with local interactions, called as (colored) Motzkin spin chain, in which entanglement entropy for its ground state grows as a square root of the volume. Since this is beyond logarithmic behavior in the ordinary critical systems, it is important to study such a model to reveal novel features of quantum entanglement. We explain how the model yields the extraordinary growth of the entanglement entropy, and then compute the Rényi entropy for the same model. As a result, we find a new phase transition with respect to the Rényi parameter, which has been never seen in any other spin chain studied so far.
PACS: 02.30.Ik, 03.65.Ud, 03.67.Bg, 05.30.-d, 12.40.Ee

1. Introduction

Quantum entanglement is one of the most surprising features of quantum theory, and gives correlations that cannot be explained in classical mechanics. This comes from the fact that superpositions of states are possible in quantum mechanics. As a measure of the entanglement, entanglement entropy (EE) is defined as follows. First, we divide a total system S into a subsystem A and the rest B. Starting with the density matrix of the total system ρ, we obtain the reduced density matrix of A by tracing out ρ by the Hilbert space belonging to B: $\rho_A = \mathrm{Tr}_B\, \rho$. Then, the EE is defined as the von Neumann entropy of ρ_A:

$$S_A = -\mathrm{Tr}_A\, \rho_A \ln \rho_A. \tag{1}$$

In case that the system S has a unique ground state, its density matrix is a pure state ($\rho = |\mathrm{GS}\rangle\langle\mathrm{GS}|$) and its von Neumann entropy is zero. However, the reduced density matrix ρ_A can become a mixed state and the EE can

* The author is grateful to all the organizers and participants of the 10th Mathematical Physics Meeting (MPHYS10) for their warm hospitality and discussions. He also thanks Vladimir Korepin and Pramod Padmanabhan for collaborations and discussions.

† e-mail address: fusugino@gmail.com

be nonzero. The mixed state reflects information of interactions between A and B, some of which is measured by the EE. Thus, different behavior of the EE implies difference of the structure of the ground state, i.e., different phase structure, and the EE can be regarded as a new order parameter of the system.

Let us consider ground states of quantum many-body systems with local interactions. Typically, the EE of such systems grows proportionally to the area of the boundary of A and B ("Area law") [1]. For systems with gap, it is naturally understood since correlation length is finite and interactions contributing to the EE localize around the boundary. A mathematical proof of the area law has been given in gapped systems in one spatial dimension [2]. Actually, in conformal field theory (CFT) in $(1+1)$ dimensions, which is gapless, logarithmic violation of the area law has been observed [3, 4, 5], where the EE grows as $\ln L$ with L length scale of the subsystem A [1].

For gapless systems in general spatial dimensions (D), it had been believed that the EE violates the area law ($S_A = O(L^{D-1})$) by at most logarithmic corrections ($S_A = O(L^{D-1}\ln L)$). Recently, Movassagh and Shor has constructed a spin chain model that gives a counterexample for the belief. In their model, called as colored Motzkin spin chain, the EE violates the area law by a square-root correction ($S_A = O(\sqrt{L})$) that is much larger than the logarithmic behavior [6]. Since this behavior is beyond the behavior for the ordinary critical systems, they called it as supercritical entanglement.

In this article, we explain the spin chain given by Movassagh and Shor, and discuss how such a large correction arises in the EE. Then, we compute the Rényi entropy of the spin chain and find a new phase transition that has not been observed in any other spin chain.

This paper is organized as follows. We introduce the Motzkin spin chain in section 2. and its colored version in section 3. with computing the EE for each case. After introducing the Rényi entropy in section 4., we compute the Rényi entropy for colorless and colored Motzkin spin chains in sections 5. and 6.. Section 7. is devoted to a summary and possible future directions.

2. Motzkin spin chain

Let us consider a quantum spin chain defined on sites of the one-dimensional lattice with length $2n$: $\{1, 2, \cdots, 2n\}$. The local Hilbert space at each site consists of spin-1 degrees of freedom, up $|u\rangle$, zero $|0\rangle$ and down $|d\rangle$, which we can correspond to up, flat and down steps, respectively. Then, each spin configuration corresponds to a length $2n$ walk on the two-dimensional (x, y)-plane. Fig. 1 shows an example for length $2n = 6$.

[1] In one spatial dimension, the boundary of A and B is point(s), and the area law means that the EE asymptotically approaches to some finite constant.

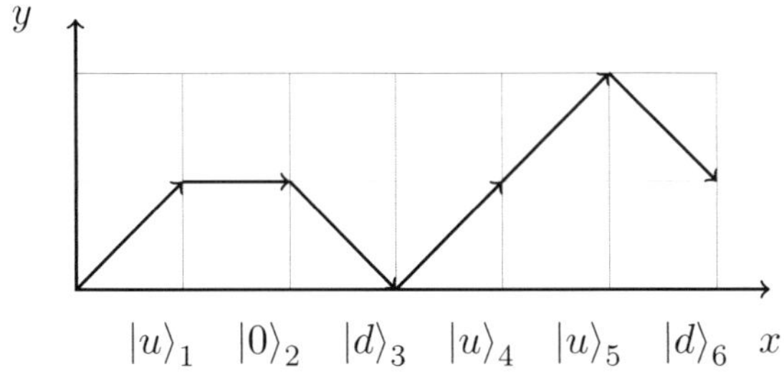

Figure 1: An example of spin configurations for length 6, and its corresponding walk. Indices of the state vectors indicate sites.

The Motzkin spin chain (or colorless Motzkin spin chain) is defined by the Hamiltonian [7]

$$H_{\mathrm{M}} = \sum_{j=1}^{2n-1} \left\{ |D\rangle_{j,j+1}\langle D| + |U\rangle_{j,j+1}\langle U| + |F\rangle_{j,j+1}\langle F| \right\} + |d\rangle_1\langle d| + |u\rangle_{2n}\langle u| \tag{2}$$

with

$$|D\rangle = \frac{1}{\sqrt{2}}\left(|0\rangle|d\rangle - |d\rangle|0\rangle\right), \qquad |U\rangle = \frac{1}{\sqrt{2}}\left(|0\rangle|u\rangle - |u\rangle|0\rangle\right),$$
$$|F\rangle = \frac{1}{\sqrt{2}}\left(|0\rangle|0\rangle - |u\rangle|d\rangle\right). \tag{3}$$

In (2), three terms inside of the sum are projection operators acting to adjacent sites j and $j+1$ as indices indicate, whereas the last two terms are boundary terms acting to sites 1 and $2n$. The Hamiltonian consists of a sum of local projection operators. Since projection operators are positive-semi definite operators, the Hamiltonian is also. Therefore, if we find a zero-energy eigenstate for (2), it must be a ground state of the system. Such a ground state is a zero-eigenstate of each local projection operator in (2). Let us express such a zero-energy ground state in terms of walks. First, the boundary terms forbid a down (up) step at the left (right) edge. Second, the projection operators with respect to the three states (3) mean that a sum of walks corresponding to the ground state should be invariant under the three moves at any adjacent two steps:

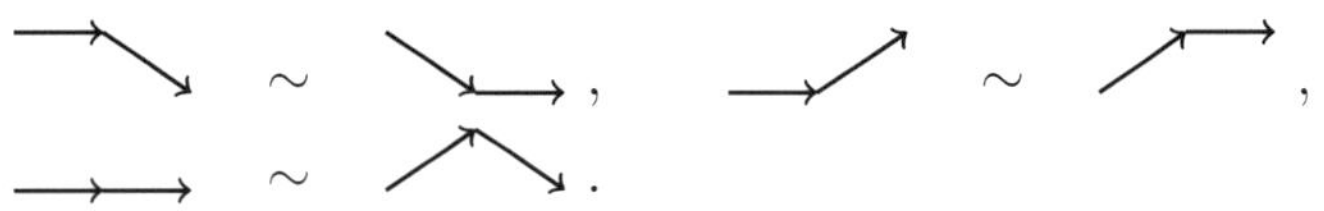
(4)

From these, it turns out that the zero-energy ground state is unique and corresponds to random walks from $(0,0)$ to $(2n,0)$ without entering the region $y < 0$. Such a kind of random walks is called as Motzkin walk, which is the origin of the name of the spin chain. For example, the ground state for length $2n = 4$ is

$$\begin{aligned} |P_4\rangle &= \frac{1}{\sqrt{9}}[|0000\rangle + |ud00\rangle + |0ud0\rangle + |00ud\rangle \\ &\quad +|u0d0\rangle + |0u0d\rangle + |u00d\rangle \\ &\quad +|udud\rangle + |uudd\rangle] \end{aligned} \tag{5}$$

and the corresponding Motzkin walk is given as

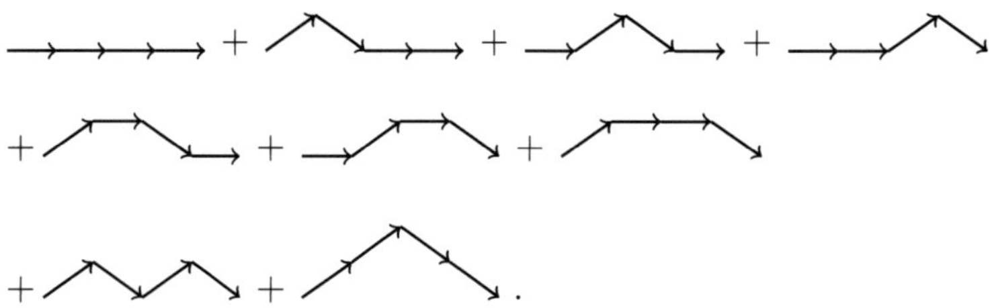

(6)

We pick a left-half of the system $\{1, 2, \cdots, n\}$ as a subsystem A and compute the EE. The ground state is decomposed as a linear combination of the tensor product of a state belonging to A and a state belonging to the rest $B = \{n+1, \cdots, 2n\}$ (Schmidt decomposition):

$$|P_{2n}\rangle = \sum_{h=1}^{n} \sqrt{p_{n,n}^{(h)}} \left|P_n^{(0\to h)}\right\rangle \otimes \left|P_n^{(h\to 0)}\right\rangle, \tag{7}$$

where $\left|P_n^{(0\to h)}\right\rangle$ is a sum of states, corresponding to length-n walks from $(0,0)$ to (n,h) and belonging to A, while $\left|P_n^{(h\to 0)}\right\rangle$ belongs to B. By using the number of the walks of $|P_{2n}\rangle$ denoted by M_{2n} and that of $\left|P_n^{(0\to h)}\right\rangle$ by $M_n^{(h)}$, $p_{n,n}^{(h)}$ in the coefficient is expressed as $p_{n,n}^{(h)} = \left(M_n^{(h)}\right)^2 / M_{2n}$. From combinatorics, the numbers are given by

$$\begin{aligned} M_{2n} &= \sum_{k=0}^{n} \frac{1}{k+1} \binom{2k}{k} \binom{2n}{2k}, \\ M_n^{(h)} &= \sum_{r=0}^{n-h} \frac{1+(-1)^{n-r+h}}{2} \binom{n}{r} \frac{h+1}{\frac{n-r+h}{2}+1} \binom{n-r}{\frac{n-r+h}{2}}, \end{aligned} \tag{8}$$

and $p_{n,n}^{(h)}$ asymptotically behaves as

$$p_{n,n}^{(h)} \sim \frac{3\sqrt{6}}{\sqrt{\pi}} \frac{(h+1)^2}{n^{3/2}} e^{-\frac{3}{2}\frac{(h+1)^2}{n}} \times \left[1 + O(n^{-1})\right] \qquad (n \to \infty). \tag{9}$$

From the density matrix of the ground state $\rho = |P_{2n}\rangle\langle P_{2n}|$, the EE is obtained as

$$S_A = -\sum_{h=0}^{n} p_{n,n}^{(h)} \ln p_{n,n}^{(h)}, \tag{10}$$

whose asymptotic form becomes

$$S_A \sim \frac{1}{2}\ln n + \frac{1}{2}\ln\frac{2\pi}{3} + \gamma - \frac{1}{2} \tag{11}$$

with γ being the Euler constant. We suppressed terms vanishing as $n \to \infty$.

Although (11) violates the area law logarithmically, the system cannot be described by relativistic CFT. Another investigation shows that the gap of the system behaves like $O(n^{-z})$ with $z \geq 2$, whereas the gap of the relativistic CFT scales as $O(n^{-1})$ [6]. We will present another evidence for this issue in computing Rényi entropy.

3. Colored Motzkin spin chain

In this section, we discuss a spin chain obtained by adding color degrees of freedom to the up and down spins in the model in the previous section. Let us add s kinds of color degrees of freedom in the up and down spins, namely $|u^k\rangle$ and $|d^k\rangle$ $(k = 1, \cdots, s)$. As we will see, this contributes to the square-root violation of the area law.

The Hamiltonian is defined as a sum of projection operators:

$$\begin{aligned} H_{\mathrm{cM},\,s} \;=\; & \sum_{j=1}^{2n-1}\sum_{k=1}^{s}\left\{\left|D^k\right\rangle_{j,j+1}\left\langle D^k\right| + \left|U^k\right\rangle_{j,j+1}\left\langle U^k\right| + \left|F^k\right\rangle_{j,j+1}\langle F|\right\} \\ & + \sum_{j=1}^{2n-1}\Pi^{\mathrm{cross}}_{j,j+1} + \sum_{k=1}^{s}\left\{\left|d^k\right\rangle_1\left\langle d^k\right| + \left|u^k\right\rangle_{2n}\left\langle u^k\right|\right\}, \end{aligned} \tag{12}$$

where $|D^k\rangle$, $|U^k\rangle$ and $|F^k\rangle$ are given in (3) with d and u replaced by d^k and u^k, respectively. The term

$$\Pi^{\mathrm{cross}}_{j,j+1} \equiv \sum_{k\neq k'}\left|u^k\right\rangle_j\left|d^{k'}\right\rangle_{j+1}\left\langle u^k\right|_j\left\langle d^{k'}\right|_{j+1} \tag{13}$$

is a peculiar to the colored model, which realizes color-matched up-down pairs in zero-energy states.

From similar argument to the previous section, we can see that (12) has a unique zero-energy eigenstate corresponding to colored Motzkin walk.

For length $2n = 4$ case, the ground state and the corresponding colored Motzkin walk are

$$|P_{4,\,s}\rangle = \frac{1}{\sqrt{1+6s+2s^2}}\left[|0000\rangle + \sum_{k=1}^{s}\left\{\left|u^k d^k 00\right\rangle + \left|0u^k d^k 0\right\rangle + \left|00u^k d^k\right\rangle \right.\right.$$
$$\left. + \left|u^k 0 d^k 0\right\rangle + \left|0u^k 0 d^k\right\rangle + \left|u^k 00 d^k\right\rangle\right\}$$
$$\left. + \sum_{k,k'=1}^{s}\left\{\left|u^k d^k u^{k'} d^{k'}\right\rangle + \left|u^k u^{k'} d^{k'} d^k\right\rangle\right\}\right] \quad (14)$$

and

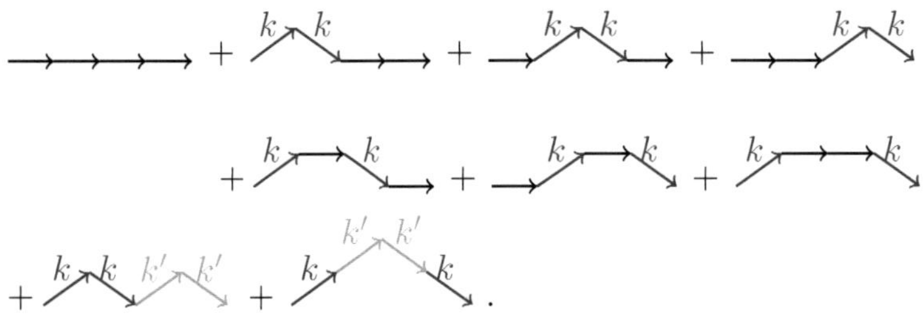

(15)

The number in the normalization $1 + 6s + 2s^2$ gives the number of the colored Motzkin walks of length 4.

In order to compute the EE of the half chain, let us consider the Schmidt decomposition of the ground state $|P_{2n,\,s}\rangle$. We divide any path of the colored Motzkin walks from $(0,0)$ to $(2n,0)$ into two length-n paths, corresponding to the subsystems A and B. For the midpoint (n,h) in the division, the left-half path from $(0,0)$ to (n,h) has colors of h up-steps unmatched inside A, which should be matched to colors of h down steps in the right-half path from (n,h) to $(2n,0)$. We show an example of length $2n = 8$ and $h = 2$ in Fig. 2.

κ_m denotes a color of the unmatched up-step from height $m-1$ to m in the left-half. By expressing the left-half (right-half) path with unmatched colors $\kappa_1, \cdots, \kappa_h$ as $P_{n,\,s}^{(0\to h)}(\{\kappa\})$ ($P_{n,\,s}^{(h\to 0)}(\{\kappa\})$), the Schmidt decomposition has the form

$$|P_{2n,\,s}\rangle = \sum_{h=0}^{n}\sum_{\kappa_1=1}^{s}\cdots\sum_{\kappa_h=1}^{s}\sqrt{p_{n,n,\,s}^{(h)}}\left|P_{n,\,s}^{(0\to h)}(\{\kappa\})\right\rangle \otimes \left|P_{n,\,s}^{(h\to 0)}(\{\kappa\})\right\rangle. \quad (16)$$

The number of paths in the colored Motzkin walks of length $2n$ is denoted by $M_{2n,\,s}$, and the number of $P_{n,\,s}^{(0\to h)}(\{\kappa\})$ by $\tilde{M}_{n,\,s}^{(h)}$. Then, $p_{n,n,\,s}^{(h)} =$

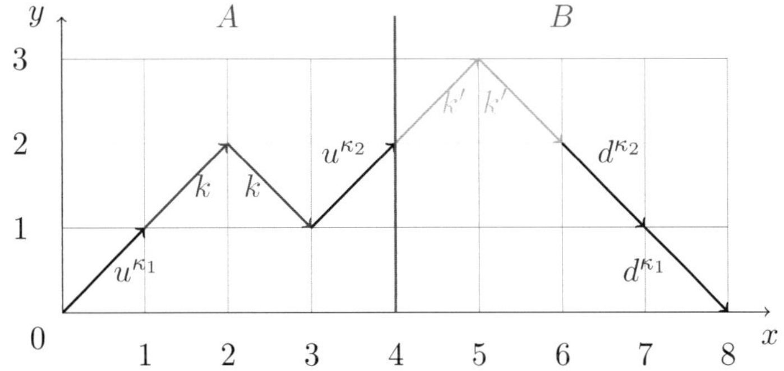

Figure 2: Division of a path with length $2n = 8$ and $h = 2$. The colors k and k' are matched inside A and B respectively, whereas the colors κ_1 and κ_2 are matched across the boundary.

$\left(\tilde{M}_{n,s}^{(h)}\right)^2 / M_{2n,s}$. The numbers are obtained by combinatorics as

$$\begin{aligned} M_{2n,s} &= \sum_{k=0}^{n} \frac{1}{k+1} \binom{2k}{k} \binom{2n}{2k} s^k, \\ M_{n,s}^{(h)} &= \sum_{r=0}^{n-h} \frac{1+(-1)^{n-r+h}}{2} \binom{n}{r} \frac{h+1}{\frac{n-r+h}{2}+1} \binom{n-r}{\frac{n-r+h}{2}} s^{\frac{n-r+h}{2}}. \end{aligned} \tag{17}$$

Assuming $n, r, n-r \pm h \gg 1$, we evaluate the sums by the saddle point method, and obtain

$$\begin{aligned} p_{n,n,s}^{(h)} &\sim \frac{s^{-h}}{\sqrt{\pi}\, s^{1/4}} \frac{(2n)^{3/2}}{(2\sqrt{s}+1)^{2n+\frac{3}{2}}} \frac{n^{2n+1}}{r_0^{2n+3}} \frac{(h+1)^2}{[4sn^2-(4s-1)h^2]^{1/2}} \\ &\quad \times \left(\frac{n-r_0-h}{n-r_0+h}\right)^{h+1} \times \left[1+O(n^{-1})\right], \end{aligned} \tag{18}$$

where the saddle point value of r is $r_0 + O(n^0)$ with

$$r_0 \equiv \frac{n}{4s-1}\left[-1+\sqrt{4s-(4s-1)\frac{h^2}{n^2}}\right]. \tag{19}$$

When, $h \le O(n^{1/2})$, (18) reduces to

$$p_{n,n,s}^{(h)} \sim \frac{\sqrt{2}\, s^{-h}}{\sqrt{\pi}\,(\sigma n)^{3/2}} (h+1)^2 e^{-\frac{(h+1)^2}{2\sigma n}} \times \left[1+O(n^{-1})\right] \tag{20}$$

with $\sigma = \sqrt{s}/(2\sqrt{s}+1)$. For the density matrix of the ground state $\rho = |P_{2n,\,s}\rangle\langle P_{2n,\,s}|$, the EE is obtained as

$$S_A = -\sum_{h=0}^{n} s^h p_{n,n,\,s}^{(h)} \ln p_{n,n,\,s}^{(h)}, \tag{21}$$

where the factor s^h arises from the sums of κ's in (16). Here, since $h \leq O(n^{1/2})$ dominantly contributes to the sum, we may use (20) in evaluating the EE. The result is

$$S_A \sim (2\ln s)\sqrt{\frac{2\sigma n}{\pi}} + \frac{1}{2}\ln n + \frac{1}{2}\ln(2\pi\sigma) + \gamma - \frac{1}{2} - \ln s \tag{22}$$

up to terms vanishing as $n \to \infty$. The first term grows as $O(\sqrt{n})$ and provides a new violation of the area law much greater than the logarithm that have ever seen. This reproduces the result in the colorless case (11) in the limit $s \to 1$. We can see that the leading term of S_A comes from the expectation value of $h\ln s$ under the weight $s^h p_{n,n,\,s}^{(h)} \sim e^{-\frac{(h+1)^2}{2\sigma n}}$. Due to the Gaussian distribution, the expectation value scales as $\sqrt{n}\,\ln s$.

For spin-1/2 case, a model exhibiting the same square-root violation has been constructed [8, 9]. By introducing a parameter that controls the randomness of the walks, models realizing the extensive EE that scales as $O(n)$ have also been discovered [10, 11, 12].

4. Rényi entropy

Rényi entropy is a genralization of the EE and defined by [13]

$$S_{A,\,\alpha} = \frac{1}{1-\alpha}\ln \mathrm{Tr}_{\,A}\,\rho_A^{\alpha}, \tag{23}$$

where α is a positive number not equal to 1. It is easy to see that (23) reduces to (1) in the limit $\alpha \to 1$. The Rényi entropy has further importance than the EE, because the whole spectrum (entanglement spectrum) of ρ_A or equivalently of the entanglement Hamiltonian

$$H_{\mathrm{ent},A} = -\ln\rho_A \tag{24}$$

can be obtained once the Rényi entropy is known as a function of α.

$S_{A,\,\alpha}$ $(0 < \alpha < 1)$ for gapped systems in one space-dimension is proven to exhibit the area law [14]. For CFT in $(1+1)$ dimensions, the Rényi entropy behaves as [3, 15]

$$S_{A,\,\alpha} \sim \frac{c}{6}\left(1+\frac{1}{\alpha}\right)\ln L, \tag{25}$$

where c is the central charge of the CFT, and L is a length scale of the subsystem A.

In terms of (24), the Rényi entropy takes a form analogous to the "thermal free energy" with the "temperature" $1/\alpha$:

$$S_{A,\alpha} = \frac{1}{1-\alpha} \ln \mathrm{Tr}_A \, e^{-\alpha H_{\mathrm{ent},\,A}}. \tag{26}$$

In the next section, we compute the Rényi entropy in the colorless and colored Motzkin spin chains, and observe a new phase transition with respect to the parameter α [16].

5. Rényi entropy of colorless Motzkin spin chain

First, let us compute asymptotic behavor of the Rényi entropy (23) as $n \to \infty$ for colorless case $s = 1$. Since ρ_A is diagonal, (23) becomes

$$S_{A,\alpha} = \frac{1}{1-\alpha} \ln \sum_{h=0}^{n} \left(p_{n,n}^{(h)}\right)^{\alpha} \tag{27}$$

with $p_{n,n}^{(h)}$ given in (9). The sum $\sum_{h=0}^{n} \left(p_{n,n}^{(h)}\right)^{\alpha}$ can be evaluated by converting to an integral. Elementary calculations lead to

$$\begin{aligned} S_{A.,\alpha} &= \frac{1}{2}\ln n + \frac{1}{1-\alpha}\ln\Gamma\left(\alpha+\frac{1}{2}\right) \\ &\quad -\frac{1}{2(1-\alpha)}\left\{(1+2\alpha)\ln\alpha + \alpha\ln\frac{\pi}{24} + \ln 6\right\} \end{aligned} \tag{28}$$

up to terms vanishing as $n \to \infty$. This logarithmically grows as the CFT case (25), but the dependence of α is different, which shows that the colorless Motzkin spin chain cannot be described by CFT. The result is consistent with the computation in [17].

6. Rényi entropy of colored Motzkin spin chain

Next, we consider the colored case ($s > 1$). What we compute is the asymptotic behavior of

$$S_{A,\alpha} = \frac{1}{1-\alpha} \ln \sum_{h=0}^{n} \left(s^h p_{n,n,\,s}^{(h)}\right)^{\alpha} \tag{29}$$

with (20). Note that the summand has a factor $s^{(1-\alpha)h}$. Due to this, the summand exponentially grows as h increases for $0 < \alpha < 1$, whereas it exponentially decays for $\alpha > 1$.

6.1. $0 < \alpha < 1$ case

For $0 < \alpha < 1$, we find a saddle point value of the sum with (18) as

$$h_* = n\frac{s^{1/(2\alpha)} - s^{1-1/(2\alpha)}}{s^{1/(2\alpha)} + s^{1-1/(2\alpha)} + 1} + O(n^0). \tag{30}$$

Since $h_* = O(n)$, here we cannot reduce (18) to (20). Saddle point analysis around (30) provides the result

$$S_{A,\alpha} = n\frac{2\alpha}{1-\alpha}\ln\left[\sigma\left(s^{\frac{1-\alpha}{2\alpha}} + s^{-\frac{1-\alpha}{2\alpha}} + s^{-1/2}\right)\right] + \frac{1+\alpha}{2(1-\alpha)}\ln n \tag{31}$$

up to $O(n^0)$ terms. The leading term linearly grows with respect to the volume n. We should note that the limit $\alpha \to 1$ or $s \to 1$ does not commute with the large-n limit. When $\alpha \to 1$ or $s \to 1$, the leading term of $O(n)$ in (30) vanishes, and thus the saddle point analysis cannot be trusted.

6.2. $\alpha > 1$ case

For $\alpha > 1$, it can be seen that $h \lesssim 1/((\alpha - 1)\ln s) = O(n^0)$ dominantly contributes to the sum (29). Using (20), we end up with

$$S_{A,\alpha} = \frac{3\alpha}{2(\alpha-1)}\ln n + O(n^0). \tag{32}$$

This grows logarithmically. Again, the limit $\alpha \to 1$ or $s \to 1$ does not commute with the $n \to \infty$ limit.

6.3. Phase transition

We saw that $S_{A,\alpha}$ grow as as $O(n)$ for $0 < \alpha < 1$, while as $O(\ln n)$ for $\alpha > 1$. There is nonanalyticity at $\alpha = 1$. As discussed in section 4., α can be regarded as the inverse temperature. From this point of view, our result means a phase transition takes place at the inverse temperature $\alpha = 1$. The transition point itself form a phase, where the EE behaves as a square-root of the volume. We summarize the result in Fig. 3.

7. Conclusions

We have mainly discussed quantum entanglement of the ground state for the Motzkin spin chain and its colored version, the latter of which significantly violates the area law by the square-root correction in spite of local interactions of the model. Then, we proceed with the computation of the Rényi entropy of the same model, and find a phase transition at the parameter $\alpha = 1$. This is a new phase transition never seen in any other spin chain studied so far.

Following future directions seem to be worth pursuing:

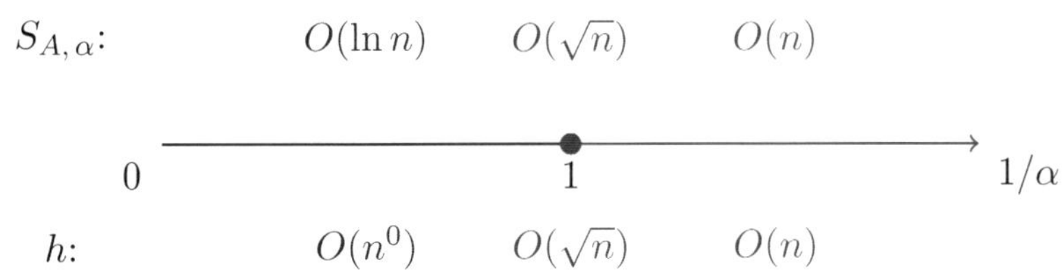

Figure 3: Phase diagram along the "temperature" $1/\alpha$. In high temperature, the Rényi entropy linearly grow with respect to n, to which $h = O(n)$ dominantly contributes. On the other hand, in low temperature, the Rényi entropy logarithmically grows, to which $h = O(n^0)$ mainly contributes. Finally, the third phase appears at the transition point $\alpha = 1$, where main contribution comes from $h = O(\sqrt{n})$ to give the EE growing as $\sqrt{n}$.

- The behavior of the EE is beyond the logarithmic growth at the ordinary critical point. So, it is interesting to investigate if the colored Motzkin spin chain is described as any quantum field theory. If so, it would have exotic properties that are not seen in the conventional quantum field theory.
- We also constructed models exhibiting the same square-root violation of the EE by using symmetric inverse semigroups [18, 19]. It would be interesting to compute the Rényi entropy for the systems.
- If we construct a higher-dimensional analog of the model, its ground state would be expressed as a sum of higher-dimensional objects, namely a sum of surfaces for a two-dimensional model. It could have some connection to string theory.
- From the viewpoint of gauge/gravity duality or holography, it would be interesting to interpret the square-root violation of the EE as a geodesic of some bulk space geometry. Since the same square-root scaling has been discovered for a geodesic in a two-dimensional fluctuating surface, i.e., two-dimensional quantum gravity [20, 21], This direction might provide a new formulation of quantum gravity.

References

[1] J. Eisert, M. Cramer, M.B. Plenio, *Rev. Mod. Phys.* **82** (2010) 277 [arXiv:arXiv:0808.3773 [quant-ph]].

[2] M.B. Hastings, J. Stat. Mech. **0708** (2007) P08024 [arXiv:0705.2024 [quant-ph]].

[3] C. Holzhey, F. Larsen, F. Wilczek, *Nucl. Phys.* **B [FS] 424** (1994) 443.

[4] V.E. Korepin, *Phys. Rev. Lett.* **92** (2004) 096402 [arXiv:cond-mat/0311056].

[5] P. Calabrese, J. Cardy, *J. Phys.* **A42** (2009) 504005 [arXiv:0905.4013 [cond-mat,stat-mech]].

[6] R. Movassagh, P.W. Shor, Counterexample to the area law for quantum matter", *Proc. Natl. Acad. Sci.* **113** (2016) 13278–13282 [arXiv:1408.1657 [quant-ph]].

[7] S. Bravyi, L. Caha, R. Movassagh, D. Nagaj, P.W. Shor, *Phys. Rev. Lett.* **109** (2012) 207202 [arXiv:1203.5801 [quant-ph]].

[8] O. Salberger, V. Korepin, *Rev. Math. Phys.* **29** (2017) 1750031 [arXiv:1605.03842 [quant-ph]].

[9] L. Dell'Anna, O. Salberger, L. Barbiero, A. Trombettoni, V.E. Korepin, *Phys. Rev.* **B 94** (2016) 155140 [arXiv:1604.08281 [cond-mat.str-el]].

[10] Z. Zhang, I. Klich, *J. Phys.* **A50** (2017) 425201 [arXiv:1702.03581 [cond-mat,stat-mech]].

[11] O. Salberger, T. Udagawa, Z. Zhang, H. Katsura, I. Klich, V. Korepin, *J. Stat. Mech. Theory Exp.* **1706** (2017) 063103 [arXiv:1611.04983 [cond-mat,stat-mech]].

[12] Z. Zhang, A. Ahmadain, I. Klich, *Proc. Natl. Acad. Sci.* **114**, 5142–5146 (2017) [arXiv:1606.07795 [quant-ph]].

[13] A. Rényi, *Probability Theory*, North Holland, Amsterdam, 1970.

[14] Y. Huang, (2015) *UC Berkeley Electronic Theses and Dissertations* https://escholarship.org/uc/item/3ct7d8tf

[15] E. Perlmutter, *J. High Energy Phys.* **03** (2014) 117 [arXiv:1308.1083 [hep-th]].

[16] F. Sugino, V. Korepin, *Int. J. Mod. Phys. B* **32**, no. 28 (2018) 1850306 [arXiv:1806.04049 [hep-th]].

[17] R. Movassagh, *J. Math. Phys.* **58** (2017) 031901 (2017) [arXiv:1602.07761 [quant-ph]].

[18] F. Sugino, P. Padmanabhan, *J. Stat. Mech. Theory Exp.* **1801** (2018) 013101 [arXiv:1710.10426 [quant-ph]].

[19] P. Padmanabhan, F. Sugino, V. Korepin, *Quantum Inf Process* (2019) 18: 69 [arXiv:1804.00978 [quant-ph]].

[20] H. Kawai, N. Kawamoto, T. Mogami and Y. Watabiki, *Phys. Lett. B* **306** (1993) 19 [hep-th/9302133].

[21] S. S. Gubser and I. R. Klebanov, *Nucl. Phys.* **B416** (1994) 827 [hep-th/9310098].

Two type of contraction of conformal algebra and the gravity limit

Michał Szcząchor[*]
Capstone Institute for Theoretical Research

Abstract

Based on the results presented in previous work [1, 2] two types of Inönü-Wigner contraction [3] is presented. After going to gravitational limits [4] the same Lagrangian is obtained in both cases i.e. Holst action with cosmological constant. This simplification dose not requite additional constrains to obtain torsionless condition. One of the contracted algebra is extended by dilation generator, and because of this the final solution is invariant under conformal resacling.

1. Introduction

In the previous work [1, 2] we presented the constructions of MacDowell–Mansouri theory in BF formulation (A similar approch has been also presented e.g. in [4, 5]). We obtained the action which can be interpreted as an action of two Lorentzian co-frame one-forms. As a solution of equation of motion one could obtain the linear relation between two of them. Based of this one soldering form can be distinguished. Another consequence of field equation is that field associated with dilation generator is identically equal zero, thus it can be set to zero. The torsion does not vanish by field equation, therefore the additional constrain is required as discussed in [6]. After preceding the above steps you can stay with metric dependent Weyl action with Holst extension and the set of boundary terms [7, 8]. The final theory is invariant under conformal rescaling of the metric. The aim of this short paper is to check if the Inönü-Wigner contraction lead to the proper gravity action, as it has been done originally i.e the theory constructed based on contraction of anti de–Sitter symmetry $so(3,2)$ algebra to the Poincaré algebra. The result which can be expected is a gravity action, which can be invariant under rescaling. This is also the behavior of the final theory presented in [1, 9]. More over all terms containing scalar field disappear from theory because of contraction of algebra. Also the torsionless condition dose not have to be added as a separate constrain but it comes directly as solution of equation of motion. Just for simplicity the gravitation limit has been chosen as the simplest one i.e. It was assumed that two co-frame one-forms are equal.

[*] e-mail address: ms@capstone-itr.eu

2. Construction of the conformal action

The construction of conformal invariant action starts from choosing $so(4,2)$ as a group of symmetry. It is well know that this Lie algebra is isomorphic to the conformal symmetry group $Conf(\mathbb{R}^{3,1})$ of the Minkowski space–time $\mathbb{R}^{3,1}$ by the relation [10, 11]:

$$J_{ij} = \mathcal{M}_{ij}, \qquad J_{4,5} = \mathcal{D}, \tag{1}$$

$$J_{4,j} = \frac{1}{2}(\mathcal{P}_j - \mathcal{K}_j) \qquad J_{5,j} = \frac{1}{2}(\mathcal{P}_j + \mathcal{K}_j)\,. \tag{2}$$

The $so(4,2)$ algebra is described by the following commutation relations:

$$[J_{IJ}, J_{KL}] = -i(\eta_{IK}J_{JL} + \eta_{JL}J_{IK} - \eta_{JK}J_{IL} - \eta_{IL}J_{JK})\,, \tag{3}$$

where $I,J = 0,\ldots,5$ and η_{IJ} denotes a pseudo-Euclidean metric of the signature $(-,+,+,+,+,-)$. In matrix representation it can written as

$$J_{IJ} = \begin{pmatrix} \begin{pmatrix} \mathcal{M}_{ij} \end{pmatrix} & \begin{pmatrix} -\frac{1}{2}(\mathcal{P}_i - \mathcal{K}_i) \end{pmatrix} & \begin{pmatrix} -\frac{1}{2}(P_i + K_i) \end{pmatrix} \\ \begin{pmatrix} \frac{1}{2}(\mathcal{P}_i - \mathcal{K}_i) \\ \frac{1}{2}(\mathcal{P}_i + \mathcal{K}_i) \end{pmatrix} & \begin{matrix} 0 \\ \mathcal{D} \end{matrix} & \begin{matrix} -\mathcal{D} \\ 0 \end{matrix} \end{pmatrix}. \tag{4}$$

The choise of Anti de–Sitter algebra allows to use MacDowell–Mansouri mechanism [12] as it was done before by several authors. The mechanism is based on a gauge field with $so(4,2)$ gauge symmetry, which is intended to be broken down to Loretnz symmetry.

As a starting point one can construct a physical viable gauge connection field

$$A \equiv A^{IJ}J_{IJ} = \frac{1}{2}\omega^{ij}J_{ij} + \frac{1}{\ell}f_1^iJ_{4i} + \frac{1}{\ell}f_2^iJ_{5i} + \frac{1}{\ell}\phi J_{45}\,, \tag{5}$$

where $i,j = 0\ldots3$. Pulling-back this form to a four-dimensional space–time manifold M^4, one should notice that f_1^i, f_2^i can be interpreted as two Lorentzian co-frame one-forms. Additionally ϕ has interpretation as Lorentzian scalar one-form.

In the presented construction it is useful to rescale the generators by λ,κ i.e. const. complex number

$$J_{ij} = M_{ij}, \qquad J_{4i} = (\lambda-\kappa)R_{1i}, \qquad J_{5i} = (\lambda+\kappa)R_{2i}, \qquad J_{45} = \kappa D\,, \tag{6}$$

then the connection (5) becomes

$$A = A^{IJ}J_{IJ} = \frac{1}{2}\omega^{ij}M_{ij} + \frac{1}{\ell(\lambda-\kappa)}f_1^iR_{1i} + \frac{1}{\ell(\lambda+\kappa)}f_2^iR_{2i} + \frac{1}{\ell\kappa}\phi D\,. \tag{7}$$

The action is a MacDowell–Mansouri action [12, 13, 14, 15] on a 4-dimensional manifold M^4 in formulation of the BF theory [16, 17], and it takes the form

$$S = \int_{M^4} B^{IJ} \wedge F_{IJ} - \frac{\beta}{2} B^{IJ} \wedge B_{IJ} - \frac{1}{2} U^{MN} \epsilon_{IJKLMN} B^{IJ} \wedge B^{KL} , \quad (8)$$

where $I, J = 0 \dots 5$; $F \equiv F^{IJ} J_{IJ}$ is a curvature two-form for A (c.f. (5)) and the two-forms B_{IJ} are the components of a Lie algebra valued two-form $B \equiv B^{IJ} J_{IJ}$. To break symmetry one can assume the components of a scalar-valued matrix $U \equiv U^{MN} J_{MN}$ in the form:

$$U^{MN} = \begin{pmatrix} 0 & \cdots & \cdots & 0 \\ \vdots & \ddots & & \vdots \\ \vdots & & \ddots & \frac{\alpha}{2} \\ 0 & \cdots & -\frac{\alpha}{2} & 0 \end{pmatrix}$$

where α is a real, positive number.

The field equations for the auxiliary fields B^{ij}, B^{i4}, B^{i5} and B^{45} can be solved and put back to the action

$$S = \int_{M^4} \frac{1}{2} (Z_{ijkl})^{-1} F^{ij} \wedge F^{kl} + \frac{1}{\beta} (F^{i4} \wedge F_{i4} - F^{i5} \wedge F_{i5} - F^{45} \wedge F_{45}) \quad (9)$$

where $Z_{ij}{}^{kl} = (\beta \delta_{ij}{}^{kl} + \frac{\alpha}{2} \epsilon_{ij}{}^{kl})$. One can note that at acion (9) the Immirzi parameter [18, 19] has been introduced and it has sense as far as Immirzi parameter $\gamma^2 \neq -1$. At this approach parameters $\alpha, \beta, \ell, \gamma$ are related to physical (gravitational and cosmological) constants by the following relations [16, 17]

$$\frac{1}{\ell^2} = -\frac{\Lambda}{3}, \qquad \alpha = \frac{G\Lambda}{3(1+\gamma^2)}, \qquad \beta = \frac{\gamma G \Lambda}{3(1+\gamma^2)} . \quad (10)$$

Straightforward calculation can show that $\alpha \sim 10^{-120}$.

Using algebraic connection field (7) it is possible to calculate the curvature tensor explicitly in analogues to[9, 20]

$$F^{ij} M_{ij} = \Big(R^{ij} - \frac{1}{\ell^2 (\lambda - \kappa)^2} f_1^i \wedge f_1^j + \frac{1}{\ell^2 (\lambda + \kappa)^2} f_2^i \wedge f_2^j \Big) M_{ij} \quad (11)$$

$$F^{45} D = \Big(\frac{1}{\ell \kappa} d\phi - \frac{1}{\ell^2 (\lambda^2 - \kappa^2)} f_1^j \wedge f_{2j} \Big) D \quad (12)$$

$$F^{i4} R_{1i} = \Big(\frac{1}{\ell (\lambda - \kappa)} D^\omega f_1^i - \frac{1}{\ell^2 \kappa (\lambda + \kappa)} \phi \wedge f_2^i \Big) R_{1i} \quad (13)$$

$$F^{i4} R_{2i} = \Big(\frac{1}{\ell (\lambda + \kappa)} D^\omega f_2^i - \frac{1}{\ell^2 \kappa (\lambda - \kappa)} \phi \wedge f_1^i \Big) R_{2i} \quad (14)$$

where two-form R_{ij} is a space–time curvature of the corresponding Riemann-Cartan connection (see e.g. [13, 14, 15]). After substituting these curvature tensors into the action (9) one can obtain

$$\begin{aligned}
S = \int_{M^4} & \frac{1}{4}\Big(\frac{\alpha}{\alpha^2+\beta^2}(\frac{\beta}{\alpha}\delta_{ijkl} - \epsilon_{ijkl})\Big)(R^{ij}\wedge R^{kl} - \frac{2}{\ell^4(\lambda^2-\kappa^2)^2} f_1^i \wedge f_1^j \wedge f_2^k \wedge f_2^l) \\
& - \frac{2}{\ell^2(\lambda-\kappa)^2} R^{ij}\wedge f_1^k \wedge f_1^l + \frac{2}{\ell^2(\lambda+\kappa)^2} R^{ij}\wedge f_2^k \wedge f_2^l \\
& + \frac{1}{\ell^4(\lambda-\kappa)^4} f_1^i \wedge f_1^j \wedge f_1^k \wedge f_1^l + \frac{1}{\ell^4(\lambda+\kappa)^4} f_2^i \wedge f_2^j \wedge f_2^k \wedge f_2^l \\
& - \frac{1}{\beta\ell^2}(\frac{1}{\kappa^2} d\phi\wedge d\phi - \frac{2}{\ell(\lambda^2-\kappa^2)\kappa} d\phi \wedge f_1^i \wedge f_{2i} - \frac{1}{\ell^2(\lambda^2-\kappa^2)^2} f_1^i \wedge f_{2i} \wedge f_1^j \wedge f_{2j}) \\
& + \frac{1}{\beta\ell^2(\lambda-\kappa)^2}(D^\omega f_1^i \wedge D^\omega f_{1i} - \frac{2}{l(\lambda^2-\kappa^2)\kappa} D^\omega f_1^i \wedge \phi \wedge f_{2i}) \\
& - \frac{1}{\beta\ell^2(\lambda+\kappa)^2}(D^\omega f_2^i \wedge D^\omega f_{2i} - \frac{2}{l(\lambda^2-\kappa^2)\kappa} D^\omega f_2^i \wedge \phi \wedge f_{1i}) \,. \qquad (15)
\end{aligned}$$

Similar form of the action has already appeared in [9, 20]. Some algebraic calculation can separate the boundary terms (i.a. Euler, Pontryagin and Nieh-Yan) form the action

$$\begin{aligned}
S = \int_{M^4} & E_4 + P_4 + NY_4(f_1) - NY_4(f_2) \\
& + \frac{\epsilon_{ijkl}}{2G(\lambda+\kappa)^2} R^{ij}\wedge f_2^k \wedge f_2^l + \frac{\epsilon_{ijkl}}{4\ell^2 G(\lambda+\kappa)^4} f_2^i \wedge f_2^j \wedge f_2^k \wedge f_2^l \\
& - \frac{\epsilon_{ijkl}}{2G(\lambda-\kappa)^2} R^{ij}\wedge f_1^k \wedge f_1^l + \frac{\epsilon_{ijkl}}{4\ell^2 G(\lambda-\kappa)^4} f_1^i \wedge f_1^j \wedge f_1^k \wedge f_1^l \\
& - \frac{\epsilon_{ijkl}}{2G\ell^2(\lambda^2-\kappa^2)^2} f_1^i \wedge f_1^j \wedge f_2^k \wedge f_2^l \\
& + \frac{1}{2G\ell^2(\lambda^2-\kappa^2)^2} f_1^i \wedge f_1^j \wedge f_{2i} \wedge f_{2j} \\
& + \frac{1}{\beta\ell^4(\lambda^2-\kappa^2)^2} f_1^i \wedge f_{2i} \wedge f_1^j \wedge f_{2j} \\
& + \frac{1}{G\gamma(\lambda+\kappa)^2} R^{ij}\wedge f_{2i}\wedge f_{2j} - \frac{1}{G\gamma(\lambda-\kappa)^2} R^{ij}\wedge f_{1i}\wedge f_{1j} \\
& + \frac{2}{\beta\ell^3(\lambda^2-\kappa^2)\kappa^2} D^\omega f_1^i \wedge f_{2i}\wedge\phi - \frac{2}{\beta\ell^3(\lambda^2-\kappa^2)\kappa^2} D^\omega f_2^i \wedge f_{1i}\wedge\phi \\
& - \frac{1}{\beta\ell^2\kappa^2} d\phi\wedge d\phi + \frac{2}{\beta\ell^3(\lambda^2-\kappa^2)\kappa^2} d\phi\wedge f_1^i \wedge f_{2i} \,. \qquad (16)
\end{aligned}$$

At that point it is more convenient to use a conformal groups isomorphism and to change the basis in the algebra by the following transformation:

$$\frac{1}{(\lambda-\kappa)}f_1^i=\frac{1}{\lambda}e^i-\frac{1}{\kappa}f^i \tag{17}$$

$$\frac{1}{(\lambda+\kappa)}f_2^i=\frac{1}{\lambda}e^i+\frac{1}{\kappa}f^i\,. \tag{18}$$

One has to note that in conformal algebra some rescaling is made. It will turn out later that at gravitational limit this rescaling makes sens.

This choice of rescaling can be equivalent with taking rescaling of conformal algebra generators:

$$D\to\kappa D,\qquad P\to\lambda P,\qquad K\to\kappa K \tag{19}$$

A Lagrangian for the action (16) reads now as

$$\begin{aligned}S=\int_{M^4}&E_4+P_4+NY_4(e,f)\\&+\frac{2\epsilon_{ijkl}}{G}R^{ij}\wedge e^k\wedge f^l+\frac{4\epsilon_{ijkl}}{\ell^2G}e^i\wedge f^j\wedge e^k\wedge f^l\\&+\frac{2}{G\ell^2}e^i\wedge e^j\wedge f_i\wedge f_j-\frac{4}{\beta\ell^4}e^i\wedge e^j\wedge f_i\wedge f_j\\&+\frac{4}{G\gamma}R^{ij}\wedge e_i\wedge f_j\\&+\frac{2}{\beta\ell^3\kappa}D^\omega(e^i\wedge f_i)\wedge\phi\\&-\frac{1}{\beta\ell^2\kappa^2}d\phi\wedge d\phi+\frac{4}{\beta\ell^3\kappa^2}d\phi\wedge e^i\wedge f_i\,.\end{aligned} \tag{20}$$

one can notices that the equation (20) has been already obtained within a bigravity approach, as presented in [21]. Since the co–frame e^i is associated with the Poincaré translation generators (soldering form), the solution of equation of motion shows the linear realtion between fields e^i and f^i [9] i.e.

$$f^i_\mu=-6R^i_\mu+Re^i_\mu=\mathcal{F}\big(e(x)\big)e^i_\mu\,. \tag{21}$$

Equation (21) describe a special choice of gravitational limits i.e. $f^i=\mathcal{F}(x)e^i$, where $\mathcal{F}$ is some function of space–time. [1] It is also possible to make a simpler choice of function $\mathcal{F}$ i.e. $\mathcal{F}(x)=1$ [4], and take limit $\kappa\to\infty$

[1] Some rescaling of field $f^i_\mu=-3R^i_\mu+\frac{1}{2}Re^i_\mu$ can give some impression that r.h.s. of this eq. is similar to r.h.s of Ricci flow equation.

the action (20) reads

$$
\begin{aligned}
S = \int_{M^4} & E_4 + P_4 + NY_4 \\
& + \frac{2\epsilon_{ijkl}}{G} R^{ij} \wedge e^k \wedge e^l + \frac{4\epsilon_{ijkl}}{\ell^2 G} e^i \wedge e^j \wedge e^k \wedge e^l \\
& + \frac{4}{G\gamma} R^{ij} \wedge e_i \wedge e_j \,.
\end{aligned}
$$

The equation of motion are standard Einstein equation and torsionless equation. The following gravity limit as well as limit of $\kappa \to \infty$ leads to two modification of conformal algebra for $\lambda = 1$

$$[\mathcal{D}, \mathcal{P}_i] = 0 \,, \tag{22}$$
$$[\mathcal{P}_i, \mathcal{P}_j] = -2i\eta_{ij}\mathcal{D} \,, \tag{23}$$
$$[\mathcal{M}_{ij}, \mathcal{P}_k] = -i(\eta_{ki}\mathcal{P}_j - \eta_{kj}\mathcal{P}_i) \,, \tag{24}$$
$$[\mathcal{M}_{ij}, \mathcal{M}_{kl}] = -i(\eta_{ik}\mathcal{M}_{jl} + \eta_{jl}\mathcal{M}_{ik} - \eta_{jk}\mathcal{M}_{il} - \eta_{il}\mathcal{M}_{jk}) \,, \tag{25}$$

and for $\lambda = \kappa$

$$[\mathcal{D}, \mathcal{P}_i] = 0 \,, \tag{26}$$
$$[\mathcal{P}_i, \mathcal{P}_j] = 0 \,, \tag{27}$$
$$[\mathcal{M}_{ij}, \mathcal{P}_k] = -i(\eta_{ki}\mathcal{P}_j - \eta_{kj}\mathcal{P}_i) \,, \tag{28}$$
$$[\mathcal{M}_{ij}, \mathcal{M}_{kl}] = -i(\eta_{ik}\mathcal{M}_{jl} + \eta_{jl}\mathcal{M}_{ik} - \eta_{jk}\mathcal{M}_{il} - \eta_{il}\mathcal{M}_{jk}) \,. \tag{29}$$

Algebra (26-29) is a Poincaré algebra as algebra (22-25) is Poincaré algebra extended by generator $\mathcal{D}$. Some one can note that Lagrangian dose not depend on λ any more, therefore it both cases Lagrangian is invariant under symmetries induced from both chose of algebras. Especially in case $\lambda = 1$ it is invariant under symmetry generated by $\mathcal{D}$.

3. Summary

The construction from [1] has been taken as starting point. But then the two types of very simple contractions has been presented on algebra level as well as action level. The obtained theory has a form of gravity with two co-frame feild without interaction term between them. Then gravitational limit has been proposed i.e. to make this two co-frame equal. The final theory is exactly the same action as obtained in [16, 17] i.e. Holst action with cosmological constant and boundary term [7, 8]. The main difference is that one of contraction contains trivial extension by dilation generator. Gauge symmetry base on it is conserved.

References

[1] A. Borowiec, M. Szczachor, to be published in AIPCP20-AR-TIM2019-00001

[2] M. Szczachor, Gravity and Supergravity as a BF theory of MacDowell-Mansouri type, PhD Thesis, Wrocław University, 27-10-2015.

[3] E. Inonu and E. P. Wigner, Proc. Nat. Acad. Sci. **39** (1953) 510.

[4] T. G. Zlosnik and H. F. Westman, arXiv:1601.00567 [gr-qc].

[5] A.B. Arbuzov, D.J. Cirilo-Lombardo, Int.J.Geom.Meth.Mod.Phys. 15 (2017) 1850005 (2017-10-05); arXiv:1708.00334 [hep-th]

[6] P. K. Townsend and P. van Nieuwenhuizen, Phys. Rev. D **19** (1979) 3166.

[7] S. Holst, Phys. Rev. D **53**, 5966 (1996). [gr-qc/9511026].

[8] D. J. Rezende and A. Perez, Phys. Rev. D **79**, 064026 (2009) [arXiv:0902.3416 [gr-qc]].

[9] M. Kaku, P. K. Townsend and P. van Nieuwenhuizen, Phys. Rev. D **17** (1978) 3179.

[10] G. Mack and A. Salam, Annals Phys. **53** (1969) 174.

[11] R. R. Metsaev, Mod. Phys. Lett. A **10** (1995) 1719.

[12] S. W. MacDowell and F. Mansouri, Phys. Rev. Lett. **38** (1977) 739 [Phys. Rev. Lett. **38** (1977) 1376].

[13] P. C. West, Phys. Lett.B7 1978, 6569.

[14] K.S. Stelle, P. C. West, J. Phys. A 12 (1979), no. 8, L205–L210.

[15] K. S. Stelle, P. C. West, Phys. Rev.D211,1980, 466.

[16] L. Freidel and A. Starodubtsev, hep-th/0501191.

[17] L. Smolin and A. Starodubtsev, hep-th/0311163.

[18] J. F. Barbero G., Phys. Rev. D **51** (1995) 5507 [gr-qc/9410014].

[19] G. Immirzi, Class. Quant. Grav. **14** (1997) L177 [gr-qc/9612030].

[20] M. Kaku, P. K. Townsend and P. van Nieuwenhuizen, Phys. Rev. Lett. **39** (1977) 1109.

[21] A. Borowiec, J. Kowalski-Glikman, M. Szczachor, Recent Developments in Theoretical and Experimental General Relativity, Astrophysics, and Relativistic Field Theories, Part B, Ed. by: K. Rosquist, R. T Jantzen, R. Ruffini, The Thirteenth Marcel Grossmann Meeting, pp. 1146-1148, World Scientific Publishing 2015; arXiv:1212.6661 [gr-qc]

Can information criteria fix the problem of degeneration in cosmology?

Marek Szydłowski*
Astronomical Observatory, Jagiellonian University,
Orla 171, 30-244 Krakow, Poland

and

Mark Kac Complex Systems Research Centre,
Jagiellonian University, Łojasiewicza 11, 30-348 Kraków, Poland

Adam Krawiec†
Institute of Economics, Finance and Management,
Jagiellonian University, Łojasiewicza 4, 30-348 Krakow, Poland

Paweł Tambor‡
Faculty of Philosophy, Catholic University of Lublin,
Al. Racławickie 12, 20-950 Lublin, Poland

Abstract

The information criteria are widely used to select the simple and proper specification of a model. In cosmology these criteria are used to select among the models with dark energy. These models are divided into two groups: the models with substantive dark energy and the models with a modified Friedmann equation. We show the advantages of using the consistent AIC over AIC in the problem considered. The cosmological model with cosmological constant – LCDM model – is favoured in the light of this criterion.

1. Introduction

Two trends are present, principally, in modern cosmology [1]. The first trend is called theoretical where different models are elaborated to explain observational facts, like acceleration, nature of dark matter and dark matter. All these attempts can be divided into three groups. First, we are looking for dark energy as a special form of substance (particles, fields). Second, we claim that theory of gravity which constitutes explanatory background should be modified. Third, we can postulate the violation of the cosmological principle as a source of acceleration (non-homogenous universe).

* e-mail address: marek.szydlowski@uj.edu.pl
† e-mail address: adam.krawiec@uj.edu.pl
‡ e-mail address: pawel.tambor@gmail.com

In the second trend, observational, we use observational methods to obtain astronomical data in order to estimate parameters of the theoretical models and probe theoretical hypothesis.

If we confront both trends, then we obtain many theoretical admissible models, which are in good agreement with observational data. This cognitive situation is called the degeneration problem. This degeneration problem has many faces in cosmology.

The same terms in model equations can mimic different effects. Objects of different ontologies produce the same term in the equations. For example, dark degeneracy means inability to resolve whether nature of dark energy is dynamical or is an effect of interaction between dark energy and dark matter [2]. This kind of degeneration is called the theoretical degeneration.

On the other side the observational degeneracy is a consequence of inability to discriminate between theoretical models on the ground of observational data. The statistical methods cannot resolve this problem definitely, but can restrict a class of admissible models. Here information criteria play their role [3, 4, 5, 6, 7, 8].

2. Choosing the right model

We can use some criteria in scientific practice to choose the best model:

- the oldest, the Middle Ages rule: Occam's razor principle — if two models describe the observations equally well choose the simpler one. This principle has aesthetical, as well as empirical justification.
- statistical criteria
  — goodness of fit (χ^2): favour the models with more parameters
  — information criteria (AIC and its generalization): realisation of Occam'srazor principle, because two models fit data equally well, the Akaike rule distinguishes model with smaller number of parameters.

Let us assume that we have N pairs of measurements (y_i, x_i) and that we want to find the relation between the y and x quantities. Suppose that we can postulate k possible relations $y \equiv f_i(x, \bar{\theta})$, where $\bar{\theta}$ is the vector of unknown model parameters and $i = 1, \ldots, k$. With the assumption that our observations come with uncorrelated gaussian errors with mean $\mu_i = 0$ and standard deviation σ_i, the goodness of fit for the theoretical model is measured by the χ^2 quantity given by

$$\chi^2 = \sum_{i=1}^{N} \frac{(f_l(x_i, \bar{\theta}) - y_i)^2}{2\sigma_i^2} = -2 \ln L,$$

where L is the likelihood function. For the particular family of models f_l the best one minimize the χ^2 quantity, which we denote $f_l(x, \hat{\bar{\theta}})$. The best model from our set of k models $f_1(x, \hat{\bar{\theta}}), \ldots, f_k(x, \hat{\bar{\theta}})$ could be the one with the smallest value of χ^2 quantity. But this method could give us misleading results. Generally speaking, for more complex model the value of χ^2 is smaller, thus the most complex one will be choosen as the best from our set under consideration.

3. Akaike information criteria

In the information theory there are no true models. There is only reality which can be approximated by models, which depend on some number of parameters. The best one from the set under consideration should be the best approximation to the truth. The information lost, when truth is approximated by model under consideration, is measured by so called Kullback-Leibler (KL) information, so the best one should minimize this quantity. It is impossible to compute the KL information directly, because it depends on truth, which is unknown. Akaike found approximation to the KL quantity, which is called the Akaike information criterion (AIC) and is given by [9]

$$\mathrm{AIC} = -2\ln L + 2d,$$

where L is the maximum of the likelihood function and d is the number of model parameters.

Model, which is the best approximation to the truth from the set under consideration, has the smallest value of the AIC quantity. It is convenient to evaluate the differences between the AIC quantities computed for the rest of models from our set and the AIC for the best one. Those differences $\Delta\mathrm{AIC}$

$$\Delta\mathrm{AIC}_i = \mathrm{AIC}_i - \mathrm{AIC}_{\mathrm{min}}$$

are easy to interpret and allow a quick 'strength of evidence' for considered model with respect to the best one. The models with

- $0 \leq \Delta\mathrm{AIC} \leq 2$ have substantial evidence as the best model,
- $4 < \Delta\mathrm{AIC} \leq 7$ have considerably less support than the best model,
- $\Delta\mathrm{AIC} > 10$ have essentially no support with respect to the best model.

The AIC is not consistent because the true model among models considered is not always pointed out. Therefore, Bozdogan proposed the consistent version of AIC [10]

$$\mathrm{cAIC} = -2\ln L + d(\ln N + 1)$$

where L is a maximum of the likelihood function, d is a number of model parameters, N is a number of observations.

Taking into account the number of observations apart from a number of parameters, the cAIC "penalizes" stronger models with high number of parameters.

4. Models of dark energy

We consider 10 representative models divided into 2 groups. There are 5 models with substantial form of dark energy and 5 models with modified gravity. The former are presented in table 1 and the latter are presented in table 2.

Table 1: Models with a substantial form of dark energy

No.	Cosmological model
1	**ΛCDM** $w_{\rm X} = -1$ $\frac{H^2(z)}{H_0^2} = \Omega_{\rm m,0}(1+z)^3 + (1-\Omega_{\rm m,0})$
2	**constant E.Q.S.** $w_{\rm X} = w_0 < -1$ $\frac{H^2(z)}{H_0^2} = \Omega_{\rm m,0}(1+z)^3 + (1-\Omega_{\rm m,0})(1+z)^{3(1+w_0)}$
3	**dynamic E.Q.S.** $w_{\rm X} = w_0 + w_1(1-a)$ $\frac{H^2(z)}{H_0^2} = \Omega_{\rm m,0}(1+z)^3 + (1-\Omega_{\rm m,0})(1+z)^{3(w_0+w_1+1)}\exp[-\frac{3w_1 z}{1+z}]$
4	**quintessence** $\bar{w}_{\rm X}(a) = \int w_{\rm X}(a)\mathrm{d}(\ln a)/\int \mathrm{d}(\ln a) \equiv w_0 a^\alpha$ $\frac{H^2(z)}{H_0^2} = \Omega_{\rm m,0}(1+z)^3 + (1-\Omega_{\rm m,0})(1+z)^{3(1+w_0(1+z)^{-\alpha})}$
5	**oscillating E.Q.S.** $w_{\rm X}(z) = -1 + (1+z)^3\left\{C\cos(\ln(1+z))\right\}$ $\frac{H^2(z)}{H_0^2} = \Omega_{\Lambda,0}\exp\Big((1+z)^3 D_2\cos(\ln(1+z))\Big) + \Omega_{\rm m,0}(1+z)^3$

Table 2: Models with modified gravity

No.	Cosmological model
6	**Interacting DE & DM** $\frac{H^2(z)}{H_0^2} = \Omega_{\rm m,0}(1+z)^3 + \Omega_{\rm int,0}(1+z)^n + 1 - \Omega_{\rm m,0} - \Omega_{\rm int,0}$
7	**Bounce ΛCDM** $\frac{H^2(z)}{H_0^2} = \Omega_{\rm m,0}(1+z)^3 - \Omega_{n,0}(1+z)^n + 1 - \Omega_{\rm m,0} + \Omega_{n,0}$
8	**Cardassian** $\frac{H^2(z)}{H_0^2} = \Omega_{\rm r,0}(1+z)^4 + \Omega_{\rm m,0}(1+z)^4\left[\frac{1}{1+z} + (1+z)^k\left(\frac{\Omega_{\rm C,0}}{\Omega_{\rm m,0}}\right)E(z)\right]$
9	**DGP** $\frac{H^2(z)}{H_0^2} = \left[\sqrt{\Omega_{\rm m,0}(1+z)^3 + \Omega_{\rm rc,0}} + \sqrt{\Omega_{\rm rc,0}}\right]^2, \quad \Omega_{\rm rc,0} = \frac{(1-\Omega_{\rm m,0})^2}{4}$
10	**Sahni-Shtanov brane I** $\frac{H^2(z)}{H_0^2} = \Omega_{\rm m,0}(1+z)^3 + \Omega_{\sigma,0} + 2\Omega_{l,0} - 2\sqrt{\Omega_{l,0}}P(z)$

To compare the above models we use the SNIa data, constraints from the CMB shift parameter, constraints from the SDSS parameter A as well as $H(z)$ observational data.

For the SNIa the likelihood function has the following form

$$L_{SN} \propto \exp\left[-\frac{1}{2}\left(\sum_{i=1}^{N_1} \frac{(\mu_i^{\text{theor}} - \mu_i^{\text{obs}})^2}{\sigma_i^2}\right)\right], \tag{1}$$

where N_1 is the number of supernovae, σ_i is known, $\mu_i^{\text{obs}} = m_i - M$ (m_i– apparent magnitude, M–absolute magnitude of SNIa), $\mu_i^{\text{theor}} = 5\log_{10} D_{Li} + \mathcal{M}$, $\mathcal{M} = -5\log_{10} H_0 + 25$ and $D_{Li} = H_0 d_{Li}$, where d_{Li} is the luminosity distance, which with the assumption $k = 0$ is given by $d_{Li} = (1+z_i)c\int_0^{z_i} \frac{dz'}{H(z')}$.

We also include information obtained from the CMB data. Here the likelihood function has the following form

$$L_R \propto \exp\left[-\frac{(R^{\text{theor}} - R^{\text{obs}})^2}{2\sigma_R^2}\right],$$

where R is the so called shift parameter, $R^{\text{theor}} = \sqrt{\Omega_{\text{m},0}}\int_0^{z_{dec}} \frac{H_0}{H(z)}dz$, and $R^{\text{obs}} = 1.70 \pm 0.03$ for $z_{\text{dec}} = 1089$.

As the third observational data we use the measurement of the baryon acoustic oscillations (BAO) from the SDSS luminous red galaxies. In this case the likelihood function has the following form

$$L_A \propto \exp\left[-\frac{(A^{\text{theor}} - A^{\text{obs}})^2}{2\sigma_A^2}\right],$$

where $A^{\text{theor}} = \sqrt{\Omega_{\text{m},0}}\left(\frac{H(z_A)}{H_0}\right)^{-\frac{1}{3}}\left[\frac{1}{z_A}\int_0^{z_A}\frac{H_0}{H(z)}dz\right]^{\frac{2}{3}}$ and $A^{\text{obs}} = 0.469 \pm 0.017$ for $z_A = 0.35$.

Finally, we used the observational $H(z)$ data. This data based on the differential ages ($\frac{dt}{dz}$) of the passively evolving galaxies which allow to estimate the relation $H(z) \equiv \frac{\dot{a}}{a} = -\frac{1}{1+z}\frac{dz}{dt}$. Here the likelihood function has the following form

$$L_H \propto \exp\left(-\frac{1}{2}\left[\sum_{i=1}^{N_2} \frac{(H(z_i) - H_i(z_i))^2}{\sigma_i^2}\right]\right),$$

where N_2 is a number of observations, $H(z)$ is the Hubble function, H_i, z_i are observational data.

The final likelihood function is given by

$$L = L_{SN}L_RL_AL_H.$$

Table 3: Results of AIC and cAIC

Model	AIC	cAIC
1	217.88	**226.51**
2	219.89	232.83
3	**217.02**	234.27
4	217.49	234.74
5	220.01	237.26
6	219.16	236.41
7	221.88	239.13
8	218.88	231.82
9	226.87	235.50
10	221.88	239.13

The AIC and cAIC have been calculated for all models considered. The results of both information criteria are presented in table 3. In this context we can see the advantage of using the consistent AIC for discrimination of the best model. While the AIC indicated substantial support for four models: 3rd, 4th, 1st and 8th, the cAIC point out only the 1st model—the ΛCDM model.

Moreover, the cAIC exposes more models to have no support with respect to the best model. These are 5th, 7th and 10th by the cAIC, and only 5th by the AIC. The rest models have less support than the best model in the respective categories.

5. Conclusions

Our investigation shows that the degeneration problem of type I (theoretical) can be removed because the cAIC favoured standard cosmological model (ΛCDM model). Note that the "true" model of the Universe may be absent from the sample of models used in our analysis. Therefore, we see how important the theoretical trend is.

The detailed results are the following.

- The AIC shows that 4 models has substantial evidence (with model 3 being the best, models 1, 4 and 8 have AIC difference less than 2 with respect to model 3) — our degeneration problem is still not solved.
- The cAIC shows model 1 — the ΛCDM model — as the best model with no substantial support for any other model — the degeneration problem is solved.

References

[1] S. Dodelson, *Modern Cosmology* (Academic Press, Amsterdam, 2003).

[2] R. von Marttens, L. Lombriser, M. Kunz, V. Marra, L. Casarini, and J. Alcaniz, (2020), 2020.100490.

[3] A. R. Liddle, Mon. Not. Roy. Astron. Soc. **351**, L49 (2004), astro-ph/0401198.

[4] A. R. Liddle, Mon. Not. Roy. Astron. Soc. **377**, L74 (2007), astro-ph/0701113.

[5] M. Szydlowski and W. Godlowski, Phys. Lett. **B633**, 427 (2006), astro-ph/0509415.

[6] M. Szydlowski, A. Kurek, and A. Krawiec, Phys. Lett. **B642**, 171 (2006), astro-ph/0604327.

[7] M. Y. J. Tan and R. Biswas, Mon. Not. Roy. Astron. Soc. **419**, 3292 (2012), 1105.5745.

[8] M. Szydlowski, A. Krawiec, A. Kurek, and M. Kamionka, Eur. Phys. J. **C75**, 5 (2015), 0801.0638.

[9] H. Akaike, IEEE Trans. Auto. Control **19**, 716 (1974).

[10] H. Bozdogan, Journal of Mathematical Psychology **44**, 62 (2000).

A quantum-classical duality and emergent space-time*

Vitaly Vanchurin†
Department of Physics, University of Minnesota,
Duluth, Minnesota, 55812

Abstract

We consider the quantum partition function for a system of quantum spinors and then derive an equivalent (or dual) classical partition function for some scalar degrees of freedom. The coupling between scalars is non-trivial (e.g. a model on 2-sphere configuration space), but the locality structure of the dual system is preserved, in contrast to the imaginary time formalism. We also show that the measure of integration in the classical partition function can be formally expressed through relativistic Green's functions which suggests a possible mechanism for the emergence of a classical space-time from anti-commutativity of quantum operators.

1. Introduction

It is well known that a D-dimensional quantum field theory at finite temperature can be mapped to a D+1-dimensional classical (or statistical) field theory. This quantum-to-classical correspondence is due to a simple observation that at equilibrium quantum systems are described by a quantum partition function

$$\mathcal{Z}_{\mathrm{q}}[\beta] = Tr\left[\exp\left(\beta \hat{H}_{\mathrm{q}}\right)\right] \tag{1}$$

where the Hamiltonian operator $\hat{H}_{\mathrm{q}}$ can be formally interpreted as a generator of evolution in imaginary time β and the trace implies that the evolution must be periodic [1, 2]. As a result the D-dimensional quantum partition function can be expressed as a D+1-dimensional classical partition function on a space with one extra dimension (see Appendix A for details).

The imaginary time formalism proved to be very useful for analyzing quantum field theories at finite temperature [3, 4, 5], but there are certain limitations that do not permit the quantum-classical duality to be applied to more general systems. First of all, the inverse temperature parameter

* The work was supported in part by Foundational Questions Institute (FQXi).
† e-mail address: vvanchur@d.umn.edu

β on the quantum side does not correspond to an inverse temperature parameter on the classical side, but to a parameter which appears inside of a classical Hamiltonian (more precisely β is the size of extra dimension). Secondly, the classical system usually has many more degrees of freedom then the corresponding quantum system. For example, if the quantum system has a finite number of localized degrees of freedom, then in the corresponding classical system each of these degrees of freedom is represented by many copies that are spread out in the extra dimension. This is, of course, not very useful if one wishes to study a possible emergence of quantum mechanics from classical systems with *local* hidden variables. And finally, the imaginary time formalism works well as a mapping from quantum to classical systems, but not in the opposite direction since the dual classical systems are very restrictive (e.g. at least one dimension must be periodic).

In this paper we shall extend the quantum-classical correspondence

$$\mathcal{Z}_{\rm q}[\beta] = \mathcal{Z}_{\rm c}[\beta] \tag{2}$$

to mappings which preserve the locality structure of both quantum and classical theories. More precisely, we shall study quantum systems whose partition function $\mathcal{Z}_{\rm q}[\beta]$ can be expressed as a classical partition function $\mathcal{Z}_{\rm c}[\beta]$ with the same number of local degrees of freedom. The price that we will have to pay is that the discrete quantum degrees of freedom (e.g. spinors) would be mapped to continuous classical degrees of freedom (e.g. scalars). In terms of field theories we shall describe how certain D-dimensional quantum systems can be mapped to D-dimensional classical systems with parameter β playing the role of an inverse temperature in both quantum and classical systems. An extra dimension on the classical side of the duality will emerge, but it will be Lorentzian and non-periodic, in contrast to the imaginary time formalism.

The paper is organized as follows. In the next section we define a quantum system and introduce the abstract-indices notations that will be used throughout the paper. In Sec. 3. we construct a classical dual of the system in the limit when all of the degrees of freedom commute. In Sec. 4. we show that the anti-commutativity of quantum operators gives rise to a non-trivial measure of integration in the classical partition function and in Sec. 5. we show that the measure can be formally expressed as a causal (retarded) propagator on an emergent space-time. In Sec. 6. we derive a sufficient condition for the existence of a classical dual and in Sec. 7. we provide an example of the dual with a separable measure of integration. In Sec. 8. we summarize and discuss the main results.

2. Quantum system

Consider a quantum system of N fermionic subsystems (which we call spinors) described by operators $\hat{\gamma}^j_a$, satisfying the following commutation relation

$$[\hat{\gamma}_a^j, \hat{\gamma}_b^k] = 0 \tag{3}$$

if $a \neq b$ and anti-commutation relation

$$\{\hat{\gamma}_a^j, \hat{\gamma}_a^k\} = 2\delta^{jk}\hat{I} \tag{4}$$

where $a, b \in \{1, ..., N\}$ and $j, k \in \{1, ..., D\}$. The spinor operators are also assumed to be Hermitian (or Majorana)

$$\hat{\gamma}_a^j = \hat{\gamma}_a^{j\dagger}, \tag{5}$$

and to satisfy a tracelessness condition

$$Tr\left[\left(\hat{\gamma}_{a_1}^{j_{1,1}} ... \hat{\gamma}_{a_1}^{j_{1,d_1}}\right)\left(\hat{\gamma}_{a_2}^{j_{2,1}} ... \hat{\gamma}_{a_2}^{j_{2,d_2}}\right) ... \left(\hat{\gamma}_{a_n}^{j_{n,1}} ... \hat{\gamma}_{a_n}^{j_{n,d_n}}\right)\right] = 0. \tag{6}$$

where $1 \leq a_1 < ... < a_n \leq N$ and $1 \leq j_{k,1} < j_{k,2} < ... < j_{k,d_k} \leq D$ for all k. For example, a collection of spinor operators for $D = 2$ can be represented by tensor products of Pauli $\hat{X}$ and $\hat{Y}$ matrices,

$$\begin{aligned}
\hat{\gamma}_1^1 &= \hat{X} \otimes \hat{I} \otimes ... \otimes \hat{I} \\
\hat{\gamma}_1^2 &= \hat{Y} \otimes \hat{I} \otimes ... \otimes \hat{I} \\
\hat{\gamma}_2^1 &= \hat{I} \otimes \hat{X} \otimes ... \otimes \hat{I} \\
\hat{\gamma}_2^2 &= \hat{I} \otimes \hat{Y} \otimes ... \otimes \hat{I} \\
&... \\
\hat{\gamma}_N^1 &= \hat{I} \otimes \hat{I} \otimes ... \otimes \hat{X} \\
\hat{\gamma}_N^2 &= \hat{I} \otimes \hat{I} \otimes ... \otimes \hat{Y}.
\end{aligned} \tag{7}$$

Similarly, the tensor products of euclidean Dirac matrices (which can be constructed from tensor products of all three Pauli matrices) can be used to represent the spinor operators for $D = 4$. Although the dimensionality D of the subsystems is kept arbitrary the two cases with $D = 1$ and $D = 3$ are of particular importance as the respective quantum systems will turn out to be dual to simple classical models on S^0 (in Sec. 3.) and on S^2 (in Sec. 7.) configuration/target spaces.

From the spinor operators we construct a Hamiltonian operator

$$\hat{H}_q = \sum_{j_1...j_N \in \{0,...,D\}} H_{j_1...j_N} \hat{\gamma}_1^{j_1} ... \hat{\gamma}_N^{j_N}. \tag{8}$$

where $\hat{\gamma}_a^0 \equiv \hat{I}$ and all of the components $H_{j_1...j_N}$ are real numbers. Note that the Hamiltonian is not only non-homogeneous and non-local, but also non-k-local as it may include terms with an arbitrary number of $\hat{\gamma}_a^k$'s operators.

The corresponding quantum partition function can be expanded in powers of inverse temperature

$$\begin{aligned}\mathcal{Z}_{\mathrm{q}}[\beta] &= Tr\left[\exp\left(\beta\hat{H}_{\mathrm{q}}\right)\right] \\ &= \sum_{n=0}^{\infty}\frac{\beta^n}{n!}Tr\left[\left(\sum_{j_1...j_N\in\{0,...,D\}} H_{j_1...j_N}\hat{\gamma}_1^{j_1}...\hat{\gamma}_N^{j_N}\right)^n\right]\end{aligned} \tag{9}$$

and then each power of Hamiltonian operator into a sum over all combinations of terms from the Hamiltonian

$$Tr\left[\left(\sum_{j_1...j_N\in\{0,...,D\}} H_{j_1...j_N}\hat{\gamma}_1^{j_1}...\hat{\gamma}_N^{j_N}\right)^n\right] = \sum_A h_A\, Tr\left[\hat{\Gamma}^A\right] \tag{10}$$

where h_A's represent products of $H_{j_1...j_N}$'s components and $\hat{\Gamma}^A$'s the corresponding products of the spinor operators. By upper-case letters (i.e. $A, B, C, ...$) we denote abstract-indices which include information about a particular order of terms from the Hamiltonian (8). Note, however, that two different products h_A and h_B of the same collections of $H_{j_1...j_N}$ components would be the equal, but the corresponding products of operators $\hat{\Gamma}^A$ and $\hat{\Gamma}^B$ may differ by a sign due to anti-commutativity condition (4). For example, if A represents $\left(H_{03}\hat{\gamma}_2^3\right)\left(H_{02}\hat{\gamma}_2^2\right)$ and B represents $\left(H_{02}\hat{\gamma}_2^2\right)\left(H_{03}\hat{\gamma}_2^3\right)$, then $h_A = H_{03}H_{02} = h_B$, but $\hat{\Gamma}^A = \hat{\gamma}_2^3\hat{\gamma}_2^2 = -\hat{\gamma}_2^2\hat{\gamma}_2^3 = -\hat{\Gamma}^B$.

It is convenient to define an ordered product of operators $:\hat{\Gamma}^A:$ (not to confuse with normal ordering) so that $\hat{\gamma}_a^k$ appears to the left of $\hat{\gamma}_b^j$ if either $a < b$ or if $a = b$ and $k < j$ (as for example in (6)). Then

$$\sum_A h_A Tr\left[\hat{\Gamma}^A\right] = \sum_A \theta(\hat{\Gamma}^A) h_A Tr\left[:\hat{\Gamma}^A:\right] \tag{11}$$

where

$$\theta(\hat{\Gamma}^A) = \begin{cases} 1 & \text{if} \quad :\hat{\Gamma}^A: = \hat{\Gamma}^A \\ -1 & \text{if} \quad :\hat{\Gamma}^A: = -\hat{\Gamma}^A. \end{cases} \tag{12}$$

Furthermore, if we define $\sigma(A)$ as a set of all abstract-indices which are equivalent to A up to different combinations of terms from the Hamiltonian, then

$$\sum_A \theta(\hat{\Gamma}^A) h_A Tr\left[:\hat{\Gamma}^A:\right] = \sum_A \mu(A) h_A Tr\left[:\hat{\Gamma}^A:\right] \tag{13}$$

where

$$\mu(A) = \frac{1}{|\sigma(A)|}\sum_{B\in\sigma(A)}\theta(\hat{\Gamma}^B) \tag{14}$$

and then (10) can be rewritten as

$$Tr\left[\left(\sum_{j_1\ldots j_N\in\{0,\ldots,D\}} H_{j_1\ldots j_N}\hat{\gamma}_1^{j_1}\ldots\hat{\gamma}_N^{j_N}\right)^n\right] = \sum_A \mu(A) h_A Tr\left[:\hat{\Gamma}^A:\right]. \quad (15)$$

For example, if A represents $\left(H_{02}\hat{\gamma}_2^2\right)\left(H_{30}\hat{\gamma}_1^3\right)$, then

$$\begin{aligned} h_A &= H_{02}H_{30} \\ \hat{\Gamma}^A &= \hat{\gamma}_2^2\hat{\gamma}_1^3 \\ :\hat{\Gamma}^A: &= \hat{\gamma}_1^3\hat{\gamma}_2^2 \\ \theta(\hat{\Gamma}^A) &= 1 \\ \mu(A) &= \frac{1}{2}(1+1) = 1, \end{aligned} \quad (16)$$

but if A represents $\left(H_{03}\hat{\gamma}_2^3\right)\left(H_{02}\hat{\gamma}_2^2\right)$, then

$$\begin{aligned} h_A &= H_{03}H_{02} \\ \hat{\Gamma}^A &= \hat{\gamma}_2^3\hat{\gamma}_2^2 \\ :\hat{\Gamma}^A: &= \hat{\gamma}_2^2\hat{\gamma}_2^3 \\ \theta(\hat{\Gamma}^A) &= -1 \\ \mu(A) &= \frac{1}{2}(1-1) = 0. \end{aligned} \quad (17)$$

In general, $\mu(A)$ can be calculated using commutation (3) and anti-commutation (4) relations, but the analysis is greatly simplified when all operators either commute (see Sec. 3.) or anti-commute (see Sec. 4.).

3. Commuting operators

Consider a system of N quantum spinors (8) with $D = 1$ whose quantum partition function is given by

$$\mathcal{Z}_q[\beta] = \sum_{n=0}^{\infty}\frac{\beta^n}{n!} Tr\left[\left(\sum_{j_1\ldots j_N\in\{0,1\}} H_{j_1\ldots j_N}\hat{\gamma}_1^{j_1}\ldots\hat{\gamma}_N^{j_N}\right)^n\right] \quad (18)$$

where $\hat{\gamma}^0 \equiv \hat{I}$, and a system of N classical scalars x_a whose classical partition function is

$$\mathcal{Z}_c[\beta] = \mathcal{N}\int\left(\prod_a dx_a\rho(x_a)\right)\sum_{n=0}^{\infty}\frac{\beta^n}{n!}\left(\sum_{j_1\ldots j_N\in\{0,1\}} H_{j_1\ldots j_N} x_1^{j_1}\ldots x_N^{j_N}\right)^n \quad (19)$$

where $x_a^1 \equiv x_a$ and $x_a^0 \equiv 1$. The classical system (19) is dual to the quantum system (18) if and only if the two partition functions are equal (2), but since the equality must be satisfied for all β, an equivalent condition is

$$\mathcal{N} \int \left(\prod_a dx_a \rho(x_a) \right) \left(\sum_{j_1 ... j_N \in \{0,1\}} H_{j_1 ... j_N} \, x_1^{j_1} ... x_N^{j_N} \right)^n =$$

$$= Tr \left[\left(\sum_{j_1 ... j_N \in \{0,1\}} H_{j_1 ... j_N} \, \hat{\gamma}_1^{j_1} ... \hat{\gamma}_N^{j_N} \right)^n \right]$$

or using the abstract-indices notation (15)

$$\mathcal{N} \int \left(\prod_a dx_a \rho(x_a) \right) \sum_A h_A \, X^A = \sum_A \mu(A) h_A Tr \left[: \hat{\Gamma}^A : \right]. \tag{20}$$

where X^A is the corresponding products of the scalars. Since all of the spinor operators $\hat{\gamma}_a^1$'s commute the products of spinors are such that $: \hat{\Gamma}^A :=\hat{\Gamma}^A$ and $\mu(A) = 1$ for all A. And then by matching individual terms we get

$$\mathcal{N} \int \left(\prod_a dx_a \rho(x_a) \right) X^A = Tr \left[: \hat{\Gamma}^A : \right]. \tag{21}$$

The ordered product of operators $: \hat{\Gamma}^A :$ either contains an even number of $\hat{\gamma}_a^1$ operators for every a, or there is at least one a for which there is an odd number of $\hat{\gamma}_a^1$'s. In the latter case equation (6) implies that $\hat{\Gamma}^A$ is traceless and so does not contribute to the partition function, and in the former case equation (4) implies that $\hat{\Gamma}^A = \hat{I}$. If we take the trace of identity to be

$$\mathcal{N} \equiv Tr \left[\hat{I} \right] = D^N, \tag{22}$$

then the measure of integration $\rho(x_a)$ should be such that all odd statistical moments vanish and all even statistical moment are the same, i.e

$$\int (x_a)^n \rho(x_a) dx_a = \begin{cases} 1 \text{ if } n \text{ is even} \\ 0 \text{ if } n \text{ is odd.} \end{cases} \tag{23}$$

But this is can be easily achieved with

$$\rho(x_a) = \frac{\delta(x_a - 1) + \delta(x_a + 1)}{2} \tag{24}$$

which corresponds to a classical partition function

$$\mathcal{Z}_{\rm c}[\beta] = \mathcal{N} \int \left(\prod_a dx_a \frac{\delta(x_a - 1) + \delta(x_a + 1)}{2} \right) \exp\left(\beta \sum_{j_1 ... j_N \in \{0,1\}} H_{j_1 ... j_N} \, x_1^{j_1} ... x_N^{j_N} \right) \tag{25}$$

or simply

$$\mathcal{Z}_{\rm c}[\beta] = \mathcal{N} \sum_{x_1 ... x_N \in \{1,-1\}} \exp\left(\beta \sum_{j_1 ... j_N \in \{0,1\}} H_{j_1 ... j_N} \, x_1^{j_1} ... x_N^{j_N} \right) . \tag{26}$$

Therefore we conclude that the partition function of a quantum system whose Hamiltonian is built out of only $\hat{\gamma}_a^1$ operators can always be expressed as a classical partition function with Hamiltonian,

$$H_{\rm c} = \sum_{j_1 ... j_N \in \{0,1\}} H_{j_1 ... j_N} \, x_1^{j_1} ... x_N^{j_N} \tag{27}$$

where x_a are the classical spinors which take values plus or minus one, or equivalently classical scalars on S^0 target space. This also suggests that the eigenvalues of the Hamiltonian $\hat{H}_{\rm q}$ are simply related to its components $H_{j_1 j_2 ... j_N}$, i.e.

$$E_x = \sum_{j_1 ... j_N \in \{0,1\}} H_{j_1 j_2 ... j_N} \, x_1^{j_1} x_2^{j_2} ... x_N^{j_N} , \tag{28}$$

where $x \in \{-1, 1\}^N$.

4. Anti-commuting operators

Next consider a single quantum spinor, i.e. $N = 1$ but with arbitrary $D \geq 1$, whose quantum partition function is

$$\mathcal{Z}_{\rm q}[\beta] = \sum_{n=0}^{\infty} \frac{\beta^n}{n!} \, Tr \left[\left(\sum_{j \in \{1,...,D\}} H_j \, \hat{\gamma}^j \right)^n \right] \tag{29}$$

where $\hat{\gamma}^0 \equiv \hat{I}$, and a classical system of D scalars whose classical partition function is

$$\mathcal{Z}_{\rm c}[\beta] = \mathcal{N} \int d^D x \rho(x) \sum_{n=0}^{\infty} \frac{\beta^n}{n!} \left(\sum_{j \in \{1,...,D\}} H_j \, x^j \right)^n . \tag{30}$$

where by upper indices j we now denote different scalars x^j. To find $\rho(x)$ such that the duality condition (2) is satisfied, we must once again match individual terms in the expansions of partition functions, i.e.

$$\mathcal{N}\int d^D x\rho(x)\left(\sum_{j\in\{1,\dots,D\}} H_j\, x^j\right)^n = Tr\left[\left(\sum_{j\in\{1,\dots,D\}} H_j\,\hat{\gamma}^j\right)^n\right]$$

or using the abstract-indices notation (15)

$$\begin{aligned}\mathcal{N}\int d^D x\rho(x)\sum_A h_A\, X^A &= \sum_A \mu(A) h_A Tr\left[:\hat{\Gamma}^A:\right],\\ \mathcal{N}\int d^D x\rho(x) X^A &= \mu(A) Tr\left[:\hat{\Gamma}^A:\right]\end{aligned} \tag{31}$$

where X^A is a given product of scalars x^k's and $:\hat{\Gamma}^A:$ is the corresponding ordered product of operators $\hat{\gamma}^k$'s. But since all odd statistical moments must vanish (because the trace of the corresponding product of operators would vanish (6)) we obtain an equivalent condition

$$\begin{aligned}\mathcal{N}\int d^D x\rho(x)\prod_k (x^k)^{2n_k} &= \mu(A) Tr\left[\prod_k (\hat{\gamma}^k)^{2n_k}\right]\\ \int d^D x\rho(x)\prod_k (x^k)^{2n_k} &= \mu(A)\end{aligned} \tag{32}$$

for some even integers $2n_k$'s which represent the number of operators $\hat{\gamma}^k$'s (or of scalars x^k's) in the product of operators $\hat{\Gamma}^A$ (or in the product of scalars X^A). And so to calculate $\mu(A)$ (and then $\rho(x_a)$) we must sum over all combinations of anti-commuting spinor operators (14) which are conveniently described by different terms in the multinomial expansions.

Consider the following two multinomials: a sum of commuting scalars raised to some even power

$$\begin{aligned}&\left(x^1 + x^2 + \dots + x^D\right)^{2K} =\\ &= \sum_{m_1+\dots+m_D=2K} \frac{(m_1+\dots+m_D)!}{(m_1)!\dots(m_D)!}(x^1)^{m_1}(x^2)^{m_2}\dots(x^D)^{m_D}\\ &= \sum_{m_1+\dots+m_D=2K} \frac{(\sum_k m_k)!}{\prod_k (m_k)!}(x^1)^{m_1}(x^2)^{m_2}\dots(x^D)^{m_D}\end{aligned} \tag{33}$$

and a sum of anti-commuting operators raised to the same even power

$$\left(\hat{\gamma}^1 + \hat{\gamma}^2 + \dots + \hat{\gamma}^D\right)^{2K} =$$

$$\begin{aligned}
&= \left((\hat{\gamma}^1)^2 + (\hat{\gamma}^2)^2 + ... + (\hat{\gamma}^D)^2\right)^K \\
&= \sum_{n_1+...+n_D=K} \frac{(n_1 + ... + n_D)!}{(n_1)!...(n_D)!} (\hat{\gamma}^1)^{2n_1} (\hat{\gamma}^2)^{2n_2} ... (\hat{\gamma}^D)^{2n_D}. \\
&= \sum_{n_1+...+n_D=K} \frac{(\sum_k n_k)!}{\prod_k (n_k)!} (\hat{\gamma}^1)^{2n_1} (\hat{\gamma}^2)^{2n_2} ... (\hat{\gamma}^D)^{2n_D}. \qquad (34)
\end{aligned}$$

Separate terms in the expansion of operators (34) represent products of $\hat{\gamma}^k$'s applied in different orders (or combinations $\sigma(A)$) and we are interested in products of $2n_1$ of $\hat{\gamma}^1$'s, $2n_2$ of $\hat{\gamma}^2$'s, etc (see Eq. (32)) The total number of such products (or $|\sigma(A)|$) is given by a multinomial coefficient in the expansion of scalars (33) with $m_k = 2n_k$, i.e.

$$|\sigma(A)| = \frac{(\sum_k 2n_k)!}{\prod_k (2n_k)!}, \qquad (35)$$

but not all of them come with the same sign when compared to an operator ordered as $(\hat{\gamma}^1)^{2n_1} (\hat{\gamma}^2)^{2n_2} ... (\hat{\gamma}^D)^{2n_D}$. And according to (14) the sum of these signs (or $\theta(B)$'s) over all combinations of operators (or $B \in \sigma(A)$) is what determines $\mu(A)$,

$$\begin{aligned}
\mu(A) &= |\sigma(A)|^{-1} \sum_{B \in \sigma(A)} \theta(B) \qquad (36) \\
&= \frac{\prod_k (2n_k)!}{(\sum_k 2n_k)!} \frac{(\sum_k n_k)!}{\prod_k n_k!} \qquad (37)
\end{aligned}$$

as is evident from (34).

This can be seen directly if we integrate (33) weighted by $\mathcal{N}\rho(x)$ and then equate it to the trace of (34),

$$\mathcal{N} \int d^D x \rho(x) \left(x^1 + x^2 + ... + x^D\right)^{2K} = Tr\left[\left(\hat{\gamma}^1 + \hat{\gamma}^2 + ... + \hat{\gamma}^D\right)^{2K}\right]$$

$$\sum_{m_1+...+m_D=2K} \frac{(\sum_k m_k)!}{\prod_k m_k!} \int d^D x \rho(x) \prod_k (x^k)^{m_k} = \sum_{n_1+...+n_D=K} \frac{(\sum_k n_k)!}{\prod_k n_k!}. \qquad (38)$$

Since the correct measure of integration $\rho(x)$ must be such that all odd moments vanish the above condition can be rewritten as

$$\sum_{2n_1+...+2n_D=2K} \frac{(\sum_k 2n_k)!}{\prod_k (2n_k)!} \int d^D x \rho(x) \prod_k (x^k)^{2n_k} = \sum_{n_1+...+n_D=K} \frac{(\sum_k n_k)!}{\prod_k n_k!} \qquad (39)$$

where on the left hand side we substituted m's for n's using $m_k = 2n_k$. Then, in order for individual terms on both sides of (39) to be equal the measure $\rho(x)$ should be such that

$$\int d^D x\, \rho(x) \prod_k (x^k)^{2n_k} = \frac{\prod_k (2n_k)!}{(\sum_k 2n_k)!} \frac{(\sum_k n_k)!}{\prod_k n_k!} = \mu(A) \qquad (40)$$

where A can represent an arbitrary product of terms with $2n_1$ of $\hat{\gamma}^1$'s, $2n_2$ of $\hat{\gamma}^2$'s, etc.

The moments generating function of $\rho(x)$ can be obtained directly from (40),

$$\begin{aligned} M(p_1, ..., p_D) &= \sum_{n_1,...,n_D} \left(\int d^D x \rho(x) \prod_k (x^k)^{2n_k} \right) \frac{p_1^{2n_1} ... p_D^{2n_D}}{\prod_k (2n_k)!} \\ &= \sum_{n_1,...,n_D} \frac{\prod_k (2n_k)!}{(\sum_k 2n_k)!} \frac{(\sum_k n_k)!}{\prod_k n_k!} \frac{p_1^{2n_1} ... p_D^{2n_D}}{\prod_k (2n_k)!} \\ &= \sum_{n_1,...,n_D} \frac{1}{(2\sum_k n_k)!} \frac{(\sum_k n_k)!}{\prod_k n_k!} p_1^{2n_1} ... p_D^{2n_D} \\ &= \sum_K \frac{1}{(2K)!} \sum_{n_1+...+n_D=K} \frac{K!}{\prod_k n_k!} p_1^{2n_1} ... p_D^{2n_D} \\ &= \sum_K \frac{1}{(2K)!} \left(p_1^2 + ... + p_D^2\right)^K \\ &= \cosh\left(\sqrt{p_1^2 + ... + p_D^2}\right) \end{aligned} \qquad (41)$$

and then the corresponding characteristic function is

$$M(ip_1, ..., ip_D) = \cos\left(\sqrt{p_1^2 + ... + p_D^2}\right) = \cos\left(\sqrt{\sum_k p_k^2}\right) \qquad (42)$$

and its inverse Fourier transform gives us the desired measure of integration

$$\rho(x) = \int \frac{d^D p}{(2\pi)^D} \cos\left(\sqrt{\sum_k p_k^2}\right) \exp\left(i \sum_k x^k p_k\right). \qquad (43)$$

5. Emergent space-time

The measure of integration (43) for $D = 1$ can be easily calculated,

$$\begin{aligned} \rho(x) &= \int \frac{dp}{2\pi} \cos\left(\sqrt{p^2}\right) \exp(ixp) \\ &= \int \frac{dp}{2\pi} \cos(p) \exp(ixp) \\ &= \frac{1}{2} \int \frac{dp}{2\pi} \left(\exp(ip + ixp) + \exp(-ip + ixp)\right) \\ &= \frac{1}{2} \left(\delta(x+1) + \delta(x-1)\right), \end{aligned} \tag{44}$$

which is in agreement with (24). As we will see shortly this result is due to the fact that the time derivative of a retarded Green's function of 1+1-dimensional d'Alembert operator is given by a sum of two delta-functions propagating in opposite directions on an emergent space-time. It turns out that the same statement is true in higher dimensions, i.e. $D > 1$ or $D+1 > 2$, but the form of the Green's functions (and of the corresponding measures $\rho(x)$) is of course different.

To evaluate the integral in (43) for arbitrary D we note that

$$\varphi(x^\mu) = \varphi(\vec{x}, x^0) \equiv \int \frac{d^D p}{(2\pi)^D} \cos\left(x^0 \sqrt{\sum_k (p_k)^2}\right) \exp\left(i \sum_k p_k x^k\right) \tag{45}$$

solves a D+1-dimensional wave equation,

$$\left((\partial_0)^2 - \sum_k (\partial_k)^2\right) \varphi(x^\mu) = 0, \tag{46}$$

with initial conditions

$$\varphi(\vec{x}, 0) = \delta^{(D)}(\vec{x}) \tag{47}$$

$$\partial_0 \varphi(\vec{x}, 0) = 0 \tag{48}$$

where

$$\partial_0 \equiv \frac{\partial}{\partial x^0} \tag{49}$$

$$\partial_k \equiv \frac{\partial}{\partial x^k}. \tag{50}$$

Then the solution of the D+1-dimensional wave equation is given by

$$\varphi(x^\mu) = \int d^D y \, \partial_0 G(\vec{x}, x^0; \vec{y}, 0) \delta^{(D)}(\vec{y}) = \partial_0 G(\vec{x}, x^0) \tag{51}$$

where

$$G(x^\mu;y^\mu) = G(x^\mu - y^\mu) = G(\vec{x}-\vec{y}, x^0 - y^0) \tag{52}$$

is the retarded Green's function (or retarded propagator) of D+1-dimensional d'Alembert operator.

The retarded propagator can be expressed as a D-dimensional integral

$$G(x^\mu;y^\mu) = \int \frac{d^D p}{(2\pi)^D} \frac{\sin\left((x^0-y^0)\,\omega(p)\right)}{\omega(p)} \exp\left(i\sum_k p_k(x^k - y^k)\right) \tag{53}$$

where

$$\omega(p) \equiv \sqrt{\sum_k (p_k)^2} \tag{54}$$

or as a more symmetric D+1-dimensional integral

$$G(x^\mu;y^\mu) = \int \frac{d^{D+1}p}{(2\pi)^{D+1}} \frac{\exp\left(i\sum_\mu p_\mu(x^\mu - y^\mu)\right)}{\sum_\mu p_\mu p^\mu} \tag{55}$$

with appropriately chosen contour of integration (i.e. both poles shifted downwards) and $p^0 \equiv -p_0$. Then the solution is

$$\varphi(x^\mu) = \partial_0 G(\vec{x}, x^0) = \int \frac{d^{D+1}p}{(2\pi)^{D+1}} \frac{ip_0}{\sum_\mu p_\mu p^\mu} \exp\left(i\sum_\mu p_\mu x^\mu\right) \tag{56}$$

and we can define an extended (into emergent "temporal" direction T) classical partition function

$$\begin{aligned} \mathcal{Z}_\mathrm{c}[\beta, T] &= \mathcal{N}\int d^D x\, \varphi(\vec{x}, T) \exp\left(\beta \sum_{j\in\{1,\dots,D\}} H_j\, x^j\right) \\ &= \mathcal{N}\int d^D x\, \partial_0 G(\vec{x}, T) \exp\left(\beta \sum_{j\in\{1,\dots,D\}} H_j\, x^j\right) \end{aligned} \tag{57}$$

which satisfies the desired duality condition

$$\mathcal{Z}_\mathrm{c}[\beta, 1] = \mathcal{Z}_\mathrm{q}[\beta] \tag{58}$$

and also normalization conditions

$$\mathcal{Z}_\mathrm{c}[\beta, 0] = \mathcal{Z}_\mathrm{c}[0, T] = \mathcal{N}. \tag{59}$$

6. Existence of duality

In the previous sections we obtained the correct measure of integration of the classical partition function for either commuting (i.e. $D = 1$) or anti-commuting (i.e. $N = 1$) degrees of freedom. Next we shall consider a more general system (8) with both commuting and anti-commuting terms (i.e. $N > 1$ and $D > 1$) described by the quantum partition function

$$\mathcal{Z}_{\mathrm{q}}[\beta] = \sum_{n=0}^{\infty} \frac{\beta^n}{n!} \, Tr \left[\left(\sum_{j_1 \ldots j_N \in \{0,\ldots,D\}} H_{j_1 \ldots j_N} \, \hat{\gamma}_1^{j_1} \ldots \hat{\gamma}_N^{j_N} \right)^n \right]. \tag{60}$$

The corresponding classical partition function (for ND classical scalars x_a^k) can be defined in a similar manner

$$\mathcal{Z}_{\mathrm{c}}[\beta] = \mathcal{N} \int \left(\prod_a d^D x_a \right) \rho(x) \sum_{n=0}^{\infty} \frac{\beta^n}{n!} \left(\sum_{j_1 \ldots j_N \in \{0,1\}} H_{j_1 \ldots j_N} \, x_1^{j_1} \ldots x_N^{j_N} \right)^n , \tag{61}$$

but it is no longer clear if (or when) the duality condition (2) could be satisfied for some measure of integration

$$\rho(x) = \rho(x_1^1, \ldots x_1^D, \ldots, x_N^1, \ldots x_N^D). \tag{62}$$

Indeed, by matching separate terms in (60) and (61) we obtain

$$\mathcal{N} \int \left(\prod_a d^D x_a \right) \rho(x) \left(\sum_{j_1 \ldots j_N \in \{0,\ldots,D\}} H_{j_1 \ldots j_N} \, x_1^{j_1} \ldots x_N^{j_N} \right)^n$$

$$= Tr \left[\left(\sum_{j_1 \ldots j_N \in \{0,\ldots,D\}} H_{j_1 \ldots j_N} \, \hat{\gamma}_1^{j_1} \ldots \hat{\gamma}_N^{j_N} \right)^n \right]$$

or using the abstract-indices notation (15)

$$\mathcal{N} \int \left(\prod_a d^D x_a \right) \rho(x) \sum_A h_A \, X^A = \sum_A \mu(A) h_A \, Tr \left[: \hat{\Gamma}^A : \right]. \tag{63}$$

This suggests that the duality condition (2) can be satisfied only if individual terms in (63) are equal,

$$\mathcal{N} \int \left(\prod_a d^D x_a \right) \rho(x) X^A = \mu(A) Tr \left[: \hat{\Gamma}^A : \right]. \tag{64}$$

Whenever the product of spinor operators $\hat{\Gamma}^A$ contains an even number $2n_k^a$ of $\hat{\gamma}_a^k$ for all a and k the trace is

$$Tr\left[:\hat{\Gamma}^A:\right] = \mathcal{N} \tag{65}$$

and thus the even statistical moments must be given by

$$\begin{aligned} \mathcal{N}\int\left(\prod_a d^D x_a\right) \;\; \rho(x) \;\; X^A \qquad &= \mathcal{N}\mu(A) \\ \int\left(\prod_a d^D x_a\right) \;\; \rho(x) \;\; \prod_{a,k}\left(x_a^k\right)^{2n_k^a} &= \mu(A) \end{aligned} \tag{66}$$

and all odd statistical moments must vanish. However, since the measure is uniquely determined from the statistical moments and the statistical moments are to be determined from $\mu(A)$, the measure can only exist if $\mu(A)$ is uniquely determined by non-negative integers n_k^a's. This puts a restriction on the quantum system (60), whose classical dual is (61), which can be expressed as the following condition:

$$:\hat{\Gamma}^A:=:\hat{\Gamma}^B: \quad \Rightarrow \qquad \mu(A) = \mu(B). \tag{67}$$

In other words, even if A and B are not in the same combination class, i.e. $\sigma(A) \neq \sigma(B)$, but the corresponding products of operators are the same, i.e. $:\hat{\Gamma}^A:=:\hat{\Gamma}^B:$, the statistical moments must also be the same, i.e. $\mu(A) = \mu(B)$. Note that (67) is only a sufficient condition as there might still be other dual classical systems which are not described by (61) (see for example imaginary time formalism in Appendix A).

7. Separable measure

In the previous section we argued that the quantum system (60) will have a classical dual (61) if the condition (67) is satisfied. In this section we are going to study a particular example of the quantum system for which the classical dual not only exists, but the integration measure is also separable, i.e.

$$\rho(x) = \rho(x_1^1, ...x_1^D, ..., x_N^1, ...x_N^D) = \prod_a \rho\left(x_a^1, ..., x_a^D\right). \tag{68}$$

Consider a Hamiltonian operator (8) with components which can be expressed as

$$H_{j_1 \dots j_N} = \sum_{k_1 \dots k_N \in \{0,1\}} \mathcal{H}_{k_1 \dots k_N} \eta_{1,j_1}^{k_1} \dots \eta_{N,j_N}^{k_N} \tag{69}$$

where we assume that

$$\eta^0_{a,j} = \delta_{j0} \tag{70}$$
$$\eta^k_{a,0} = \delta_{0k} \tag{71}$$

and, without loss of generality, $H_{0...0} = 0$. Then the Hamiltonian operator can be rewritten as

$$\begin{aligned} \hat{H}_{\rm q} &= \sum_{j_1...j_N \in \{0,...,D\}} H_{j_1...j_N} \hat{\gamma}_1^{j_1}...\hat{\gamma}_N^{j_N} \\ &= \sum_{k_1...k_N \in \{0,1\}} \mathcal{H}_{k_1...k_N} \left(\sum_{j_1 \in \{0,...,D\}} \eta^{k_1}_{1,j_1} \hat{\gamma}_1^{j_1} \right) ... \left(\sum_{j_N \in \{0,...,D\}} \eta^{k_N}_{N,j_N} \hat{\gamma}_N^{j_N} \right) \\ &= \sum_{k_1...k_N \in \{0,1\}} \mathcal{H}_{k_1...k_N} \hat{\eta}_1^{k_1}...\hat{\eta}_N^{k_N} \end{aligned} \tag{72}$$

where the combined operators $\hat{\eta}_a$'s are defined as linear combinations of spin operators

$$\hat{\eta}_a = \hat{\eta}^1_a = \sum_{j \in \{0,...,D\}} \eta^1_{a,j} \hat{\gamma}^j_a = \sum_{j \in \{1,...,D\}} \eta^1_{a,j} \hat{\gamma}^j_a \tag{73}$$

and

$$\hat{\eta}^0_a = \sum_{j \in \{0,...,D\}} \eta^0_{a,j} \hat{\gamma}^j_a = \sum_{j \in \{0,...,D\}} \delta_{0j} \hat{\gamma}^j_a = \hat{\gamma}^0_a = \hat{I}. \tag{74}$$

Since the combined operators satisfy a commutation relation

$$[\hat{\eta}_a, \hat{\eta}_b] = 0 \tag{75}$$

we can essentially follow the analysis of Sec. 3. with classical partition function

$$\begin{aligned} \mathcal{Z}_{\rm c}[\beta] &= \\ &= \mathcal{N} \int \left(\prod_a d^D x_a \rho(x_a) \right) \sum_{n=0}^{\infty} \frac{\beta^n}{n!} \left(\sum_{j_1...j_N \in \{0,...,D\}} H_{j_1...j_N} x_1^{j_1}...x_N^{j_N} \right)^n \\ &= \mathcal{N} \int \left(\prod_a d^D x_a \rho(x_a) \right) \sum_{n=0}^{\infty} \frac{\beta^n}{n!} \left(\sum_{k_1...k_N \in \{0,1\}} \mathcal{H}_{k_1...k_N} \chi_1^{k_1}...\chi_N^{k_N} \right)^n \end{aligned} \tag{76}$$

where

$$\chi_a = \chi^1_a = \sum_{j \in \{1,...,D\}} \eta^1_{a,j} x^j_a \tag{77}$$
$$\chi^0_a = 1. \tag{78}$$

Indeed, the duality condition (2) implies

$$\mathcal{N}\int\left(\prod_a d^D x_a \rho(x_a)\right)\left(\sum_{k_1...k_N\in\{0,1\}} \mathcal{H}_{k_1...k_N}\chi_1^{k_1}...\chi_N^{k_N}\right)^n$$

$$= Tr\left[\left(\sum_{k_1...k_N\in\{0,1\}} \mathcal{H}_{k_1...k_N}\hat{\eta}_1^{k_1}...\hat{\eta}_N^{k_N}\right)^n\right]$$

which can be satisfied (due to commutativity of the combined operators (75)) only if

$$\mathcal{N}\int d^D x_a \rho(x_a)\,(\chi_a)^n = Tr\left[(\hat{\eta}_a)^n\right] \tag{79}$$

or, equivalently, if

$$\mathcal{N}\int d^D x_a \rho(x_a)\left(\sum_{j\in\{1,...,D\}} \eta_j^1 x_a^j\right)^n = Tr\left[\left(\sum_{j\in\{1,...,D\}} \eta_j^1 \hat{\gamma}_a^j\right)^n\right] \tag{80}$$

for all a and n. But since all of $\hat{\gamma}_a^j$ operators in (80) anti-commute the corresponding measure is the same as in Sec. 4. which was shown to be given by a retarded Green's function in Sec. 5.. Therefore, for a quantum system whose Hamiltonian components can be expressed as in (69), the dual classical system is described by a classical partition function

$$\mathcal{Z}_c[\beta,T] = \mathcal{N}\int\left(\prod_a d^D x_a\,\partial_0 G(\vec{x}_a,T_a)\right)$$

$$\exp\left(\beta\sum_{j_1...j_N\in\{0,1\}} H_{j_1...j_N}\,x_1^{j_1}...x_N^{j_N}\right), \tag{81}$$

where $G(\vec{x}_a,T_a)$ is the retarded Green's function of $D{+}1$-dimensional d'Alembert operator (see Sec. 5.). Note that the measure of integration is already normalized,

$$\int\left(\prod_a d^D x_a\,\partial_0 G(\vec{x}_a,T_a)\right) = 1, \tag{82}$$

but it can be interpreted as a probability density only if it takes non-negative values. For example, when $D=1$

$$\mathcal{Z}_c[\beta,T] = \mathcal{N}\int\prod_a\left(\frac{dx_a}{2}\left(\delta(x_a^1 - T_a) + \delta(x_a^1 + T_a)\right)\right)\exp\left(\beta H_c\right), \tag{83}$$

in agreement with Sec 3., or when $D = 3$

$$\mathcal{Z}_c[\beta, T] = \mathcal{N} \int \prod_a \left(\frac{d^3 x_a}{4\pi T_a^2} \delta \left(\sum_k \left(x_a^k \right)^2 - T_a^2 \right) \right) \exp\left(\beta H_c\right). \tag{84}$$

Of course there is no reason to expect that the measure will remain positive for more general quantum systems and then the dual system defined in the similar manner would not be classical *per se* as the corresponding partition function would involve integration over configurations with negative probabilities.

8. Discussion

In this paper we took a step towards extending the quantum-classical duality (2) beyond the imaginary time formalism. In particular we showed that the quantum partition function for a system of quantum spinors (8) with non-homogeneous, non-local, and non-k-local interactions can be described as a classical partition for some scalar degrees of freedom. The measure of integration in the classical partition functions is non-trivial, but the locality structure of the dual theories was preserved, in contrast to the imaginary time formalism. We derived a general sufficient condition for the existence of the duality (67) and gave three examples of the duality for quantum systems with only commuting (see Sec. 3.), with only anti-commuting (see Sec. 4.), and with both commuting and anti-commuting (see Sec. 7.) degrees of freedom.

An interesting byproduct of our analysis was the realization that the non-trivial measure of integration in the classical partition function can be described using relativistic Green's functions (Sec. 5.). This suggest a possible (and quite general) mechanism for the emergence of a classical space-time from anti-commutativity of quantum operators which deserves a separate study (see, however, Refs. [6, 7, 9, 8] for other recent attempts to derive space-time from quantum mechanics). But since the duality mapping supposedly works both ways (from quantum to classical and from classical to quantum) the very same result can be interpreted as a possible mechanism for the emergence of anti-commutativity of quantum operators from Lorentzian symmetry of a classical space-time. And then it would be interesting to see if the phenomena can be responsible for the emergence of quantum mechanics from classical/statistical mechanics (see, for example, Refs. [10, 11, 12, 13] for some recent attempts to derive quantum mechanics).

The stumbling block for any classical theories with local hidden variables are the Bell's inequalities [14, 15]. It is hard to see how quantum effects, such as entanglement between qubits in an EPR pair, can be described using classical hidden variables that are also local [16]. However, we have already seen that the quantum-to-classical mapping between equilibrium systems can preserve locality, and so it would be important to see if the locality can also be preserved away from the equilibrium. In Sec. 5. we derived an extended partition function (57) with a temporal parameter T which describes the dynamics in an emergent space-time. For the equilibrium classical partition function the parameter T had to be set to one, but the physical meaning of other values of T remains unclear. Could it be that the extended partition function describes a non-equilibrium dynamics of the system from some zero entropy state at $T = 0$ towards some maximum entropy state at $T = 1$? And if so does this evolution follow the principle of the stationary entropy production that was recently proposed in [13]? We leave these questions for future work.

A Imaginary time formalism

Consider a quantum system with a preferred tensor product factorization of Hilbert space into N factors. The quantum partition function for such system can be written as

$$\mathcal{Z}_{\mathrm{q}}[\beta] = Tr\left[\exp\left(\beta \hat{H}_{\mathrm{q}}\right)\right] = \sum_{j} \langle j| \exp\left(\beta \hat{H}_{\mathrm{q}}\right) |j\rangle \tag{85}$$

where the sum is taken over some set of orthonormal basis vectors

$$|j\rangle \equiv \bigotimes_{a=1}^{N} |j_a\rangle \tag{86}$$

and

$$j_a \in \{1, ..., D\}. \tag{87}$$

This partition function can be evaluated as a path integral over imaginary time with periodic boundary conditions. By splitting the imaginary time β into T intervals we obtain

$$\mathcal{Z}_{\mathrm{q}}[\beta] =$$

$$= \sum_{j(1)} \sum_{j(2)} ... \sum_{j(T)} \langle j(T)| e^{\frac{\beta}{T}\hat{H}_{\mathrm{q}}} |j(1)\rangle \langle j(1)| e^{\frac{\beta}{T}\hat{H}_{\mathrm{q}}} |j(2)\rangle ... \langle x(T-1)| e^{\frac{\beta}{T}\hat{H}_{\mathrm{q}}} |j(T)\rangle \tag{88}$$

But this may be also interpreted as the classical partition function with summation taken over all configurations $j(1), j(2), ..., j(T)$ (or discretized paths) weighted by

$$p\left(j(1), j(2) ... j(T)\right) =$$

$$= \langle j(T)| e^{\frac{\beta}{T}\hat{H}_{\mathrm{q}}} |j(1)\rangle \langle j(1)| e^{\frac{\beta}{T}\hat{H}_{\mathrm{q}}} |j(2)\rangle ... \langle j(T-1)| e^{\frac{\beta}{T}\hat{H}_{\mathrm{q}}} |j(T)\rangle. \tag{89}$$

Then what we have is two systems one quantum, described by partition function (85), and another one classical (or more precisely statistical), described by

$$\begin{aligned} \mathcal{Z}_{\mathrm{c}}[\beta] &= \sum_{j(1)}\sum_{j(2)}...\sum_{j(T)} p\left(j(1), j(2)...j(T)\right) \\ &= \sum_{j(1)}\sum_{j(2)}...\sum_{j(T)} \exp\left(H_{\mathrm{c}}(\beta, j)\right), \end{aligned} \tag{90}$$

where

$$H_{\mathrm{c}}(\beta, j) = \sum_{t=1}^{T} \log\left(\left\langle j(t) \left| \exp\left(\frac{\beta}{T}\hat{H}_{\mathrm{q}}\right) \right| j(t+1) \right\rangle\right) \tag{91}$$

such that the duality condition (2) is satisfied.

For sufficiently large T the exponential can be approximated by a linear term,

$$\exp\left(\frac{\beta}{T}\hat{H}_{\mathrm{q}}\right) \approx \hat{I} + \frac{\beta}{T}\hat{H}_{\mathrm{q}} \tag{92}$$

and then

$$\left\langle i \left| \exp\left(\frac{\beta}{T}\hat{H}_{\mathrm{q}}\right) \right| j \right\rangle \approx \delta_{ij} + \frac{\beta}{T}\langle i|\hat{H}_{\mathrm{q}}|j\rangle. \tag{93}$$

Equation (93) describes what is known as transfer matrix, but if we want the product of elements of these matrices (89) to represent probabilities or, equivalently, the classical Hamiltonian (91) to be real, certain restrictions must apply to its from. In particular, what we want is to choose basis vectors $|i\rangle$ so that the off-diagonal terms of the Hamiltonian $\langle i|\hat{H}_{\mathrm{q}}|j\rangle$ are non-negative and then to choose T large enough so that the diagonal terms of the transfer matrix are also non-negative.

References

[1] Bloch, F. (1932), "Zur Theorie des Austauschproblems und der Remanenzerscheinung der Ferromagnetika," Z. Phys. 74 (5-6): 295-335.

[2] Jean Zinn-Justin (2002), " Quantum Field Theory and Critical Phenomena," Oxford University Press

[3] D.A. Kirznits and A.D. Linde, "Macroscopic Consequences of the Weinberg Model," Phys. Lett. B42 (1972) 471

[4] Weinberg, S. (1974), "Gauge and Global Symmetries at High Temperature," Phys. Rev. D. 9 (12): 3357-3378.

[5] L. Dolan, and R. Jackiw (1974), "Symmetry behavior at finite temperature," Phys. Rev. D. 9 (12): 3320-3341.

[6] M. Van Raamsdonk, "Building up spacetime with quantum entanglement," Gen. Rel. Grav. **42**, 2323 (2010)

[7] C. Cao, S. M. Carroll and S. Michalakis, "Space from Hilbert Space: Recovering Geometry from Bulk Entanglement," Phys. Rev. D 95, 024031 (2017)

[8] V. Vanchurin, "Information Graph Flow: a geometric approximation of quantum and statistical systems," Found.Phys. 48 (2018) no.6, 636-653

[9] M. Noorbala, "SpaceTime from Hilbert Space: Decompositions of Hilbert Space as Instances of Time, " Fortsch.Phys. 66 (2018) no.2, 1800002

[10] S. Adler, "Quantum Theory as an Emergent Phenomenon" (Cambridge UP, Cambridge, 2004)

[11] A. Caticha, "Entropic Inference and the Foundations of Physics" (EBEB 2012, São Paulo, Brazil)

[12] G. 't Hooft, "The Cellular Automaton Interpretation of Quantum Mechanics" (Springer, 2016)

[13] V. Vanchurin, "Entropic Mechanics," arXiv:1901.07369 [cond-mat.stat-mech].

[14] S.J. Freedman; J.F. Clauser (1972). "Experimental test of local hidden-variable theories". Phys. Rev. Lett. 28 (938): 938-941

[15] Alain Aspect; Philippe Grangier; Gérard Roger (1981). "Experimental Tests of Realistic Local Theories via Bell's Theorem" Phys. Rev. Lett. 47 (7): 460-3

[16] Bell, John (1964). "On the Einstein Podolsky Rosen Paradox". Physics. 1 (3): 195-200

Transformation properties of nonlinear evolution equations in 1+1 dimensions*

Olena Vaneeva[†]
Institute of Mathematics of the National
Academy of Sciences of Ukraine,
3 Tereshchenkivs'ka Str., 01024 Kyiv, Ukraine

Abstract

We study transformation properties of the general class of evolution equations $u_t = H(t, x, u, u_x, u_{xx})$, $H_{u_{xx}} \neq 0$ and construct a chain of its nested normalized subclasses. A special attention is paid to a class of variable coefficient equations of reaction-diffusion–convection type. The obtained equivalence groups can further be used in group analysis of such classes of nonlinear evolution equations including the construction of exact solutions and local conservation laws.

1. Introduction

A number of mathematical models in physics and biology are represented by (1+1)-dimensional second-order nonlinear evolution equations. Such models are used in such diverse fields as quantum field theory [1, pp. 294–341], physics of nanowire semiconductor devices [2], and population genetics [3].

Many nonlinear partial differential equations (PDEs) that are important for applications are parameterized by arbitrary elements (constants or functions) and constitute classes of PDEs. An important task is to study transformation properties of such classes, i.e. to describe explicitly nondegenerate point transformations that link fixed equations from the class (it is also possible to extend the consideration to contact transformations). If two PDEs are connected by such a transformation, then associated objects like exact solutions, local conservation laws, and various kinds of symmetries of these equations are also related by the respective transformation. Such connected equations are called equivalent or similar [4]. In particular, the equivalence method allows one to construct exact solutions for variable coefficient PDEs using known exact solutions for their constant coefficient

* The author would like to thank the Organizing Committee of the 10th Mathematical Physics Meeting for the provided hospitality and support. She is also grateful to Roman Popovych and Christodoulos Sophocleous for useful discussions and ongoing collaboration.

† e-mail address: vaneeva@imath.kiev.ua

counterparts, see, e.g., [5, 6, 7]. At the same time, nondegenerate point transformations appear to be a useful tool not only for finding exact solutions but also for exhaustive solving group classifications problems (see, e.g., [5, 6, 7, 8, 9, 10]), design of physical parameterization schemes [11], and study of integrability [12, 13, 14, 15].

The systematic investigation of transformation properties of classes of nonlinear PDEs was initiated in [16, 17]. Roughly speaking, an admissible transformation is a triple consisting of two fixed equations from a class and a transformation that links these equations. The set of admissible transformations considered with the standard operation of composition of transformations has a groupoid structure and therefore it is also called the equivalence groupoid [11]. Equivalence transformations, which are invaluable tools of group analysis of differential equations, generate a subset in a set of admissible transformations and always form a group. An equivalence transformation applied to any equation from the class always maps it to another equation from the same class, while an admissible transformation may exist only for a specific pair of equations from the class under consideration.

The usual equivalence group consists of the nondegenerate point transformations of the independent and dependent variables and of the arbitrary elements of the class, where transformations for independent and dependent variables are projectable on the space of these variables [4]. If the transformations for independent and/or dependent variables involve arbitrary elements, then the corresponding equivalence group is called generalized one. If new arbitrary elements appear to depend on old ones in a nonlocal way (e.g., new arbitrary elements are expressed via integrals of old ones), then the corresponding equivalence group is called extended one. Simultaneously weakening the conditions of locality and projectability, leads to the notion of extended generalized equivalence group.

If any admissible transformation in a given class is induced by a transformation from its usual equivalence group, then this class is called *normalized* in the usual sense. In analogous way, the notions of normalization of a class in generalized, extended and extended generalized senses are formulated [8]. It is important that normalized classes of differential equation have simpler transformation properties than non-normalized ones. Moreover, once it is proved that certain class is normalized, finding the equivalence groupoids of its subclasses becomes essentially simpler since they are always subgroupoids of the equivalence groupoid of the initial class.

We aim to study transformation properties of the general class of nonlinear second-order evolution equations of the form

$$u_t = H(t, x, u, u_x, u_{xx}), \quad H_{u_{xx}} \neq 0, \tag{1}$$

to construct a chain of its nested normalized subclasses and to find their equivalence groupoids. Then the obtained results is used to find equivalence groups for non-normalized classes of reaction–diffusion equations that are important for applications.

2. Admissible transformations of evolution equations

Any nondegenerate point transformation $\mathcal{T}$ relating two fixed equations $u_t = H$ and $\tilde{u}_{\tilde{t}} = \tilde{H}$ from the class (1) has the form $\tilde{t} = T(t)$, $\tilde{x} = X(t,x,u)$, $\tilde{u} = U(t,x,u)$ with $T_t(X_xU_u - X_uU_x) \neq 0$ [16, 18]. The partial derivatives are transformed as follows:

$$\tilde{u}_{\tilde{t}} = \frac{D_tUD_xX - D_xUD_tX}{T_tD_xX}, \quad \tilde{u}_{\tilde{x}} = \frac{D_xU}{D_xX}, \quad \tilde{u}_{\tilde{x}\tilde{x}} = \frac{1}{D_xX}D_x\left(\frac{D_xU}{D_xX}\right),$$

where $D_t = \partial_t + u_t\partial_u + u_{tt}\partial_{u_t} + u_{tx}\partial_{u_x} + \ldots$ and $D_x = \partial_x + u_x\partial_u + u_{tx}\partial_{u_t} + u_{xx}\partial_{u_x} + \ldots$ are operators of the total differentiation with respect to t and x. Moreover, it was proved in [19] that class (1) is normalized. For further consideration we use the following statement.

Theorem 1 ([19]). *Class (1) is normalized in the usual sense. Its equivalence group is formed by the transformations*

$$\tilde{t} = T(t), \quad \tilde{x} = X(t,x,u), \quad \tilde{u} = U(t,x,u), \tag{2}$$

$$\tilde{H} = \frac{X_xU_u - X_uU_x}{T_tD_xX}H + \frac{U_tD_xX - X_tD_xU}{T_tD_xX}, \tag{3}$$

where $T_t(X_xU_u - X_uU_x) \neq 0$,

Formula (3) implies that the subclass of class (1) singled out by the condition $H_{u_{xx}u_{xx}} = 0$ has the same equivalence transformation components for variables. The following statement is true.

Theorem 2. *The class of quasilinear second-order evolution equations,*

$$u_t = G(t,x,u,u_x)u_{xx} + F(t,x,u,u_x), \quad G \neq 0, \tag{4}$$

is normalized in the usual sense. Its equivalence group is formed by the transformation components for variables (2) and the transformations for arbitrary elements

$$\tilde{G} = \frac{(D_xX)^2}{T_t}G, \quad \tilde{F} = \frac{X_xU_u - X_uU_x}{T_tD_xX}F + \frac{U_tD_xX - X_tD_xU}{T_tD_xX} +$$
$$\frac{(X_{xx} + 2X_{xu}u_x + X_{uu}u_x^2)D_xU - (U_{xx} + 2U_{xu}u_x + U_{uu}u_x^2)D_xX}{T_tD_xX}G.$$

The transformation component for G implies that, if G does not depend on u_x, then $X_u = 0$. We formulate more generally Lemma 1 from [20].

Theorem 3. *The class*

$$u_t = G(t,x,u)u_{xx} + F(t,x,u,u_x), \quad G \neq 0, \tag{5}$$

is normalized in the usual sense. Its equivalence group comprises the transformations

$$\tilde t = T(t),\quad \tilde x = X(t,x),\quad \tilde u = U(t,x,u),\quad \tilde G = \frac{X_x^2}{T_t}G, \tag{6}$$

$$\tilde F = \frac{U_u}{T_t}F + \frac{U_tX_x - X_tD_xU}{T_tX_x} + \frac{X_{xx}D_xU - (U_{xx} + 2U_{xu}u_x + U_{uu}u_x^2)X_x}{T_tX_x}G,$$

where $T_tX_xU_u \neq 0$.

Particular important subclass of class (5) is one, where the function F is polynomial in u_x and especially when it is quadratic or linear in u_x. We formulate separate statements for each of these cases.

Theorem 4. *The class*

$$u_t = G(t,x,u)u_{xx} + \sum_{k=0}^{n} F^k(t,x,u)u_x^k,\ n\geq 2,\quad G\neq 0, \tag{7}$$

is normalized in the usual sense. Its equivalence group consists of the transformations (6) and the transformation components for the arbitrary elements F^k, $k=0,\dots,n$, *are found as solutions of the algebraic system resulting from the splitting of the following equation with respect to different powers of* u_x

$$\sum_{k=0}^{n} \tilde F^k\left(\frac{U_u}{X_x}u_x + \frac{U_x}{X_x}\right)^k = \frac{1}{T_tX_x}\big[X_xU_u\sum_{k=0}^{n}F^ku_x^k + U_tX_x - X_tD_xU +$$
$$\big(X_{xx}D_xU - (U_{xx} + 2U_{xu}u_x + U_{uu}u_x^2)X_x\big)\,G\big].$$

Theorem 5. *The class*

$$u_t = G(t,x,u)u_{xx} + F^2(t,x,u)u_x^2 + F^1(t,x,u)u_x + F^0(t,x,u),\quad G\neq 0, \tag{8}$$

is normalized in the usual sense. Its equivalence group is formed by the transformations

$$\tilde t = T(t),\quad \tilde x = X(t,x),\quad \tilde u = U(t,x,u),$$

$$\tilde G = \frac{X_x^2}{T_t}G,\quad \tilde F^2 = \frac{X_x^2}{T_tU_u^2}\left(U_uF^2 - U_{uu}G\right),\quad \tilde F^1 = \frac{1}{T_tU_u}\bigg[X_xU_uF^1 +$$
$$2\frac{X_xU_x}{U_u}(U_{uu}G - U_uF^2) - X_tU_u + (X_{xx}U_u - 2U_{xu}X_x)G\bigg],$$

$$\tilde F^0 = \frac{1}{T_t}\left[\frac{U_x^2}{U_u}F^2 - U_xF^1 + U_uF^0 + U_t + \left(2\frac{U_x}{U_u}U_{xu} - U_{xx} - \frac{U_x^2}{U_u^2}U_{uu}\right)G\right],$$

where $T_tX_xU_u \neq 0$.

If we consider the subclass of class (8) singled out by the condition $F^2 = 0$, then its equivalence group is a proper subgroup of the equivalence group of class (8). The constraints for the transformations are derived by setting $\tilde{F}^2 = 0$ and $F^2 = 0$ in Theorem 5. This implies that $U_{uu} = 0$. The following statement is true.

Theorem 6. *The class*

$$u_t = G(t,x,u)u_{xx} + F^1(t,x,u)u_x + F^0(t,x,u), \quad G \neq 0, \tag{9}$$

is normalized in the usual sense. Its equivalence group comprises the transformations

$$\tilde{t} = T(t), \quad \tilde{x} = X(t,x), \quad \tilde{u} = U^1(t,x)u + U^0(t,x), \quad T_t X_x U^1 \neq 0, \tag{10}$$

$$\tilde{G} = \frac{X_x^2}{T_t}G, \quad \tilde{F}^1 = \frac{1}{T_t U^1}\left(X_x U^1 F^1 - X_t U^1 + (X_{xx}U^1 - 2U_x^1 X_x)G\right),$$

$$\tilde{F}^0 = \frac{1}{T_t}\Bigg[U^1 F^0 - (U_x^1 u + U_x^0)F^1 + U_t^1 u + U_t^0 +$$
$$\left(2\frac{U_x^1}{U^1}(U_x^1 u + U_x^0) - U_{xx}^1 u - U_{xx}^0\right)G\Bigg].$$

Consider one more subclass of class (8) for which the condition $U_{uu} = 0$ holds for admissible transformations. This is the subclass singled out by the condition $F^2 = G_u$,

$$u_t = (G(t,x,u)u_x)_x + K(t,x,u)u_x + P(t,x,u), \quad G \neq 0. \tag{11}$$

This class can be written in the form

$$u_t = Gu_{xx} + G_u u_x^2 + (G_x + K)u_x + P,$$

where the connections between arbitrary elements of the latter class and class (8) are given by the formulas $F^2 = G_u$, $F^1 = G_x + K$, $F^0 = P$. It contains derivatives of G and they naturally appear in transformation components for arbitrary elements of the equivalence group. This group can be considered as usual one if we extend the tuple of arbitrary elements by new elements G_u and G_x. Then the transformations of these variables take the form

$$\tilde{G}_{\tilde{x}} = \frac{X_x}{T_t}G_x + 2\frac{X_{xx}}{T_t}G, \quad \tilde{G}_{\tilde{u}} = \frac{X_x^2}{T_t U^1}G_u.$$

It is easy to see that the group is really usual one, representing the above class in the form

$$u_t = Gu_{xx} + G^1 u_x^2 + (G^2 + K)u_x + P$$

with additional arbitrary elements $G^1 = G_u$, and $G^2 = G_x$.

Theorem 7. *Reparameterized class (11) is normalized in the usual sense. Its equivalence group is formed by the transformations*

$$\tilde t = T(t),\quad \tilde x = X(t,x),\quad \tilde u = U^1(t,x)u + U^0(t,x),\quad T_tX_xU^1 \neq 0,$$
$$\tilde G = \frac{X_x^2}{T_t}G,\quad \tilde K = \frac{X_x}{T_t}\left[K - \left(\frac{X_{xx}}{X_x} + 2\frac{U^1_x}{U^1}\right)G - 2(U^1_xu + U^0_x)\frac{G_u}{U^1} - \frac{X_t}{X_x}\right],$$
$$\tilde P = \frac{1}{T_t}\left[U^1P + \frac{(U^1_xu + U^0_x)^2}{U^1}G_u - (U^1_xu + U^0_x)(G_x + K) + U^1_tu + U^0_t + \right.$$
$$\left.\left(2\frac{U^1_x}{U^1}(U^1_xu + U^0_x) - U^1_{xx}u - U^0_{xx}\right)G\right].$$

The subclass of class (11) singled out by the condition $K = 0$,

$$u_t = (G(t,x,u)u_x)_x + P(t,x,u),\quad G \neq 0, \tag{12}$$

is not normalized anymore in contrast to its covering classes considered above.

Constraints for its admissible transformations are derived setting $K = 0$ and $\tilde K = 0$ in the transformations adduced in the previous theorem, which results in the equation

$$2\left(\frac{U^1_x}{U^1}u + \frac{U^0_x}{U^1}\right)G_u + \left(\frac{X_{xx}}{X_x} + 2\frac{U^1_x}{U^1}\right)G + \frac{X_t}{X_x} = 0.$$

The further constraints on forms of X, U^1 and U^0 depend on values of the function G. If G does not satisfy the equation of the form $(au + b)G_u + cG + d = 0$, where a, b, c and d are functions of t and x, then the point transformations between equations from this class necessarily satisfy the conditions $X_t = X_{xx} = U^1_x = U^0_x = 0$, and such a subclass of class (12) will be normalized in the usual sense. The whole class (12) is not normalized. We adduce its equivalence group in the following statement.

Theorem 8. *The equivalence group of class (12) is comprised of the transformations*

$$\tilde t = T(t),\quad \tilde x = \delta_1x + \delta_2,\quad \tilde u = U^1(t)u + U^0(t),\quad T_tU^1\delta_1 \neq 0,$$
$$\tilde G = \frac{\delta_1^2}{T_t}G,\quad \tilde P = \frac{1}{T_t}\left(U^1P + U^1_tu + U^0_t\right).$$

The description of the entire equivalence groupoid of class (12) needs additional study (see Conclusion).

Classes of evolution equations with variable coefficients of u_t often appear in applications. That is why we additionally consider the generalization of equations (11) of the form

$$S(t,x)u_t = (G(t,x,u)u_x)_x + K(t,x,u)u_x + P(t,x,u),\quad SG \neq 0. \tag{13}$$

In particular, the classes of variable-coefficient diffusion–reaction equations $f(x)u_t = (g(x)A(u)u_x)_x + h(x)B(u)$ and diffusion–convection equations $f(x)u_t = (g(x)A(u)u_x)_x + h(x)B(u)u_x$ ($fgA \neq 0$) are subclasses of this class.

In principle, the coefficient $S(t,x)$ can be gauged to one by the family of point transformation

$$\tilde{t} = t, \quad \tilde{x} = \int_{x_0}^{x} S(t,y)\,\mathrm{d}y, \quad \tilde{u} = u.$$

Nevertheless, we will consider class (13) separately since its transformational properties become more complicated in comparison with those of class (11). The following statement is true.

Theorem 9. *Any point transformation between two fixed equations from class* (13) *has the form* (10). *Then the respective values of the arbitrary elements are related via the formulas*

$$\frac{\tilde{K}+\tilde{G}_{\tilde{x}}}{\tilde{S}} = \frac{X_x}{T_t S}\left[K + G_x + \left(\frac{X_{xx}}{X_x} - 2\frac{U^1_x}{U^1}\right)G - 2(U^1_x u + U^0_x)\frac{G_u}{U^1} - \frac{X_t}{X_x}S\right],$$

$$\frac{\tilde{G}}{\tilde{S}} = \frac{X_x^2}{T_t}\frac{G}{S}, \quad \frac{\tilde{P}}{\tilde{S}} = \frac{1}{T_t S}\left[U^1 P + \frac{(U^1_x u + U^0_x)^2}{U^1}G_u + (U^1_t u + U^0_t)S + \left(2\frac{U^1_x}{U^1}(U^1_x u + U^0_x) - U^1_{xx}u - U^0_{xx}\right)G - (U^1_x u + U^0_x)(K + G_x)\right].$$

It is obvious that transformation properties of class (13) become more complicated in comparison with those of class (11). Transformations are defined only for fractions of arbitrary elements. It is explained by the fact that this class admits peculiar gauge equivalence transformation (an equivalence transformation for which independent and dependent variables do not transform but only arbitrary elements). This is the transformation

$$\tilde{S} = Z(t,x,S), \quad \tilde{G} = \frac{G}{S}Z, \quad \tilde{K} = \frac{K}{S}Z - G\left(\frac{Z}{S}\right)_x, \quad \tilde{P} = \frac{P}{S}Z,$$

where Z is an arbitrary smooth function of its variables with $Z_S \neq 0$.

Theorem 10. *The equivalence group of class* (13) *comprises the transfor-*

mations

$$\tilde t = T(t),\quad \tilde x = X(t,x),\quad \tilde u = U^1(t,x)u + U^0(t,x),\quad T_tX_xU^1 \neq 0,$$
$$\tilde S = Z(t,x,S),\quad \tilde G = \frac{X_x^2}{T_t}\frac{G}{S}Z,$$
$$\tilde K = \frac{X_xZ}{T_tS}\left[K - \left(\frac{X_{xx}}{X_x} + 2\frac{U^1_x}{U^1}\right)G - 2\mathcal{U}\frac{G_u}{U^1} - \frac{X_t}{X_x}S\right] - \frac{X_x}{T_t}G\left(\frac{Z}{S}\right)_x,$$
$$\tilde P = \frac{Z}{T_tS}\left[U^1P + \frac{\mathcal{U}^2}{U^1}G_u - \mathcal{U}(K+G_x) + (U^1_tu + U^0_t)S + \left(2\frac{U^1_x}{U^1}\mathcal{U} - U^1_{xx}u - U^0_{xx}\right)G\right],$$

where $\mathcal{U} = U^1_xu + U^0_x$.

Class (13) can be regarded as normalized in the usual sense since we can present it in the form $Su_t = Gu_{xx} + G^1u_x^2 + G^2u_x + P$ with additional arbitrary elements $G^1 = G_u$, and $G^2 = K + G_x$.

We note that the subclass of class (13) singled out by the condition $K = 0$, i.e. the class

$$S(t,x)u_t = (G(t,x,u)u_x)_x + P(t,x,u),\quad SG \neq 0, \tag{14}$$

is not normalized. In contrast to the case of class (13) the coefficient S is essential for class (14). The gauge equivalence transformations are quite simple in this case, namely, each coefficient can be multiplied by a nonvanishing smooth function of t. The equivalence group of class (14) is wider than the equivalence group of its subclass with $S = 1$ (cf. Theorem 8). The following statement is true.

Theorem 11. *The equivalence group of class* (14) *consists of the transformations*

$$\tilde t = T(t),\quad \tilde x = X(x),\quad \tilde u = U^1(t)u + U^0(t),\quad T_tX_xU^1 \neq 0,$$
$$\tilde S = \psi(t)\frac{T_t}{X_x}S,\quad \tilde G = \psi(t)X_xG,\quad \tilde P = \frac{\psi(t)}{X_x}\left(U^1P + (U^1_tu + U^0_t)S\right),$$

where $\psi(t)$ *is a nonvanishing smooth function of its variable.*

3. Conclusions

The chain of nested subclasses of the general class of (1+1)-dimensional second-order nonlinear evolution equations is constructed. Transformation properties of the derived subclasses are investigated. For all the subclasses the widest possible equivalence groups are found. For those subclasses that are proved to be normalized the found equivalence groups give the exhaustive description of the respective equivalence groupoids of these classes. So, we firstly proved that classes (4), (5), (7)–(9), (11) and (13) are normalized and then looked for their equivalence groups, which lead to the complete description of the equivalence groupoids of these classes.

Finding the equivalence groupoids for the non-normalized classes is a difficult task since the determining equations for components of admissible transformations are nonlinear ones. That is why for such classes other techniques are needed like partition of the class into normalized subclasses and the method of furcate splitting [21]. In future work we plan to use these methods to find the entire equivalence groupoids of classes (12) and (14), that are of special interest for further applications. There are many models with application in physics and biology which are members of these classes, e.g. variable coefficient Fisher and Newell–Whitehead–Segel equations. The results of this paper will be used, in particular, to complete the group classification of the class of variable coefficient reaction–diffusion equations $f(x)u_t = (g(x)A(u)u_x)_x + h(x)B(u)$, $fgA_u \neq 0$.

References

[1] L. D. Faddeev, *40 years in Mathematical Physics*, World Scientific Publishing, 1995.

[2] J.-P. Colinge and J. C. Greer, *Nanowire Transistors: Physics of Devices and Materials in One Dimension*, Cambridge University Press, 2016.

[3] J. D. Murray, *Mathematical Biology I: An Introduction*, Springer, 2002.

[4] L. V. Ovsiannikov, *Group Analysis of Differential Equations*, Academic Press, 1982.

[5] O. O. Vaneeva, *Commun. Nonlinear Sci. Numer. Simul.* **17** (2012) 611.

[6] O. O. Vaneeva, in: Dobrev V. (eds) *Lie Theory and Its Applications in Physics. Springer Proc. Math. Stat.* **36** (2013) 451.

[7] O. Vaneeva, V. Boyko, A. Zhalij and C. Sophocleous, *J. Math. Anal. Appl.* **474** (2019) 264.

[8] R. O. Popovych, M. Kunzinger and H. Eshraghi, *Acta Appl. Math.* **109** (2010) 315.

[9] S. Opanasenko, V. Boyko and R. O. Popovych, *J. Math. Anal. Appl.* **484** (2020) 123739.

[10] C. Kurujyibwami, P. Basarab-Horwath and R. O. Popovych, *Acta Appl. Math.* **157** (2018) 171.

[11] R.O. Popovych and A. Bihlo, *J. Math. Phys.* **53** (2012) 073102.

[12] T. Brugarino, *J. Math. Phys.* **30** (1989) 1013.

[13] N. Joshi, *Phys. Lett. A* **125** (1987) 456.

[14] A. Kundu, *Phys. Rev. E* **79** (2009) 015601.

[15] O. O. Vaneeva, R. O. Popovych and C. Sophocleous, *Phys. Scr.* **89** (2014) 038003.

[16] J. G. Kingston, *J. Phys. A: Math. Gen.* **24** (1991) L769.

[17] J. G. Kingston and C. Sophocleous, *Phys. Lett. A* **155** (1991) 15.

[18] J. G. Kingston and C. Sophocleous, *J. Phys. A: Math. Gen.* **31** (1998) 1597.

[19] R. O. Popovych and A. M. Samoilenko, *J. Phys. A: Math. Theor.* **41** (2008) 362002.

[20] N. M. Ivanova, R. O. Popovych and C. Sophocleous, *Lobachevskii J. Math.* **31** (2010) 100.

[21] O. O Vaneeva, A. Bihlo and R. O. Popovych, Preprint arXiv:2002.08939.

Talks not included in the Proceedings

Ignatios Antoniadis
Inflation from supersymmetry breaking

Irina Aref'eva
Holography for heavy-ion collision

Sudipto Bhattacharjee
Role of particle creation mechanism on the collapse of a massive star

Asmus Bisbo
Polynomial representations of the Lie superalgebra $osp(1|2n)$

David Edward Bruschi
On the weight of entanglement

Martin Cederwall
Extended geometry and tensor hierarchy algebras

Radu Constantinescu
Special methods for solving nonlinear differential equations through polynomial expansion

Ion Cotaescu
Leptogenesis in a spatially flat Milne-type universe

Branislav Cvetković
Entropy in Poincare gauge theory: Hamiltonian approach

Ivan Dimitrijević
Cosmological solutions of a nonlocal square root gravity

Dragoljub Gočanin
Noncommutative $OSp(4|2)$ supergravity

Alexey Golovnev
Spin connections, local Lorentz transformations and cosmological perturbations in modified teleparallel theories of gravity

Nikola Konjik
Angular noncommutative scalar field phenomenology in flat and curved space

Alexey Koshelev
On quantum properties of analytic infinite derivative field theories

Abhijit Mandal
Thermodynamic study of Reissner-Nordstrom quintessence black hole

Milan Milošević
Slow-roll parameters in extended RSII model

Sergei D. Odintsov
The universe acceleration in modified gravity: an overview

Anna Pachol
Digital quantum geometries

Shibesh Kumar Jas Pacif
An accelerating cosmological model from a parametrization of Hubble parameter

Gauranga Charan Samanta
Wormhole modeling in general relativity

Ciprian Sporea
Fermionic quasibound states for Reissner-Nordstrom black holes

Alexei Starobinsky
Talk 1: Pre-inflation, inflation and post-inflationary heating in $f(R)$ and scalar-tensor modified gravity
Talk 2: The mixed Higgs-R^2 inflationary model

Mykola Stetsko
Black holes in the theory with nonminimal derivative coupling and some aspects of their thermodynamics

Francesco Toppan
New issues in superconformal quantum mechanics

Mihai Visinescu
Sasaki-Ricci flow on Sasaki-Einstein space $T_{1,1}$

Xiaoning Wu
The null-timelike boundary problems of linear wave equations in asymptotically anti-de Sitter space

Naqing Xie
Construction of vacuum initial data by the conformally covariant split system

List of Participants

1. Ignatios Antoniadis (Bern, Switzerland)
2. Irina Arefeva (Moscow, Russia)
3. Sudipto Bhattacharjee (Kolkata, India)
4. Asmus Bisbo (Gent, Belgium)
5. Duško Borka (Belgrade, Serbia)
6. Vesna Borka Jovanović (Belgrade, Serbia)
7. Danijela Branković (Belgrade, Serbia)
8. David Edward Bruschi (Vienna, Austria)
9. Maja Burić (Belgrade, Serbia)
10. Martin Cederwall (Goteborg, Sweden)
11. Diego Cirilo-Lombardo (Buenos Aires, Argentina)
12. Sanja Ćirković (Belgrade, Serbia)
13. Radu Constantinescu (Craiova, Romania)
14. Ion Cotaescu (Timisoara, Romania)
15. Mihailo Čubrović (Belgrade, Serbia)
16. Branislav Cvetković (Belgrade, Serbia)
17. Ljubica Davidović (Belgrade, Serbia)
18. Aleksandra Dimić (Belgrade, Serbia)
19. Ivan Dimitrijević (Belgrade, Serbia)
20. Marija Dimitrijević Ćirić (Belgrade, Serbia)
21. Goran Đorđević (Nis, Serbia)
22. Branko Dragovich (Belgrade, Serbia)
23. Stefano Gregorio Giaccari (Holon, Israel)
24. Dragoljub Gočanin (Belgrade, Serbia)
25. Alexey Golovnev (Saint Petersburg, Russia)
26. Ilija Ivanišević (Belgrade, Serbia)

27. Predrag Jovanović (Belgrade, Serbia)
28. Nikola Konjik (Belgrade, Serbia)
29. Alexey Koshelev (Covilha, Portugal)
30. Bojan Kovačević (Banja Luka, Bosnia and Herzegovina)
31. Abhijit Mandal (Kolkata, India)
32. Nenad Manojlović (Faro, Portugal)
33. Veselin Manojlović (Oxford, United Kingdom)
34. Žarko Mijajlović (Belgrade, Serbia)
35. Aleksandar Miković (Lisbon, Portugal)
36. Milan Milošević (Nis, Serbia)
37. Đorđe Minić (Blacksburg, USA)
38. Nataša Mišić (Belgrade, Serbia)
39. Biljana Nikolić (Belgrade, Serbia)
40. Bojan Nikolić (Belgrade, Serbia)
41. Emil Nissimov (Sofia, Bulgaria)
42. Dušan Novičić (Belgrade, Serbia)
43. Danijel Obrić (Belgrade, Serbia)
44. Sergei Odintsov (Barcelona, Spain)
45. Svetlana Pacheva (Sofia, Bulgaria)
46. Anna Pachol (London, United Kingdom)
47. Shibesh Kumar Jas Pacif (Vellore, India)
48. Dragan Prekrat (Belgrade, Serbia)
49. Tijana Radenković (Belgrade, Serbia)
50. Voja Radovanović (Belgrade, Serbia)
51. Viktor Radović (Belgrade, Serbia)
52. Zoran Rakić (Belgrade, Serbia)
53. Igor Salom (Belgrade, Serbia)
54. Gauranga Charan Samanta (Vasco da Gama, India)

55. Branislav Sazdović (Belgrade, Serbia)
56. Dejan Simić (Belgrade, Serbia)
57. Đorđe Šijački (Belgrade, Serbia)
58. Paul Sorba (Annecy, France)
59. Ciprian Sporea (Timisoara, Romania)
60. Jelena Stanković (Belgrade, Serbia)
61. Alexei Starobinsky (Moscow, Russia)
62. Mykola Stetsko (Lviv, Ukraine)
63. Ovidiu Cristinel Stoica (Bucharest, Romania)
64. Fumihiko Sugino (Daejeon, South Korea)
65. Michal Szczachor (Wroclaw, Poland)
66. Marek Szydlowski (Krakow, Poland)
67. Francesco Toppan (Rio de Janeiro, Brasil)
68. Dejan Urošević (Belgrade, Serbia)
69. Vitaly Vanchurin (Minnesota, USA)
70. Olena Vaneeva (Kyiv, Ukraine)
71. Mihai Visinescu (Bucharest, Romania)
72. Marko Vojinović (Belgrade, Serbia)
73. Xiaoning Wu (Beijing, China)
74. Naqing Xie (Shanghai, China)
75. Aleksandar Zejak (Belgrade, Serbia)